Fundamentals of Organic Chemistry

Herman G. Richey, Jr.

The Pennsylvania State University

Prentice-Hall, Inc., Englewood Cliffs, New Jersey 07632

Library of Congress Cataloging in Publication Data

RICHEY, HERMAN G. (Herman Glenn), (date)
Fundamentals of organic chemistry.

Includes index.
1. Chemistry, Organic. I. Title.
QD2512.R5   547   82-7520
ISBN 0-13-341511-2   AACR2

Permission for the publication herein of Sadtler Standard Spectra®
has been granted, and all rights are reserved, by Sadtler
Research Laboratories, Division of Bio-Rad Laboratories, Inc.

Editorial/production supervision by Ellen W. Caughey and Joan Semels
Interior design by Dawn Stanley and Ellen W. Caughey
Cover photo by Laimute E. Druskis
Cover design by Anne T. Bonanno
Manufacturing buyer: John Hall

The cover photograph is of a model of the alkaloid reserpine (see Sec. 8.11). Reserpine, isolated
from the Indian snakeroot plant, has been used medically both for reducing elevated blood pressure
and for its tranquilizing effects. The model is constructed from CPK space-filling molecular models
produced by the Ealing Corporation.

Printed in the United States of America
10 9 8 7 6 5 4 3 2 1

ISBN 0-13-341511-2

Prentice-Hall International, Inc., *London*
Prentice-Hall of Australia Pty. Limited, *Sydney*
Editora Prentice-Hall do Brazil Ltda, Rio de Janeiro
Prentice-Hall of Canada Inc., *Toronto*
Prentice-Hall of India Private Limited, *New Delhi*
Prentice-Hall of Japan, Inc., *Tokyo*
Prentice-Hall of Southeast Asia Pte. Ltd., *Singapore*
Whitehall Books Limited, *Wellington, New Zealand*

To Jane, Susan, Anna, and Daniel

Contents

2

Alkanes   19

3

Alkenes and Alkynes   51

4

Aromatic Compounds 86

Alcohols, Ethers, and Phenols 126

Chiral Compounds 165

7

Organic Halides   194

Amines   217

Aldehydes and Ketones   246

Carbohydrates   279

11

Carboxylic Acids   300

12

Derivatives of Carboxylic Acids: Esters and Amides   318

13

Lipids   352

14

Amino Acids and Proteins   372

15

Nucleic Acids   402

16

Spectroscopy of Organic Compounds   418

Preface

I have written this book for students who need a knowledge of organic chemistry but who cannot devote the time required for the lengthy courses taken by chemistry majors. Such students plan professional careers in areas as diverse as agriculture, nutrition, forestry, medical technology, ecology, microbiology, fuel technology, pharmacy, veterinary medicine, nursing, and dentistry. Many learn organic chemistry because of a need to understand the functioning of living systems at the molecular level, but some because of the role of organic compounds in industry, the environment, or energy generation. This book focuses on those fundamentals that will be of greatest value to these students.

The fundamentals of organic chemistry are not inherently difficult. To assimilate in a short time what at first may seem a strange language and a large assortment of facts, however, is a formidable challenge to students. The challenge to instructors and to the author of a text is to ease as much as possible the task facing beginning students.

The organization of this book was guided by my experience of what is helpful to students. I have avoided confronting students at the start of the book with an overdose of review information, concepts, and theories that are isolated from the material to which they will be applied. As much as possible, these are deferred until they can be directly related to behavior of organic compounds. An effort is made to present fundamental ideas thoroughly—not with unusual depth or detail—but with ample explanations and examples. Names associated with some reactions, trivial nomenclature close to oblivion, and other jargon of marginal importance to non-chemists are used only sparingly. Finally, the chapters are carefully organized into sections in a way that will be useful to instructors who want to make assignments that emphasize or omit particular topics, and to students when studying and reviewing.

I have incorporated several study aids. Most students find that studying with pencil and paper at hand—continually writing structures, names, and reactions of organic compounds—greatly accelerates learning organic chemistry. To aid in this practice, two kinds of problems are included. **Problems** scattered throughout the earlier chapters are related specifically to fundamental information that immediately precedes them. These problems generally are simple and can be done quickly. I urge students to stop and do these problems as they read through a chapter; answers to them are given at the back of the book. **Problem sets** at the end of each chapter provide many straightforward problems as well as some that are more challenging. These problem sets are unusually extensive, not because students will have time to work them all, but rather to permit an instructor to make problem assignments suitable for a particular course without recourse to a supplementary problem book. Answers to these problems are given in a **solutions manual** prepared by Jane Richey.

Many students find **molecular models** to be useful in studying aspects of organic structure. **Exercises with models** in some chapters are for those who wish to use molecular models while studying. Access to virtually any simple model set will be adequate for doing these exercises. Finally, **summaries** at the end of each chapter review all important definitions, ideas, and reactions.

Due to the trenchant comments of one reviewer, this book adheres more closely than most to current IUPAC guidelines for nomenclature. Where several procedures are accepted by the IUPAC, those most likely to be encountered have been selected. The last chapter contains a discussion of spectroscopic methods that should be sufficient for most courses that incorporate such material. The sections on these methods can be introduced at any time after Chapter 4 has been started.

Many persons contributed, directly or indirectly, to the preparation of this book. I am grateful to the following reviewers of the manuscript for many helpful comments: Professor Winfield M. Baldwin, Jr., of the University of Georgia; Professor Irving Lillien of Miami-Dade Community College; Professor Kenneth L. Marsi of California State University at Long Beach; Professor Charles B. Rose of the University of Nevada at Reno; Professor Kevin Smith of the University of California at Davis; and Professor Walter S. Trahanovsky of Iowa State University. I owe a great debt to my colleagues in the Chemistry Department at Penn State from whom I have learned a great deal about chemistry and about approaches to teaching. I am grateful to Sandy Williams for her dedication and care in typing the manuscript through all of the revisions. Finally, I want particularly to thank Jane Richey for suggestions, corrections, and assistance at every stage of the preparation of this book.

H. G. R.

1

Some Simple Organic Compounds

Life on our earth depends on the formation, interconversion, and degradation of compounds of carbon. It was once believed that the carbon-containing compounds found in living organisms could arise only through the operation of a "vital force" residing in living cells. Because of this belief, such compounds were designated **organic**. During the first half of the last century it became evident that organic compounds, like any others, can be synthesized in the laboratory. In fact, the number of carbon-containing compounds synthesized by chemists now greatly exceeds the number isolated from natural sources. We continue, however, to apply the name organic to carbon-containing compounds.

The mystical aura of organic compounds has faded but not their importance. Except for water and salts, the compounds of living systems are organic. The increasing effort to understand living systems at the molecular level requires an understanding of the nature of these compounds and their chemical transformations. Unique features of the chemistry of organic compounds seem central to life as we know it. In fact, scientists searching for evidence of life on the moon and Mars have looked particularly for the presence of specific types of organic compounds.

The importance of organic compounds extends beyond their presence in living systems. Much of what we ingest—foodstuffs, medicines, vitamins—consists of organic compounds. Our major sources of energy are the organic compounds that constitute natural gas, petroleum, and coal. About 95% of the energy used in the United States is currently derived from these materials. Important structural materials—wood, plastics, rubber—are composed of organic compounds. So are fibers, papers, paints, dyes, detergents, scents, and adhesives. Unfortunately, the widespread use of organic compounds has also led to some of the environmental problems that confront us.

The study of these carbon-containing compounds so vitally connected with our lives has developed into the extensive branch of science known as **organic chemistry**. More than 5 million organic compounds have already been synthesized or isolated from natural sources. The number that can be synthesized in the future is virtually limitless. Organic molecules present an enormous diversity in size, shape, complexity, properties, and reactions. Comprehending and organizing any significant part of what chemists have learned about this vast array of organic compounds may seem overwhelming. Fortunately, some common features in the properties and behavior of organic compounds greatly simplify our task. Even in a single volume of this size, we can make a significant start in understanding some of the most important aspects of organic chemistry.

1.1
METHANE

Methane is an excellent compound with which to begin our study of organic chemistry. The methane molecule is particularly simple, containing only one carbon atom and four hydrogen atoms. Nevertheless, methane is the organic compound used in largest amounts. Annual consumption of methane in the United States exceeds *2 tons per person.*

Methane, a product of microbial decay of the organic material of plants in the absence of air, is the gas seen bubbling out of marshes. It has recently been detected in interstellar space in the constellation Orion. More important to us, it is the major constituent of natural gas.

Methane is colorless and odorless; the smell sometimes associated with the gas distributed to homes is due to odorous compounds that are added to make leaks evident. At atmospheric pressure methane boils at $-162°$. In mammoth ocean-going tankers that carry up to 100,000 tons of methane, it is maintained at that temperature in carefully insulated compartments to avoid the need for construction that could withstand high pressures.

The fate of most of the methane that we use is combustion.

$$CH_4 + 2\,O_2 \longrightarrow CO_2 + 2\,H_2O$$

The value of this reaction is not due to the carbon dioxide and water that are products—these compounds are abundant in our environment. The significant product is heat—213 kilocalories for every mole (16 grams) of methane.

In spite of the use of enormous quantities of methane, it is a relatively inert compound. Methane does not react readily with acids and bases or oxidizing and reducing agents. In fact, we know that a mixture of methane and air must be heated by a flame or spark to initiate the combustion reaction. Once this reaction has started, the heat produced warms the reactants sufficiently to maintain the reaction.

At very high temperatures (in the absence of oxygen) methane can be converted to acetylene (C_2H_2) and hydrogen.

$$2\,CH_4 \xrightarrow{\;1500°\;} C_2H_2 + 3\,H_2$$

Acetylene is an important intermediate in the chemical industry. Because it burns with a particularly hot flame, it is used in oxyacetylene torches for cutting and welding metals.

Treatment of methane with chlorine leads to replacement of a hydrogen atom by a chlorine atom.

$$CH_4 + Cl_2 \longrightarrow CH_3Cl + HCl$$

High temperatures or ultraviolet light are necessary to initiate and maintain the reaction. Since further replacement of hydrogens occurs readily, the reaction furnishes a mixture of compounds.

$$CH_4 \xrightarrow[\substack{\text{heat or} \\ \text{ultraviolet light}}]{Cl_2} CH_3Cl \quad +$$

Methyl chloride
(chloromethane)

$$CH_2Cl_2 \quad + \quad CHCl_3 \quad + \quad CCl_4$$

Methylene chloride   Chloroform   Carbon tetrachloride
(dichloromethane)   (trichloromethane)   (tetrachloromethane)

Methylene chloride is used for degreasing metal parts and in paint removers. Chloroform and carbon tetrachloride are common solvents.

Most of the hydrogen used in the chemical industry is produced by heating methane with water at a high temperature.

$$CH_4 + H_2O \longrightarrow CO + 3\,H_2$$

A significant portion of this hydrogen is used to produce ammonia by reaction with nitrogen at high temperature and pressure.

$$N_2 + 3\,H_2 \xrightarrow{\;\text{iron catalyst}\;} 2\,NH_3$$

Most of the ammonia is used directly as a fertilizer or is converted to other compounds also used as fertilizers. Therefore the cost and availability of nitrogen-containing fertilizers are linked closely to the price and supply of methane.

1.2
THE COVALENT BOND

An understanding of how atoms are attached together to form molecules provides a basis for comprehending the vast array of organic molecules. Section 1.1 presented the molecular formulas of a variety of simple molecules. **A molecular formula** indicates the types of atoms and the number of each in a molecule. In this section we consider the structural formulas of the same molecules. **A structural formula** is a representation of a molecule that indicates which atoms are bonded to which. To draw structural formulas, we must consider briefly the nature of the chemical bonds that link atoms together in molecules. Chemical bonding will be considered again in Chapter 4 after we have become more familiar with organic molecules.

Most chemical bonds in organic molecules are of the type called **covalent**. All bonds of the simple molecules in Section 1.1 are covalent. Before considering covalent bonds, we need to review some material found in any general chemistry course.

Recall that an atom contains a small, positively charged nucleus. The nucleus consists of protons, which are positively charged, and neutrons, which are neutral. The atomic number is the number of protons in the nucleus—and hence the magnitude of the positive charge of the nucleus. The atomic number characterizes the atoms of a particular element; all hydrogen atoms have atomic number 1, all carbon atoms 6, and all chlorine atoms 17. Electrons, which are negatively charged, surround the nucleus. In a neutral atom the number of electrons equals the atomic number. The electrons are arranged in different shells. Only the electrons in the outermost or **valence** shell are usually involved in the formation of chemical bonds. The number of valence electrons is 1 for hydrogen, 4 for carbon, 5 for nitrogen, 6 for oxygen, and 7 for chlorine. It is sometimes convenient to indicate the valence electrons by dots.

$$\text{H}\cdot \qquad \cdot\overset{\displaystyle\cdot}{\underset{\displaystyle\cdot}{\text{C}}}\cdot \qquad \cdot\overset{\displaystyle\cdot}{\text{N}}\text{:} \qquad \cdot\overset{\displaystyle\cdot}{\underset{\displaystyle\cdot}{\text{O}}}\text{:} \qquad \cdot\overset{\displaystyle\cdot}{\underset{\displaystyle\cdot}{\text{Cl}}}\text{:}$$

A **covalent bond** results when atoms share an electron pair. Consider the hydrogen molecule, H_2. Its covalent bond results from the hydrogen atoms sharing the only two electrons of this simple molecule. Heat is liberated on formation of the hydrogen molecule from hydrogen atoms.

$$\text{H}\cdot + \text{H}\cdot \longrightarrow \text{H:H} + \text{heat}$$

Application of an equivalent amount of energy (104 kcal/mole of hydrogen molecules) is required to reverse this reaction. The hydrogen molecule is more stable than two hydrogen atoms because each electron is attracted by two positive nuclei rather than one. Although moving constantly, electrons spend more time in the region between the nuclei than elsewhere. The sum of the attractive forces between the positive nuclei and negative electrons is greater in a hydrogen molecule than in two isolated hydrogen atoms.

In the chlorine and hydrogen chloride molecules one pair of electrons also is shared.

$$: \overset{..}{\underset{..}{Cl}} : \overset{..}{\underset{..}{Cl}} : \qquad H : \overset{..}{\underset{..}{Cl}} :$$

In the water and ammonia molecules one electron pair is shared between each hydrogen and the oxygen or nitrogen.

$$\begin{matrix} H : \overset{..}{\underset{..}{O}} : & \quad & H : \overset{..}{N} : H \\ H & & H \end{matrix}$$

Note that not all valence electrons are necessarily involved in covalent bonds.

How many shared electron pairs surround an atom? In stable compounds the total number of valence electrons (including all shared electrons) surrounding an atom is usually the maximum number of electrons that its valence shell can hold. This number is 2 for H and 8 for C, N, O, and Cl. To achieve these numbers of electrons, the number of shared electron pairs in neutral molecules is 1 for H and Cl (and also for F, Br, and I), 2 for O, 3 for N, and 4 for C.

It is customary to represent the shared electron pair of a covalent bond by a line instead of a pair of dots. Therefore the structural formulas of molecules considered previously can be written

$$\begin{matrix} H{-}H & \quad H{-}\overset{..}{\underset{..}{Cl}} : & \quad H{-}\overset{..}{O} : & \quad H{-}\overset{..}{N}{-}H \\ & & | & | \\ & & H & H \end{matrix}$$

Structural formulas are usually further simplified by omitting any unshared electrons (although sometimes it will be important for us to remember that they are present).

$$\begin{matrix} H{-}Cl & \quad H{-}O & \quad H{-}N{-}H \\ & | & | \\ & H & H \end{matrix}$$

In the methane molecule one electron pair is shared between each hydrogen and the carbon.

$$\begin{matrix} & H & & & H & \\ & \overset{..}{\underset{..}{}} & & & | & \\ H : & C & : H & \quad H{-} & C & {-}H \\ & H & & & | & \\ & & & & H & \end{matrix}$$

To be surrounded by eight valence electrons, carbon atoms in other organic compounds, such as methyl chloride and chloroform, also have four covalent bonds.

$$\begin{matrix} & Cl & & & Cl & \\ & | & & & | & \\ H{-} & C & {-}H & \quad Cl{-} & C & {-}Cl \\ & | & & & | & \\ & H & & & H & \end{matrix}$$

A carbon dioxide molecule has only three atoms and thus cannot possibly have four atoms bonded to carbon. The carbon is surrounded, however, by eight valence electrons, because carbon and each oxygen share *two electron pairs*.

$$:\ddot{O}::C::\ddot{O}: \qquad O=C=O$$

Note that each oxygen atom also is surrounded by eight valence electrons. Each nitrogen atom of the nitrogen molecule also is surrounded by eight electrons, because the nitrogen atoms share three electron pairs.

$$:N:::N: \qquad N\equiv N$$

In the acetylene molecule each carbon atom shares three electron pairs with the other carbon atom and one electron pair with a hydrogen atom.

$$H:C:::C:H \qquad H-C\equiv C-H$$

---

problem 1.1   Using dots to represent electrons, draw a structural formula for the poisonous gas HCN (hydrogen cyanide).

A covalent bond that results when two atoms share one electron pair is called a **single bond**. Covalent bonds that result from the sharing of more than one electron pair are called **multiple bonds**. A multiple bond is a **double bond** if two pairs are shared and a **triple bond** if three pairs are shared. Note that the total number of bonds to an atom is the same in compounds that have double or triple bonds as in compounds having only single bonds. The carbon atoms of methane, carbon dioxide, and acetylene all have four bonds. We would have arrived at the structural formulas of the compounds considered earlier simply by fitting together, by trial and error, atoms surrounded by the appropriate numbers of covalent attachments:

$$H- \qquad Cl- \qquad -O- \qquad -N- \qquad -\overset{\displaystyle |}{\underset{\displaystyle |}{C}}-$$

---

problem 1.2   Draw structural formulas for the following compounds.
(a) CH_2O (b) C_2H_4

Charges sometimes appear in structural formulas. The ammonium ion, obtained by protonation of ammonia, is a familiar example.

$$H:\overset{\displaystyle ..}{N}:H + H^+ \longrightarrow H:\overset{\displaystyle \overset{H}{..}}{\underset{..}{N}}:H$$
$$\overset{}{\underset{\displaystyle H}{}} \qquad\qquad\qquad\qquad \overset{}{\underset{\displaystyle H}{}}$$

Because it can be formed by combination of a neutral molecule and a positive ion, the ammonium ion obviously has a positive charge. From its electron dot structure, we would also know that it is positive and that it is reasonable to place the charge on nitrogen.

An unbonded atom is neutral if the number of electrons around its nucleus equals the number of protons in its nucleus (the atomic number). The negative charge of the one electron of a hydrogen atom, $H\cdot$, balances the positive charge of the one proton in its nucleus. H^+ lacks the electron, however, and hence is positively charged.

It is convenient to consider that a covalently bonded atom owns one-half the electrons shared with other atoms. On this basis, each atom in the compounds considered earlier owns the same number of electrons that surround it when it is unbonded and neutral. The nitrogen in NH_3, for example, owns five valence electrons—the two unshared electrons and one electron of each of the three shared electron pairs. The nitrogen atom of NH_4^+, however, owns one electron from each of the four shared electron pairs, a total of only four electrons. By this manner of counting, the nitrogen of NH_4^+ has one electron less than a neutral nitrogen atom and hence has a positive charge. The charges of covalently bonded atoms determined by this bookkeeping process are called **formal charges**.

1.3
ELECTRONEGATIVITY

In such molecules as H_2 or Cl_2 identical nuclei equally attract and share the electron pair that forms the covalent bond. In HCl or H_2O, however, the atoms bonded together are not identical and their nuclei attract unequally a shared electron pair. In HCl, for example, the shared electron pair is attracted somewhat more by the chlorine nucleus than by the hydrogen nucleus. As a result, the shared electrons are somewhat nearer to chlorine than would be the case for equal sharing.

$$H :\overset{..}{\underset{..}{Cl}}:$$

The chlorine has somewhat more than its share of electrons and therefore is partially negative; conversely, the hydrogen has somewhat less than its share of electrons and is partially positive. The charge separation is sometimes shown by using the symbol δ (delta) to indicate partial charges.

$$^{\delta^+}H—Cl^{\delta^-}$$

Bonds in which electron pairs are shared unequally are called **polar covalent bonds**.

The tendency of an atom to attract shared electrons is called its **electronegativity**. Electronegativity can be expressed on a numerical scale. One widely used set of electronegativity values is listed in Figure 1.1. Electronegativity increases from left to right in a horizontal row of the periodic table because electrons are attracted by an increasing nuclear charge. It decreases in going down a group because the valence

H 2.2						
Li 1.0	Be 1.6	B 2.0	C 2.5	N 3.0	O 3.4	F 4.0
Na 0.9	Mg 1.3	Al 1.6	Si 1.9	P 2.2	S 2.6	Cl 3.2

Figure 1.1 Electronegativities of some atoms.

electrons are shielded from the nucleus by an increasing number of inner-shell electrons.

The greater the difference in electronegativity between two atoms, the greater the difference between their abilities to attract a shared electron pair and the more polar the bond between them. The polarity of bonds due to unequal sharing of electrons influences some properties of organic molecules. Note that carbon and hydrogen have nearly equal electronegativities. Carbon-hydrogen bonds, so common in organic compounds, are nonpolar.

1.4
SHAPES OF MOLECULES

Methane does not have the planar geometry implied by the structural formulas that we have drawn for this molecule. Each hydrogen of the methane molecule is separated from the other three hydrogens by the same distance. The H—C—H angles all are 109°28′. This is the angle, sometimes called the **tetrahedral angle**, that results when the carbon is placed in the middle and the hydrogens at the vertices of a regular tetrahedron.

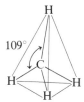

The following is an alternate representation of the tetrahedral geometry of methane.

Similar geometries are generally observed for other compounds in which carbon is joined by single bonds to four groups. Carbon tetrachloride, for instance, also has a tetrahedral geometry.

$$\underset{\underset{Cl}{|}}{\overset{\overset{Cl}{|}\;109°}{\underset{\displaystyle Cl}{C}}} \text{—Cl}$$

Even in compounds, such as methyl chloride or chloroform, in which the groups bonded to carbon are not all the same, deviations of the bond angles from 109° are usually small.

Molecules that have doubly or triply bonded carbons exhibit significantly different geometries. Acetylene, for example, is linear.

$$\overset{180°}{H \text{—} C \equiv C \text{—} H}$$

We face a problem when we wish to represent three-dimensional molecules on the two-dimensional surface of a sheet of paper. Various expedients can be used to suggest three-dimensional geometry. We will sometimes use a convention illustrated earlier: bonds in the plane of the paper are represented by solid lines (—), bonds pointing out of the page (toward the reader) by wedges (◄), and bonds pointing behind the page by broken lines (---). Usually, however, we will represent all bonds with solid lines and must mentally supply the geometry.

1.5
MOLECULAR MODELS

The geometries of molecules are more apparent if the molecules are represented by appropriate models. In Figure 1.2 methane is represented by three kinds of models in common use. In **ball-and-stick** models the balls represent atoms and the rods serve as the chemical bonds linking them. In **framework** models the atoms are understood to be at the intersections and ends of the rods. **Space-filling** models represent more accurately the relative sizes of atoms and the volumes occupied by molecules. As atoms that are not bonded to each other are pushed closely together,

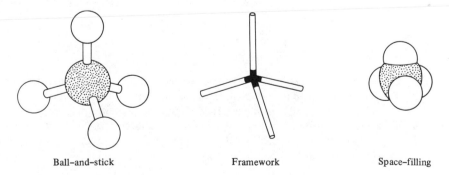

Ball–and–stick   Framework   Space–filling

Figure 1.2   Three molecular models of methane.

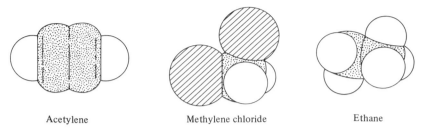

Acetylene   Methylene chloride   Ethane

Figure 1.3   Space-filling models of acetylene, methylene chloride, and ethane.

they begin to repel due to repulsion of their electron clouds. For this reason, even the application of high pressures reduces the volumes of liquids relatively little. A space-filling model of a molecule indicates the approximate boundaries within which other molecules cannot easily penetrate. Figure 1.3 shows space-filling models of several other molecules. Although space-filling models more accurately represent relationships involving size, ball-and-stick or framework models are often more useful in showing geometrical relationships between atoms.

---

exercise with models 1.1   Construct a model of methane. Note that the environments of the four hydrogens are identical. Convince yourself that if two carbon-hydrogen bonds are held in the plane of a sheet of paper, then the other two bonds are respectively above and below the plane of the paper, as indicated by the wedge and the broken line in the following drawing.

$$\underset{H}{\overset{H}{H-\overset{|}{\underset{|}{C}}-H}}$$

exercise with models 1.2   Construct a model of methylene chloride. Construct other models of the same molecule and note that they are always identical. Drawings such as

$$Cl-\overset{\overset{\displaystyle H}{|}}{\underset{\underset{\displaystyle H}{|}}{C}}-Cl \quad \text{and} \quad Cl-\overset{\overset{\displaystyle Cl}{|}}{\underset{\underset{\displaystyle H}{|}}{C}}-H$$

might lead us incorrectly to believe that geometrically different dichloromethanes exist. These drawings must represent the same molecule. The observation that only one compound of molecular formula CH_2Cl_2 could be prepared provided some of the earliest evidence that four groups bonded to carbon could not lie in a plane.

Convince yourself by making models that only one geometry is possible for methyl chloride or chloroform.

Ethane, a colorless and odorless gas of boiling point $-89°$, has the molecular formula C_2H_6. In many of its properties it resembles methane. Ethane, for example, is relatively inert but burns to carbon dioxide and water with the evolution of heat. In a reaction initiated by heat or ultraviolet light ethane reacts with chlorine to give products in which hydrogen atoms have been replaced by chlorine atoms: C_2H_5Cl, $C_2H_4Cl_2$, $C_2H_3Cl_3$ . . . .

The two carbons of ethane are linked by a single bond, and each carbon also is bonded by single bonds to three hydrogens.

$$
\begin{array}{cc}
\text{H H} & \text{H H} \\
\text{H:C:C:H} & \text{H—C—C—H} \\
\text{H H} & \text{H H}
\end{array}
$$

No other structure for C_2H_6 places four bonds at each carbon and one at each hydrogen (eight valence electrons around each carbon and two around each hydrogen).

Ethane is not a flat molecule. As in other compounds in which carbons are bonded to four groups, the groups attached to each carbon of ethane are arranged tetrahedrally. The tetrahedral arrangement around each carbon does not specify one aspect of geometry. An infinite number of geometrical relationships between groups attached to one carbon and those attached to the other carbon are possible. As one extreme, ethane can have a **staggered** geometry, the hydrogens attached to one carbon lying exactly between those attached to the other carbon. As another extreme, ethane can have an **eclipsed** geometry, the hydrogens on one carbon lying directly behind those on the other carbon. A multitude of possible geometries lie between these extremes. Three alternative representations useful for indicating this aspect of geometry are shown in Figure 1.4. The third representation of each geometry, called a Newman projection, is a view directly down the carbon-carbon bond. Bonds to the front carbon intersect at the center of the circle. Bonds to the back carbon are drawn only to the edge of the circle. Different geometrical arrangements of a molecule, such as the staggered and eclipsed forms of ethane, that can be interconverted by rotation around single bonds are called **conformations**.

The staggered conformation is the most stable and, at any instant, most molecules in a sample of ethane have geometries close to staggered. The eclipsed conformation is the least stable, and few molecules have geometries close to eclipsed. A hydrogen atom is so small that the instability of the eclipsed conformation is not due to hydrogens bonded to one carbon pushing directly against those bonded to the other carbon. Instead it arises from a preference for maintaining the maximum possible distance between the electron pairs surrounding one carbon and those surrounding the other carbon. Although the staggered conformation is more stable,

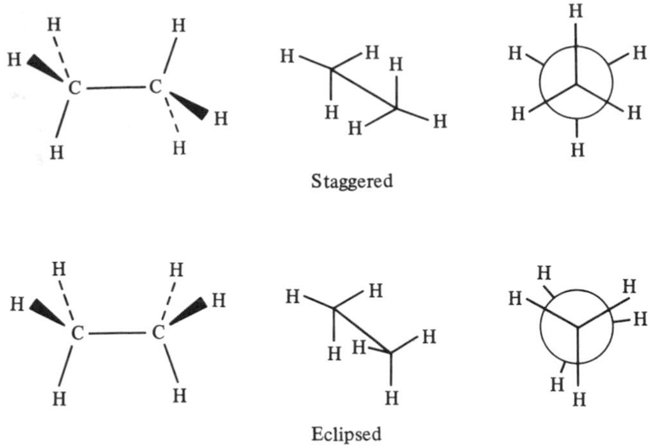

Staggered

Eclipsed

Figure 1.4 Three representations of the staggered and eclipsed conformations of ethane. The eclipsed conformation pictured in the right-hand representation is slightly distorted so that all bonds to hydrogens will be visible.

relatively little energy is required to reach the eclipsed conformation. At normal temperatures most molecules possess enough energy for frequent rotation (thousands of times a second) around the carbon-carbon bond, a motion that requires passing through the eclipsed conformation.

---

exercise with models 1.3 Construct a model of ethane. Rotate one carbon relative to the other, noting that identical staggered conformations are formed every one-third of a turn. Convince yourself that the hydrogens on adjoining carbons are as distant as possible when ethane is in the staggered conformation.

Note that (regardless of the conformation) all hydrogens in a given ethane molecule have identical environments. Convince yourself that replacement of any hydrogen atom by a chlorine atom leads to the same molecule; there is only one molecule having the molecular formula C_2H_5Cl.

1.7
PROPANE

Propane, a compound of molecular formula C_3H_8, is related to methane and ethane. Its boiling point is $-42°$, so it is a gas at ordinary temperatures and pressures. Because it can be compressed more readily than methane, it is often used as a bottled gas. Three carbons and eight hydrogens can combine in only one way to give a structure with four bonds at each carbon and one bond at each hydrogen.

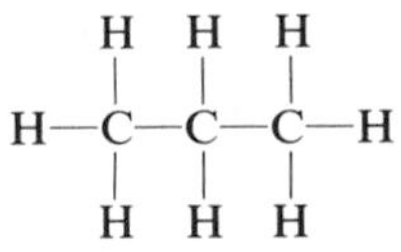

$$H-\underset{\underset{H}{|}}{\overset{\overset{H}{|}}{C}}-\underset{\underset{H}{|}}{\overset{\overset{H}{|}}{C}}-\underset{\underset{H}{|}}{\overset{\overset{H}{|}}{C}}-H$$

exercise with models 1.4   Construct a model of propane. Adjust it so that the groups around both carbon-carbon bonds are staggered. This is the most stable conformation of propane.

1.8
BUTANES

The molecules methane (CH_4), ethane (C_2H_6), and propane (C_3H_8) form a series in which each member contains one carbon and two hydrogens more than the preceding member. The next molecular formula in this series is C_4H_{10}. Two compounds having this formula are known. The properties of these compounds, although similar in many respects, are not identical. For example, the compound called butane has a boiling point of $0°$ and the compound called isobutane a boiling point of $-12°$. Compounds that have the same molecular formula but are different are called **isomers**.

If we try to draw structures that place four bonds at each carbon and one bond at each hydrogen, we find two possibilities for C_4H_{10}.

$$
\begin{array}{cc}
 & \begin{array}{c} H \\ | \\ H-C-H \end{array} \\
\begin{array}{c} H \ \ H \ \ H \ \ H \\ | \ \ | \ \ | \ \ | \\ H-C-C-C-C-H \\ | \ \ | \ \ | \ \ | \\ H \ \ H \ \ H \ \ H \end{array} &
\begin{array}{c} H \ \ \ \ \ H \\ | \ \ \ \ \ | \\ H-C-C-C-H \\ | \ \ | \ \ | \\ H \ \ H \ \ H \end{array} \\
\text{Butane} & \text{Isobutane}
\end{array}
$$

One structure has a continuous chain of four carbon atoms; the other structure has a chain of only three carbon atoms with the other carbon linked to the central carbon of the chain. Isomers, such as butane and isobutane, that differ in which atoms are bonded to which are called **constitutional isomers** (the designation structural isomers is also widely used).

As was the case with ethane, the butanes most commonly have conformations in which arrangements around the carbon-carbon bonds are approximately staggered. As illustrated in Figure 1.5, however, butane has more than one staggered arrangement around the central carbon-carbon bond. The conformation with the —CH_3 groups (called methyl groups) as far apart as possible is named **anti**; the conformations with these groups nearer are named **gauche**.

At room temperature about 80% of the molecules in a sample of butane have the anti conformation and about 20% a gauche conformation. The slightly lesser stability of a gauche conformation is due to a slight crowding together of the methyl groups, which is avoided in the anti conformation. The anti and gauche conformations are interconverted rapidly by rotation around the central carbon-carbon bond.

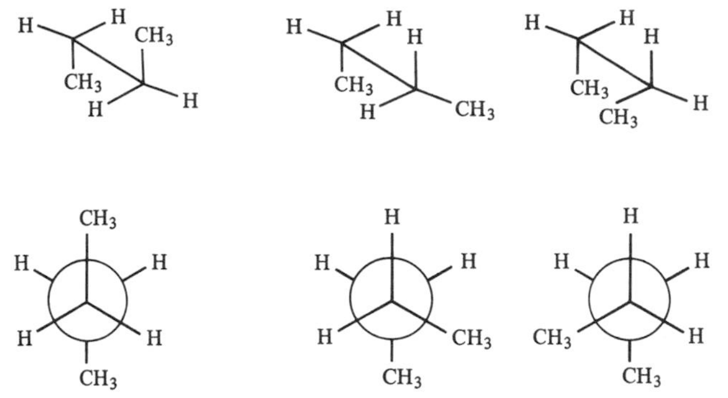

Anti conformation   Gauche conformations

Figure 1.5 Two representations of the anti and gauche conformations of butane.

As a result, samples of butane cannot readily be separated into portions containing molecules with only anti or only gauche conformations. Under any given conditions all samples of butane contain the same mixture of conformations. Therefore we do not consider various conformations to be different compounds, just as we do not consider coiled and stretched snakes to be different creatures.

---

exercise with models 1.5   Construct a model of the molecule

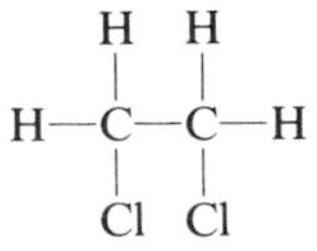

$$
\begin{array}{ccc}
& \text{H} & \text{H} \\
& | & | \\
\text{H}- & \text{C}-\text{C} & -\text{H} \\
& | & | \\
& \text{Cl} & \text{Cl}
\end{array}
$$

If the balls or rods used to represent chloro groups are imagined to be methyl groups, then this model also represents butane. Place the molecule in the conformation with the chloro (or methyl) groups as far apart as possible. This is the anti conformation. Note that it is a staggered conformation. Now rotate one carbon by exactly one-third of a turn. The staggered conformation that results is a gauche conformation. Rotate the carbon another one-third of a turn in the same direction. This step again produces a gauche conformation. Another one-third of a turn returns the model to the anti conformation.

1.9
THE PROBLEM OF DRAWING STRUCTURES

Deciding whether two structural drawings represent the same or different compounds can give us some puzzling moments when we first encounter organic structures. This problem arises because three-dimensional molecules are represented on the two-dimensional surface of a sheet of paper. The most convenient ways of drawing structural formulas have misleading aspects. The following drawings appear different.

```
                                    H
                                    |
                    H           H—C—H           H     H
    H  H  H         |               |              \   /
    |  |  |      H—C—H              H               C
 H—C—C—C—H       |               |              / \
    |  |  |      H—C—H          H—C—C—H      H—C     C—H
    H  H  H         |             / \         / \   / \
                 H—C—H          H  H         H  H  H  H
                    |
                    H
```

They are, however, equivalent structural formulas of propane. All contain the same structural information: three carbons are linked by single bonds into a chain; the other bonds at each carbon are to hydrogens. A molecule that has a tetrahedral arrangement of groups around each carbon has been squashed onto the paper. The different drawings result from positioning the propane molecule in different ways over the paper and bending the bonds in different directions in the flattening process. The resulting formulas do not represent shapes—they indicate only which atoms are bonded together.

Similarly, butane can be represented by a variety of drawings, but all will have four carbon atoms linked by single bonds into a chain. In any drawing of isobutane, the longest chain has only three carbon atoms and the remaining carbon atom is linked to the middle carbon of the chain.

---

problem 1.3   Which of the following structural formulas are of butane and which of isobutane?

(a)
```
    H  H
    |  |
 H—C—C—H
    |  |
    |  H
 H—C—H
    |
 H—C—H
    |
    H
```

(b)
```
       H
       |
    H—C—H
       |
       H
       |
 H—C—C—H
    |  |
    H  H—C—H
       |
       H
```

(c)
```
              H
             /
        H   C—H
         \ / \
      H—C    H
        / \
       H   C—H
          / \
         H   H
        H
```

(d)
```
              H
             /
     H  H   C—H
     |  |  /
  H—C—C      H
     |  \   /
     H   C
          \
           H
          H
```

(e)
```
    H   H
    |   |
 H—C———C—H
    |   |
 H—C—H
    |
 HH—C—H
    |
    H
```

(f)
```
    H  H
    |  |
 H—C—C—H
    |  |
    H  H
    |  |
 H—C—C—H
    |  |
    H  H
```

exercise with models 1.6 Construct models of butane and isobutane. Rotate each carbon relative to the others until convinced that rotations round carbon-carbon bonds never make the models identical. Construct models of compounds having molecular formula C_4H_{10} until convinced that butane and isobutane are the only possible structures.

1.10
THE IONIC BOND

Amidst the covalent bonds that predominate in organic compounds, an ionic bond occasionally lurks. This is a good point at which to review briefly what was learned in an earlier course about such bonds.

Consider the familiar ionic compound sodium chloride. In an aqueous solution of sodium chloride, sodium ions (Na^+) and chloride ions (Cl^-) are separated by solvent molecules and move about relatively independently. In solid sodium chloride the same ions are packed closely together (Fig. 1.6). Each sodium ion is surrounded equally by six chloride ions and each chloride ion by six sodium ions. The formula NaCl for this compound expresses only the fact that the numbers of sodium and chloride ions are equal—ordinarily there are no discrete NaCl molecules.

The **ionic bond** is simply the electrostatic attraction between nearby positive and negative ions. An ionic bond between two atoms results not by their sharing a pair of electrons but by transfer of one (or more than one) electron from one atom to the other to produce ions that attract each other electrostatically.

$$Na\cdot \; + \; \cdot \ddot{\underset{\cdot\cdot}{Cl}} : \quad \longrightarrow \quad Na^+ \; + \; : \ddot{\underset{\cdot\cdot}{Cl}} :^-$$

In ionic compounds, as in covalent compounds, the number of electrons around each atom often corresponds to that associated with a filled valence shell.

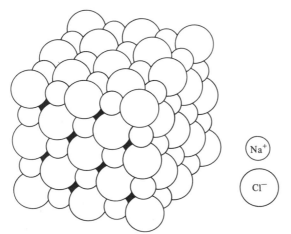

Figure 1.6 The crystal structure of sodium chloride.

In sodium chloride the chloride ion has one more electron than a neutral chlorine atom and therefore eight valence electrons. The sodium ion has lost the one electron that occupies the valence shell of a neutral sodium atom. The outermost shell that still contains electrons, however, has eight electrons and is full.

Ionic bonds are found between elements that, because of greatly different electronegativities, cannot share electrons effectively. We particularly encounter ionic bonds between an element of low electronegativity, such as sodium, potassium, or calcium, and an element of high electronegativity, such as nitrogen, oxygen, fluorine, or chlorine. Because of the intermediate electronegativity of carbon and because it must gain or lose four electrons to form an ion with a filled outer shell, its bonds to most other atoms are ordinarily covalent.

SUMMARY

Definitions and Ideas

Organic chemistry is the chemistry of compounds containing carbon. Most of the compounds of living systems are organic.

A **molecular formula** indicates the types of atoms and the numbers of each in a molecule. A **structural formula** also indicates which atoms are bonded to which.

A **covalent bond** results when two atoms share one or more pairs of electrons. The electrons of an atom involved in bonding are **valence electrons**, the electrons in the outermost shell. A **single bond** between two atoms results when they share one electron pair. Two pairs of electrons are shared in a **double bond** and three in a **triple bond**.

Different atoms bonded together generally attract a shared electron pair unequally. The tendency of an atom to attract a shared electron pair is called its **electronegativity**. Covalent bonds in which sharing is unequal are called **polar covalent bonds**.

In most stable compounds that contain covalent bonds the number of valence electrons is 2 around each H and 8 around each C, N, O, or halogen (F, Cl, Br, I). As a result, in neutral molecules the number of covalent bonds is 1 for H and the halogens, 2 for O, 3 for N, and 4 for C:

$$\text{H—} \quad \text{F—} \quad \text{Cl—} \quad \text{Br—} \quad \text{I—} \quad \text{—O—} \quad \text{—N—} \quad \text{—C—}$$

When carbon is bonded to four groups, the bond angles are about 109° (the **tetrahedral angle**).

Conformations are different geometrical arrangements of a molecule that can be interconverted by rotation around single bonds. Rotation around carbon-carbon single bonds is usually rapid.

A **staggered** conformation is more stable than an **eclipsed** conformation. The **anti** conformation of butane is more stable than a **gauche** conformation.

Isomers are compounds that have the same molecular formula but differ in some way. Isomers, such as butane and isobutane, that differ in which atoms are bonded to which are called **constitutional isomers** (or structural isomers).

An **ionic bond** results from attraction between positive and negative ions.

PROBLEMS

1. Using dots to represent valence electrons, draw structural formulas for the following compounds.
   (a) CCl_4
   (b) H_2O_2
   (c) $COCl_2$
   (d) CH_4O
   (e) N_2H_4
   (f) CH_5N
   (g) C_2H_5Cl
   (h) CH_2O_2

2. Using lines to represent shared electron pairs and omitting unshared electrons, draw structural formulas for the compounds in problem 1.

3. Using dots to represent valence electrons, draw structural formulas for the following ions.
   (a) OH^-
   (b) $HCO_3{}^-$
   (c) H_3O^+

4. What, if anything, is wrong with the following molecular formulas?
   (a) CH_2Cl
   (b) C_3H_8
   (c) C_2H_3
   (d) C_2H_7Cl
   (e) C_3H_7Cl
   (f) C_4H_9

5. What, if anything, is wrong with the following structural formulas?

   (a)
   $$H-\underset{\underset{H}{|}}{C}=\underset{\underset{H}{|}}{C}-H$$

   (b)
   $$H-\overset{H}{\underset{H}{C}}-H$$

   (c)
   $$H-\underset{\underset{H}{|}}{\overset{\overset{H}{|}}{C}}-H-\underset{\underset{H}{|}}{\overset{\overset{H}{|}}{C}}-H$$

   (d)
   $$\overset{H}{\underset{H}{\diagdown}}C\underset{\underset{H}{|}}{\overset{\overset{Cl}{|}}{-}}\underset{\underset{H}{|}}{\overset{\overset{H}{|}}{C}}-H$$

   (e)
   $$H-\overset{H}{\underset{\underset{H-\underset{\underset{H}{|}}{C}-\underset{\diagdown H}{C}-H}{|}}{\overset{\diagup H}{C}}}$$

   (f)
   $$H-\underset{\underset{H}{|}}{\overset{\overset{H}{|}}{C}}-\underset{\underset{H}{|}}{\overset{\overset{H}{|}}{C}}-\overset{\overset{H}{|}}{C}-H$$

6. Using the ordinary valences for each atom ($H = 1$, $Cl = 1$, $O = 2$, $N = 3$, $C = 4$), draw structural formulas for all compounds that have the following molecular formulas.
   (a) C_2H_6O
   (b) C_2H_7N
   (c) C_3H_6
   (d) $C_2H_4Cl_2$
   (e) C_3H_7Cl
   (f) C_5H_{12}

7. The gauche and anti conformations of butane differ as much in shape as do butane and isobutane. Why, then, are butane and isobutane considered distinct compounds but the gauche and anti conformations of butane the same compound?

8. When $AgNO_3$ is added to an aqueous solution of NaCl, a precipitate of AgCl forms instantaneously. No precipitate forms, however, when $AgNO_3$ is shaken with CH_3Cl. Explain why these simple chlorine-containing compounds behave so differently.

9. Describe the bonding in ammonium chloride (NH_4Cl).

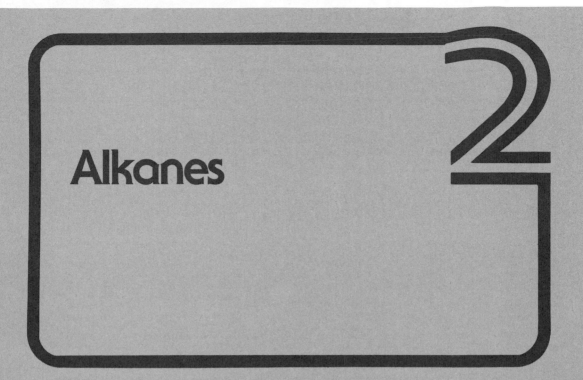

Alkanes

Methane, ethane, propane, and the butanes are members of the family of compounds called **alkanes**. Alkanes are one group of **hydrocarbons**, organic molecules that contain only carbon and hydrogen atoms. An alkane is a hydrocarbon that contains only single bonds.

Alkanes are a convenient family to examine first in our survey of organic chemistry. By studying them, we will become acquainted with many of the ways in which carbons are linked together to form the skeletons of organic molecules. Moreover, alkanes are the principal constituents of natural gas and petroleum and therefore our most significant energy source and the major raw material for the organic chemical industry.

The alkanes considered so far have molecular formulas that can be expressed by the equation C_nH_{2n+2}, where n is 1, 2, 3 . . . . As indicated by the last column of Table 2.1, the number of isomers in this series increases rapidly as the number of carbon atoms increases. Because many alkanes often have the same molecular formula, structural formulas are essential in considering these compounds.

Table 2.1   Names, structural formulas, boiling points, and densities of linear alkanes, and the numbers of compounds having the same molecular formulas

Name	Molecular formula	Structural formula	Boiling point (°C)	Density at 20°C	Number of compounds having this molecular formula
Methane	CH_4	CH_4	-162		1
Ethane	C_2H_6	CH_3CH_3	-89		1
Propane	C_3H_8	$CH_3CH_2CH_3$	-42		1
Butane	C_4H_{10}	$CH_3(CH_2)_2CH_3$	0		2
Pentane	C_5H_{12}	$CH_3(CH_2)_3CH_3$	36	.626	3
Hexane	C_6H_{14}	$CH_3(CH_2)_4CH_3$	69	.659	5
Heptane	C_7H_{16}	$CH_3(CH_2)_5CH_3$	98	.684	9
Octane	C_8H_{18}	$CH_3(CH_2)_6CH_3$	126	.703	18
Nonane	C_9H_{20}	$CH_3(CH_2)_7CH_3$	151	.718	35
Decane	$C_{10}H_{22}$	$CH_3(CH_2)_8CH_3$	174	.730	75
Undecane	$C_{11}H_{24}$	$CH_3(CH_2)_9CH_3$	196	.740	159
Dodecane	$C_{12}H_{26}$	$CH_3(CH_2)_{10}CH_3$	216	.749	355
Tetradecane	$C_{14}H_{30}$	$CH_3(CH_2)_{12}CH_3$	254	.763	1,858
Hexadecane	$C_{16}H_{34}$	$CH_3(CH_2)_{14}CH_3$	287	.773	10,359

2.1
PENTANES

There are three isomeric pentanes.

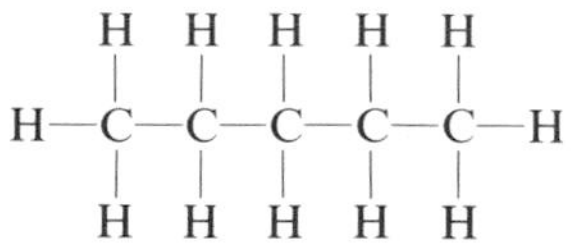

Pentane	Isopentane	Neopentane
bp 36°	bp 28°	bp 9°

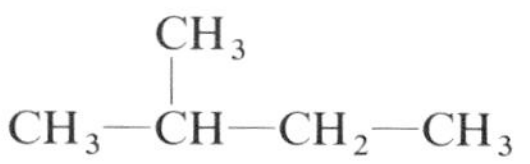

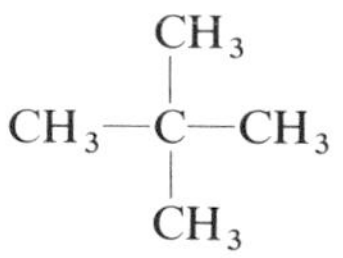

In the representations of isopentane and neopentane, the lines representing carbon-hydrogen bonds are omitted and the hydrogens attached to each carbon are written immediately after it. As we become familiar with structures of alkanes, we will generally use such condensed representations since they are simpler to write and occupy less space.

problem 2.1   Draw representations of isopentane and neopentane that show each bond.

problem 2.2   Draw some C_5H_{12} structural formulas that look different than the three given at the beginning of this section. Determine which pentane isomer is depicted by each of your drawings.

2.2
HEXANES

There are five isomeric hexanes.

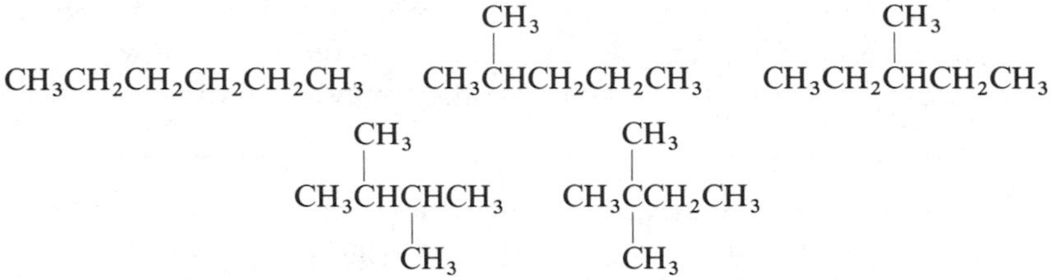

$$CH_3CH_2CH_2CH_2CH_2CH_3 \qquad CH_3\overset{\overset{\displaystyle CH_3}{|}}{C}HCH_2CH_2CH_3 \qquad CH_3CH_2\overset{\overset{\displaystyle CH_3}{|}}{C}HCH_2CH_3$$

$$CH_3\overset{\overset{\displaystyle CH_3}{|}}{\underset{\underset{\displaystyle CH_3}{|}}{C}}HCHCH_3 \qquad CH_3\overset{\overset{\displaystyle CH_3}{|}}{\underset{\underset{\displaystyle CH_3}{|}}{C}}CH_2CH_3$$

These representations are further simplified by omitting horizontal lines representing carbon-carbon bonds. When viewing such structural formulas, we must remember that the bonds to each carbon from the hydrogens written after it and from adjoining carbons should total four. With some practice, we can mentally supply the missing bonds.

A further simplification, used in Table 2.1, is to place a repeated unit in parentheses, followed by a number to indicate the number of repetitions. Thus pentane can be represented by $CH_3(CH_2)_3CH_3$. Another simplification, sometimes used so that a structural formula will fit onto a single line, is illustrated in the following example.

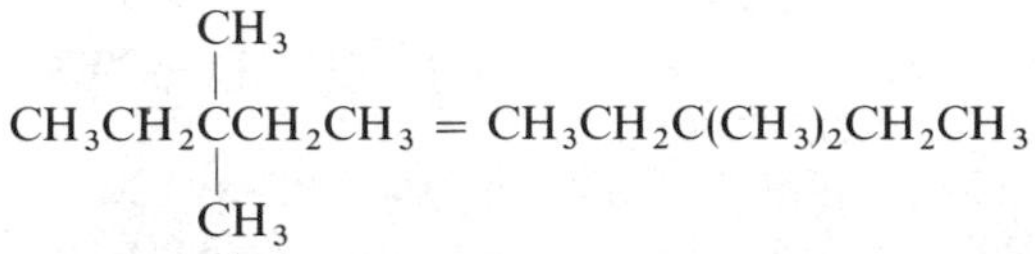

$$CH_3CH_2\overset{\overset{\displaystyle CH_3}{|}}{\underset{\underset{\displaystyle CH_3}{|}}{C}}CH_2CH_3 = CH_3CH_2C(CH_3)_2CH_2CH_3$$

---

problem 2.3 Which isomeric hexane shown at the beginning of this section does each of the following structural formulas represent?

(a) 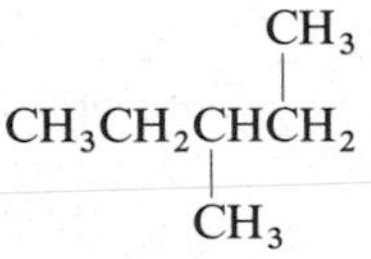
$$CH_3CH_2\overset{\overset{\displaystyle CH_3}{|}}{\underset{\underset{\displaystyle CH_3}{|}}{C}}HCH_2$$

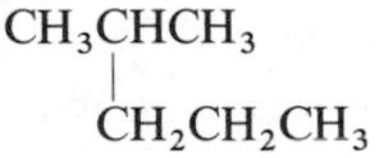

(b) CH_3CHCH_3
$\quad\quad\ CH_2CH_2CH_3$

(c) $CH_3 \quad CH_2CH_3$
$\quad\ \ CH_2CH_2CH_2$

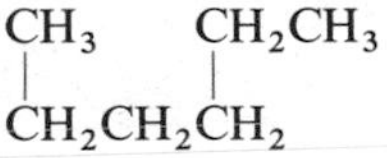

problem 2.4 Redraw the structural formulas of the compounds in Problem 2.3 to show every bond.

2.3
LARGER ALKANES

The number of isomeric compounds having a particular molecular formula increases dramatically as the number of carbon atoms becomes larger. For example, 366,319 reasonable structural formulas can be drawn for the molecular formula $C_{20}H_{42}$ and

36,797,588 for $C_{25}H_{52}$. Most isomeric alkanes containing up to ten carbon atoms and many with more carbons already are known compounds. Almost any one of the larger alkanes could be prepared if it became important to do so. Even though we are considering compounds that contain only carbon-carbon and carbon-hydrogen single bonds, the numbers are astronomical. The bonding together of many carbon atoms and the possibility of the linkages between them occurring in a variety of ways make possible a limitless number of organic molecules.

---

exercise with models 2.1   Construct a model of dodecane (the linear C_{12} alkane) and twist it into different shapes. Convince yourself that even when restricted to staggered arrangements around each carbon-carbon bond, this molecule has an extremely large number of conformations.

2.4
NOMENCLATURE OF ALKANES

As organic chemistry first developed, it was natural to assign a new name to each compound as though it were a newly discovered star or island. This procedure became cumbersome, however, as the number of known compounds grew. The butanes and pentanes were distinguished by using prefixes (iso-, neo-), but there is not sufficient time in most organic chemistry courses to devise and remember different prefixes for the 10,359 $C_{16}H_{34}$ alkanes. The need for a more systematic way of naming compounds is evident.

Rules for naming organic compounds have been codified by the International Union of Pure and Applied Chemistry (IUPAC) and are now in use throughout the world. The IUPAC accepts a few alkane names, such as isobutane, isopentane, and neopentane, that are entrenched by long use. These simple alkanes, however, and more complex alkanes can all be named in a more systematic fashion by applying a few rules.

   1.   Alkanes with linear (unbranched) chains are given names ending in *-ane*.

The names assigned to the most commonly encountered linear alkanes are listed in Table 2.1. It is important to learn these names, for they will be used to form names for branched alkanes and for members of many other families of compounds.

   2.   To name a compound with a branched chain, first select its longest continuous chain of carbon atoms. Use the name of the linear alkane with that number of carbon atoms as the basis for the name.

   3.   Add as a prefix the name of an atom or group of atoms (other than a hydrogen) that is attached to the longest chain of carbons. Number the carbons of the longest chain in sequence, starting from the end of the chain that results in the lower number for the carbon bearing the atom or group. Place the number of the carbon to which it is attached before the name of the atom or group.

Consider the following compound.

$$CH_3$$
$$|$$
$$CH_3CH_2CHCH_2CH_3$$

| 1 | 2 | 3 | 4 | 5 |
| 5 | 4 | 3 | 2 | 1 |

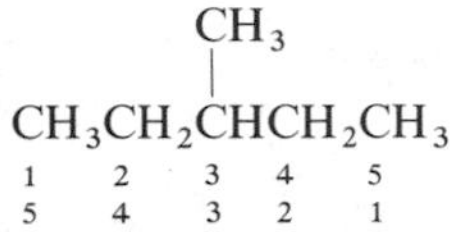

3-Methylpentane

Because its longest chain has five carbons, this compound is named as a pentane. That name does not include the —CH_3 group attached to the middle carbon of the chain. We have already learned that a —CH_3 group is called methyl. If we number the chain from either end, the methyl group is attached to carbon number 3. Therefore the name is 3-methylpentane. Note that the following compound is also a methylpentane.

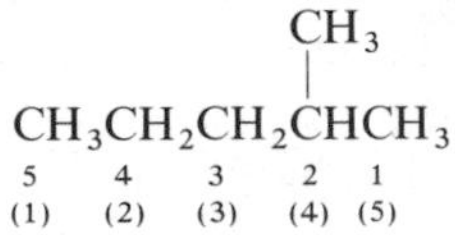

$$CH_3$$
$$|$$
$$CH_3CH_2CH_2CHCH_3$$

| 5 | 4 | 3 | 2 | 1 |
| (1) | (2) | (3) | (4) | (5) |

2-Methylpentane

The methyl group is at carbon number 2 if the chain is numbered from right to left, and this compound is named 2-methylpentane. If we had numbered in the other direction, the methyl group would have been at carbon number 4 and the name would have been 4-methylpentane. We always number the chain from the end that results in the lower number.

An atom or group, such as methyl, that can be considered to replace a hydrogen atom of some simpler compound is called a **substituent**.

4. When more than one kind of substituent is present, name and number each substituent. Number the chain carbons from the end that gives the lowest number to the substituent that is nearest to an end of the chain.

Consider the following examples.

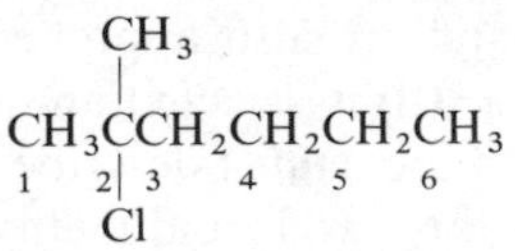

$$CH_3$$
$$|$$
$$CH_3CCH_2CH_2CH_2CH_3$$
1 2| 3   4   5   6
$$Cl$$

2-Chloro-2-methylhexane

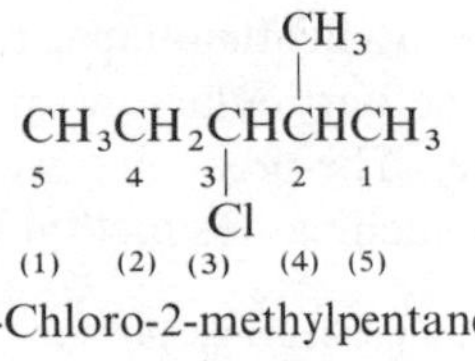

$$CH_3$$
$$|$$
$$CH_3CH_2CHCHCH_3$$
5   4   3| 2   1
$$Cl$$
(1) (2) (3) (4) (5)

3-Chloro-2-methylpentane

Note that a —Cl substituent is called chloro; similarly, F is fluoro, Br is bromo, and I is iodo. Substituents are generally placed in alphabetical order.* Therefore the name shown in the second example is preferred over 2-methyl-3-chloropentane. The chain in this example is numbered from right to left because that makes the lowest number assigned to a substituent a 2 instead of a 3.

5. When the same substituent occurs more than once, indicate how many times by using di-, tri-, tetra-, penta-, and so on. Indicate the position of each substituent by the appropriate number.

This rule is illustrated in the following examples.

$$\underset{\substack{1 \quad 2| \ 3 \quad 4 \quad 5 \\ CH_3}}{\overset{\overset{\displaystyle CH_3 \quad CH_3}{|\qquad |}}{CH_3CCH_2CHCH_3}}$$

$$\underset{\substack{7 \quad 6 \quad 5| \ 4 \quad 3 \quad 2| \ 1 \\ Br \qquad Br \\ (1) \ (2) \ (3) (4) \ (5) \ (6) \ (7)}}{\overset{\overset{\displaystyle CH_3}{|}}{CH_3CH_2CCH_2CH_2CHCH_3}}$$

2,2,4-Trimethylpentane   2,5-Dibromo-5-methylheptane

Note that *each* substituent is assigned a number. Because it has a lower number for the substituent nearest the end of the chain, the name given above for the dibromo compound is used instead of 3,6-dibromo-3-methylheptane, the name that would result from numbering the chain from the other end.

---

problem 2.5 Name the following compounds.

(a) $CH_3-CH_2-CH_2-\underset{\underset{Cl}{|}}{CH}-CH_2-CH_3$ (b) $Cl-\underset{\underset{CH_3}{|}}{CH}-CH_2-CH_2-\underset{\overset{|}{CH_2-CH_3}}{CH}-CH_3$

Names formed by the preceding rules are called **substitutive names**. In forming substitutive names, some simpler compound is selected as the parent compound and every substituent on that parent compound is indicated in the name.

When we discussed the chlorination of methane (Sec. 1.1), two names were assigned to each product. We can now recognize the names in parentheses (chloro-methane, dichloromethane, etc.) as substitutive names formed by the preceding rules. Names like methyl chloride for CH_3Cl and methylene chloride for CH_2Cl_2 are called **radicofunctional names** and are also approved by the IUPAC. In such names one particular group (here —Cl) is selected and becomes the final word of the name. The preceding word (or words) must describe the remainder of the molecule. In such names methyl stands for —CH_3 and methylene for —CH_2—.

---

* In alphabetizing, prefixes like tri that indicate numbers of groups and hyphenated prefixes like *tert-* (Sec. 2.5) are usually ignored.

As did isopentane and neopentane, the names chloroform for $CHCl_3$ and carbon tetrachloride for CCl_4 originated long ago and belong to neither system of nomenclature. The IUPAC tries to discourage the use and hasten the demise of such nonsystematic or "trivial" names. Yielding, however, to our tendency to cling to the familiar, the IUPAC permits use of these and some other trivial names that are firmly entrenched.

It is sometimes useful to classify carbons according to the number of other carbon atoms to which they are directly bonded. A carbon bonded to only one other carbon is called primary (1°). Carbons bonded to 2, 3, or 4 carbons are called secondary (2°), tertiary (3°), or quaternary (4°). The carbons of the following structure are designated in this manner.

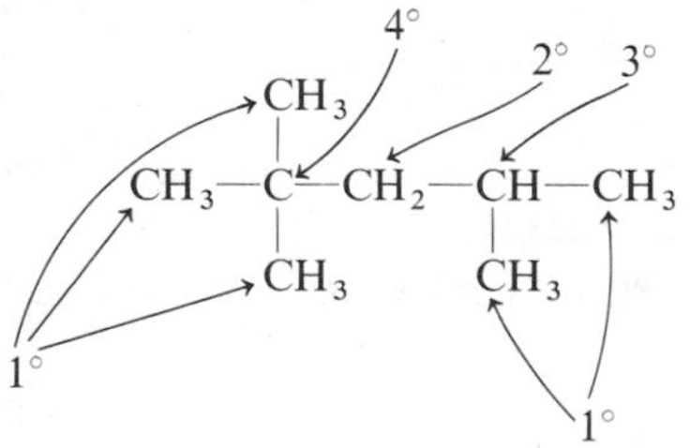

Hydrogens (and other groups) can be classified according to the nature of the carbon to which they are attached. Those attached to a primary carbon, for instance, are primary hydrogens.

2.5
ALKYL GROUPS

How do we name carbon-containing substituents larger than methyl? Consider the following compound.

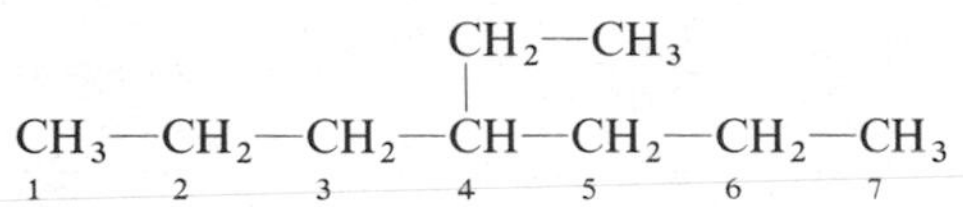

It will be named as a heptane, but we need a name for the two-carbon substituent at carbon number 4.

The $—CH_3$ group corresponds to methane minus a hydrogen and we already know that it is called methyl. The $—CH_2CH_3$ group corresponds to ethane minus a hydrogen and is called ethyl.

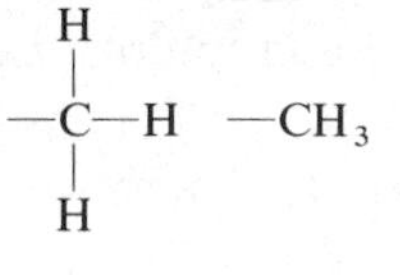

Methyl 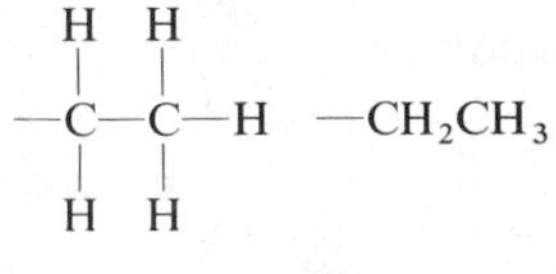   Ethyl

Therefore the preceding alkane is named 4-ethylheptane. Groups that correspond to an alkane minus one hydrogen are called **alkyl** groups. They are named by changing the -*ane* ending of an alkane to -*yl*. Alkyl groups have no independent existence; they are simply structural units of larger molecules to which we assign names that are useful for nomenclature.

The following compounds both have seven-carbon chains to which a three-carbon substituent is attached.

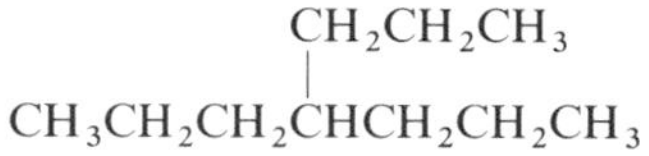

$$CH_2CH_2CH_3$$
$$|$$
$$CH_3CH_2CH_2CHCH_2CH_2CH_3$$

4-Propylheptane

$$CH_3$$
$$|$$
$$CHCH_3$$
$$|$$
$$CH_3CH_2CH_2CHCH_2CH_2CH_3$$

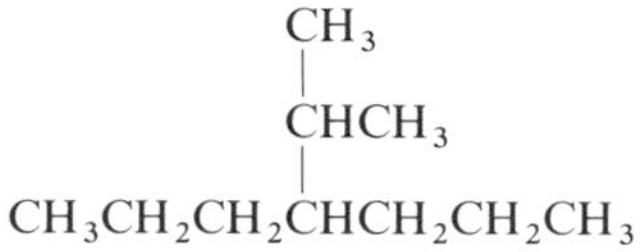

4-Isopropylheptane

The substituents are different, however, and we need some way of designating each. The group with the attachment at the end carbon is called propyl;* that with the attachment at the central carbon is called isopropyl.

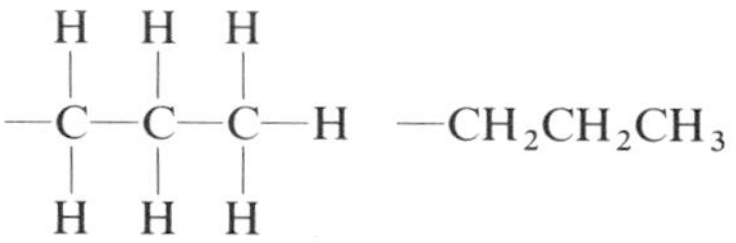

$$\begin{array}{ccc} H & H & H \\ | & | & | \\ -C & -C & -C-H \\ | & | & | \\ H & H & H \end{array} \quad -CH_2CH_2CH_3$$

Propyl

$$\begin{array}{ccc} H & H & H \\ | & | & | \\ H-C & -C & -C-H \\ | & & | \\ H & & H \end{array} \quad CH_3CHCH_3$$

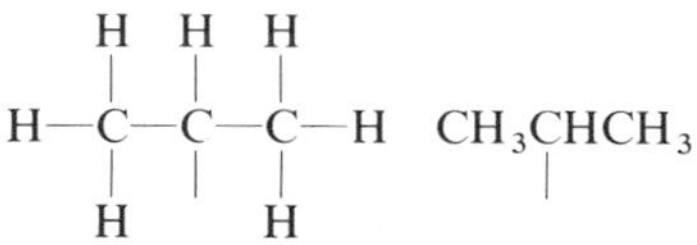

Isopropyl

The propyl group is propane minus a hydrogen from one of the end carbons, and the isopropyl group is propane minus the hydrogen from the middle carbon.

---

exercise with models 2.2 Construct models of C_3H_7 alkyl groups until convinced that propyl and isopropyl are the only possible structures.

There are four isomeric butyl groups.

$$CH_2-CH_2-CH_2-CH_3$$

Butyl

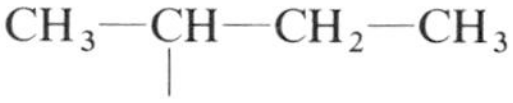

$$CH_3-\overset{}{C}H-CH_2-CH_3$$
$$|$$

sec-Butyl

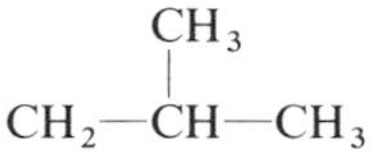

$$\overset{\textstyle CH_3}{\overset{|}{CH_2-CH-CH_3}}$$

Isobutyl

$$\overset{\textstyle CH_3}{\overset{|}{CH_3-\underset{|}{C}-CH_3}}$$

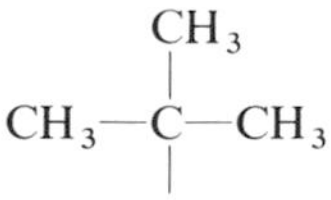

tert-Butyl

Butyl and *sec*-butyl (secondary butyl) have four-carbon chains. Attachment is to an end carbon of the chain in butyl but to an internal carbon in *sec*-butyl. These

---

* You may sometimes see this group called *n*-propyl. The prefix *n* (for normal) was once commonly used to indicate a straight chain.

groups correspond to butane minus a hydrogen. Isobutyl and *tert*-butyl (tertiary butyl) have branched chains. Isobutyl corresponds to isobutane minus a (primary) hydrogen from one of the methyl groups and *tert*-butyl to isobutane minus the (tertiary) hydrogen of the central carbon. It may help in remembering the names of these butyl groups to note that attachment to *sec*-butyl is at a secondary carbon and to *tert*-butyl at a tertiary carbon. Any C_4H_9 alkyl group that you draw will turn out to be identical in structure to one of the four preceding groups.

The names of the one- to four-carbon alkyl groups are widely used in nomenclature and it is essential to learn them. Only rarely will we encounter alkyl groups with more than four carbon atoms, and in this text we will not learn the systematic nomenclature used to name them.

problem 2.6   Name the following alkane.

$$CH_3{-}CH_2{-}CH{-}CH_3$$
$$CH_3{-}CH_2{-}CH_2{-}CH_2{-}\overset{|}{CH}{-}CH_2{-}CH_2{-}CH_2{-}CH_3$$

problem 2.7   Draw a structural formula for 4-*tert*-butyl-2-chloroheptane.

2.6
PHYSICAL PROPERTIES OF ALKANES

Physical properties are characteristics such as solubility or boiling point that can be observed without chemically altering a compound. Alkanes dissolve readily in liquid hydrocarbons as well as in many other organic liquids. They are, however, virtually insoluble in water. As indicated by the data in Table 2.1, alkanes that are liquids are somewhat less dense than water. You know already that "oil and water don't mix" and probably have seen how even a small amount of oil or gasoline does not dissolve in water but forms a slick on the surface. Alkanes are the major constituents of gasoline and most lubricating oils.

The boiling point is another important property of an alkane. As Table 2.1 shows, the boiling points of linear alkanes increase regularly with an increasing number of carbon atoms. In a group of isomeric alkanes boiling points decrease somewhat with increasing chain branching. Note, for example, the boiling points of pentane (36°), isopentane (28°), and neopentane (9°). It is often useful to make a rough estimate of the boiling point of a compound. Because of the regularity of their boiling points, it is easy to do this for alkanes by remembering just a few values. It is convenient to remember that butane boils at 0°, heptane at about 100°, and undecane at about 200°. The alkanes in Table 2.1 are liquids at room temperature. Some larger alkanes are solids, however. Eicosane ($C_{20}H_{42}$), for instance, melts at 36°.

2.7
REACTIONS OF ALKANES

As is methane, alkanes are relatively inert. They are attacked rapidly only by very reactive chemical species or when vigorous reaction conditions are used.

The largest amounts of alkanes are consumed in combustion reactions to provide thermal or mechanical energy. As is general for organic compounds containing carbon-carbon and carbon-hydrogen bonds, alkanes react with oxygen to form carbon dioxide and water. Combustion is exothermic, producing approximately 12 kcal/g of alkane, but requires elevated temperatures to occur rapidly.

$$CH_3CH_2CH_2CH_2CH_2CH_2CH_3 + 11\,O_2 \longrightarrow 7\,CO_2 + 8\,H_2O + 1150\ \text{kcal/mole}$$
(molecular weight = 100)

Another important large-scale reaction of alkanes called cracking is discussed in the next chapter.

Alkanes also react with halogens. Although this reaction is less important commercially than combustion or cracking, we will use it to introduce a new problem—how reactions occur.

2.8
HALOGENATION

Alkanes react with chlorine and bromine in reactions that must be initiated by heat or ultraviolet light. Chlorination of ethane first produces a product in which one hydrogen has been replaced by a chlorine.

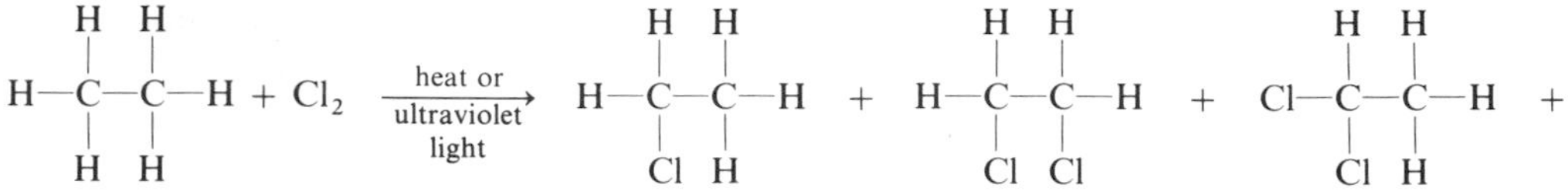

Ethyl chloride   1,2-Dichloroethane   1,1-Dichloroethane
Chloroethane

trichloroethanes + tetrachloroethanes + pentachloroethane + hexachloroethane

The substitutive name of that compound is chloroethane, but it is more often called ethyl chloride, a radicofunctional name. Further substitution of hydrogens by chlorine occurs readily so that products containing more chlorines are also formed. Reactions, such as halogenation of alkanes, in which one atom or group is replaced by another are called **substitution reactions**.

Bromination of propane produces two monobromopropanes as well as products containing two or more bromines.

$$
\begin{array}{c}
\underset{\substack{| \ \ | \ \ |\\ H \ \ H \ \ H}}{\overset{\substack{H \ \ H \ \ H\\ | \ \ | \ \ |}}{H-C-C-C-H}} + Br_2
\end{array}
\xrightarrow[\substack{\text{ultraviolet}\\ \text{light}}]{\text{heat or}}
\begin{array}{c}
\underset{\substack{| \ \ | \ \ |\\ H \ \ H \ \ H}}{\overset{\substack{H \ \ H \ \ H\\ | \ \ | \ \ |}}{Br-C-C-C-H}}
\end{array}
+
\begin{array}{c}
\underset{\substack{| \ \ | \ \ |\\ H \ \ Br \ \ H}}{\overset{\substack{H \ \ H \ \ H\\ | \ \ | \ \ |}}{H-C-C-C-H}}
\end{array}
$$

Propyl bromide   Isopropyl bromide
1-Bromopropane   2-Bromopropane

We have already learned that there are just two kinds of propyl groups. Substitution of a single H of a propane molecule must produce a product that contains one of these groups. The monobromo products are named by using either substitutive or radicofunctional nomenclature.

---

exercise with models 2.3   Construct models of compounds having molecular formula C_3H_7Br until convinced that just two structures, propyl bromide and isopropyl bromide, are possible.

---

problem 2.8   Draw a structural formula and give two names for each monochloro product obtained from chlorination of butane and isobutane.

2.9
MECHANISM OF HALOGENATION

How does the halogenation of alkanes actually occur? Knowing the products of a chlorination or bromination reaction does not explain the role of ultraviolet light or heat in initiating the reaction or how the replacement of hydrogen by halogen takes place. A detailed description of the pathway leading from the reactants to the products of a reaction is called a **mechanism**. Here we consider specifically the chlorination of methane, but halogenation of other alkanes is essentially identical.

The reaction is initiated by cleavage of the chlorine molecule into chlorine atoms.

$$:\!\overset{..}{\underset{..}{Cl}}\!:\!\overset{..}{\underset{..}{Cl}}\!: \longrightarrow 2 \cdot\overset{..}{\underset{..}{Cl}}\!: $$

Even though the Cl—Cl bond is weaker than other bonds in the reactants or products, to break it requires a substantial input of energy. This energy can be supplied by ultraviolet light, which is absorbed by the chlorine, or by heat. A species, such as a chlorine atom, that contains an unpaired electron is called a **free radical**. The chlorine atom has only seven valence electrons and is extremely reactive.

When a chlorine atom collides with a methane molecule, it often rips away a hydrogen atom to form a molecule of hydrogen chloride (step 1).

$$
\text{Step 1} \quad
\underset{\underset{\displaystyle H}{\displaystyle |}}{\overset{\displaystyle H}{H:\overset{\cdot\cdot}{C}:H}} + \cdot \overset{\cdot\cdot}{\underset{\cdot\cdot}{C}}l: \quad \longrightarrow \quad
\underset{\underset{\displaystyle H}{\displaystyle |}}{\overset{\displaystyle H}{H:\overset{\cdot\cdot}{C}\cdot}} + H:\overset{\cdot\cdot}{\underset{\cdot\cdot}{C}}l:
$$

$$
\text{Step 2} \quad
\underset{\underset{\displaystyle H}{\displaystyle |}}{\overset{\displaystyle H}{H:\overset{\cdot\cdot}{C}\cdot}} + :\overset{\cdot\cdot}{\underset{\cdot\cdot}{C}}l:\overset{\cdot\cdot}{\underset{\cdot\cdot}{C}}l: \quad \longrightarrow \quad
\underset{\underset{\displaystyle H}{\displaystyle |}}{\overset{\displaystyle H}{H:\overset{\cdot\cdot}{C}:\overset{\cdot\cdot}{\underset{\cdot\cdot}{C}}l:}} + \cdot \overset{\cdot\cdot}{\underset{\cdot\cdot}{C}}l:
$$

$$
\text{Step 1 + Step 2} \qquad CH_4 + Cl_2 \quad \longrightarrow \quad CH_3Cl + HCl
$$

The chlorine atom now is surrounded by eight electrons. This step, however, forms another very reactive free radical, the methyl radical. When a methyl radical collides with a chlorine molecule, it often pulls away a chlorine atom to form a molecule of methyl chloride (step 2). This step also forms a chlorine atom that can then react with another methane molecule (step 1). Adding steps 1 and 2 together (and canceling species that appear on both sides of the arrow) gives the balanced equation for the reaction. Although essential for the reaction to take place, the reactive chlorine and methyl radicals are present only transiently and do not appear in this equation.

Chlorination occurs by a sequence of steps, each generating a reactive species necessary for the next. Such a reaction is called a **chain reaction**. A chain reaction requires a **chain-initiating** step to create a reactive species. Cleavage of the chlorine molecule into chlorine atoms is the chain-initiating step of chlorination. The repeating steps of a chain reaction that lead to formation of products are called **chain-propagating** steps. Once chlorine atoms are formed, the chain-propagating steps (1 and 2) can continue over and over. Thousands of methyl chloride molecules can be formed for each dissociation of a chlorine molecule into chlorine atoms.

Ultimately some step occurs that removes a reactive free radical, breaking a reaction chain. A step of a chain reaction in which a reactive species is consumed (without generating a new one) is called a **chain-terminating** step. Three chain-terminating steps are important in the chlorination of methane.

$$
2 :\overset{\cdot\cdot}{\underset{\cdot\cdot}{C}}l\cdot \quad \longrightarrow \quad :\overset{\cdot\cdot}{\underset{\cdot\cdot}{C}}l:\overset{\cdot\cdot}{\underset{\cdot\cdot}{C}}l:
$$

$$
2\,CH_3\cdot \quad \longrightarrow \quad CH_3:CH_3
$$

$$
CH_3\cdot + \cdot\overset{\cdot\cdot}{\underset{\cdot\cdot}{C}}l: \quad \longrightarrow \quad CH_3:\overset{\cdot\cdot}{\underset{\cdot\cdot}{C}}l:
$$

Because their concentrations are very low, the probability of the transient free radicals colliding together is much less than of their colliding with methane or chlorine molecules. Therefore the chain-terminating steps occur relatively infrequently.

Chemists strive to learn as much as possible about the mechanisms of reactions. Once we understand its mechanism, we can often control a reaction more effectively or make it more useful. From what we have learned about the mechanism of chlorination, for example, we know that removal of a single free radical can prevent the formation of many thousands of product molecules. Chlorination could be drastically slowed by the addition of compounds that would combine with the free radicals.   A substance that even in small amounts slows a reaction is called an **inhibitor**. When we want to chlorinate an alkane, we must exclude even small amounts of impurities that are effective inhibitors.

2.10
CYCLOALKANES

Some alkanes have fewer hydrogens per carbon than the alkanes already discussed. The simplest is cyclopropane, a potent gaseous anesthetic. Cyclopropane has the molecular formula C_3H_6 and a cyclic structure.

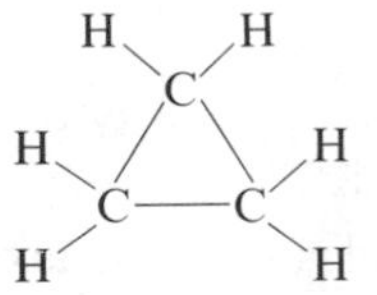

Cyclopropane

The other alkanes with fewer hydrogens per carbon also contain rings of carbons. They are called **cycloalkanes** to distinguish them from the **acyclic** (not cyclic) alkanes already considered.

A cycloalkane ring is named by adding the prefix *cyclo-* to the name of the acyclic alkane having the same number of carbon atoms.

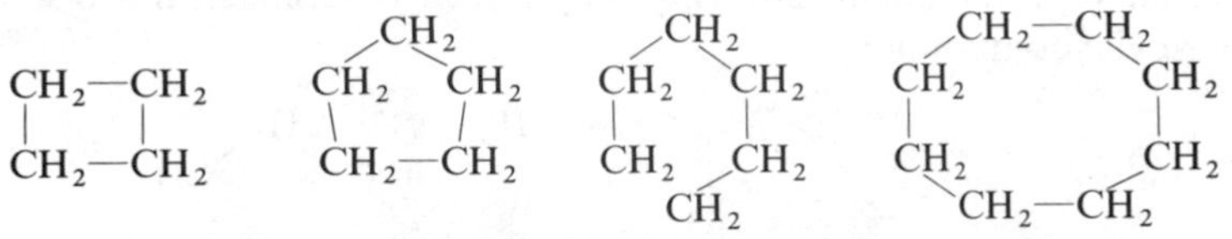

Cyclobutane   Cyclopentane   Cyclohexane   Cyclooctane

Substituents attached to the ring must be named. If there is only one substituent, its position is not indicated by a number, since there is only one monosubstituted cycloalkane.

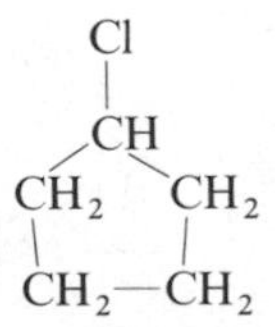

Chlorocyclopentane
Cyclopentyl chloride

Note that a cyclic alkane minus one hydrogen can also be named as an alkyl group.

When more substituents are present, the position of each must be indicated by a number. The ring is numbered so that as many substituents as possible are at carbon number 1 and the nearest carbon bearing another substituent has the lowest possible number.

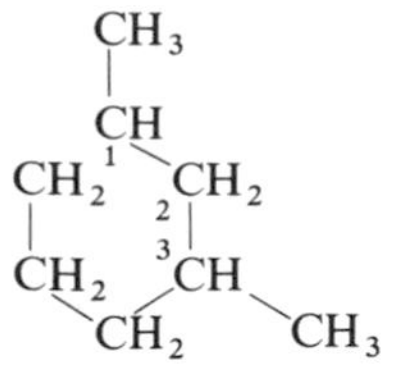

1,3-Dimethyl-
cyclohexane

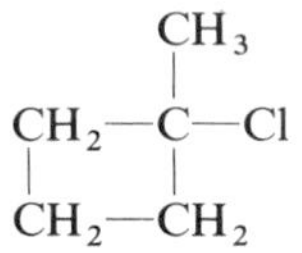

1-Chloro-1-
methylcyclobutane

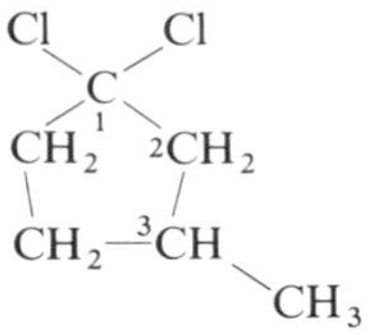

1,1-Dichloro-3-
methylcyclopentane

(not 1,1-dichloro-4-
methylcyclopentane or
3,3-dichloro-1-methyl-
cyclopentane)

The structural formulas of cyclic alkanes are often simplified by omitting the carbon atoms of the rings and the hydrogens attached to them. The foregoing compounds can be represented by the following simplified structural formulas.

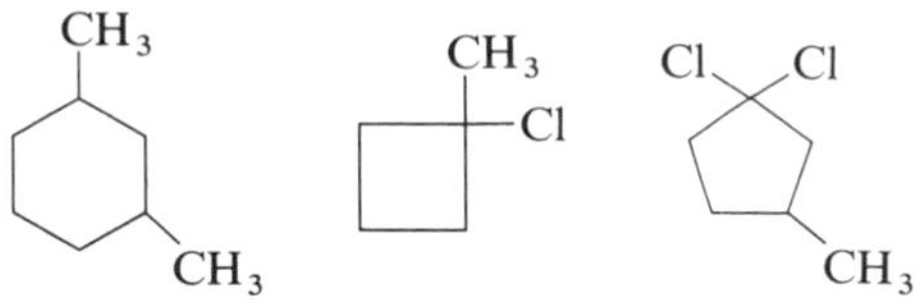

Similar representations showing only the carbon skeleton are sometimes used even for acylic compounds.

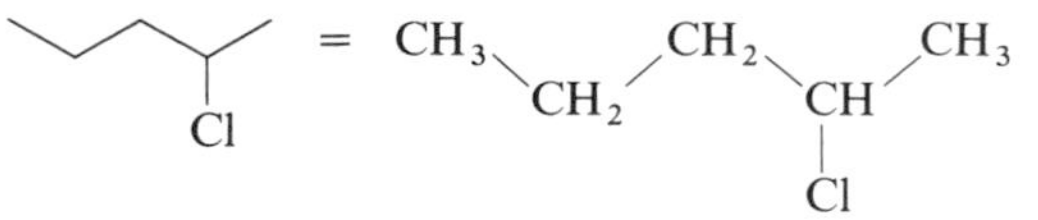

2-Chloropentane

---

problem 2.9 Draw expanded structural formulas (showing every bond) and give names for the following compounds.

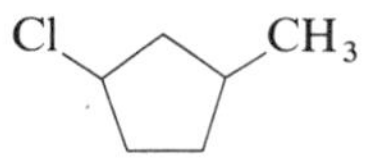

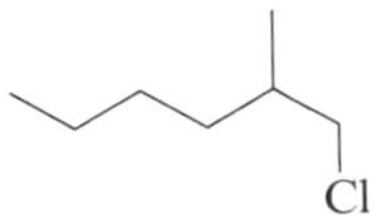

problem 2.10 Draw a structural formula for 1-ethyl-2-methylcyclohexane.

Cycloalkanes have properties similar to those of acyclic alkanes. Cycloalkanes, for example, are virtually insoluble in water, have boiling points close to those of comparable acyclic alkanes, and react with chlorine to form chlorine-substituted cycloalkanes. That cycloalkanes and acyclic alkanes have similar properties is not surprising since they contain the same structural features: carbon-carbon and carbon-hydrogen single bonds.

2.11
SHAPES OF CYCLOALKANES

Carbon atoms attached to four groups usually have bond angles that are near the tetrahedral angle of 109°, for this geometry leads to the strongest bonds. All carbon-carbon bond angles of cyclopropane, however, are 60°.

As a result, the carbon-carbon bonds in cyclopropane are unusually weak, and cyclopropane undergoes some chemical reactions, such as cleavage by strong acids, that are not characteristic of alkanes.

$$\triangle + HBr \longrightarrow CH_3{-}CH_2{-}CH_2{-}Br$$

The instability of some cyclic compounds due to deviations of bond angles from the tetrahedral angle is called **angle strain**. Cyclobutane also has considerable angle strain. The carbons lie nearly in a plane and the bond angles are about 90°. Cyclopentane has angles near the tetrahedral angle, not surprising because a regular pentagon has angles of 108°. Although a regular hexagon has angles of 120°, cyclohexane has exactly tetrahedral bond angles. It achieves these angles by adopting a nonplanar geometry that is considered in the next section. Larger rings also are nonplanar and have bond angles near the tetrahedral angle.

---

exercise with models 2.4   Construct models of cyclopropane and cyclobutane. If your model set has carbon atoms to which attachments can be only at the tetrahedral angle, then the connectors that serve as bonds must be flexible to permit forming these models.

Construct a model of cyclopentane. This model can be put together readily with the tetrahedral carbons of a model set.

2.12
CONFORMATIONS OF CYCLOHEXANE

Compounds containing cyclohexane rings are encountered frequently and their geometries are important to understand. Cyclohexane can assume two geometries, **chair** and **boat** (see Figure 2.1), that have tetrahedral angles at each carbon and hence no angle strain. The chair and boat can be interconverted by rotation around carbon-carbon bonds (just as can conformations of butane) and thus are conformations of cyclohexane.

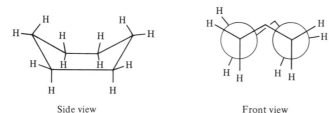

Side view   Front view

Chair conformation

Side view   Front view

Boat conformation

Figure 2.1 Two views of the chair and boat conformations of cyclohexane. The boat conformation pictured in the right-hand representation is slightly distorted so that all bonds to the back carbons will be visible.

The chair is the most stable conformation of cyclohexane and of most substituted cyclohexanes. Most molecules in a sample of cyclohexane have the chair conformation. As can be seen from the front view of the chair conformation (looking down two of the carbon-carbon bonds), the groups attached to adjoining carbons are perfectly staggered.

Although the boat conformation also has tetrahedral angles—and hence no angle strain—it is considerably less stable than the chair conformation. One source of the lesser stability of the boat (seen clearly in the front view) is the eclipsed arrangement of groups attached to the carbon-carbon bonds along the sides ("gunwales"). Another source is the crowding together of the "flagpole" hydrogens.

The interaction between groups forced too closely together is called **steric repulsion**.

The chair conformation has two types of hydrogens. Six hydrogens that lie around the circumference of the ring of carbons are called **equatorial** and six hydrogens that are perpendicular to this ring are called **axial**.

Bonds to the six axial hydrogens are indicated by thick lines and to the six equatorial hydrogens by thin lines.

A chair conformation is converted into a boat conformation by bending up one carbon, but a chair is again produced by bending down the opposite carbon.

| Chair | Boat | Chair |

As a result of this transformation, all hydrogens that were originally equatorial become axial and vice versa. Such synchronous rotations around carbon-carbon bonds that interconvert chair forms occur thousands of times a second at room temperature.

---

exercise with models 2.5   Construct a model of the chair conformation of cyclohexane. Note that this model has tetrahedral bond angles. Observe that the six axial hydrogens are all identical (have identical environments), as are also the six equatorial hydrogens. Note that the hydrogens on any carbon are staggered with respect to those on either adjoining carbon.

Construct a model of the boat conformation of cyclohexane. Note that this model also has tetrahedral bond angles. Observe that the hydrogens on the "gunwale" (starred) cabons ( ) are eclipsed and that the "flagpole" hydrogens ( ) are close together.

Construct a model of the chair conformation of cyclohexane, using one color for the six axial hydrogens and another for the six equatorial hydrogens. Convert this model into a boat by lifting up one carbon and then into another chair by bending down the opposite carbon ( ). Note that all six hydrogens that were originally axial are now equatorial. Convince yourself by similar twistings involving other carbons that in a chair conformation these six hydrogens are all axial or all equatorial.

exercise with models 2.6   Construct a model of cyclodecane. Convince yourself that this molecule can assume a large number of conformations, although none in which the hydrogens attached to each carbon are exactly staggered with respect to those on the adjoining carbons. Analysis of preferred shapes is more difficult for such large rings than for cyclohexane rings.

2.13
CONFORMATIONS OF SUBSTITUTED CYCLOHEXANES

Due to the rapid interconversion of chair conformations, the methyl substituent of methylcyclohexane alternately occupies equatorial and axial positions.

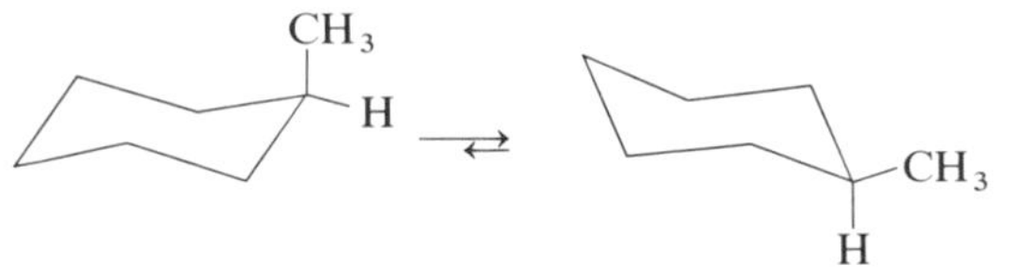

CH$_3$

H $\rightleftarrows$ CH$_3$

H

Methyl group axial   Methyl group equatorial

The conformation with the equatorial methyl group is more stable. The lesser stability of an axial methyl group is due to steric repulsion between this group and the axial hydrogens on the same side of the ring.

H H   CH$_3$

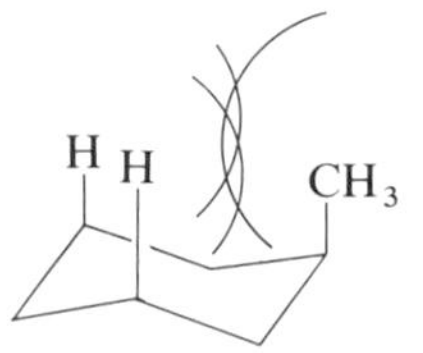

Therefore a methylcyclohexane molecule is most often in the chair conformation in which the methyl substituent is equatorial. At room temperature about 95% of methylcyclohexane molecules have equatorial methyl groups. For groups that are larger than methyl, the preference for the equatorial position is greater. In more than 99.9% of *tert*-butylcyclohexane molecules, the substituent occupies an equatorial position.

---

exercise with models 2.7 Construct a model of methylcyclohexane. By motions of the sort that you did in Exercise 2.5, interconvert chair conformations in which the methyl group occupies an axial or an equatorial position. Note that an axial methyl group is close to the two axial hydrogens on the same side of the cyclohexane ring.

Replace the axial methyl group by an axial *tert*-butyl group. Note that the steric repulsion between the substituent and the axial hydrogens on the same side of the ring must be much larger in *tert*-butylcyclohexane than in methylcyclohexane.

2.14
STEREOISOMERISM IN CYCLOALKANES

There are two isomeric 1,3-dichlorocyclobutanes.

cis-1,3-Dichlorocyclobutane   *trans*-1,3-Dichlorocyclobutane

The chlorines are on the same side of the ring in the isomer called **cis** but on opposite sides in the isomer called **trans**. The two 1,3-dichlorocyclobutanes have the same atoms bonded together and hence are not constitutional isomers. They differ only in the spatial arrangements of the atoms. We cannot indicate the difference between these isomers without referring in some way to their geometries. Isomers that differ only in the spatial arrangements of atoms are called **stereoisomers**. The spatial arrangement of atoms that characterizes a particular stereoisomer is called its **configuration**.

Many other examples of such isomerism are known.

cis-1-Chloro-2-methylcyclopropane*   *trans*-1-Chloro-2-methylcyclopropane*

cis-1,4-Dimethylcyclohexane   *trans*-1,4-Dimethylcyclohexane

Stereoisomers of this sort are sometimes called **cis-trans isomers**. Just as do constitutional isomers, cis-trans isomers have different properties.

---

problem 2.11   Draw and name all isomeric dimethylcyclobutanes.

---

exercise with models 2.8   Construct models of *cis*- and *trans*-1,3-dichlorocyclobutane. Note that they cannot be interconverted without breaking and re-forming bonds.

exercise with models 2.9   Both for convenience in drawing and for ease in recognizing cis-trans isomers, we drew the rings of the 1,4-dimethylcyclohexane isomers as though they were flat. The fact that the rings are nonplanar and assume more than one conformation makes it more difficult to distinguish stereoisomers but does not alter the conclusion that there are two (and only two) isomeric 1,4-dimethylcyclohexanes.

Construct two models of chair 1,4-dimethylcyclohexane. In one model (A) put both methyl groups in axial positions. By flipping the ring, you can produce another chair conformation in which both methyl groups are equatorial. As you would expect, the diequatorial conformation of A is considerably more stable and therefore most molecules have this conformation.

In the other model (B) put one methyl group in an axial position and the other in an equatorial position. Flipping the ring produces an identical chair conformation in which the methyl groups have reversed positions.

* This compound has an isomer that is its mirror image. This additional aspect of stereoisomerism is considered in Chapter 6.

Convince yourself that no ring flipping of models A and B ever interconverts them. Therefore the models represent isomeric compounds. Isomer A is called trans and isomer B cis. Note that in the more important (diequatorial) conformation of the trans isomer, the methyl groups are not very obviously on opposite sides of the ring. However, one group occupies the upper position and the other group the lower position at the carbons to which they are bonded. In a cis isomer, both groups have upper (or lower) positions at their respective carbons.

2.15
ALKANES WITH MORE COMPLEX RING STRUCTURES

Many compounds contain more than one ring of carbon atoms. Some compounds have independent rings bonded together.

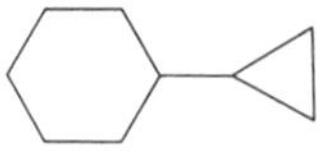

Cyclopropylcyclohexane

Other compounds have structures in which some carbons are shared by rings. The decalin ring system that appears in many compounds contains **fused rings**, rings that share two adjacent carbon atoms.

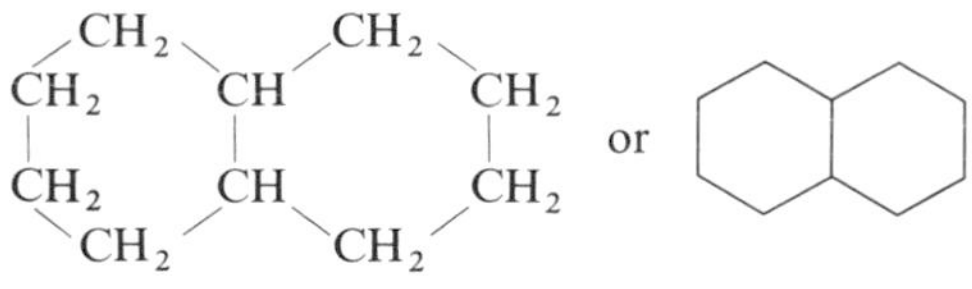

Decalin

Steroids (Sec. 9.15), important naturally occurring compounds, have a skeleton containing four fused rings.

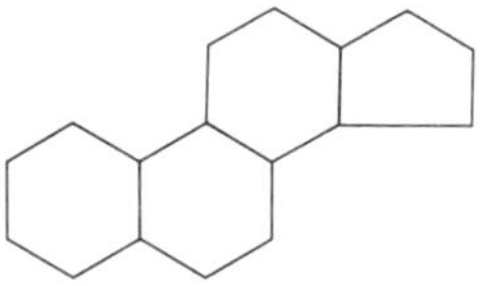

The ring skeleton that is usually found in steroids.

Norbornane has rings joined at nonadjacent atoms.

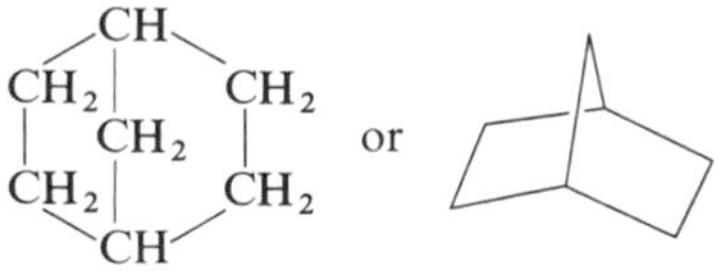

Norbornane

The two rings of spiropentane have only one carbon in common.

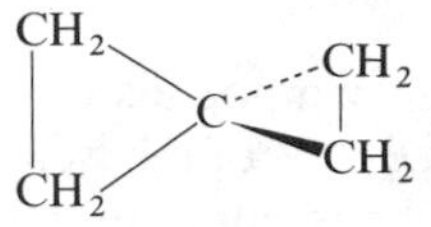

$$
\begin{array}{c}
CH_2 \\
| \quad\quad\quad C \cdots CH_2 \\
CH_2 \quad\quad CH_2
\end{array}
$$

Spiropentane

Each carbon of cubane is located at the corner of a cube.

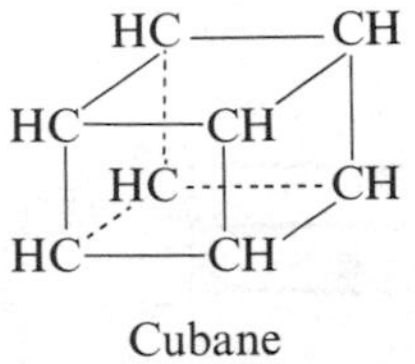

$$
\begin{array}{c}
HC\text{------}CH \\
HC\text{------}CH \\
HC\text{------}CH \\
HC\text{------}CH
\end{array}
$$

Cubane

A diamond is a giant molecule of carbons, each attached to four others by bonds arranged at the regular tetrahedral angle; every ring has the chair conformation.

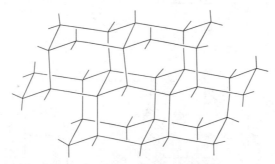

A portion of a diamond

Chemists have synthesized compounds with catenane and rotaxane structures. A catenane contains rings that are interlocked but not chemically bonded together. The linear chain of a rotaxane is threaded through a ring but cannot escape because of bulky end groups.

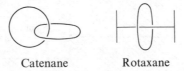

Catenane   Rotaxane

Carbon atoms can link together by stable covalent bonds to form an endless variety of chain and ring skeletons. Moreover, carbon also forms strong covalent bonds to several other atoms of low atomic number, especially hydrogen, nitrogen, and oxygen. The combination of these two properties is unique to carbon and responsible for the diversity of stable organic compounds.

2.16
THE CARBON CYCLE

The importance of organic compounds is not due to their being abundant. Carbon atoms constitute less than 0.03% of the weight of the earth's outer crust, oceans, and atmosphere. Of these carbons, more than 99% are in sedimentary deposits, chiefly as carbonate ion (CO_3^{2-}) in rocks such as limestone ($CaCO_3$) and marble. It is because of their extreme concentration in living systems and fossil fuels that organic compounds are so significant. Important interconversions of carbon compounds are outlined in Figure 2.2.

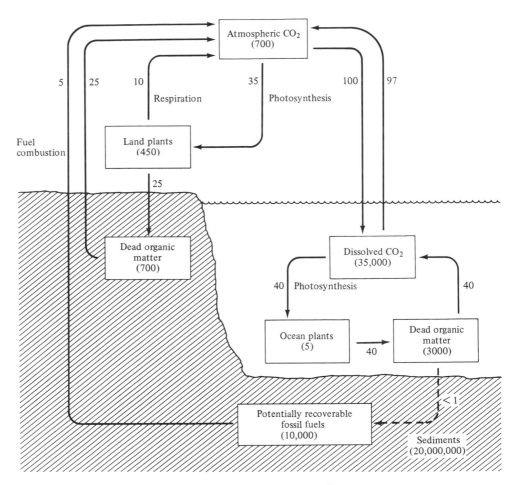

Figure 2.2 Major parts of the carbon cycle. Numbers in parentheses represent estimated quantities of carbon stored in different reservoirs and numbers over arrows the estimated annual flows between reservoirs. The unit is 10^9 metric tons. Land and ocean animals eat a portion of the plants but also ultimately become dead organic matter. [Data taken from R. Bolin, *Sci. Amer.*, **223**, no. 3, 125 (1970).]

The reduction of carbon dioxide by ocean and land plants is critical to life.

$$\text{Energy} + n\,CO_2 + n\,H_2O \xrightarrow{\text{photosynthesis}} (CH_2O)_n + n\,O_2$$

A complex series of reactions called **photosynthesis** produces carbohydrates, organic molecules that for the moment we represent by a general molecular formula $(CH_2O)_n$. Chapter 10 is devoted to these important biological molecules. Photosynthesis is endothermic. The necessary energy is provided by sunlight absorbed by plants.

Photosynthesis is the ultimate source of the organic compounds and the energy required by living organisms. The fundamental biological compounds used to construct the varied organic compounds in living systems arise from reactions of carbohydrates. Oxidation to carbon dioxide of carbohydrates and other biological compounds produced from them is exothermic. These oxidations, carried out in complex series of reactions, provide the energy needed for life.

Photosynthesis is also the source of most energy used for human activities. A small amount of the organic material of living organisms escapes oxidation by deposition in environments where there is little oxygen, principally in underwater sediments. Petroleum, natural gas, and coal are formed by slow transformations of these sediments. Most of the energy that we use is derived from the combustion of these fossil fuels, which are also the principal raw materials for the organic chemical industry. Current use of fossil fuels far exceeds the slow rate of their formation, and we are rapidly consuming deposits that were created over many millions of years.

As a result of combustion of enormous amounts of fossil fuels, the concentration of carbon dioxide in the atmosphere is increasing by about 0.3% per year. Because carbon dioxide effectively absorbs certain frequencies of infrared radiation, its rising concentration will tend to reduce infrared radiation from the earth to space. This reduction could alter the present balance between the visible and ultraviolet radiation that arrives from the sun and the radiation, mainly infrared, that leaves the earth. Many are concerned that the average surface temperature of the earth may increase, leading to significant changes in global climate.

2.17
PETROLEUM AND NATURAL GAS

The annual use of petroleum and natural gas in the United States is about 7 tons per person. These materials are thought to arise in oxygen-deficient environments from the action of bacteria and of some heat and pressure on organic materials largely derived from marine plankton. The compounds produced slowly in this way are sometimes concentrated by geological action into porous rock formations. Only when significant amounts have accumulated in localized areas are drilling and extraction economically feasible.

Natural gas consists mainly of methane (typically about 90%) but also contains varying amounts of other volatile alkanes. Natural gas is found sometimes in conjunction with petroleum but often in separate deposits.

Petroleum, a viscous liquid, is a complex mixture of organic compounds, most of which fall in the C_1 to C_{40} range. Hydrocarbons are the most plentiful components, although organic compounds containing nitrogen, oxygen, and sulfur also are present. The most abundant hydrocarbons usually are acyclic alkanes, both linear and branched, and cycloalkanes with cyclopentane and cyclohexane rings.

The first step in the processing of crude petroleum in a refinery is distillation to achieve a separation into fractions having different boiling point ranges. Typical distillation fractions are listed in Table 2.2. Asphalt is an even less volatile material.

Gasoline, diesel fuel, and other marketable products rarely come directly from the distillation of petroleum. Fractions distilled from petroleum undergo processing in a refinery to remove undesirable materials (e.g., sulfur-containing compounds) and to modify the composition of compounds. Then processed materials are blended to give products with properties suitable for particular applications.

About one-half of all petroleum is used to produce gasoline. Gasoline is a mixture of hydrocarbons, mainly alkanes, boiling between about 25° and 200°. Butanes are added to increase volatility and ease of starting in areas where winters are cold.

Combustion of alkanes is a complex process that proceeds by free-radical chain reactions. An automobile engine generates a maximum of mechanical work when the reacting flame front advances steadily through the gasoline–air mixture in the combustion chamber. Spontaneous or too rapid combustion, called **knocking**, reduces efficiency and increases engine wear. An increase in the compression ratio of engines increases efficiency but also the tendency toward knocking.

The octane rating of a gasoline indicates its tendency to produce knocking. Heptane, a compound particularly prone to cause knocking, is assigned an octane rating of 0. Isooctane (a nonsystematic name for 2,2,4-trimethylpentane) is given a rating of 100 because of its low tendency toward knocking. Methylcyclohexane is assigned an octane rating of 75 because its tendency to produce knocking is the same as that of a mixture of 75% isooctane and 25% heptane. Octane ratings are

Table 2.2   Distillation fractions from petroleum

Fraction	Approximate boiling range (°C)	Approximate size range of molecules
Liquefied petroleum gas (LPG)	below 25	C_3–C_4
Straight-run gasoline	25–200	C_5–C_{12}
Kerosene	200–275	C_{12}–C_{16}
Fuel oils	275–400	C_{15}–C_{25}
Lubricating oils, waxes, greases	400–500	C_{25}–C_{35}

generally lowest for linear alkanes, higher for cyclic alkanes, even higher for very branched alkanes, and highest for aromatic hydrocarbons (discussed in Chapter 4).

Straight-run gasolines, obtained directly from distillation of petroleum, have low octane ratings. Moreover, not enough of the distillate from petroleum is in the boiling point range of gasoline. Much of the processing in a petroleum refinery involves chemical transformations that convert higher-boiling fractions into gasoline and increase the proportions of types of compounds that have high octane ratings. The octane rating can also be increased by addition of small amounts of compounds that are particularly effective in interrupting some of the free-radical chains. Tetra-ethyllead (TEL) is the most widely used additive.

$$
\begin{array}{c}
CH_2CH_3 \\
| \\
CH_3CH_2-Pb-CH_2CH_3 \\
| \\
CH_2CH_3
\end{array}
$$

Tetraethyllead

Addition of only 3 mL of TEL to a gallon of gasoline can increase its octane rating by as much as 10 to 15 units.

2.18
ENVIRONMENTAL PROBLEMS ACCOMPANYING
THE USE OF PETROLEUM PRODUCTS

Our large-scale use of petroleum has created obvious environmental problems. The gasoline engine is the greatest offender.

The rapid cycling in a gasoline engine prevents complete combustion of the hydrocarbon fuel. As a result, the exhaust contains significant amounts of unreacted hydrocarbons and of carbon monoxide, a product of partial combustion.

The toxicity of carbon monoxide stems from its action on hemoglobin, the molecule that functions as an oxygen carrier in our bodies. Carbon monoxide combines so effectively with hemoglobin that the combination with oxygen, necessary for oxygen transport, is blocked. The amount of carbon monoxide released by automobiles is relatively small compared to that formed by natural processes, but where automobile density is high, local concentrations of carbon monoxide can be dangerously high.

Most hydrocarbons are not themselves very toxic. In the presence of nitrogen dioxide and sunlight, however, hydrocarbons are involved in a complex group of reactions responsible for the photochemical smog associated with Los Angeles and other cities.

Nitrogen dioxide (NO_2), a brown gas responsible for the color of smog, results from an atmospheric reaction of nitrous oxide (NO) with oxygen.

$$2\,NO + O_2 \longrightarrow 2\,NO_2$$

Some nitrous oxide arises from natural sources, but it also results from the reaction of nitrogen and oxygen at high temperatures.

$$N_2 + O_2 \xrightarrow{\text{high temperature}} 2\,NO$$

Nitrous oxide formation accompanies hydrocarbon combustion, since nitrogen is the principal component of the air used as the source of oxygen.

Emission-control systems in automobiles are designed to reduce the concentrations of these pollutants. Recycling a portion of the exhaust back through the engine provides further opportunity for combustion of carbon monoxide and unreacted hydrocarbons. In the catalytic converters installed on many automobiles, the exhaust gases first pass over a catalyst that aids in reducing NO to N_2. Then the exhaust gases, mixed with additional air, pass over a catalyst that promotes further oxidation of the carbon monoxide and remaining hydrocarbons to carbon dioxide. Cars equipped with catalytic converters cannot use gasoline containing tetraethyllead because lead compounds deactivate the catalysts. For this reason and because of concern about health hazards of lead compounds emitted in exhaust gases, governmental regulation is greatly reducing tetraethyllead use in the United States.

SUMMARY

Definitions and Ideas

Hydrocarbons are molecules containing only C and H. **Alkanes** are hydrocarbons containing only single bonds. A multitude of acyclic and cyclic alkanes can exist.

The generally accepted nomenclature for organic compounds is that fostered by the IUPAC. Alkanes are named by using **substitutive nomenclature**. This system considers a compound to be related to a simpler molecule (parent compound), some of whose hydrogens have been replaced by other atoms or groups. Atoms or groups replacing hydrogens of the parent compound are called **substituents**.

To name an acyclic alkane, pick the longest straight chain of carbons. Use the name of the linear (unbranched) alkane with that number of carbons as the parent name. Add as a prefix the name and position of attachment of each substituent. Determine positions by numbering the parent chain, starting from the end nearest a substituent.

An **alkyl group** is a portion of a molecule corresponding to an alkane minus one hydrogen. Frequently encountered alkyl groups are methyl, ethyl, propyl, isopropyl, butyl, *sec*-butyl, isobutyl, and *tert*-butyl.

In **radicofunctional nomenclature** (e.g., isopropyl chloride) one group (here —Cl) is selected and becomes the final word of the name. This word is preceded by a word (or words) to describe the remainder of the molecule.

Depending on the number of other carbons to which it is attached, a carbon is classified as **primary** (attached to one other carbon), **secondary** (two), **tertiary** (three), or **quaternary** (four).

Cyclopropane and cyclobutane have considerable **angle strain**, instability due to deviations of bond angles from the tetrahedral angle.

In both the **chair** and **boat** conformations of cyclohexane, each carbon has tetrahedral bond angles. The chair conformation is more stable, however, because the boat conformation has **steric repulsion** between two hydrogens and eclipsed relationships between groups on some adjoining carbons. A cyclohexane ring in a chair conformation has one **equatorial** and one **axial** group at each carbon. Axial and equatorial groups are rapidly interconverted by ring flipping. A substituent is more stable in an equatorial than an axial position.

Stereoisomers are isomers that differ only in the spatial arrangements of atoms. Conformations also differ in this way but are not considered isomers because they readily interconvert. The spatial arrangement that characterizes a particular stereoisomer is called its **configuration**.

Stereoisomerism is found with some cyclic compounds. Substituents on the same side of a ring are called **cis** and those on opposite sides are called **trans**.

Alkanes are virtually insoluble in water, and liquid alkanes are less dense than water. The boiling points of alkanes increase with increasing number of carbons.

Alkanes are relatively inert, reacting only with particularly reactive chemical species or when vigorous reaction conditions are used. Halogenation of alkanes is a **substitution reaction**, a reaction in which one atom or group is replaced by another.

The **mechanism** of a reaction is a detailed description of the pathway leading from the reactants to the products.

Halogenation is a **free-radical chain reaction**. A **free radical** is a species that has an unpaired electron and is usually very reactive. The mechanism of a **chain reaction** includes **chain-propagating** steps, each producing a reactive species that causes the next step to occur. A chain reaction also has **chain-initiating** and **chain-terminating** steps.

Important Reactions of Alkanes

1. *Combustion*   alkane $\xrightarrow{\text{O}_2}$ $CO_2 + H_2O$ + heat

2. *Halogenation* $-\overset{|}{\underset{|}{C}}-H$ $\xrightarrow[\substack{\text{heat or} \\ \text{ultraviolet light}}]{X_2}$ $-\overset{|}{\underset{|}{C}}-X$

   $X = Cl \text{ or } Br$

PROBLEMS

1. Draw expanded structural formulas (showing every bond) for the following compounds.
   (a) $CH_3(CH_2)_4CH_3$
   (b) $CH_3CHClCH_3$
   (c) $(CH_3)_2CHCH_2Cl$
   (d) $CH_3CH(CH_3)CH_2CH(CH_3)_2$
   (e) —Cl
   (f) $(CH_3)_3CBr$

2. Draw structural formulas for the following compounds.

(a) 2-methylhexane   (b) 2,3-dimethylbutane

(c) 2,2-dimethylheptane   (d) 2-bromobutane

(e) 1,2-dichloropropane   (f) 2,2,3,3-tetramethylbutane

(g) 3-ethylhexane   (h) 4-isopropyl-2-methylheptane

(i) 4-*tert*-butyl-5-isopropyloctane   (j) *sec*-butyl bromide

(k) butylcyclohexane   (l) 2,2,4-trimethylpentane

(m) isobutyl bromide   (n) 1-chloro-3-methylcyclopentane

3. Draw simplified structural formulas for the following compounds.

(a) $CH_3CH_2CH_2CH_2CH_2CH_3$

(b)
$$CH_3\underset{\underset{\displaystyle CH_3}{|}}{\overset{\overset{\displaystyle CH_3}{|}}{C}}CH_3$$

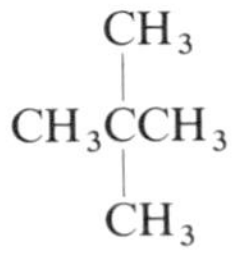

(c)
$$\begin{array}{c} CH_2 \\ CH_2 \qquad CH-CH_3 \\ CH_2 \qquad CH_2 \\ CH_2 \end{array}$$

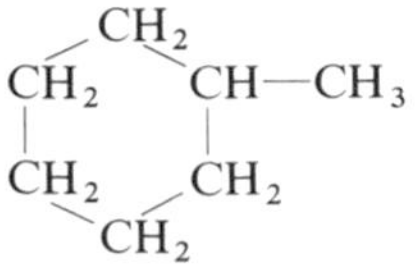

(d)
$$CH_3CHCH_2CH_2CHCH_3$$
with CH_2CH_3 on one CH and CH_3 on the other

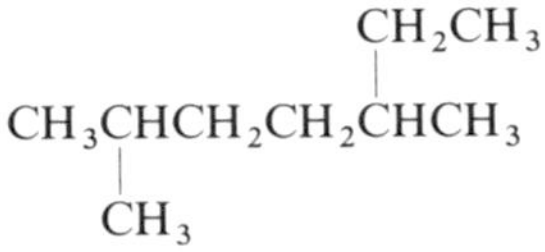

4. Which structural formulas in each group represent the same compound?

(a) 1.
$$\begin{array}{c} H \ \ H \ \ H \\ | \ \ | \ \ | \\ H-C-C-C-H \\ | \ \ | \ \ | \\ H \ \ H \ \ Br \end{array}$$

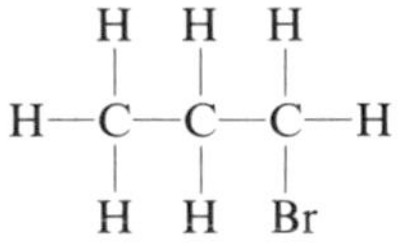

2. $CH_2CH_2CH_3$
    $\ |$
    Br

3. $\ \ \ \ Br$
    $\ \ \ \ |$
    CH_3CHCH_3

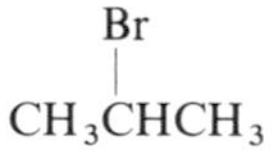

4. $(CH_3)_2CHBr$

5. CH_2-Br
    $\ |$
    CH_3-CH_2

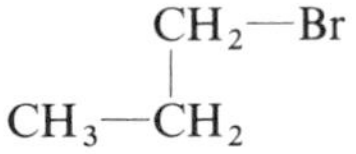

6.
$$\begin{array}{c} H \ H \\ \backslash / \ \ H \\ H \ \ C \\ | \ \ | \\ H-C-C-H \\ | \ \ | \\ Br \ H \end{array}$$

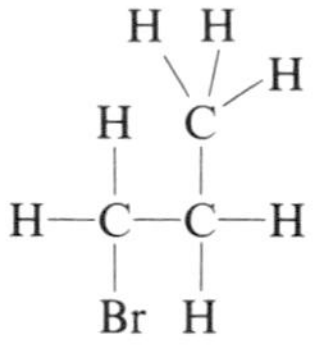

(b) 1.
$$\begin{array}{c} CH_3 \\ | \\ CH_2 \\ | \\ CH_3CHCH_2 \\ | \\ CH_3 \end{array}$$

2.
$$\begin{array}{c} CH_3 \\ | \\ CHCH_3 \\ | \\ CH_3CH_2CH_2 \end{array}$$

3. $CH_3CH_2CHCH_2CH_3$
    $\ \ \ \ \ \ \ \ \ |$
    $\ \ \ \ \ \ \ \ \ CH_3$

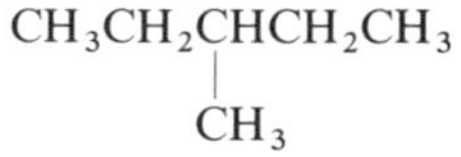

4. $CH_2-CH-CH_2$
    $\ |\ \ \ \ \ |\ \ \ \ \ |$
    $CH_3 \ CH_3 \ CH_3$

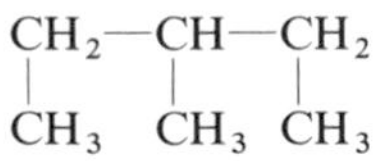

5. $CH_3 \ CH_2CH_3$
    $\ |\ \ \ \ \ \ \ |$
    $CH-CH_2$
    $\ |$
    CH_3

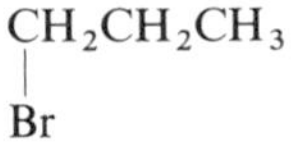

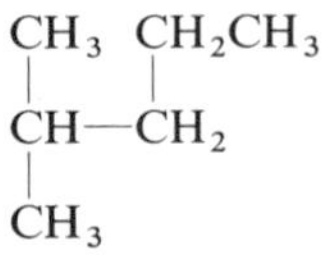

6. $\ \ \ \ \ \ \ \ \ \ \ \ \ CH_3$
    $\ \ \ \ \ \ \ \ \ \ \ \ \ |$
    CH_3-CH_2-CH
    $\ \ \ \ \ \ \ \ \ \ \ \ \ |$
    $\ \ \ \ \ \ \ \ \ \ \ \ \ CH_2$
    $\ \ \ \ \ \ \ \ \ \ \ \ \ |$
    $\ \ \ \ \ \ \ \ \ \ \ \ \ CH_3$

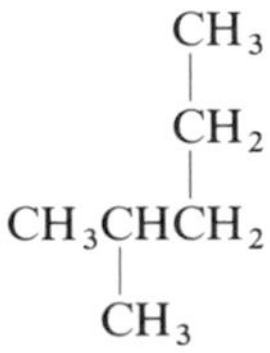

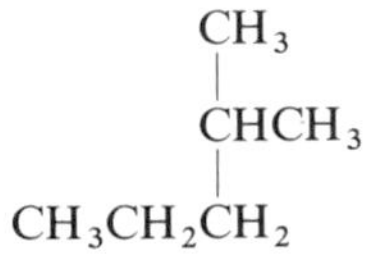

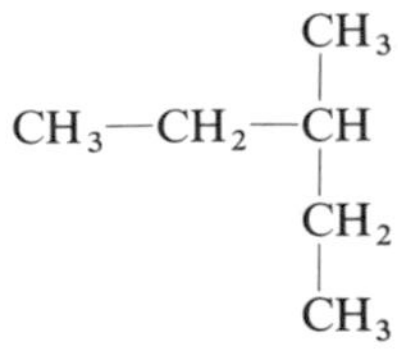

(c) 1. $CH_3CH_2C(CH_3)_2CH_2CH_2CH_3$

2. 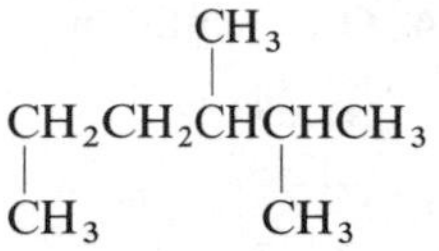
$$CH_3$$
$$CH_2CH_2CHCHCH_3$$
$$\hspace{1.2cm}|\hspace{0.9cm}|$$
$$\hspace{1.2cm}CH_3\hspace{0.5cm}CH_3$$

3.
$$CH_3CH_2$$
$$CH_3CH_2CH_2CCH_3$$
$$\hspace{2.2cm}|$$
$$\hspace{2.2cm}CH_3$$

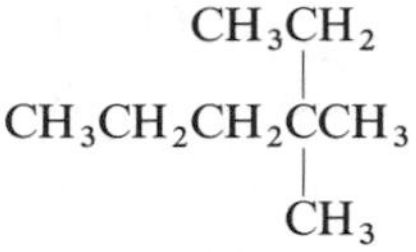

4. 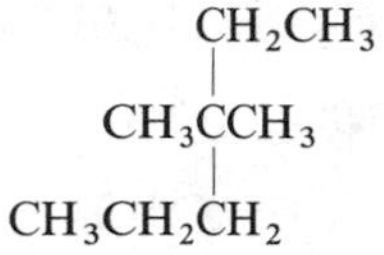
$$CH_2CH_3$$
$$CH_3CCH_3$$
$$CH_3CH_2CH_2$$

5. 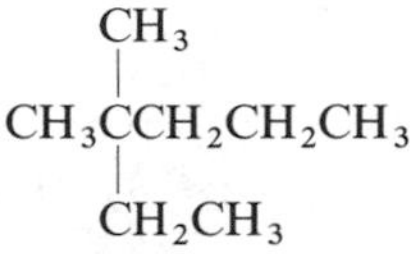
$$CH_3$$
$$CH_3CCH_2CH_2CH_3$$
$$\hspace{0.5cm}|$$
$$\hspace{0.5cm}CH_2CH_3$$

6. 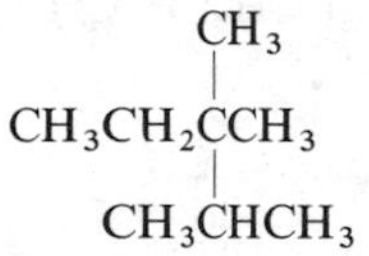
$$CH_3$$
$$CH_3CH_2CCH_3$$
$$\hspace{1.5cm}|$$
$$\hspace{1.5cm}CH_3CHCH_3$$

(d) 1. 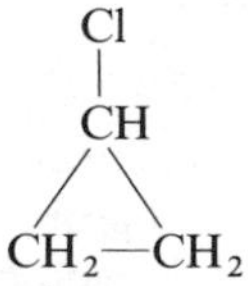
$$Cl$$
$$|$$
$$CH$$
$$\diagup\;\diagdown$$
$$CH_2{-}CH_2$$

2.
$$\triangleright{-}Cl$$

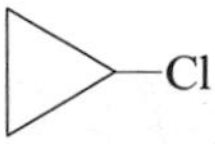

3. 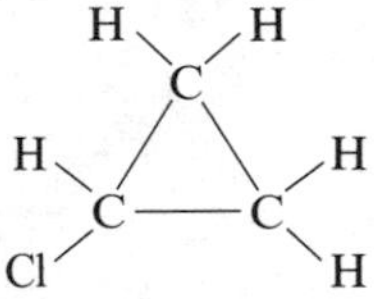
$$\hspace{0.5cm}H\hspace{0.4cm}H$$
$$\hspace{0.8cm}\diagdown\;C\;\diagup$$
$$H\hspace{1.5cm}H$$
$$\hspace{0.3cm}\diagdown C{-}C\diagup$$
$$Cl\diagup\hspace{1.0cm}\diagdown H$$

5. Give substitutive names for the following compounds.

(a) CH_3CH_2Br

(b) 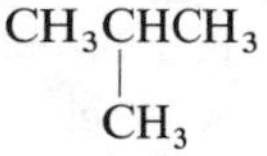 CH_3CHCH_3
$$\hspace{1.2cm}|$$
$$\hspace{1.2cm}CH_3$$

(c) 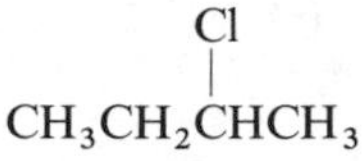
$$Cl$$
$$|$$
$$CH_3CH_2CHCH_3$$

(d) 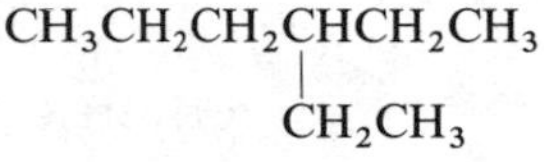 $CH_3CH_2CH_2CHCH_2CH_3$
$$\hspace{3.0cm}|$$
$$\hspace{3.0cm}CH_2CH_3$$

(e) 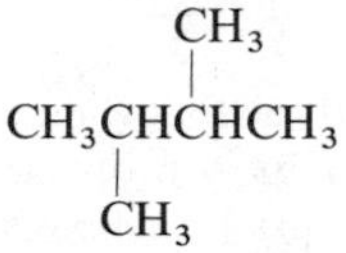
$$CH_3$$
$$|$$
$$CH_3CHCHCH_3$$
$$\hspace{1.5cm}|$$
$$\hspace{1.5cm}CH_3$$

(f) 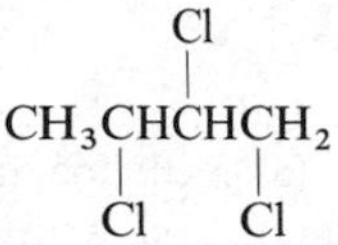
$$Cl$$
$$|$$
$$CH_3CHCHCH_2$$
$$\hspace{0.9cm}|\hspace{0.5cm}|$$
$$\hspace{0.9cm}Cl\hspace{0.5cm}Cl$$

(g) 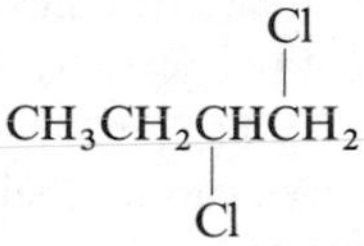
$$Cl$$
$$|$$
$$CH_3CH_2CHCH_2$$
$$\hspace{1.8cm}|$$
$$\hspace{1.8cm}Cl$$

(h) 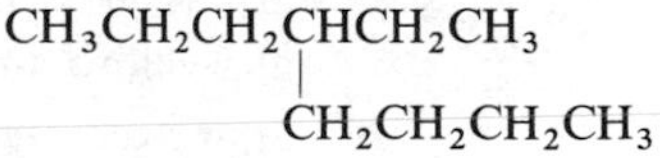 $CH_3CH_2CH_2CHCH_2CH_3$
$$\hspace{3.0cm}|$$
$$\hspace{2.5cm}CH_2CH_2CH_2CH_3$$

(i) 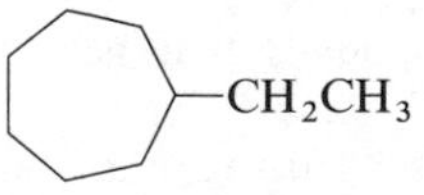 $\bigcirc{-}CH_2CH_3$

(j) 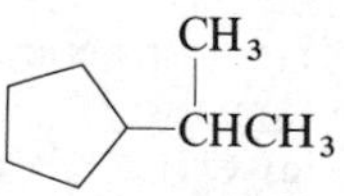
$$CH_3$$
$$|$$
$$\bigcirc{-}CHCH_3$$

(k) 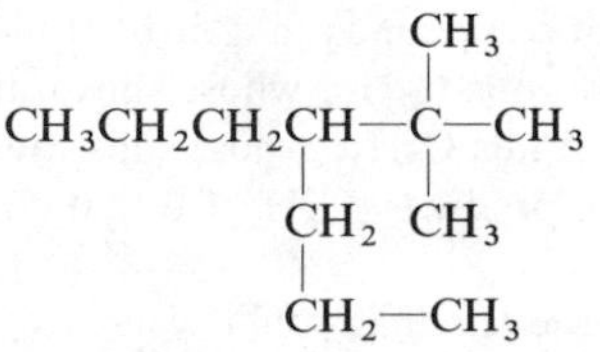
$$CH_3$$
$$|$$
$$CH_3CH_2CH_2CH{-}C{-}CH_3$$
$$\hspace{2.3cm}|\hspace{0.5cm}|$$
$$\hspace{2.3cm}CH_2\hspace{0.3cm}CH_3$$
$$\hspace{2.3cm}|$$
$$\hspace{2.3cm}CH_2{-}CH_3$$

(l) 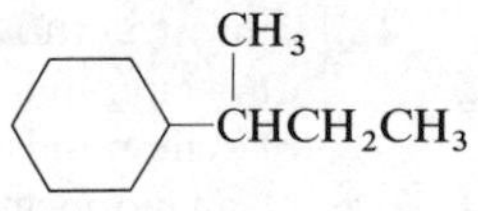
$$CH_3$$
$$|$$
$$\bigcirc{-}CHCH_2CH_3$$

6. Give substitutive and radicofunctional names for the following compounds.

(a) CH_3Br

(b) CH_3CHCH_3
  $|$
  Br

(c) CH_3CH_2Cl

(d) CH_2Br_2

(e) CH_3
     $|$
  CH_3CHCH_2Br

(f) $CH_3CH_2CH_2Cl$

(g) $CH_3CH_2CH_2CH_2I$

(h) –Cl

(i) CH_3
     $|$
  CH_3CCH_3
     $|$
     Br

(j) $CH_3CHCH_2CH_3$
  $|$
  I

7. Give another name for each of the following compounds.
  (a) ethyl bromide
  (b) isopropyl chloride
  (c) 2-methylpropane
  (d) 2-chlorobutane
  (e) 1-iodopropane
  (f) *tert*-butyl chloride
  (g) cyclobutyl chloride
  (h) 1-chloro-2-methylpropane
  (i) butyl bromide
  (j) bromoform

8. Classify each carbon of the following compounds as 1°, 2°, 3°, or 4°.

(a) CH_3 CH_3
     $|$ $|$
  $CH_3CH_2CCH_2CHCH_2CCH_3$
     $|$ $|$
     CH_3 CH_3 CH_3

(b) CH_3
     $|$

9. Indicate what, if anything, is incorrect about the following names.
  (a) 2-ethylpentane
  (b) isopropane
  (c) 3,3-dimethylbutane
  (d) 1-methyloctane
  (e) 2-isopropyl chloride
  (f) *cis*-1,4-dimethylcyclopentane
  (g) 3-chloro-4-methylpentane
  (h) 3-*sec*-butylhexane

10. Draw structural formulas for the nine isomeric heptanes. (A systematic approach to this task is to draw first the isomer with a seven-carbon chain, then any isomers with six-carbon chains, then any with five-carbon chains . . . .)

11. Draw structural formulas of all compounds that have the following molecular formulas.
  (a) $C_2H_3Cl_3$
  (b) C_4H_9Br
  (c) $C_3H_6Br_2$
  (d) C_3H_6BrCl

12. Draw structural formulas of all compounds that fit the following descriptions.
  (a) compounds of molecular formula C_7H_{16} whose substitutive names end in hexane
  (b) compounds of molecular formula C_8H_{18} whose substitutive names end in pentane
  (c) all monochloro substitution products $(C_5H_{11}Cl)$ that could result from chlorination of isopentane
  (d) all dibromo substitution products that could result from bromination of ethane

(e) all dichloro substitution products that could result from chlorination of isobutane

(f) all monochloro (C_5H_9Cl) and dichloro ($C_5H_8Cl_2$) products that could result from chlorination of cyclopentane

13. Draw a structural formula that corresponds to each of the following names. If stereoisomers exist, draw a suitable representation of each isomer and label it cis or trans.

    (a) 1-bromo-3-chlorocyclohexane   (b) 1,1-dichlorocyclohexane

    (c) cyclooctane   (d) 1,2-dimethylcyclopropane

    (e) 1-ethyl-2-methylcyclopentane   (f) isopropylcyclopentane

14. Draw suitable representations of all isomers (including stereoisomers) that fit the following descriptions.

    (a) cyclic compounds having molecular formula C_5H_{10}

    (b) dichlorocyclopropanes

    (c) dibromocyclohexanes

    (d) 1,2,3-trichlorocyclopentanes

15. Arrange each group of compounds in order of increasing boiling point.

    (a) hexane, 3-methylpentane, 2,2-dimethylbutane

    (b) 2-methylhexane, octane, heptane

16. Draw the chain-propagating steps for the formation of isopropyl chloride by chlorination of propane. Do the same for the formation of propyl chloride.

17. Why is chlorination of cyclohexane a practical synthesis of chlorocyclohexane but chlorination of hexane not a practical synthesis of 1-chlorohexane?

18. How could the reaction conditions for the chlorination of methane be adjusted to favor the formation mainly of methyl chloride instead of products that contain more than one chlorine atom? How could the formation of carbon tetrachloride be favored?

19. Chlorination of which C_5H_{12} compound furnishes only one monochloro product?

20. Draw representations of the possible chair conformations of the following compounds.

    (a) *cis*-1,2-dichlorocyclohexane   (b) *cis*-1,3-dimethylcyclohexane

21. Which member of each of the following pairs of compounds is the more stable?

    (a) *cis*- and *trans*-1,2-dichlorocyclohexane

    (b) *cis*- and *trans*-1,3-dimethylcyclohexane

22. Draw a representation of the most stable conformation of the most stable isomer of 1,2,3,4,5,6-hexachlorocyclohexane. (The insecticide Lindane is another of the stereoisomers.)

ADDITIONAL EXERCISES WITH MODELS

1. Construct models of cyclobutane and cyclopentane. Note that if the carbons of these models lie in a plane, then hydrogens on adjoining carbons are eclipsed. When considering acyclic alkanes, we learned that staggered arrangements are preferred. Bend one carbon of the cyclobutane and the cyclopentane models out of the plane of the other carbons. This motion somewhat reduces the eclipsing. Cyclobutane and cyclopentane adopt geometries that are slightly nonplanar. This puckering reduces eclipsing (although at the cost of somewhat increasing the deviations from the tetrahedral angle and hence increasing the angle strain).

2.   Construct a model of *trans*-decalin.

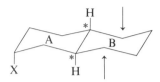

The model can be constructed by first making a chair cyclohexane ring (ring A) and then forming the other ring by attaching a four-carbon chain of atoms between the equatorial positions at the starred atoms. Note that the hydrogens on the starred carbons are both axial. Now try to flip ring A to interchange its equatorial and axial groups. You will find this impossible. The chain of carbon atoms in ring B is not long enough for the carbons marked with arrows to be axial substituents on adjacent carbons of ring A. Both rings are locked into particular chair conformations and substituent X is forever axial. Fused ring structures with such restrictions to ring flipping are found in steroids (Sec. 9.15) and many other compounds. Properties of a substituent, such as the rate of its reaction with a particular reagent or which of two possible reactions it undergoes, can depend on whether it is axial or equatorial. Therefore the position to which a substituent is constrained can be important.

Alkenes
and Alkynes

The **alkenes** are a family of hydrocarbons that contain two fewer hydrogens than alkanes with the same skeletons. In contrast to alkanes, alkenes readily undergo a variety of reactions.

Ethylene, the simplest alkene, is a gas of molecular formula C_2H_4 and boiling point $-104°$. Ethylene is used extensively; its annual industrial production in the United States is nearly 60 kg per person. Ethylene is also produced by some plants, especially by fruits. Since one of its effects is to promote ripening and loosening of fruit, compounds that slowly produce ethylene are sometimes sprayed on trees in orchards.

Ethylene has a structure in which the two carbons are linked by a double bond.

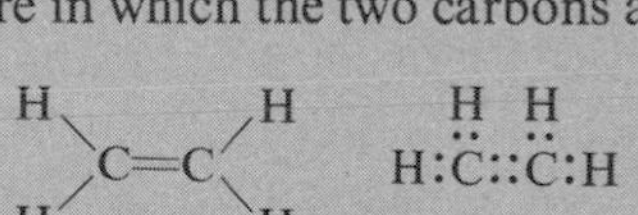

$$\begin{array}{c} H \\ \diagdown \\ C=C \\ \diagup \qquad \diagdown \\ H \qquad\qquad H \end{array} \qquad H\!:\!\overset{..}{C}\!:\!\!:\!\overset{..}{C}\!:\!H$$

This is the only structure for C_2H_4 that places four bonds (eight electrons) at each carbon and one bond (two electrons) at each hydrogen. A carbon-carbon double bond is the characteristic feature of alkenes.

3.1
STRUCTURE OF ALKENES

All atoms of ethylene lie in a plane. In this instance, the structural drawing on a sheet of paper can faithfully represent the geometry of the molecule. The bond angles at each carbon are approximately 120°.

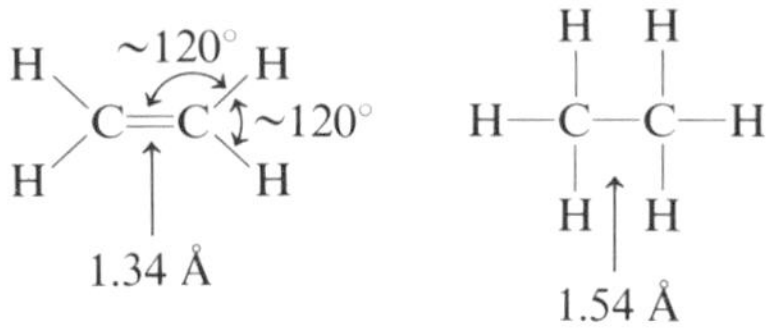
H
~120°
H
C=C
~120°
H
H
1.34 Å
H H
H—C—C—H
H H
1.54 Å

Nuclei are held closer together when bonded by two electron pairs rather than only one. In most alkenes the length of the carbon-carbon double bond (the distance between the carbon nuclei) is about 1.34 Å (1 Å = 1 angstrom unit = 10^{-10} m); the carbon-carbon single bonds of alkanes are about 1.54 Å long.

The next member of the alkene family, propene (propylene), has the molecular formula C_3H_6 and the following structure.

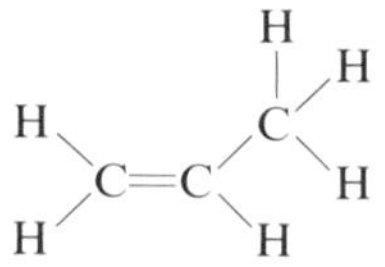

Propene
(propylene)

As was the case with ethylene, the carbons of the double bond and the four atoms attached directly to them lie in a plane. In the following drawing these atoms and the bonds linking them are darkened.

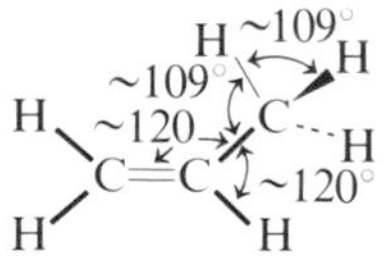
~109°
~109°
~120
~120°

The bond angles at each carbon of the double bond are approximately 120°. As is usual for carbons attached to four groups, however, the other carbon has bond angles of approximately 109°.

Following ethylene and propene, the next molecular formula is C_4H_8. Three structures can be drawn for alkenes that have this molecular formula.

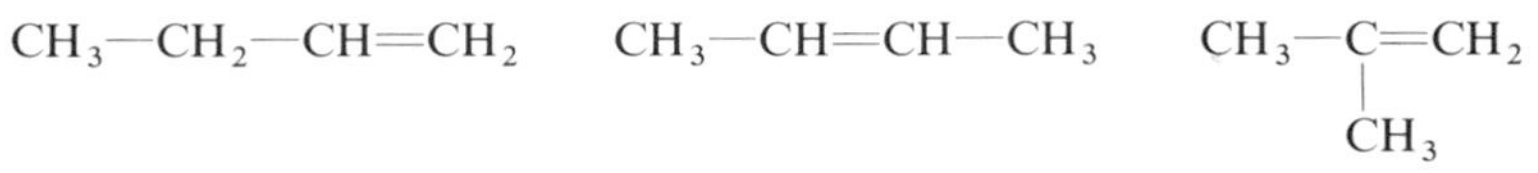
CH3—CH2—CH=CH2
CH3—CH=CH—CH3
CH3—C=CH2
CH3

1-Butene 2-Butene 2-Methyl-1-propene

These structures are drawn in the condensed fashion that omits bonds to hydrogen. The first two structures have the linear skeleton of butane; the third structure has the branched skeleton of isobutane.

Four, not three, butenes are known. From experimental evidence it is apparent that two compounds have the 2-butene structure. In considering the geometry characteristic of carbon-carbon double bonds, we find two geometrically different ways in which the methyl groups and hydrogens can be arranged.

$$CH_3 \diagdown C{=}C \diagup CH_3 \qquad CH_3 \diagdown C{=}C \diagup H$$
$$H \diagup \qquad \diagdown H \qquad H \diagup \qquad \diagdown CH_3$$

cis-2-Butene   *trans*-2-Butene

The isomer with the methyl groups on the same side of the double bond is called **cis** and the isomer with the methyl groups on opposite sides **trans**.

Rotation around the carbon-carbon double bond would interconvert *cis*-2-butene and *trans*-2-butene. However, rotation around carbon-carbon double bonds does not ordinarily occur. A sample of *cis*-2-butene or *trans*-2-butene is not converted into an equilibrium mixture of the two isomers. Therefore we consider *cis*-2-butene and *trans*-2-butene to be different compounds, just as we do butane and isobutane (or 1-butene and 2-methyl-1-propene). Because *cis*- and *trans*-2-butene have the same constitution (the same atoms bonded together) and differ only in the spatial arrangements of atoms, they are stereoisomers. In common with cyclic stereoisomers, such alkene stereoisomers are sometimes called **cis-trans isomers**.

An alkene has a stereoisomer if reversing the substituents at one carbon of the double bond gives a different geometry.

$$Cl \diagdown C{=}C \diagup Cl \qquad H \diagdown C{=}C \diagup Cl$$
$$H \diagup \qquad \diagdown H \qquad Cl \diagup \qquad \diagdown H$$

cis-1,2-Dichloroethene   *trans*-1,2-Dichloroethene

$$CH_3 \diagdown C{=}C \diagup Br \qquad H \diagdown C{=}C \diagup Br$$
$$H \diagup \qquad \diagdown H \qquad CH_3 \diagup \qquad \diagdown H$$

cis-1-Bromo-1-propene   *trans*-1-Bromo-1-propene

If, however, either carbon of the double bond is attached to two identical groups, then there is no stereoisomer. Propene, 1-butene, and 2-methyl-1-propene have no stereoisomers.

---

exercise with models 3.1   Construct models of propene that have the following arrangements.

$$CH_3 \diagdown C{=}C \diagup H \qquad H \diagdown C{=}C \diagup H$$
$$H \diagup \qquad \diagdown H \qquad CH_3 \diagup \qquad \diagdown H$$

Note that simply turning one of the models over makes it identical to the other.

Construct models of *cis*-2-butene and *trans*-2-butene. Note that these models could be interconverted only by rotating around the double bonds.

With increasing numbers of carbon atoms, the number of isomeric alkenes increases rapidly. In addition to the variations in carbon skeleton that we encountered for alkanes, the double bond often can occupy different positions and be cis or trans. The double bond can also be part of a ring.

$$
\begin{array}{c}
\text{CH}_2 \\
\text{CH}_2 \quad \text{CH} \\
| \quad\quad || \\
\text{CH}_2 \quad \text{CH} \\
\text{CH}_2
\end{array}
$$

$$
\begin{array}{c}
\text{CH}_2\!-\!\text{CH} \\
| \quad\quad || \\
\text{CH}_2\!-\!\text{CH}
\end{array}
$$

Cyclohexene   Cyclobutene

---

problem 3.1 Draw suitable representations of all alkenes (including stereoisomers) of molecular formula C_6H_{12} that have the following skeleton.

$$
\begin{array}{c}
\text{C}\!-\!\text{C}\!-\!\text{C}\!-\!\text{C}\!-\!\text{C} \\
| \\
\text{C}
\end{array}
$$

3.2
NOMENCLATURE OF ALKENES

Alkenes are usually assigned substitutive names by a procedure similar to that used for alkanes. Of course, the presence, location, and cis or trans arrangement of the double bond must also be indicated.

1. Select the longest continuous chain of carbon atoms that contains the carbons of the double bond. Form the parent name by taking the name of the linear alkane with that number of carbons and changing the *-ane* ending to *-ene*.

The name derived in this manner for the simplest alkene is ethene. However, ethylene, an old and almost universally used name, is accepted by the IUPAC. Propylene is often used instead of propene.

2. Number the chain starting at the end nearer to the double bond. Indicate the position of the double bond by the number of its lowest-numbered carbon. Indicate the configuration by cis or trans (or by *E* or *Z*, using a more general system explained in Sec. 6.3).

Consider the following pentenes.

$$
\underset{5}{\text{CH}_3}\!-\!\underset{4}{\text{CH}_2}\!-\!\underset{3}{\text{CH}_2}\!-\!\underset{2}{\text{CH}}\!=\!\underset{1}{\text{CH}_2}
$$

$$
\begin{array}{cc}
\underset{5}{\text{CH}_3}\!-\!\underset{4}{\text{CH}_2} & \underset{1}{\text{CH}_3} \\
\diagdown & \diagup \\
& \text{C}\!=\!\text{C} \\
\diagup & \diagdown \\
\text{H} \underset{3}{\quad} & \underset{2}{\quad} \text{H}
\end{array}
$$

1-Pentene   *cis*-2-Pentene

Note that the number of only one of the double-bond carbons is given.

3. Name each substituent as a prefix and, when necessary, indicate the number of the carbon to which it is attached.

Note in the following example that the longest chain including the double bond is used as the basis for the name, even though it is not the longest chain in the molecule.

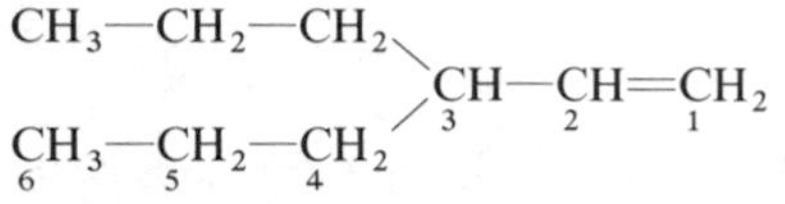

$$CH_3—CH_2—CH_2 \diagdown$$
$$\underset{3}{CH}—\underset{2}{CH}=\underset{1}{CH_2}$$
$$\underset{6}{CH_3}—\underset{5}{CH_2}—\underset{4}{CH_2} \diagup$$

3-Propyl-1-hexene

Because a chain is numbered to give the double bond the lowest possible number—whatever the consequence may be for the numbers assigned to substituents—the name of the following is not *cis*-2-chloro-3-pentene.

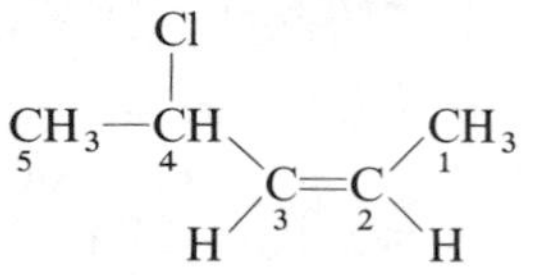

$$Cl$$
$$|$$
$$\underset{5}{CH_3}—\underset{4}{CH} \diagdown \qquad \diagup \underset{1}{CH_3}$$
$$\underset{3}{C}=\underset{2}{C}$$
$$H \diagup \qquad \diagdown H$$

cis-4-Chloro-2-pentene

In cyclic alkenes the carbons of the double bond are assigned numbers 1 and 2. Therefore the position of the double bond need not be indicated.

Cyclopentene 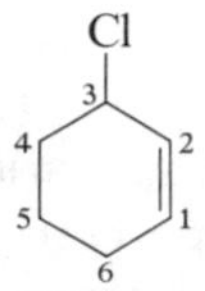   3-Chlorocyclohexene

The names **vinyl** and **allyl** are often used for the following groups that appear frequently in organic compounds.

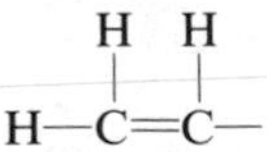

$$\overset{H}{\underset{|}{}} \overset{H}{\underset{|}{}}$$
$$H—C=C—$$

$$\overset{H}{\underset{|}{}} \overset{H}{\underset{|}{}} \overset{H}{\underset{|}{}}$$
$$H—C=C—C—$$
$$\underset{|}{\overset{}{}}$$
$$H$$

Vinyl 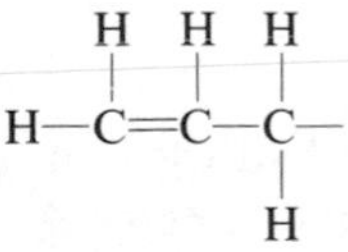   Allyl

$$CH_2=CH—Cl \qquad CH_2=CH—CH_2—\triangleleft$$

Vinyl chloride   Allylcyclopentane
(chloroethene)   (3-cyclopentyl-1-propene)*

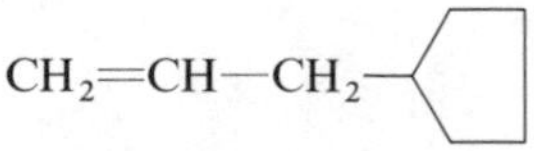

* In a few compounds, such as 2-methyl-1-propene (methylpropene) and 3-cyclopentyl-1-propene (3-cyclopentylpropene), some numbers are not essential. It is often less trouble to include them, however, than to verify that a name omitting numbers unambiguously specifies only one compound.

problem 3.2 Name the following alkenes.

(a)

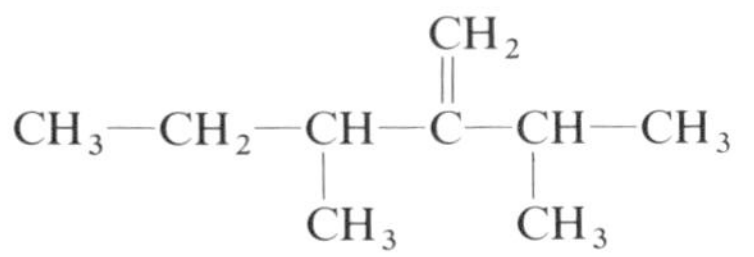

$$CH_3—CH_2—CH—\overset{\overset{\displaystyle CH_2}{\|}}{C}—CH—CH_3$$
$$\underset{\displaystyle CH_3}{|} \qquad \underset{\displaystyle CH_3}{|}$$

(b)

$$\underset{CH_3}{\overset{H}{\diagdown}}C=C\underset{H}{\overset{\overset{\displaystyle Cl}{|}}{\diagup}}\overset{CHCH_3}{}$$

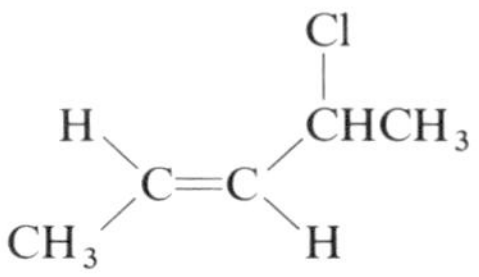

problem 3.3 Draw suitable representations of the following compounds.
(a) 4-methyl-1-pentene   (b) *cis*-1,4-dichloro-2-butene

3.3
PHYSICAL PROPERTIES OF ALKENES

The physical properties of alkenes are similar to those of the corresponding alkanes. Compare, for example, the boiling points of 1-pentene (30°) and pentane (36°) or of cyclohexene (83°) and cyclohexane (81°). As are alkanes, alkenes are practically insoluble in water but soluble in other liquid hydrocarbons and in many other organic liquids.

3.4
REACTIONS OF ALKENES

Alkenes react with a wide variety of chemical reagents, often under mild conditions. Some particularly important reactions are described in the following sections. The reactivity of alkenes, a sharp contrast with the inertness of alkanes, is due to the presence of the carbon-carbon double bond.

3.5
HYDROGENATION

Alkenes react with hydrogen in the presence of catalysts to form alkanes. Ethylene and hydrogen form ethane.

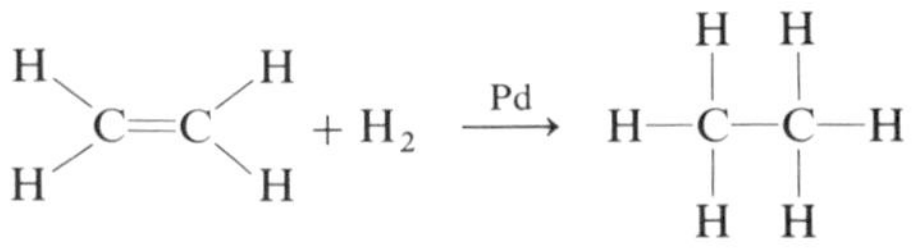

$$\underset{H}{\overset{H}{\diagdown}}C=C\underset{H}{\overset{H}{\diagup}} + H_2 \xrightarrow{\text{Pd}} H—\overset{\overset{\displaystyle H}{|}}{\underset{\underset{\displaystyle H}{|}}{C}}—\overset{\overset{\displaystyle H}{|}}{\underset{\underset{\displaystyle H}{|}}{C}}—H$$

As are many of the characteristic reactions of alkenes, this is an **addition reaction**. In an addition, atoms or groups become linked to each end of a multiple bond without the loss of any atoms or groups already present. In additions to alkenes, each carbon of the double bond becomes attached by a single bond to a portion of the reagent and one of the bonds of the double bond is lost. It may be more evident that the number of electrons involved in bonding is the same in the product as in the reactants if the relevant electrons are indicated as dots.

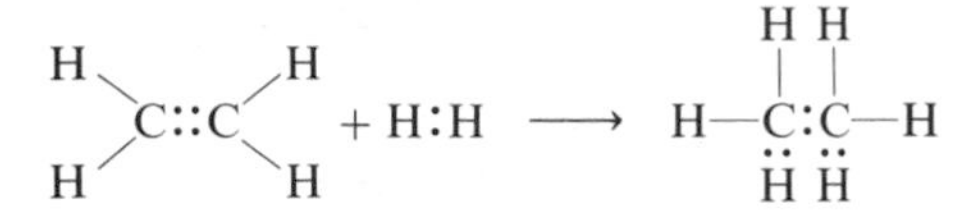

Addition of hydrogen to a multiple bond is called **hydrogenation**. Other examples of hydrogenation include the following.

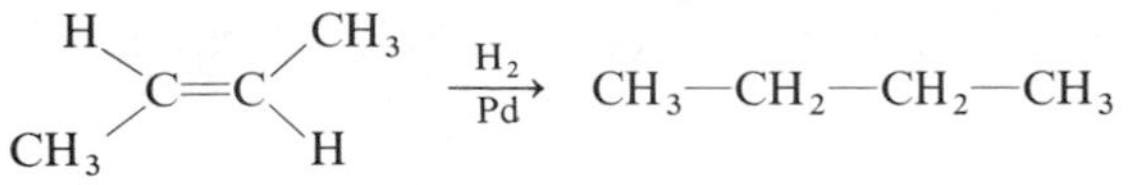

trans-2-Butene   Butane

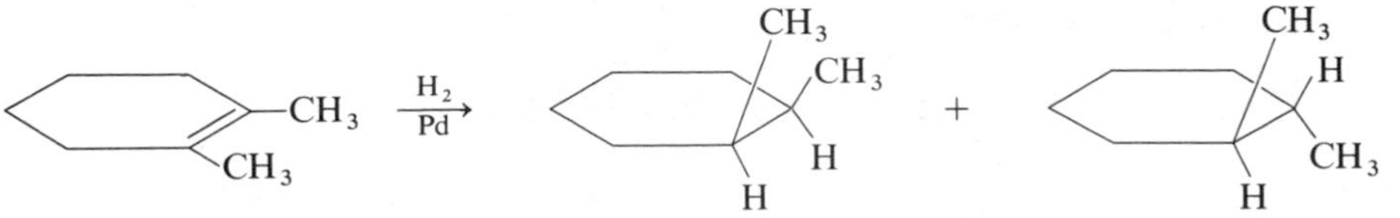

1,2-Dimethylcyclohexene   *cis*-1,2-Dimethylcyclohexane   *trans*-1,2-Dimethylcyclohexane
                                          (major product)   (minor product)

Compounds that contain a sufficient number of hydrogen atoms for each carbon to be attached to four other atoms or groups are referred to as **saturated**. Compounds with fewer hydrogens contain one or more multiple bonds and are called **unsaturated**. Therefore alkenes are unsaturated hydrocarbons, but their "saturation" with hydrogen converts them to saturated hydrocarbons. "Saturated" and "unsaturated" fats (Chapter 13) differ in carbon-carbon double-bond content.

---

problem 3.4 A compound of molecular formula C_6H_{10} can be hydrogenated, but even with severe reaction conditions one mole of the compound reacts with only one mole of H_2. How many rings and how many double bonds are in the compound?

Alkenes do not react readily with hydrogen in the absence of a catalyst. You learned in an earlier chemistry course that a **catalyst** is a substance that increases the rate of a chemical reaction without itself undergoing any permanent change. The catalysts used for hydrogenation of alkenes are metals, particularly palladium, platinum, and nickel. Hydrogen is adsorbed at the surface of these metals, and the H—H bond is broken and replaced by bonds from the hydrogen atoms to the metal.

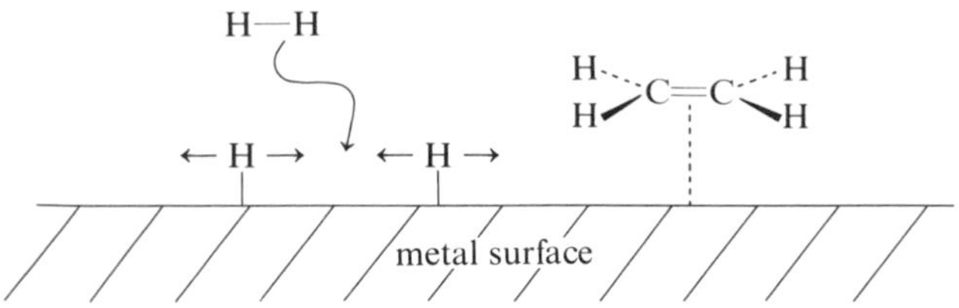

Alkenes are weakly adsorbed at the surface. The hydrogens migrate relatively freely about the surface and can bind to the carbons of a double bond when they encounter an alkene. Both hydrogens tend to add to the side of the double bond adsorbed to the catalyst surface. Therefore the major product in the hydrogenation of 1,2-dimethylcyclohexene is *cis*-1,2-dimethylcyclohexane. Such additions in which the groups become attached to the same side of the molecule are called **syn** (or cis) additions.

3.6
ADDITION OF HALOGENS

Chlorine and bromine add readily to alkenes.

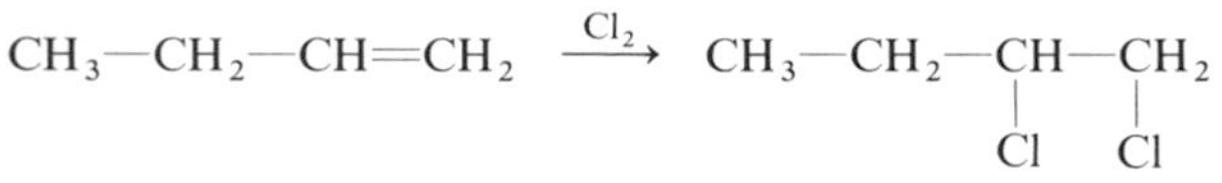

$$CH_3{-}CH_2{-}CH{=}CH_2 \xrightarrow{Cl_2} CH_3{-}CH_2{-}\underset{\underset{Cl}{|}}{CH}{-}\underset{\underset{Cl}{|}}{CH_2}$$

1,2-Dichlorobutane

trans-1,2-Dibromocyclopentane

In additions of halogens to alkenes, the halogen atoms become attached to opposite sides of the molecule. Such additions are called **anti** (or trans). Fluorine is so reactive that it does not undergo addition in an easily controlled fashion. Products from the addition of iodine revert readily to alkenes with the loss of iodine.

The rapid reactions of alkenes with bromine and chlorine are very different from the reactions of alkanes with these halogens. Initiation by heat or ultraviolet light is not needed, and the products result from addition rather than substitution. The mechanism of addition of halogens is considered in Section 3.9.

---

problem 3.5   What is the nature of a compound of molecular formula $C_{10}H_{18}$ that does not react readily with a carbon tetrachloride solution of bromine?

3.7
ADDITION OF HYDROGEN HALIDES

The hydrogen halides HCl, HBr, and HI add to alkenes.

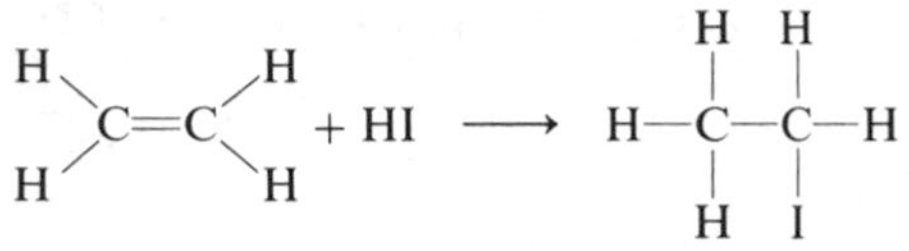

$$\underset{H}{\overset{H}{}}\!\!\diagdown C=C\diagup\!\!\underset{H}{\overset{H}{}} + HI \longrightarrow H\!-\!\underset{H}{\overset{H}{\underset{|}{\overset{|}{C}}}}\!-\!\underset{I}{\overset{H}{\underset{|}{\overset{|}{C}}}}\!-\!H$$

Ethyl iodide

The two groups of a hydrogen halide are different. If the carbons of the double bond are not identical, then two addition products are conceivable. In the addition of hydrogen chloride to propene, the chlorine atom might become attached to the middle carbon to give isopropyl chloride or to the end carbon to give propyl chloride. The reaction actually gives isopropyl chloride.

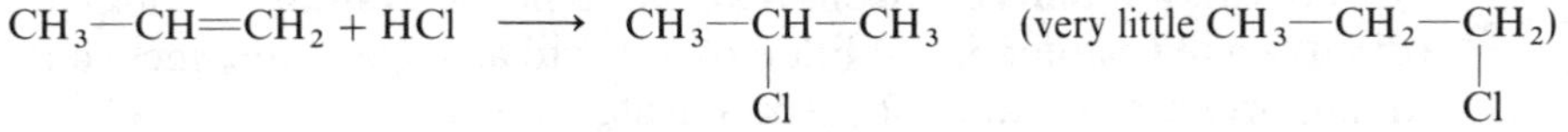

$$CH_3\!-\!CH\!=\!CH_2 + HCl \longrightarrow \underset{\underset{Cl}{|}}{CH_3\!-\!CH\!-\!CH_3} \quad (\text{very little } \underset{\underset{Cl}{|}}{CH_3\!-\!CH_2\!-\!CH_2})$$

Propene   Isopropyl   Propyl chloride
                            chloride

Similarly, addition of hydrogen bromide to 2-methyl-2-butene furnishes one of the possible products.

$$\underset{\underset{CH_3}{|}}{CH_3\!-\!C\!=\!CH\!-\!CH_3} + HBr \longrightarrow$$

2-Methyl-2-butene

$$\underset{\underset{CH_3}{|}}{\overset{\overset{Br}{|}}{CH_3\!-\!C\!-\!CH_2\!-\!CH_3}} \quad (\text{very little } \underset{\underset{CH_3}{|}}{\overset{\overset{Br}{|}}{CH_3\!-\!CH\!-\!CH\!-\!CH_3}})$$

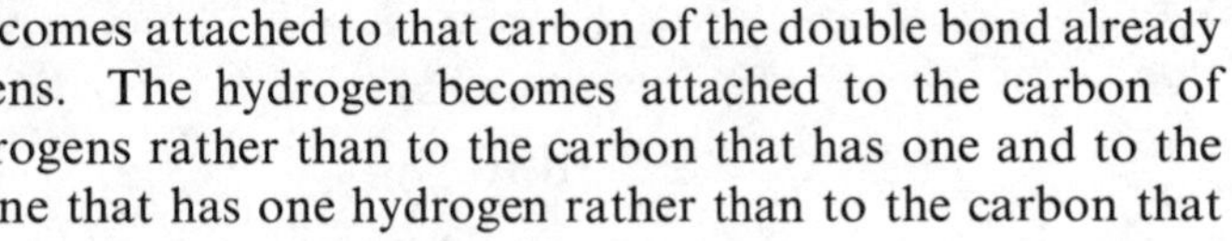

2-Bromo-2-methylbutane   2-Bromo-3-methylbutane

Results of many such additions are summarized in **Markovnikov's rule**: The hydrogen of the reagent becomes attached to that carbon of the double bond already bearing the most hydrogens. The hydrogen becomes attached to the carbon of propene that has two hydrogens rather than to the carbon that has one and to the carbon of 2-methyl-2-butene that has one hydrogen rather than to the carbon that has none.

3.8
ADDITION OF WATER

Water adds to alkenes in the presence of an acidic catalyst.

$$CH_2{=}CH_2 + H_2O \xrightarrow{\ H_3O^+\ } CH_2{-}CH_2{-}OH$$

Ethyl alcohol

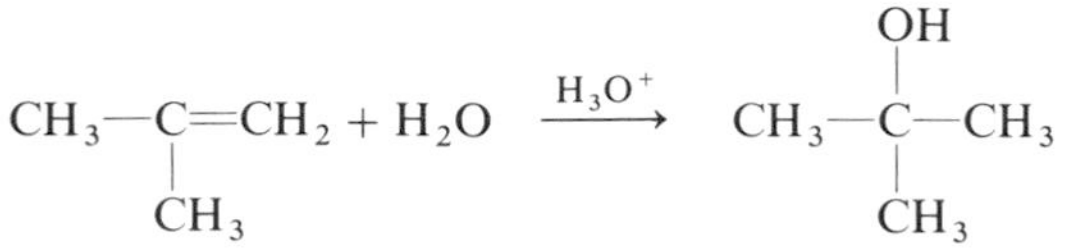

$$CH_3{-}\underset{\underset{CH_3}{|}}{C}{=}CH_2 + H_2O \xrightarrow{\ H_3O^+\ } CH_3{-}\underset{\underset{CH_3}{|}}{\overset{\overset{OH}{|}}{C}}{-}CH_3$$

tert-Butyl alcohol

The reaction is often carried out by adding the alkene to a solution of sulfuric acid in water. Note that Markovnikov's rule is followed: the hydrogen becomes attached to that carbon of the double bond of 2-methyl-1-propene that bears the most hydrogens.

The —OH group is called a **hydroxyl group**, and compounds that contain this group are called **alcohols**. Alcohols are an important family of compounds and will be discussed in Chapter 5. Addition of water to alkenes, sometimes called **hydration**, is an important procedure for preparing alcohols.

---

problem 3.6 Draw structural formulas of the products of the following reactions.
(a) hydrogenation of 1-hexene
(b) addition of bromine to 2-methyl-2-butene
(c) addition of hydrochloric acid to 2-methyl-1-propene
(d) acid-catalyzed addition of water to 1-methylcyclopentene

3.9
MECHANISM OF ADDITION OF HALOGENS, HYDROGEN HALIDES, AND WATER

Many additions to alkenes take place by two-step mechanisms. In the first step an electron-poor or **electrophilic** (electron-seeking) species becomes attached to one of the carbons of the double bond. Consider the addition of hydrogen chloride to ethylene.

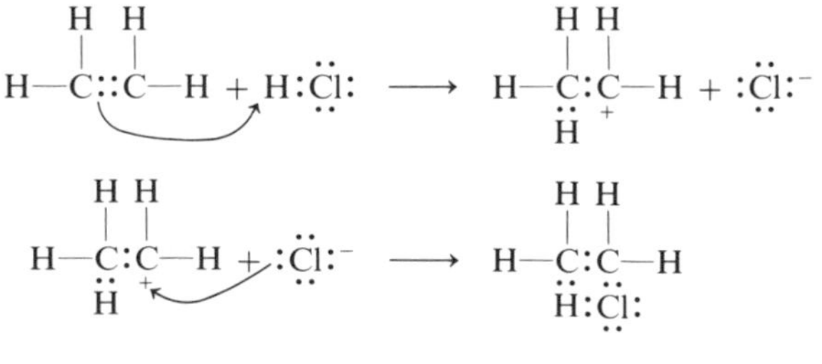

In the first step a proton becomes bonded to the alkene. Both electrons necessary for the new bond to the electrophilic proton are furnished by the alkene. The arrow is a device to help us keep track of the valence electron pairs.

The species with a positive carbon surrounded by only six electrons is called a **carbocation** (the older term carbonium ion is also still used). Carbocations are potent electrophilic reagents and extremely reactive. Their lifetimes are ordinarily short, for they rapidly undergo some reaction that will lead to eight electrons around the carbon. A common reaction is with an electron-rich or **nucleophilic** (nucleus-seeking) reagent that can furnish an electron pair to form a bond. In the addition of hydrogen chloride to ethylene, the carbocation reacts with the nucleophilic chloride ion to form the product, both electrons for the new bond coming from chloride.

The acid-catalyzed addition of water occurs in a similar fashion. We know that strong acids, such as sulfuric acid, react almost quantitatively with water to form the hydronium ion.

$$H_2SO_4 + H_2O \longrightarrow HSO_4^- + H_3O^+$$

Therefore we will represent the catalyst by that species. In the first step of the addition, the acid donates the electrophilic proton to the alkene to form a carbocation.

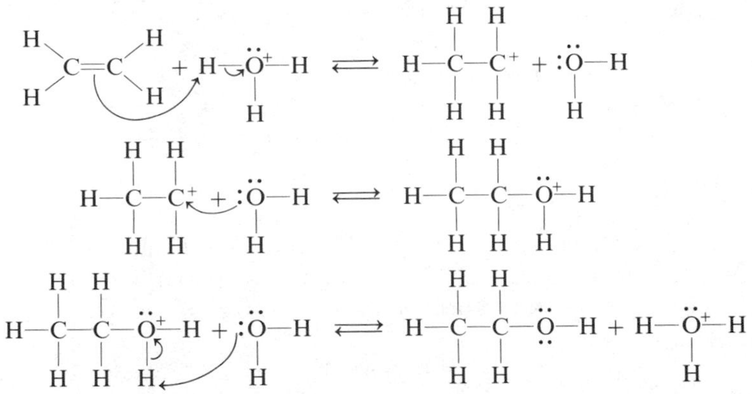

In the second step the carbocation reacts with the nucleophile water. That water can donate a pair of electrons to form a third bond to oxygen is familiar to us already from our knowledge of the hydronium ion. Transfer of a proton from the oxygen of the alcohol to the oxygen of a water molecule regenerates the H_3O^+ catalyst.

It may not be obvious that a halogen molecule can act as an electrophilic reagent. As a halogen molecule and alkene collide, however, one halogen atom can become bonded to carbon by using an electron pair furnished by the alkene.

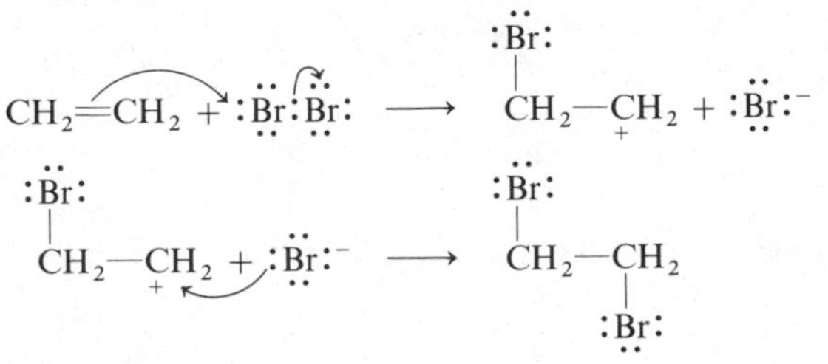

This step produces a carbocation and a halide ion. The second step is the usual combination of a carbocation with a nucleophile.*

---

problem 3.7 When ethylene reacts with bromine in an aqueous solution saturated with sodium chloride, three organic products are produced.

$$
\begin{array}{ccc}
\text{CH}_2\text{—CH}_2 & \text{CH}_2\text{—CH}_2 & \text{CH}_2\text{—CH}_2 \\
| \qquad | & | \qquad | & | \qquad | \\
\text{Br} \quad \text{Br} & \text{Br} \quad \text{Cl} & \text{Br} \quad \text{OH}
\end{array}
$$

No reaction takes place if the bromine is omitted. Explain the formation of these products.

Why does addition occur in the direction summarized by Markovnikov's rule? The step that forms a carbocation determines the direction of addition. Attachment of a proton to the end carbon of propene is much slower than to the middle carbon.

$$
\text{CH}_3\text{—CH}=\text{CH}_2
\begin{array}{l}
\xrightarrow{\text{HCl}} \overset{+}{\text{CH}_3\text{—CH—CH}_3} \xrightarrow{\text{Cl}^-} \text{CH}_3\text{—CH—CH}_3 \\
\qquad\qquad \text{A 2}^\circ \text{ carbocation} \qquad\qquad\qquad | \\
\qquad\qquad\qquad\qquad\qquad\qquad\qquad\qquad\qquad \text{Cl} \\[6pt]
\xcancel{\xrightarrow{\text{HCl}}} \text{CH}_3\text{—CH}_2\text{—}\overset{+}{\text{CH}}_2 \xrightarrow{\text{Cl}^-} \text{CH}_3\text{—CH}_2\text{—CH}_2 \\
\qquad\qquad \text{A 1}^\circ \text{ carbocation} \qquad\qquad\qquad\quad | \\
\qquad\qquad\qquad\qquad\qquad\qquad\qquad\qquad\qquad \text{Cl}
\end{array}
$$

This difference in rate is related to the stabilities of the carbocation intermediates.

Carbocations are classified as primary, secondary, or tertiary, depending on the number of alkyl groups attached to the positive carbon.

increasing stability →

$$
\underset{\substack{\text{A primary (1}^\circ) \\ \text{carbocation}}}{R\overset{+}{-}C\big\langle{}^{H}_{H}} \;<\; \underset{\substack{\text{A secondary (2}^\circ) \\ \text{carbocation}}}{R\overset{+}{-}C\big\langle{}^{H}_{R}} \;<\; \underset{\substack{\text{A tertiary (3}^\circ) \\ \text{carbocation}}}{R\overset{+}{-}C\big\langle{}^{R}_{R}}
$$

R represents any alkyl group.

* The intermediates in additions of halogens are actually better represented as cyclic structures in which the halogen is bonded to both carbons of the double bond.

$$
-\overset{|}{C}=\overset{|}{C}- \; + \; :\ddot{\text{X}}:\ddot{\text{X}}: \longrightarrow \; -\overset{|}{C}\underset{\underset{:\ddot{\text{X}}\overset{+}{:}}{\diagup\diagdown}}{}\overset{|}{C}- \; + \; :\ddot{\text{X}}:^- \longrightarrow \; -\overset{|}{\underset{:\ddot{\text{X}}:}{C}}-\overset{\overset{:\ddot{\text{X}}:}{|}}{C}- \qquad \text{X = Cl or Br}
$$

Anti (trans) addition occurs because the nucleophile attacks from the side opposite the C—X bond that will be broken.

As in the structures shown, R is often used to stand for any alkyl group. Although most carbocations are unstable and very reactive, their relative stability increases in the order primary < secondary < tertiary. The ease of formation of carbocations generally parallels their stabilities. Therefore Markovnikov's rule can be restated in a way that is related to the mechanism of addition: An electrophilic reagent adds to an alkene in the direction that forms the most stable carbocation intermediate.

The replacement of hydrogen atoms by alkyl groups makes a carbocation more stable because the electrons of bonds to an adjoining carbon are shared partially with the positive carbon.

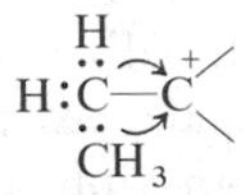

$$\begin{array}{c} H \\ H:\overset{\cdot\cdot}{\underset{\cdot\cdot}{C}}{-}\overset{+}{C} \\ CH_3 \end{array}$$

This sharing somewhat reduces the charge on the positive carbon by placing portions of it elsewhere in the molecule. A carbocation is stabilized by alkyl groups in the same way that any charged object is stabilized if its charge is dispersed.

---

problem 3.8 Would you expect hydrogen chloride to add more rapidly to ethylene or to 2-methylpropene?

Compared to the multitude of chemical reactions, the number of fundamentally different reaction steps is few. Seemingly diverse reactions may actually occur in similar ways. We have seen, for instance, that several additions to alkenes occur by initial addition of an electrophile to form a carbocation intermediate. The ease of addition to a particular alkene or the direction of addition of the reagent would seem to be isolated observations if we knew nothing about the mechanism. We now recognize, however, that these features have a common origin in the relative stabilities of carbocations.

3.10
OXIDATION

As do alkanes (and most other organic compounds), alkenes burn in air to form carbon dioxide and water. Alkenes, however, also react more selectively with a variety of oxidizing agents in reactions that involve the double bond.

The reaction that is illustrated for 1-butene occurs with potassium permanganate when mild conditions are used.

$$3\,CH_3{-}CH_2{-}CH{=}CH_2 + 2\,K^+\,MnO_4^- + 4\,H_2O \longrightarrow$$

$$3\,CH_3{-}CH_2{-}\underset{OH}{\overset{|}{CH}}{-}\underset{OH}{\overset{|}{CH_2}} + 2\,MnO_2{\downarrow} + 2\,K^+\,OH^-$$

During the course of the reaction, the purple permanganate color fades as permanganate is reduced to form MnO_2, which appears as a brown precipitate. Due to formation of a cyclic intermediate, the oxidation with permanganate proceeds in a syn fashion.

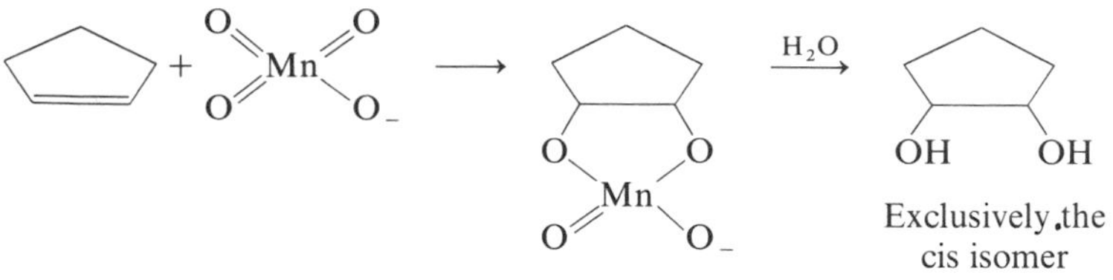

Alkenes react with ozone (O_3), a molecule that can be prepared by passing oxygen (O_2) through an electric discharge. Ozone is a constituent of photochemical smog and its reaction with alkenes plays a role in that environmental disaster. The initial product is called an ozonide.

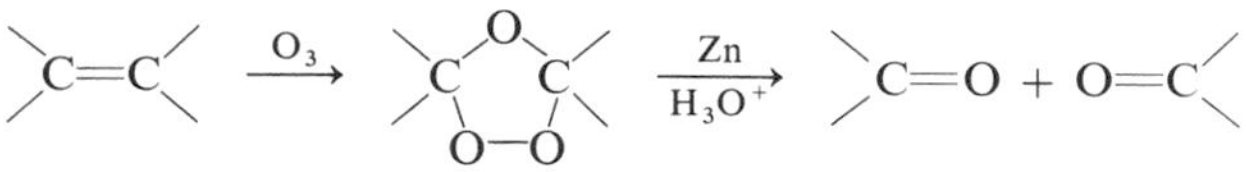

Ozonide

The ozonide is usually treated with a reducing agent, often zinc in an acidic solution, to cleave it into two fragments. In the overall reaction, called ozonolysis, the double bond is cleaved and each carbon originally part of the double bond becomes doubly bonded to an oxygen.

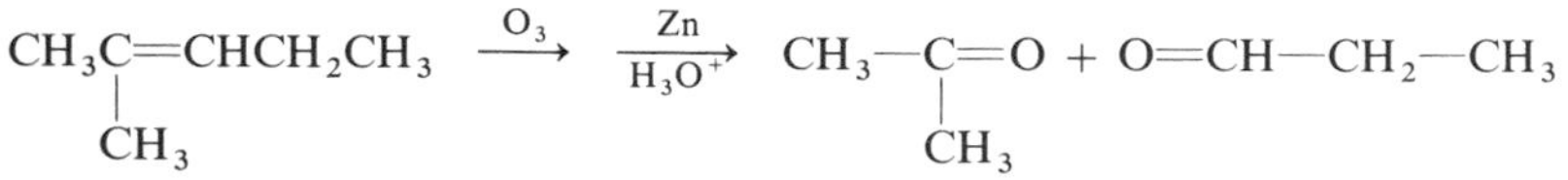

The $>C=O$ group is called a **carbonyl group**; aldehydes and ketones, important compounds containing this group, are the subject of Chapter 9.

---

problem 3.9 Ozonolysis can be used to identify the structure of an alkene if the resulting carbonyl-containing compounds can be identified.
(a) What alkenes on treatment with ozone followed by Zn and acid give only CH_3—$CH=O$?
(b) What alkene gives only $O=CH—CH_2—CH_2—CH_2—CH_2—CH=O$ on similar treatment?

3.11
POLYMERIZATION

Except for combustion, the most extensive use of organic compounds is for preparing polymers. **Polymers** are large molecules, molecular weights of several thousand to several million, that contain repeating structural units. Some of the most widely used polymers are prepared from alkenes.

To illustrate what polymers are and an important method for their preparation, we will consider a polymer called polyvinyl chloride. Annual production of this polymer in the United States exceeds 13 kg per person. It is found in a variety of articles, including phonograph records and plastic pipe. Upon addition of some high-boiling organic compounds called plasticizers that dissolve in the polymer and make it more flexible, it is used in such articles as raincoats, overshoes, garden hose, and automobile upholstery.

Polyvinyl chloride can be prepared by heating vinyl chloride with a small amount of a peroxide. A **peroxide** is a compound that contains an oxygen-oxygen single bond. Such bonds are weak and are readily cleaved on heating to produce free radicals.

$$R—\overset{..}{\underset{..}{O}}:\overset{..}{\underset{..}{O}}—R \xrightarrow{\text{heat}} 2\,R—\overset{..}{\underset{..}{O}}\cdot \qquad \text{R is some organic group.}$$

Arrows are again used to help us keep track of the valence electrons. Here, however, they are only half-headed because single electrons rather than electron pairs are involved.

The reactive free radicals add to the double bond of a vinyl chloride molecule.

$$R\overset{..}{\underset{..}{O}}\cdot + CH_2\!::\!\overset{.}{C}H \longrightarrow R\overset{..}{\underset{..}{O}}\!:\!CH_2\!:\!\overset{.}{C}H$$
$$\qquad\qquad\qquad |\qquad\qquad\qquad\qquad |$$
$$\qquad\qquad\qquad Cl\qquad\qquad\qquad\qquad Cl$$

Formation of the new bond in this addition requires only one electron from the alkene, and the addition product is another free radical.* The new free radical also adds to a vinyl chloride molecule, producing yet another free radical.

$$\begin{array}{l}
RO—CH_2—\overset{.}{C}H \xrightarrow{\;CH_2=CH,\;Cl\;} RO—CH_2—CH—CH_2—\overset{.}{C}H \xrightarrow{\;CH_2=CH,\;Cl\;}\\
\qquad\qquad\quad |\qquad\qquad\qquad\qquad\quad |\qquad\qquad\quad |\\
\qquad\qquad\quad Cl\qquad\qquad\qquad\qquad\quad Cl\qquad\qquad Cl
\end{array}$$

$$RO—CH_2—CH—CH_2—CH—CH_2—\overset{.}{C}H \longrightarrow \longrightarrow \longrightarrow$$
$$\qquad\qquad\quad |\qquad\qquad\quad |\qquad\qquad\quad |$$
$$\qquad\qquad\quad Cl\qquad\qquad Cl\qquad\qquad Cl$$

$$RO\!-\!(CH_2—CH)_n\!-\!CH_2—\overset{.}{C}H \xrightarrow{\;Z\cdot\;} RO\!-\!(CH_2—CH)\!-\!CH_2—CH—Z$$
$$\qquad\qquad\;\; |\qquad\qquad\quad |\qquad\qquad\qquad\qquad\; |\qquad\qquad\; |$$
$$\qquad\qquad\;\; Cl\qquad\qquad Cl\qquad\qquad\qquad\qquad Cl\qquad\quad Cl$$

n is a large integer.
Z· could be RO· or another growing polymer chain.

The free radical produced by each successive addition adds to another molecule of vinyl chloride, resulting ultimately in the construction of a long chain.

* The direction of addition is that producing the most stable free radical.

Table 3.1 Some important polymers prepared from alkenes

Monomer	Polymer (trade names are in parentheses)	Approximate annual use per person in United States (kg)	Some uses
$CH_2{=}CH_2$ Ethylene	$+CH_2CH_2)_{\overline{n}}$ polyethylene (Polythene)	27	pipe, film, sheeting, molded articles, electrical wire insulation
$CH_2{=}CHCH_3$ Propylene (propene)	$+CH_2CH)_{\overline{n}}$ $\quad\quad\mid$ $\quad\quad CH_3$ polypropylene (Herculon)	8	molded articles, fibers for garments and carpeting, film
$CH_2{=}CHCl$ Vinyl chloride	$+CH_2CH)_{\overline{n}}$ $\quad\quad\mid$ $\quad\quad Cl$ polyvinyl chloride (Koroseal)	13	pipe, vinyl flooring, upholstery materials, wire coatings, phonograph records
$CH_2{=}CH{-}C{\equiv}N$ Acrylonitrile	$+CH_2CH)_{\overline{n}}$ $\quad\quad\mid$ $\quad\quad C{\equiv}N$ polyacrylonitrile (Orlon, Acrilan, Creslan)	2	woollike fibers
$CF_2{=}CF_2$ Tetrafluoroethylene	$+CF_2CF_2)_{\overline{n}}$ polytetrafluoro-ethylene (Teflon)	<1	chemically resistant objects, nonstick coatings for kitchen utensils
$CH_2{=}CH{-}C_6H_5$ Styrene	$+CH_2CH)_{\overline{n}}$ $\quad\quad\mid$ $\quad\quad C_6H_5$ polystyrene (Styrofoam)	13	molded articles, foam insulation and packaging

Free-radical polymerization of vinyl chloride is another example of a chain reaction. Peroxide cleavage is the initiation step. Each addition of a free radical to vinyl chloride to produce a new free radical is a chain-propagation step. Possible chain-termination steps include coupling of the growing chain with $RO\cdot$ or with another growing chain. Because the concentrations of free radicals are low, combination of radicals is infrequent. Hundreds or thousands of chain-propagation steps may occur before the growth of a chain is terminated.

The overall reaction can be summarized by a simple equation.

$$n\ CH_2{=}\underset{\underset{Cl}{|}}{CH} \xrightarrow{\text{catalyst}} {\color{black}\text{-}}(CH_2{-}\underset{\underset{Cl}{|}}{CH}{)_n}$$

The joining together of small molecules to construct polymers is called **polymerization** and the small molecules from which a polymer is prepared are called **monomers**.

In 1974 the U.S. Occupational Safety and Health Administration (OSHA) concluded that vinyl chloride is a carcinogen (cancer-producing substance) for humans and set strict limits to exposure to this compound. Because of the large-scale use of vinyl chloride to prepare polyvinyl chloride, many persons have been exposed to it. Nevertheless, it took many years for the danger of vinyl chloride to become evident, for cancer may not appear until long after exposure.

Some other polymers prepared from alkenes are listed in Table 3.1. The nature of the $-C_6H_5$ grouping in styrene is discussed in the next chapter. All these polymers can be produced by free-radical reactions. Catalysts that result in polymerization by other mechanisms, however, are used to produce much of the polyethylene, most of the polypropylene, and many polymers not in the table.

A polymer (Saran) used for packaging film is prepared by polymerizing a mixture of 1,1-dichloroethene ($CH_2{=}CCl_2$) and vinyl chloride. Therefore it contains both $-CH_2CCl_2-$ and $-CH_2CHCl-$ units. Polymers that contain two monomer units are known as **copolymers**. By choice of the monomers and their ratio, a range of polymer properties can be obtained.

3.12
PREPARATION OF ALKENES

A carbon-carbon double bond is introduced into saturated molecules by **elimination reactions**, reactions in which atoms or groups are lost from two adjoining carbons. Eliminations are formally the reverse of additions.

$$-\underset{\underset{A}{|}}{C}{-}\underset{\underset{B}{|}}{C}{-} \underset{\xleftarrow[\text{addition}]{}}{\overset{\text{elimination}}{\rightleftharpoons}} {>}C{=}C{<} + A{-}B$$

3.13
DEHYDROHALOGENATION OF ALKYL HALIDES

Dehydrohalogenation is the elimination of the elements of a hydrogen halide.

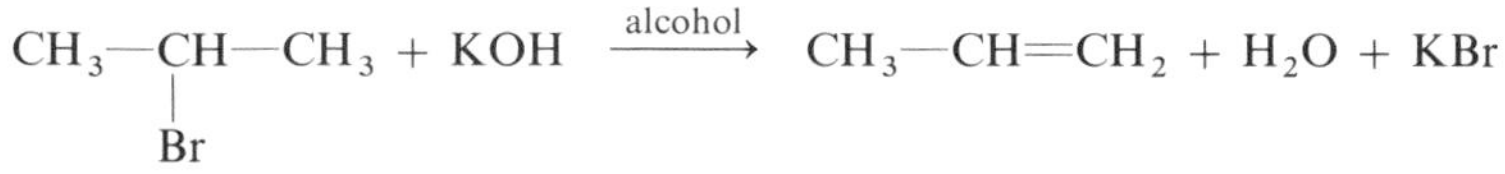

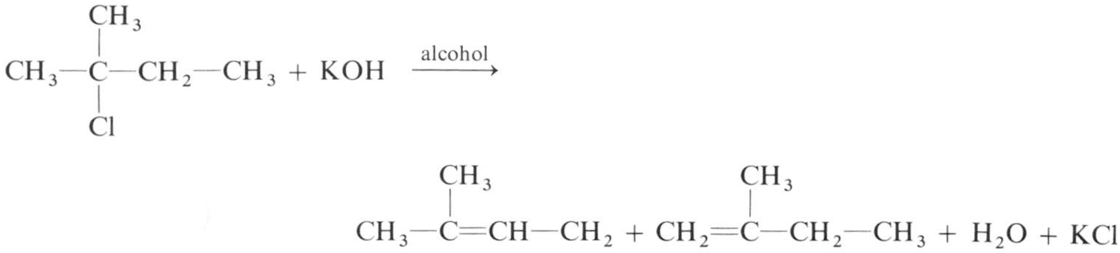

$$CH_3-CH-CH_3 + KOH \xrightarrow{\text{alcohol}} CH_3-CH=CH_2 + H_2O + KBr$$
$$| \quad \text{Br}$$

$$CH_3-\overset{\overset{\displaystyle CH_3}{|}}{C}-CH_2-CH_3 + KOH \xrightarrow{\text{alcohol}}$$
$$\underset{\displaystyle Cl}{|}$$

$$\underset{\displaystyle CH_3}{|} \qquad \underset{\displaystyle CH_3}{|}$$
$$CH_3-C=CH-CH_2 + CH_2=C-CH_2-CH_3 + H_2O + KCl$$

Major product	Minor product
(double bond has three alkyl substituents)	(double bond has two alkyl substituents)

It is carried out by using a strong base—for example, potassium hydroxide dissolved in an alcohol. The hydrogen that is eliminated must come from a carbon adjacent to the carbon bearing the halogen atom. When loss of such hydrogens can result in formation of more than one alkene, then that alkene bearing the most alkyl substituents is ordinarily the major product.

---

problem 3.10 Draw the major organic products that result from dehydrohalogenating the following alkyl halides.

(a) 2-bromobutane   (b) 3-chlorohexane

3.14
DEHYDRATION OF ALCOHOLS

Dehydration, the elimination of water, is achieved by heating an alcohol with an acid catalyst.

$$CH_3-CH_2-OH \xrightarrow[\text{heat}]{H_2SO_4} CH_2=CH_2 + H_2O$$

Ethyl alcohol

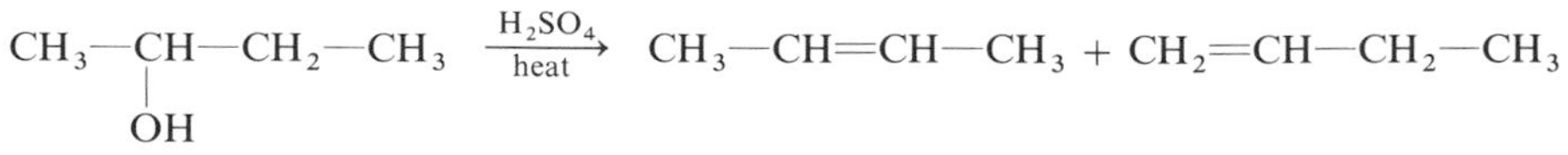

$$CH_3-\underset{\underset{\displaystyle OH}{|}}{CH}-CH_2-CH_3 \xrightarrow[\text{heat}]{H_2SO_4} CH_3-CH=CH-CH_3 + CH_2=CH-CH_2-CH_3$$

| | (cis + trans) | |
| *sec*-Butyl alcohol | Major product | Minor product |

In a laboratory the catalyst is often sulfuric or phosphoric acid. Solids, such as Al_2O_3, with acidic properties are used for large-scale industrial preparations. The hydroxyl group and hydrogen are lost from adjoining carbons. As was the case for dehydrohalogenation, that alkene bearing the most alkyl substituents is usually the major product when elimination can occur in more than one direction.

---

problem 3.11 How could you synthesize isopropyl chloride from propyl alcohol?

Dehydration of alcohols is the reverse of hydration of alkenes.

$$-\underset{\underset{H}{|}}{\overset{|}{C}}-\underset{\underset{OH}{|}}{\overset{|}{C}}- \; \underset{\xrightarrow{H_3O^+}}{\rightleftharpoons} \; \diagdown C{=}C \diagup \; + H_2O$$

The mechanism of dehydration has the same steps (but in reverse order) shown in Section 3.9 for the hydration of an alkene. Formation of a carbocation is the key step in dehydration, and the ease of dehydrating alcohols parallels the stabilities of the carbocations generated from them. Dehydration of ethyl alcohol, which involves a primary carbocation, requires more severe conditions (higher temperature, stronger acid, longer reaction time) than does dehydration of *sec*-butyl alcohol, which involves a secondary carbocation.

An equilibrium mixture of alkene, alcohol, and water is reached in a hydration or dehydration reaction. Large amounts of water increase the concentration of the alcohol in the equilibrium mixture and are used when hydration is the objective. Conversely, removal of the water or alkene produced on dehydration displaces the equilibrium and favors alkene formation. Because alkenes are more volatile than the corresponding alcohols, they can sometimes be distilled selectively from the reaction mixture as they are formed.

3.15
CRACKING OF ALKANES

The process of heating alkanes at such high temperatures that bonds are broken and smaller molecules formed is called **cracking**. As illustrated with butane, cracking produces alkenes with the same number of carbon atoms by loss of hydrogen and smaller alkenes by loss of alkanes.

$$CH_3CH_2CH_2CH_3 \xrightarrow{700\text{--}800°} CH_3CH_2CH{=}CH_2 + \quad CH_3CH{=}CHCH_3 + H_2 +$$

$$\text{(cis + trans)}$$

$$CH_3CH{=}CH_2 + CH_4 + CH_2{=}CH_2 + CH_3CH_3$$

Ethylene and propene are usually made by cracking alkanes. Cracking provides an important way of converting the higher boiling fractions of petroleum into lower boiling materials suitable for use in or conversion to gasoline. By using appropriate solid catalysts, cracking can be achieved at lower temperatures (400–500°) and is accompanied by some rearrangement of linear alkanes to branched isomers, which have higher octane ratings.

3.16
ALKYNES

The alkynes, another family of unsaturated hydrocarbons, contain four fewer hydrogens than alkanes with the same skeletons. Acetylene, the simplest alkyne, has the molecular formula C_2H_2 and a structure in which the two carbons are linked by a triple bond.

$$H:C:::C:H \qquad H-C{\equiv}C-H$$

This is the only structure for C_2H_2 that places four bonds (eight electrons) at each carbon. A carbon-carbon triple bond is the characteristic feature of alkynes.

Acetylene is linear and its carbon-carbon bond is shorter than that of ethylene.

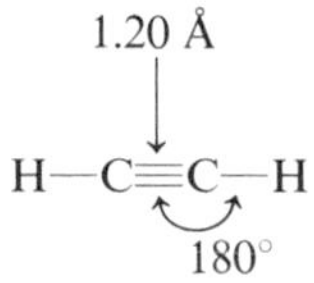

$$1.20 \text{ Å}$$
$$H-C{\equiv}C-H$$
$$180°$$

In more complex alkynes the carbons of the triple bond and the atoms attached directly to them are also arranged linearly.

Substitutive nomenclature for alkynes is identical to that for alkenes except that the ending *-yne* is used to indicate the presence of the triple bond. Acetylene, a trivial name in common use, is accepted by the IUPAC.

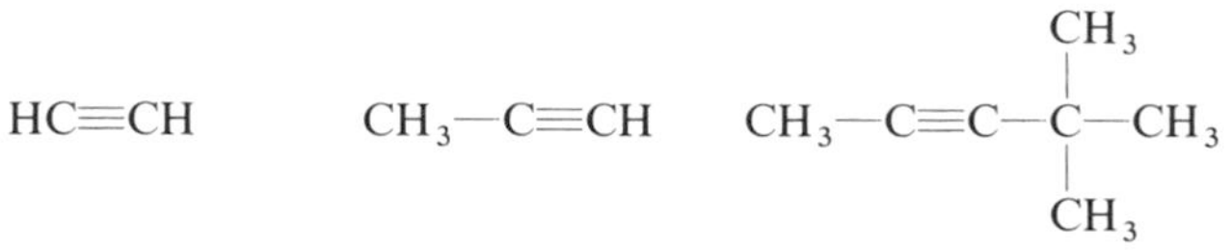

$$HC{\equiv}CH \qquad CH_3-C{\equiv}CH \qquad CH_3-C{\equiv}C-\overset{\displaystyle CH_3}{\underset{\displaystyle CH_3}{C}}-CH_3$$

| Acetylene | Propyne | 4,4-Dimethyl-2-pentyne |
| Ethyne | | |

Boiling points, solubilities, and densities of alkynes are similar to those of alkanes and alkenes.

Many chemical reactions of alkynes resemble reactions of alkenes. Alkynes, for example, form addition products with hydrogen (in the presence of a catalyst), halogens, and hydrogen halides. As illustrated for the hydrogenation of acetylene, addition of one molecule of a reagent produces a carbon-carbon double bond, which we already know can also react with the reagent.

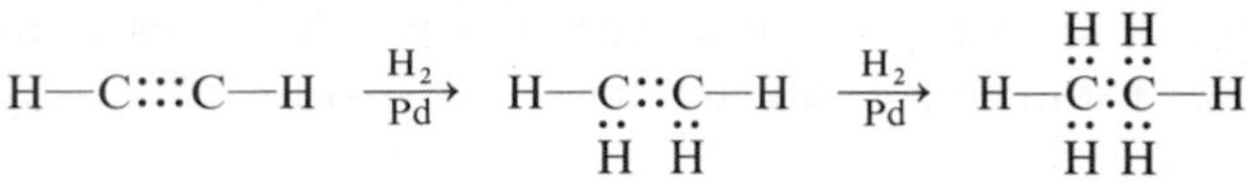

$$H—C:::C—H \xrightarrow[Pd]{H_2} H—\overset{\cdot\cdot}{\underset{\overset{|}{H}\ \underset{}{H}}{C}}::\overset{\cdot\cdot}{C}—H \xrightarrow[Pd]{H_2} H—\overset{\overset{H\ \ H}{\cdot\cdot}}{\underset{\underset{H\ \ H}{}}{C}}:\overset{\cdot\cdot}{C}—H$$

By appropriate choice of reaction conditions, it is usually possible to add either one or two molecules of most reagents.

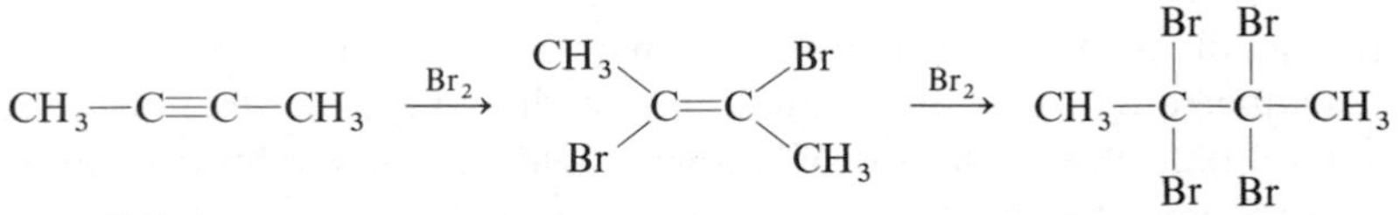

$$CH_3—C≡C—CH_3 \xrightarrow{Br_2} \underset{Br}{\overset{CH_3}{}}C=C\underset{CH_3}{\overset{Br}{}} \xrightarrow{Br_2} CH_3—\overset{\overset{Br}{|}}{\underset{\underset{Br}{|}}{C}}—\overset{\overset{Br}{|}}{\underset{\underset{Br}{|}}{C}}—CH_3$$

2-Butyne

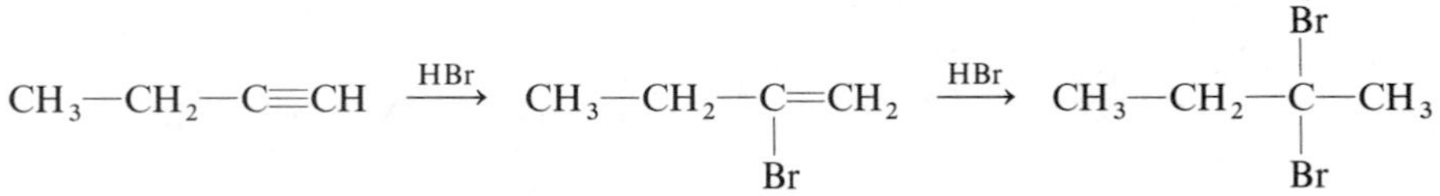

$$CH_3—CH_2—C≡CH \xrightarrow{HBr} CH_3—CH_2—\overset{}{\underset{\underset{Br}{|}}{C}}=CH_2 \xrightarrow{HBr} CH_3—CH_2—\overset{\overset{Br}{|}}{\underset{\underset{Br}{|}}{C}}—CH_3$$

1-Butyne

$$HC≡CH \xrightarrow{HCl} CH_2=\overset{}{\underset{\underset{Cl}{|}}{CH}} \xrightarrow{HCl} CH_3—\overset{\overset{Cl}{|}}{\underset{\underset{Cl}{|}}{CH}}$$

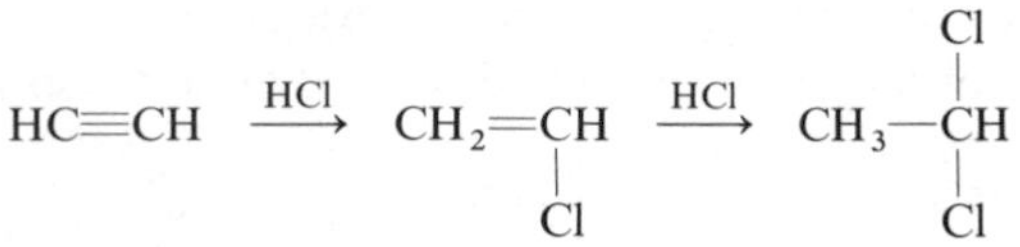

In analogy to the reactions of alkenes, addition of halogens is anti and addition of hydrogen halides follows Markovnikov's rule.

---

problem 3.12 What would you expect to be the major product from addition (using a palladium catalyst) of one mole of hydrogen to 2-butyne?

3.17
THE FUNCTIONAL GROUP

Organic chemists often regard organic molecules as carbon frameworks adorned with **functional groups**. A functional group is an atom or a particular grouping of atoms that confers characteristic chemical and physical behavior on molecules in which it is incorporated. That such behavior is relatively similar in many molecules makes feasible the task of systematizing the behavior of millions of organic molecules. Carbon-carbon double and triple bonds are examples of functional groups. Other examples encountered but not yet considered in detail are halide (—F, —Cl, —Br, —I), hydroxyl (—OH), and carbonyl ($>C=O$) groups.

Consider the carbon-carbon double bond. Chemical reactions of an alkene ordinarily involve the double bond, not the saturated portions of the molecule, and are similar to those of other alkenes. A simplified equation that specifies nothing about the structure of the alkene into which the carbon-carbon double bond is embedded summarizes literally tens of thousands of hydrogenation reactions.

$$\text{C}{=}\text{C} \xrightarrow[\text{Pt, Pd, or Ni}]{\text{H}_2} -\underset{\text{H}}{\overset{}{\text{C}}}-\underset{\text{H}}{\overset{}{\text{C}}}-$$

Instead of memorizing hydrogenation reactions of individual alkenes, we need to remember only this simple equation and the additional fact that hydrogenation is preferentially syn. Addition of halogens to alkenes is summarized by a simple general equation and the observation that addition is anti.

$$\text{C}{=}\text{C} \xrightarrow{\text{X}_2} -\overset{\text{X}}{\underset{}{\text{C}}}-\underset{\text{X}}{\overset{}{\text{C}}}- \qquad \text{X}_2 = \text{Cl}_2 \text{ or } \text{Br}_2$$

Similarly, addition of hydrogen halides or water can be summarized by a single equation and the observation that the direction of addition follows Markovnikov's rule.

$$\text{C}{=}\text{C} \xrightarrow{\text{HX}} -\underset{\text{H}}{\overset{}{\text{C}}}-\underset{\text{X}}{\overset{}{\text{C}}}- \qquad \text{HX} = \text{HCl, HBr, HI, or HOH (H}_3\text{O}^+)$$

Each alkene, of course, has some individuality. We know, for instance, that reactivity toward additions of electrophilic reagents depends on alkene structure, particularly as it affects the stability of the carbocation intermediate. Differences in size or speed among horses, however, do not negate the value of considering them a common group, quite distinct from turtles. So it is with different organic functional groups.

3.18
RESONANCE THEORY

We described covalent bonds as the result of two atoms sharing one, two, or three electron pairs. This simple idea is remarkably effective in describing a wide variety of organic compounds. Sometimes, however, such a description is misleading.

Consider the allyl cation, the carbocation that results from treating allyl alcohol with an acid.

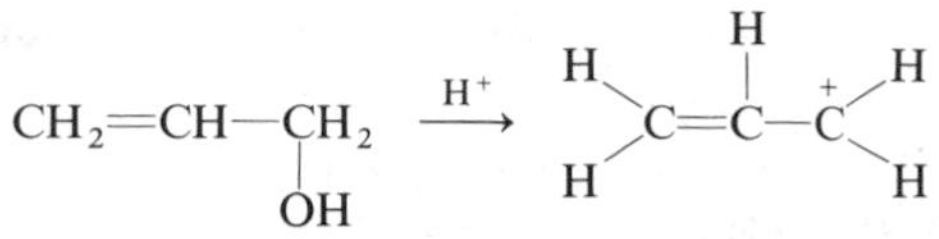

$$CH_2{=}CH{-}CH_2 \xrightarrow{\;H^+\;} \begin{array}{c} H \\ | \\ \end{array}$$

Allyl cation

The structure indicates that it is a primary carbocation. The allyl cation is considerably more stable, however, than typical primary carbocations. The structure also indicates that this ion contains a carbon-carbon single bond and a carbon-carbon double bond. Single and double bonds differ in properties. We know, for example, that double bonds are significantly shorter than single bonds. Contrary to the implications of this structure, the carbon-carbon bonds in the allyl cation are identical and their properties are intermediate between those of typical single and double bonds.

When a structure is misleading, we frequently find that we can draw one or more additional structures that differ only in the position of electrons and that also appear reasonable. In such cases, the assumption that every electron pair is shared by just two atoms is no longer a good approximation. Note that the allyl cation could just as well have been drawn as shown here.

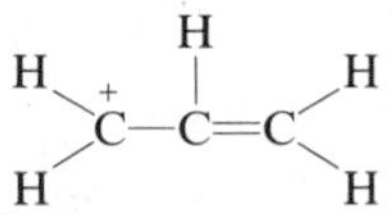

The same atoms are attached together in both structures; only an electron pair is placed differently. A particular end carbon is bonded to the center carbon with a double bond (two electron pairs) in one structure but with a single bond (one electron pair) in the other. In an earlier course you may have encountered some simple inorganic species, such as O_3, HNO_3, and $CO_3{}^{2-}$, for which two or more reasonable structures differing only in electron placement also could be drawn.

The resonance theory provides a way to use the electron-pair structures with which we have become familiar to recognize, represent, and predict the properties of chemical species in which electrons are shared by more than two atoms. When a molecule can be represented by two or more reasonable structures that have the same atoms bonded together but differ in the placement of electrons, the theory states that

1. the molecule is better represented as a hybrid of these structures than by any one of them.

2. the molecule is more stable than would be predicted for any one structure.

According to the resonance theory, the allyl cation is better represented as a hybrid of two structures than by either structure alone.

$$CH_2{=}CH{-}\overset{+}{C}H_2 \quad\longleftrightarrow\quad \overset{+}{C}H_2{-}CH{=}CH_2$$

Structures that contribute to a hybrid are called **resonance structures** and, by convention, are connected by a double-headed arrow. Since each carbon-carbon bond is single in one structure and double in the other, the resonance theory predicts that both bonds in the actual ion are equivalent and intermediate in properties between carbon-carbon single and double bonds. The theory also predicts that the ion is more stable than would be predicted on the basis of either resonance structure.

A mule is a hybrid of a horse and a donkey. It is always a mule, not some days a horse and other days a donkey. Similarly, the allyl cation is a hybrid of two resonance structures. It does not sometimes have one structure and sometimes the other. It always has the same structure, and that structure is intermediate between the individual resonance structures.

3.19
DIENES

Hydrocarbons that contain two double bonds are called **dienes**. The examples given show how the substitutive nomenclature is adapted to indicate the presence and location of two double bonds.

$$CH_2\!=\!CH\!-\!CH\!=\!CH_2 \qquad CH_2\!=\!CH\!-\!CH_2\!-\!CH\!=\!CH_2$$

1,3-Butadiene   1,4-Pentadiene

Multiple bonds are called **conjugated** if separated by only one single bond and **isolated** if separated by more than one single bond.* Isolated multiple bonds ordinarily react independently. Bromine, for example, adds to the isolated double bonds of 1,4-pentadiene in the same fashion as to simple alkenes.

$$CH_2\!=\!CH\!-\!CH_2\!-\!CH\!=\!CH_2 \xrightarrow{Br_2} \underset{\underset{Br}{|}\ \ \underset{Br}{|}}{CH_2\!-\!CH\!-\!CH_2\!-\!CH\!=\!CH_2} \xrightarrow{Br_2}$$

$$\underset{\underset{Br}{|}\ \ \underset{Br}{|}\qquad\ \underset{Br}{|}\ \ \underset{Br}{|}}{CH_2\!-\!CH\!-\!CH_2\!-\!CH\!-\!CH_2}$$

Reactions with conjugated double bonds can be more complex. Addition of one Br_2 to the conjugated double bonds in 1,3-butadiene produces two dibromo products.

$$CH_2\!=\!CH\!-\!CH\!=\!CH_2 \xrightarrow{Br_2} \underset{\underset{Br}{|}\ \ \underset{Br}{|}}{CH_2\!-\!CH\!-\!CH\!=\!CH_2} + \underset{\underset{Br}{|}\qquad\quad\ \underset{Br}{|}}{CH_2\!-\!CH\!=\!CH\!-\!CH_2}$$

                                            (cis + trans)
          "1,2-Addition"   "1,4-Addition"
             product   product

---

* Double bonds that share a carbon, as in $CH_2\!=\!C\!=\!CH_2$, are called **cumulated**. Such double bonds are encountered infrequently.

In the "1,2-addition" product, as in products of addition to simple alkenes, the bromines are attached to adjoining carbons that had formed a double bond. In the "1,4-addition" product, however, the bromines are attached to nonadjoining carbons that were parts of different double bonds; the position of the remaining double bond does not correspond to that of either double bond of the diene.

By considering the mechanism of bromine addition to alkenes, we can see how the formation of two products from a conjugated diene is reasonable. The first step of bromine addition furnishes a carbocation.

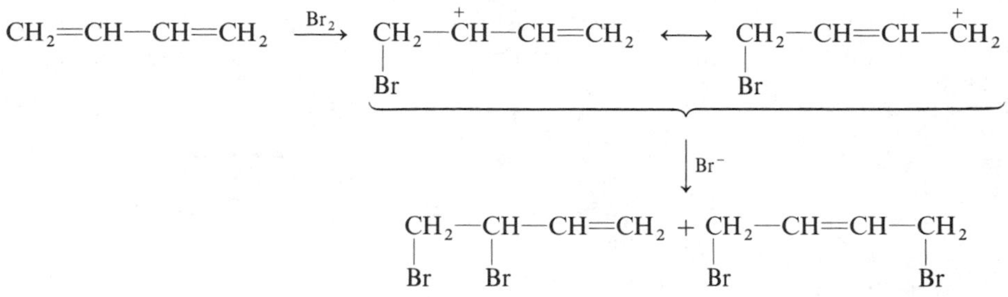

$$CH_2{=}CH{-}CH{=}CH_2 \xrightarrow{Br_2} \underset{Br}{CH_2{-}\overset{+}{C}H{-}CH{=}CH_2} \longleftrightarrow \underset{Br}{CH_2{-}CH{=}CH{-}\overset{+}{C}H_2}$$

$$\downarrow Br^-$$

$$\underset{Br\quad Br}{CH_2{-}CH{-}CH{=}CH_2} + \underset{Br\qquad\quad Br}{CH_2{-}CH{=}CH{-}CH_2}$$

This carbocation, a substituted version of the allyl cation considered in the last section, is best represented as a hybrid of two resonance structures. The second step of bromine addition, combination of the carbocation and a bromide ion, takes place at both partially positive carbons.

3.20
NATURAL AND SYNTHETIC RUBBER

Natural rubber is a high molecular weight polymer obtained from a milky fluid (latex) present in certain trees. A portion of a rubber molecule can be represented in the following ways.

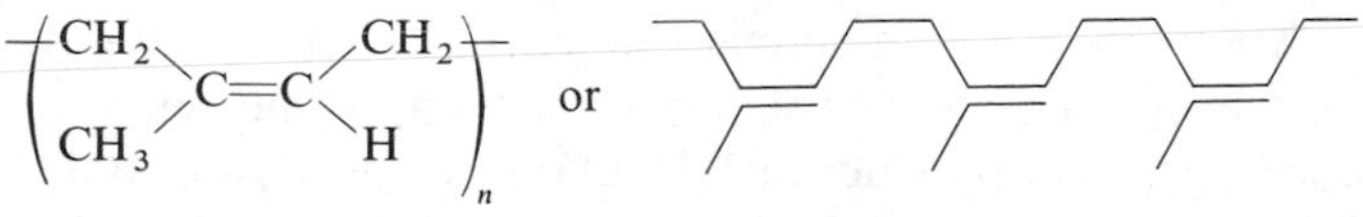

$$\left(\underset{CH_3}{\overset{CH_2}{}}{>}C{=}C{<}\underset{H}{\overset{CH_2}{}}\right)_n \quad \text{or}$$

The methylene (CH_2) groups of the chain have a cis relationship at almost every double bond.

Rubber could result from a 1,4-polymerization of isoprene.

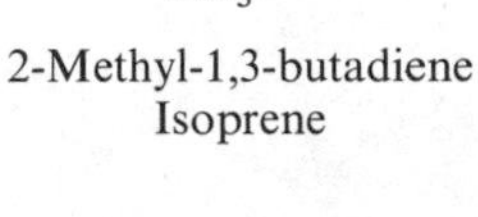

$$\underset{CH_3}{CH_2{=}C{-}CH{=}CH_2}$$

2-Methyl-1,3-butadiene
Isoprene

Although a polymer virtually identical to natural rubber can be synthesized by treating isoprene with an appropriate catalyst, plants synthesize such molecules in a different manner (see Sec. 7.9).

Double-bond configuration is critical in imparting the particular properties of rubber. Gutta percha, a polymer obtained from some plants, differs from rubber only in having a trans arrangement of the methylene groups at each double bond. It is tough and nonelastic.

Polymers like rubber that can be stretched severalfold in length but that regain their original dimensions when tension is released are called **elastomers**. The molecules of an elastomer must have the property of changing reversibly from random arrangements to more linear ones as tension is applied and released.

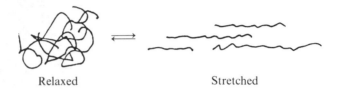

Relaxed   Stretched

Because of their trans double bonds, stretched and aligned chains of gutta percha fit together so well that they do not readily return to a more random arrangement.

When an elastomer is stretched, the chains tend to slide by one another. This sliding, which prevents a stretched rubber article from returning to its original shape, can be prevented by **vulcanization**. In this process, rubber is heated with a small amount of sulfur to introduce occasional sulfur links between the chains.

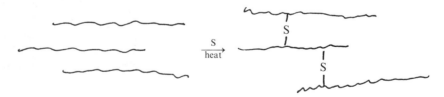

About 80% of commercial rubberlike materials are now synthetic. Most synthetic rubbers are polymers of dienes or are copolymers in which dienes are important components. A copolymer of 1,3-butadiene and styrene (CH_2=CH—C_6H_5) called SBR (styrene-butadiene rubber) is the most widely used synthetic rubber, particularly for tires. Neoprene, used in applications where resistance to gasoline, oil, or organic solvents is important, is a polymer of 2-chloro-1,3-butadiene.

$$CH_2=C-CH=CH_2 \longrightarrow \left(\begin{array}{c} CH_2 \\ | \\ Cl \end{array} C=C \begin{array}{c} H \\ | \\ CH_2 \end{array} \right)_n$$
$$\qquad\quad |$$
$$\qquad\quad Cl$$

Poly-2-chloro-1,3-butadiene
Neoprene

NATURALLY OCCURRING ALKENES; TERPENES

The carbon-carbon double bond, often in conjunction with other functional groups, is found in many compounds of living systems. The innocuous-looking alkene called muscalure affects us all.

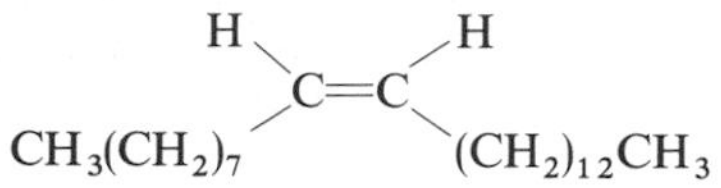

$$CH_3(CH_2)_7 \quad \overset{H}{\underset{}{C}} = \overset{H}{\underset{(CH_2)_{12}CH_3}{C}}$$

Muscalure

It is used by female houseflies to attract males. Such chemicals used between members of the same species as messages for mating, defense, trailmarking, and other activities are called **pheromones**. Because of the possibility that these natural substances could be used to control insect populations, considerable effort is being devoted to their isolation, structure determination, and synthesis.

The compounds called **terpenes** usually contain double bonds. Terpenes are constructed of five-carbon "isoprene units." Most terpenes have 10, 15, 20, 30, or 40 carbons, although rubber and gutta percha are very much larger terpenes. Many terpenes, including the following examples, are found in the organic oils obtained from pressing or distilling plant materials. Terpenes are often responsible for much of the fragrance of plants, and some are used in perfumes and flavoring agents. β-Myrcene is obtained from oil of bay, and limonene from many sources, including orange and lemon peels. Geraniol is the major component of oil of rose. α-Pinene is a major constituent of turpentine, a mixture of terpenes obtained from pine trees. Farnesol is found in citronella and lemon grass oils, and zingiberene in ginger oil.

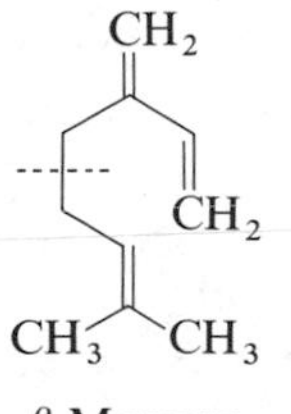

β-Myrcene 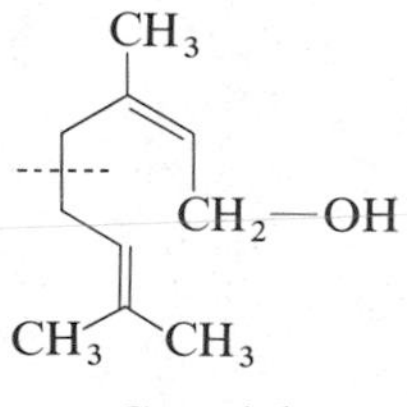   Geraniol 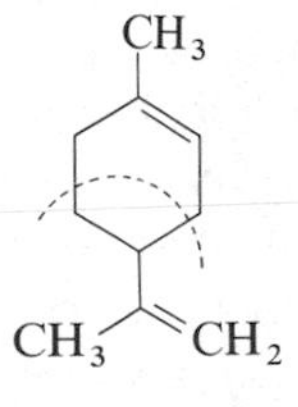   Limonene

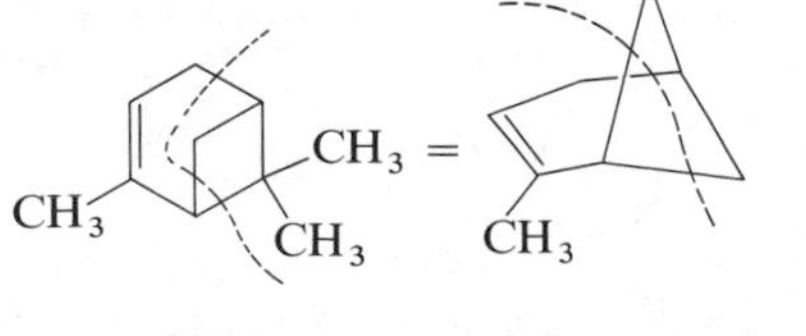

α-Pinene 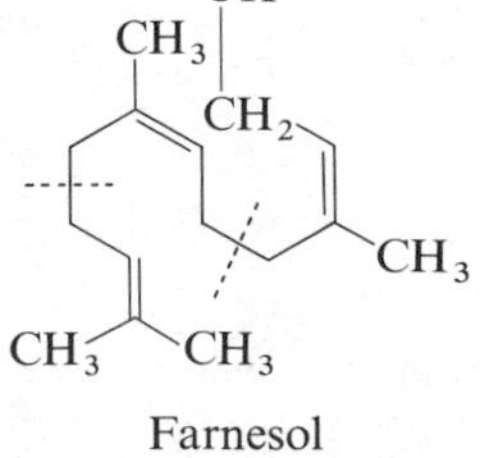   Farnesol 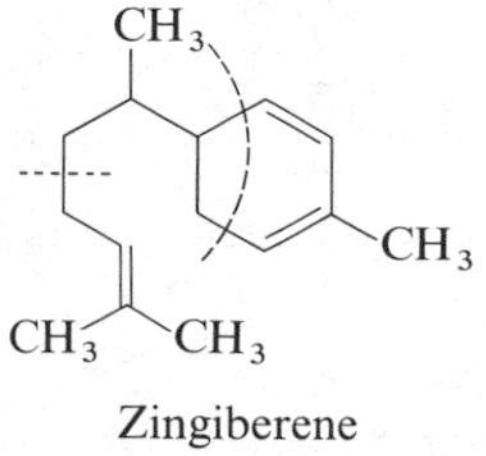   Zingiberene

Some terpenes contain hydroxyl, carbonyl, or other functional groups. Many have rings; the structures of acyclic terpenes shown were drawn in shapes that make more evident their relationship to the cyclic structures. The dotted lines in the structures help reveal the "isoprene units" ($-C-C-C-C-$) in their skeletons.
$$\begin{array}{c} | \\ C \end{array}$$

β-Carotene, an orange-red compound, occurs in carrots and is sometimes used as a food coloring.

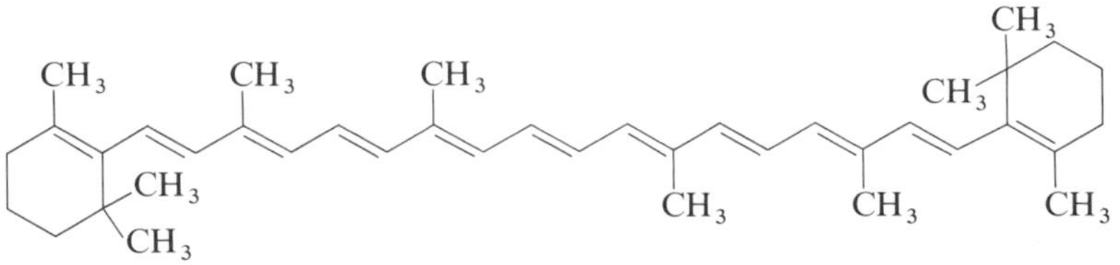

β-Carotene

It is a member of a group of C_{40} terpenes with closely related structures that often are responsible for yellow, orange, and red colors in some plants and animals. It can be cleaved in our bodies to furnish two molecules of the terpene vitamin A.

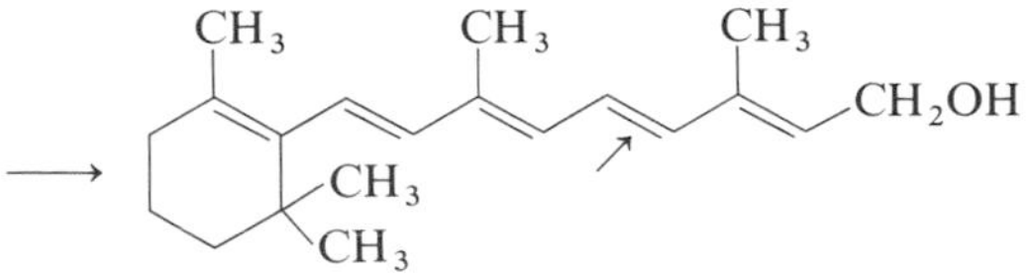

Vitamin A

A cis-trans isomerization of the bond indicated by an arrow occurs in a derivative of vitamin A on absorption of light. The change in molecular shape triggers the response responsible for our "night vision." For this reason, a deficiency of vitamin A results in "night-blindness." The ability of β-carotene and vitamin A to absorb light is due to their extensive chains of conjugated double bonds (Sec. 16.3).

Trivial rather than systematic names are used in this section for terpenes. A systematic name of vitamin A is (all-E)-3,7-dimethyl-9-(2,6,6-trimethyl-1-cyclohexen-1-yl)-2,4,6,8-nonatetraen-1-ol. This mouthful may suggest why trivial names are often used.

SUMMARY

Definitions and Ideas

An **alkene** is a hydrocarbon with a carbon-carbon double bond, and an **alkyne** a hydrocarbon with a carbon-carbon triple bond. **Saturated** compounds (e.g., ethane) contain a sufficient number of hydrogen atoms for each carbon to be attached to four other atoms or groups. **Unsaturated** compounds (e.g., ethylene, acetylene) contain fewer hydrogens and one or more multiple bonds.

A **functional group** is an atom or grouping of atoms that gives characteristic properties to the molecules in which it is incorporated. Carbon-carbon double and triple bonds, halogen groups, and hydroxyl groups are examples.

The double-bond carbons of an alkene and the four atoms bonded directly to them lie in a plane. Bond angles are about 120° at the double-bond carbons of an alkene and 180° at the triple bond carbons of an alkyne. Carbon-carbon double bonds (about 1.34 Å) are shorter than carbon-carbon single bonds (about 1.54 Å).

Stereoisomeric alkenes, such as *cis-* and *trans-*2-butene, exist when neither double-bond carbon has two identical groups. A stereoisomer is called **cis** if like groups are on the same side of the double bond and **trans** if like groups are on opposite sides.

An alkene is named by a substitutive system similar to that used for alkanes except that the presence, location, and configuration of the double bond must also be indicated. Select the longest chain of carbon atoms that contains the double-bond carbons and use the name of the corresponding alkane with the *-ane* ending changed to *-ene* as the parent name. Number the chain from the end nearer to the double bond and indicate the position of the double bond by the number of its lowest-numbered carbon. Name alkynes in a similar fashion except use the ending *-yne*.

Boiling points, solubilities, and densities of alkenes and alkynes are similar to those of comparable alkanes.

In contrast to alkanes, alkenes react readily with a variety of chemical reagents. Many of the reactions of alkenes are **additions**, reactions in which atoms or groups become linked to each end of the multiple bond without the loss of any atoms or groups already present. Each carbon of the double bond becomes attached by a single bond to some new atom or group and one of the bonds of the double bond is lost. Groups become attached to the same side of the alkene in a **syn** addition and to opposite sides in an **anti** addition. Additions of hydrogen halides and water follow **Markovnikov's rule**: The hydrogen becomes attached to that double-bond carbon already bearing the most hydrogens.

Additions of halogens, hydrogen halides, and water to alkenes proceed by two-step mechanisms. In the first step an **electrophile**, an electron-seeking species, becomes attached to one of the double-bond carbons to form a carbocation intermediate. In the second step the carbocation reacts with a **nucleophile**, a species that can furnish an electron pair. Preferential attack by the electrophile in the direction that produces the most stable carbocation is the reason for Markovnikov's rule. The ease of these additions parallels the stabilities of the carbocation intermediates.

A **carbocation** (or carbonium ion) is a species with a positive carbon surrounded by only six electrons. Carbocation stability increases in the order primary < secondary < tertiary.

Preparations of alkenes are often **eliminations**, reactions that are formally the reverse of additions. Dehydration of alcohols and addition of water to alkenes proceed by the same mechanistic steps.

Polymers are high molecular weight molecules that contain repeating structural units. The joining together of small molecules to construct polymers is called **polymerization**. The small molecules from which a polymer is prepared are called **monomers**.

Compounds containing two carbon-carbon double bonds are called **dienes**. Multiple bonds are **conjugated** if separated by only one single bond (C=C—C=C) and **isolated** if separated by more than one single bond (e.g., C=C—C—C=C).

The **resonance theory** is used when a molecule can be represented by two or more **resonance structures**, reasonable structures that have the same atoms bonded together but differ in the placement of electrons. The theory states that

1. the molecule is better represented as a hybrid of these structures than by any one of them.

2. the molecule is more stable than would be predicted for any one structure.

The application of this theory permits the use of simple structures (with lines to represent covalent bonds) to recognize, represent, and predict properties of chemical species in which the assumption that every electron pair is shared by just two atoms is not a good approximation.

Important Reactions Involving Alkenes

1. *Hydrogenation* $\text{C}=\text{C} \xrightarrow[\text{Pt, Pd, or Ni}]{\text{H}_2}$ $-\overset{|}{\underset{|}{\text{C}}}-\overset{|}{\underset{|}{\text{C}}}-$ H H

   Addition is predominantly syn.

2. *Addition of halogens* $\text{C}=\text{C} \xrightarrow{\text{X}_2}$ $-\overset{\text{X}}{\underset{|}{\text{C}}}-\overset{|}{\underset{\text{X}}{\text{C}}}-$

   Addition is anti. X = Cl or Br

3. *Addition of hydrogen halides* $\text{C}=\text{C} \xrightarrow{\text{HX}}$ $-\overset{|}{\underset{\text{H}}{\text{C}}}-\overset{|}{\underset{\text{X}}{\text{C}}}-$

   Addition follows Markovnikov's rule. X = Cl, Br, or I

4. *Addition of water (hydration)* $\text{C}=\text{C} \xrightarrow[\text{H}_3\text{O}^+]{\text{H}_2\text{O}}$ $-\overset{|}{\underset{\text{H}}{\text{C}}}-\overset{|}{\underset{\text{OH}}{\text{C}}}-$

   Addition follows Markovnikov's rule.

5. *Oxidation with potassium permanganate* $\text{C}=\text{C} \xrightarrow{\text{KMnO}_4}$ $-\overset{|}{\underset{\text{OH}}{\text{C}}}-\overset{|}{\underset{\text{OH}}{\text{C}}}-$

   Addition is syn.

6. *Ozonolysis* $\text{C}=\text{C} \xrightarrow[\text{H}_3\text{O}^+]{\text{O}_3 \quad \text{Zn}}$ $\text{C}=\text{O} + \text{O}=\text{C}$

7. *Polymerization* $\text{C}=\text{C} \xrightarrow[\text{other catalyst}]{\text{peroxide or}}$ $-(\overset{|}{\underset{|}{\text{C}}}-\overset{|}{\underset{|}{\text{C}}})_n$

8. *Preparation by dehydrohalogenation of alkyl halides*

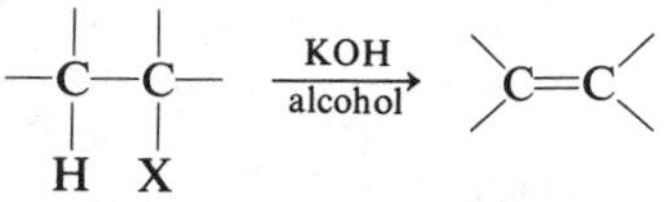

$$-\overset{|}{\underset{H}{C}}-\overset{|}{\underset{X}{C}}- \xrightarrow[\text{alcohol}]{\text{KOH}} \overset{\diagdown}{\diagup}C=C\overset{\diagup}{\diagdown}$$

This is the reverse of reaction 3.

9. *Preparation by dehydration of alcohols* $-\overset{|}{\underset{H}{C}}-\overset{|}{\underset{OH}{C}}- \xrightarrow[\text{heat}]{\text{H}^+} \overset{\diagdown}{\diagup}C=C\overset{\diagup}{\diagdown}$

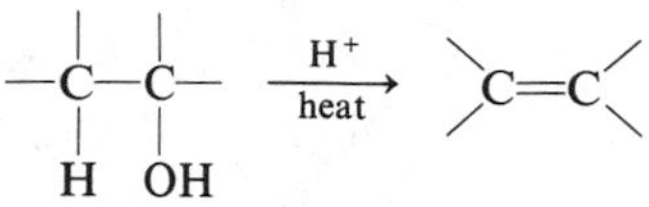

This is the reverse of reaction 4.

Important Reactions of Alkynes

Except that one or two molecules of the reagent can be added to an alkyne molecule, hydrogenation, addition of halogens, and addition of hydrogen halides are similar to the reactions (1–3) with alkenes.

PROBLEMS

1. Draw suitable representations of the following compounds.
   - (a) 1-pentene
   - (b) 2-methyl-2-butene
   - (c) cyclopentene
   - (d) *trans*-3-hexene
   - (e) *cis*-1,2-dichloroethene
   - (f) *cis*-4,4-dimethyl-2-octene
   - (g) vinyl bromide
   - (h) allyl chloride
   - (i) 1-methylcyclohexene
   - (j) 3-chlorocyclopentene
   - (k) vinylcyclohexane
   - (l) 1-pentyne
   - (m) 3-chloro-1-hexyne
   - (n) 3-methyl-1-butyne
   - (o) 1,5-hexadiene
   - (p) 2-methyl-1,3-butadiene

2. Name the following compounds.
   - (a) $CH_3CH_2CH_2CH_2CH=CH_2$

   - (b) 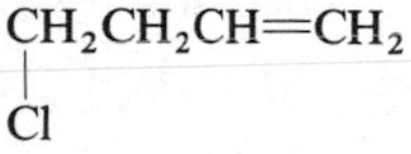 $\underset{\underset{Cl}{|}}{CH_2}CH_2CH=CH_2$

   - (c) 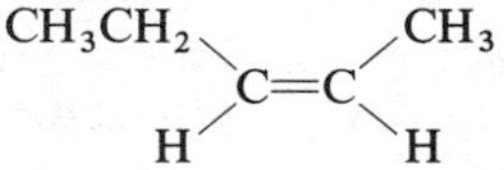 $\underset{H}{\overset{CH_3CH_2}{\diagdown}}C=C\underset{H}{\overset{CH_3}{\diagup}}$

   - (d) 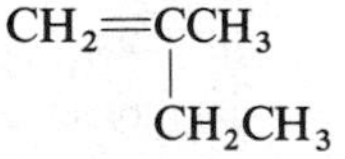 $CH_2=\underset{\underset{CH_2CH_3}{|}}{C}CH_3$

   - (e)

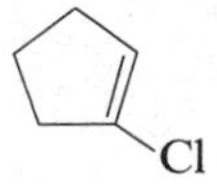

   - (f)

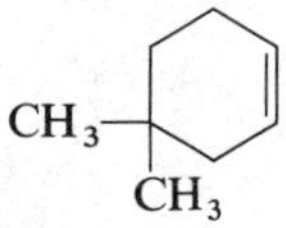

   - (g) 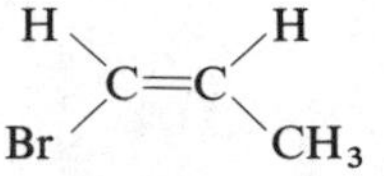 $\underset{Br}{\overset{H}{\diagdown}}C=C\underset{CH_3}{\overset{H}{\diagup}}$

   - (h) $CH_3C{\equiv}CCH_2CH_3$

(i) $CH_3CHC{\equiv}CH$
      |
      CH_3

(j) $CH_2{=}CHC{=}CH_2$
                     |
                     CH_3

3. Give two names for each of the following compounds.
   (a) $CH_2{=}CHCH_3$
   (b) $CH_2{=}CHI$
   (c) $CH_2{=}CHCH_2Br$
   (d) $CH_2{=}CH_2$

4. Give another name for each of the following compounds.
   (a) propylene
   (b) allyl iodide
   (c) vinylcyclopentane
   (d) 3-cyclohexyl-1-propene

5. What, if anything, is wrong with the following names?
   (a) 3-butene
   (b) 2-ethyl-1-propene
   (c) *cis*-2-methyl-2-hexene
   (d) 1,2-pentene
   (e) 1-chloro-2-propene
   (f) 3-methylbutene
   (g) 2-chlorocyclopentene
   (h) 1-methyl-2-cyclohexene
   (i) trans-1,2-dimethylcyclohexene
   (j) 4-pentyne

6. Draw suitable representations of the six isomeric pentenes (include stereoisomers).

7. Draw structural formulas of the isomeric pentynes.

8. Draw suitable representations of all compounds (including stereoisomers) that fit the following descriptions.
   (a) compounds of molecular formula C_4H_8 that on catalytic hydrogenation react with one mole of H_2
   (b) compounds of molecular formula C_4H_8
   (c) compounds of molecular formula C_5H_{10}
   (d) compounds of molecular formula $C_2H_2Cl_2$
   (e) all methylcyclohexenes
   (f) compounds of molecular formula C_3H_5Cl
   (g) compounds of molecular formula C_6H_{12} whose names end in hexene
   (h) alkynes of molecular formula C_6H_{10}
   (i) dienes of molecular formula C_5H_8

9. Which of the following structural formulas represent a pair of cis-trans isomers? Where such isomers exist, draw suitable representations that indicate their geometry.
   (a) $CH_3CH{=}CCl_2$
   (b) $ClCH{=}CHBr$
   (c) $ClCH{=}CHCl$
   (d) $CH_3(CH_2)_6CH{=}CH_2$
   (e) $(CH_3)_2C{=}CHCl$
   (f) $CH_3CHCH{=}CHCHCH_3$
              | |
              CH_3 CH_3

10. What are the approximate numerical values of the indicated bond angles in the actual molecules represented by the following structural formulas?

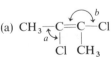

11. All cyclic alkenes illustrated in this chapter have cis stereochemistries. Such compounds as *trans*-cyclopentene and *trans*-cyclohexene are highly unstable. In what situations will trans isomers of cyclic alkenes be stable compounds? (The use of models may be helpful.)

12. Why isn't cyclohexyne a stable compound?

13. Draw structural formulas for any organic products that result from reaction of 1-butene with the following reagents.
    (a) H_2 (Pt) (b) Br_2
    (c) HCl (d) H_2O (H_3O^+)
    (e) a peroxide (f) $KMnO_4$
    (g) O_3 followed by Zn (H_3O^+)

14. Draw suitable representations of any organic products that result from the following reactions.

    (a) cyclopentene $\xrightarrow{\text{HCl}}$ (b) 1-pentene $\xrightarrow[\text{H}_3\text{O}^+]{\text{H}_2\text{O}}$

    (c) 1-bromobutane $\xrightarrow[\text{alcohol}]{\text{KOH}}$ (d) 2-bromo-2-methylbutane $\xrightarrow[\text{alcohol}]{\text{KOH}}$

    (e) 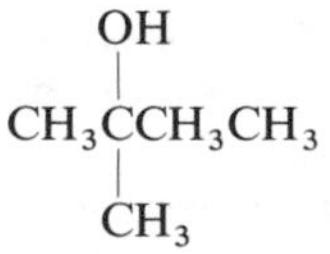 $CH_3\overset{\overset{\displaystyle OH}{|}}{\underset{\underset{\displaystyle CH_3}{|}}{C}}CH_3CH_3 \xrightarrow[\text{heat}]{\text{H}^+}$ (f) cyclopentene $\xrightarrow{\text{Br}_2}$

    (g) 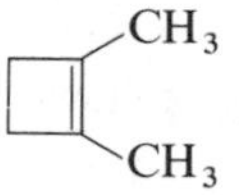 $\xrightarrow[\text{Pt}]{\text{H}_2}$ (h) 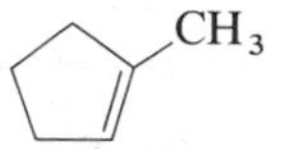 $\xrightarrow{\text{O}_3}$ $\xrightarrow[\text{H}_3\text{O}^+]{\text{Zn}}$

    (i) limonene (structure in Sec. 3.21) $\xrightarrow{\text{HCl (excess)}}$ (j) 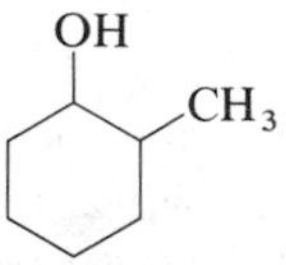 $\xrightarrow[\text{heat}]{\text{H}^+}$

    (k) 3-hexyne $\xrightarrow{\text{Cl}_2 \text{ (one mole)}}$ (l) 1-pentyne $\xrightarrow{\text{HCl (excess)}}$

    (m) 2-butyne $\xrightarrow[\text{Pt}]{\text{H}_2 \text{ (excess)}}$

    (n) 2,3-dimethyl-1,3-butadiene $\xrightarrow[\text{Pt}]{\text{H}_2 \text{ (excess)}}$

    (o) 1,3-butadiene $\xrightarrow{\text{HCl (one mole)}}$

15. Give the structural formula of the alkyl halide that on dehydrohalogenation gives only the indicated alkene.
    (a) 2-pentene (b) 4-methylcyclohexene

16. Polymerization of chlorotrifluoroethene produces a polymer known as Kel-F that has excellent stability toward many chemicals and toward high temperatures. Draw the structure of a ten-carbon segment of the chain of this polymer and also draw the structure of the repeating unit.

17. Write equations to show how the following conversions can be carried out. Each requires the combination of two reactions.

    (a) $CH_3CHCH_2CH_3 \longrightarrow$ butane
        $\overset{|}{OH}$

    (b) 1-bromopropane $\longrightarrow$ 1,2-dibromopropane

    (c) 1-chlorobutane $\longrightarrow$ 2-bromobutane

    (d) 2-butyne $\longrightarrow$ $CH_3CHCH_2CH_3$
        $\overset{|}{OH}$

    (e) $(CH_3)_2CHCH_2OH \longrightarrow$ *tert*-butyl chloride

    (f) $\overset{\text{OH}}{\diagup}$ pentagon
        $\longrightarrow$ $\overset{O}{\overset{||}{HCCH_2CH_2CH_2CH}}$

18. How many rings are in the following compounds?

    (a) the terpene menthene (molecular formula $C_{10}H_{18}$), which reacts with one mole of bromine to form a compound of molecular formula $C_{10}H_{18}Br_2$

    (b) the terpene camphene (molecular formula $C_{10}H_{16}$), a constituent of turpentine, which takes up one mole of H_2 on catalytic hydrogenation

    (c) the steroid cholestane (molecular formula $C_{27}H_{48}$), which reacts neither with hydrogen and a platinum catalyst nor with bromine

    (d) squalene (molecular formula $C_{30}H_{50}$), an important precursor of many biological compounds, which takes up six moles of H_2 on catalytic hydrogenation

19. Draw the structure of the compound that gives the following products on ozonolysis.

    (a) $CH_2{=}O + CH_3CHCH{=}O$
        $\overset{|}{CH_3}$

    (b) $\text{pentagon}{=}O + CH_3CH{=}O$

20. Draw the structure of the major product obtained from ozonolysis of natural rubber.

21. α-Terpinene, a terpene of molecular formula $C_{10}H_{16}$, reacts with hydrogen and a platinum catalyst to form a compound of molecular formula $C_{10}H_{20}$. Ozonolysis of α-terpinene gives two compounds.

    $$\overset{O \quad O}{\overset{|| \quad ||}{HC{-}CH}} \qquad \overset{O \qquad\quad O}{\overset{|| \qquad\quad ||}{CH_3CCH_2CH_2CCHCH_3}}$$
    $$\overset{|}{CH_3}$$

    Draw the structure of α-terpinene.

22. Compound X on treatment with potassium hydroxide in alcohol gives compound Y, molecular formula C_5H_{10}. Ozonolysis of compound Y gives $CH_2{=}O$ and $CH_3CH_2CH_2CH{=}O$. Draw reasonable structures for compounds X and Y.

23. Why is cyclopropene an extraordinarily reactive alkene?

24. Using appropriate conditions, 2-methyl-1-propene can be converted efficiently with a sulfuric acid catalyst into a mixture of 2,4,4-trimethyl-2-pentene and 2,4,4-trimethyl-1-pentene. Write a mechanism that accounts for the formation of these products.

25. Arrange the following alcohols in order of increasing ease of dehydration.

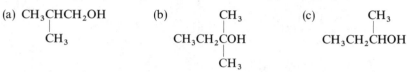

(a) CH_3CHCH_2OH
      |
      CH_3

(b) CH_3
         |
    CH_3CH_2COH
         |
        CH_3

(c) CH_3
         |
    CH_3CH_2CHOH

26. The role of the acid in an acid-catalyzed dehydration is to protonate an alcohol to form ROH_2^+. The key step is then loss of water to form a carbocation intermediate.

$$R-OH_2^+ \longrightarrow R^+ + H_2O$$

Explain why formation of the carbocation from the alcohol itself is so much less favorable.

$$R-OH \not\longrightarrow R^+ + OH^-$$

27. Write resonance structures for the following species.

    (a) ozone

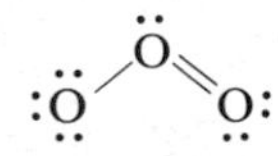

    (b) nitric acid

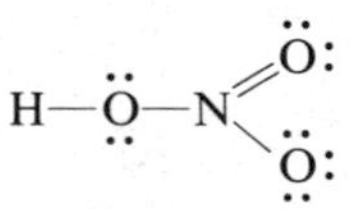

    (c) bicarbonate ion

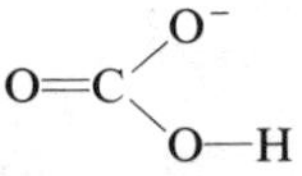

    (d) carbonate ion

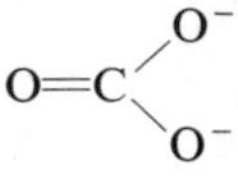

28. Draw all stereoisomers of the following.
    (a) 2,4-heptadiene   (b) 2,4-hexadiene

29. Why does addition of HBr to 1,3-butadiene produce significant amounts of $CH_3-CHBr-CH=CH_2$ and $CH_3-CH=CH-CH_2Br$ but not of $BrCH_2-CH_2-CH=CH_2$?

30. Indicate the isoprene units in β-carotene (structure in Sec. 3.21).

Aromatic Compounds

Many compounds that are unsaturated have little of the reactivity characteristic of alkenes and alkynes. Some of the first encountered were isolated from plants. Because of characteristic and often pleasant odors, they were called **aromatic**. The term aromatic is now applied more generally to all compounds that share certain properties. Benzene is the simplest, most common, and most studied of these compounds and so we will use it to introduce the special features of aromatic compounds.

Later in this chapter we again consider covalent bonding, a topic discussed briefly in Chapter 1. Now that we are familiar with the structures and properties of a variety of organic compounds, we are better prepared to learn more about ways of describing their bonds.

4.1
SOME PROPERTIES OF BENZENE

Benzene, a water-insoluble liquid boiling at 80°, has the molecular formula C_6H_6. Therefore benzene has eight fewer hydrogens than do C_6 alkanes. Because a compound has two fewer hydrogens than an acyclic alkane for each ring or double bond, the total number of rings plus double bonds in a C_6H_6 compound is four. Benzene can be hydrogenated, but conditions far more stringent are needed than for hydrogenation of alkenes.

$$\text{Benzene} + 3\,H_2 \xrightarrow{\text{Ni}} \hexagon$$

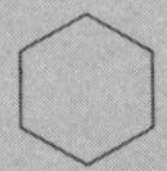

Cyclohexane

Hydrogenation results in the uptake of three moles of hydrogen per mole of benzene and the formation of cyclohexane. Therefore it is reasonable to assume that benzene has the equivalent of three double bonds in a six-membered ring and a structure like the one shown.

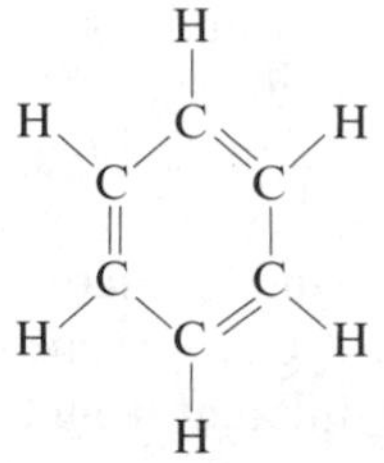

This structure, one of the first proposed for benzene, still is frequently used. Yet, the chemical properties of benzene are not what we expect for a compound having carbon-carbon double bonds. Benzene is much less reactive than a typical alkene. It is difficult to hydrogenate, gives no apparent reaction with hydrogen halides, and does not react readily with bromine. When benzene is heated with bromine in the presence of a catalyst, a reaction occurs, but the product results from substitution of an H by a Br.

$$C_6H_6 + Br_2 \xrightarrow{\text{FeBr}_3} C_6H_5Br + HBr$$

In fact, most reactions of benzene are substitutions rather than the additions characteristic of alkenes.

The geometry of benzene also is not in accord with the presence of alternating double bonds and single bonds. All carbon-carbon bonds of benzene are identical; their length (1.39 Å) is intermediate between the lengths of typical carbon-carbon single bonds (1.54 Å) and double bonds (1.34 Å).

4.2
RESONANCE DESCRIPTION OF BENZENE

Benzene is a molecule to which the resonance theory (Sec. 3.18) applies. We can draw two resonance structures for benzene that have the same atoms connected together but that differ in the arrangement of electrons.

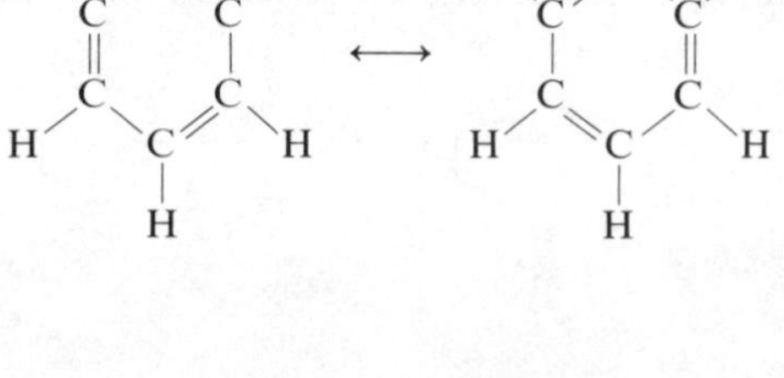

According to the resonance theory, the benzene molecule is better represented as a hybrid of these structures than by either structure alone.

Each carbon-carbon bond of benzene is a single bond in one resonance structure and a double bond in the other. Therefore the resonance theory correctly predicts that all carbon-carbon bonds are identical and intermediate in properties between those of carbon-carbon single and double bonds. The resonance theory also predicts that benzene is more stable than would be expected on the basis of either resonance structure. Because the reactivity of molecules tends to decrease with increasing stability, the resonance theory also explains the low reactivity of benzene.

Drawing two resonance structures is tedious and so benzene is generally represented in other ways. One expedient is to use some notation indicating that all the bonds are equal and different from those of alkenes.

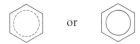

In this book we will frequently use the notation with a solid circle. It is also common to represent benzene by only one of the resonance structures.

We will sometimes do so when we want to keep track of the number of bonding electrons but must not then forget the failings of this structure.

4.3
RESONANCE ENERGY OF BENZENE

Benzene's special stability is due to some of its electrons being attracted by more than two nuclei. The magnitude of this extra stability can be estimated by measuring the changes in heat that are associated with reactions. Hydrogenation of cyclohexene to form cyclohexane evolves 28.6 kcal/mole, a value typical for hydrogenation of alkenes.

$$\text{⬡} + H_2 \longrightarrow \text{⬡} + 28.6 \text{ kcal}$$

Hydrogenation of both double bonds of 1,3-cyclohexadiene evolves 55.4 kcal/mole, approximately double the amount observed for cyclohexene.

$$\text{⬡} + 2 H_2 \longrightarrow \text{⬡} + 55.4 \text{ kcal}$$

The molecule 1,3,5-cyclohexatriene, containing three ordinary double bonds, is unknown—any efforts to produce it yield benzene. We would, however, expect complete hydrogenation of the unknown 1,3,5-cyclohexatriene to evolve approximately 3 × 28.6 or 85.8 kcal/mole.

$$\left[\bigcirc \right] + 3\,H_2 \longrightarrow \bigcirc + \begin{array}{l} 85.8\ \text{kcal} \\ \text{(estimated)} \end{array}$$

A hypothetical molecule
containing three ordinary
carbon-carbon double bonds

Hydrogenation of benzene furnishes only 49.8 kcal/mole.

$$\bigcirc + 3\,H_2 \longrightarrow \bigcirc + 49.8\ \text{kcal}$$

Figure 4.1 may make the energy relationships more evident. The 36-kcal difference between the heat evolved in the hydrogenation of benzene and that estimated for

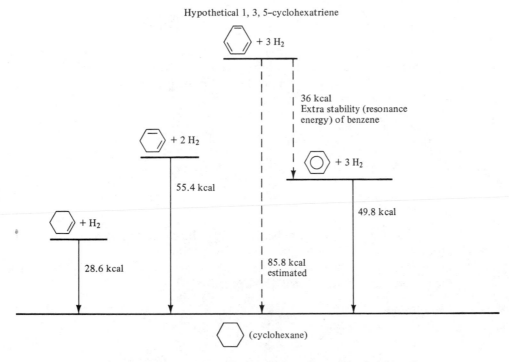

Figure 4.1   Heats evolved on hydrogenation of one mole of some cyclic compounds.

hydrogenation of a compound with three ordinary double bonds is the added stability of benzene. This added stability is sometimes called the **resonance energy,** a misleading term because it refers to energy that a molecule does *not* have.

4.4
SUBSTITUTION REACTIONS OF BENZENE

The characteristic reactions of benzene and other aromatic compounds are substitutions. In these reactions a hydrogen is replaced by some other group and the stable aromatic system is preserved. We will examine four of the most important of these reactions.

In the presence of a suitable catalyst, often a ferric halide, benzene reacts with bromine or chlorine (**halogenation**).

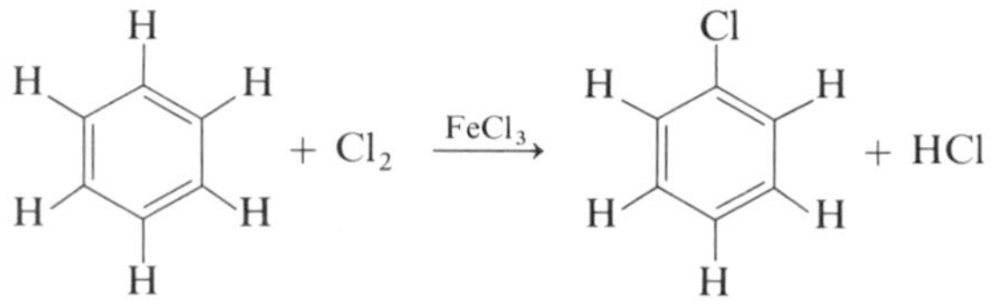

Chlorobenzene

Hydrogens are usually omitted from the structures of aromatic compounds but were included here to remind us that the halogen replaces a hydrogen.

Reaction of benzene with nitric acid in the presence of sulfuric acid introduces —NO_2, a nitro group (**nitration**).

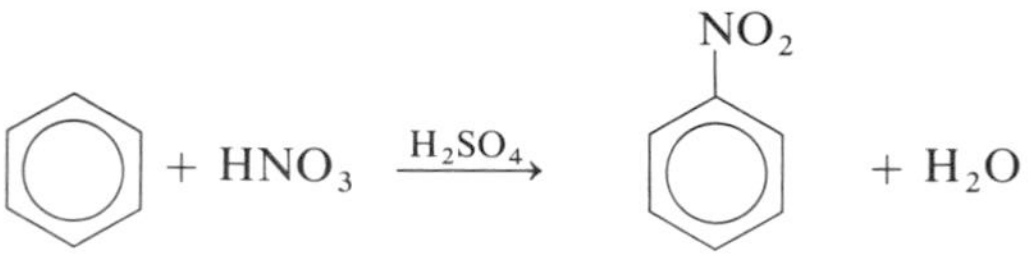

Nitrobenzene

The nitro group is important because it can be converted to many other functional groups.

When benzene is heated with strong sulfuric acid, a hydrogen is replaced by the sulfonic acid group, —SO_3H (**sulfonation**).

Benzenesulfonic acid

Benzenesulfonic acid is comparable in acidity to sulfuric acid and hence almost completely ionized in water.

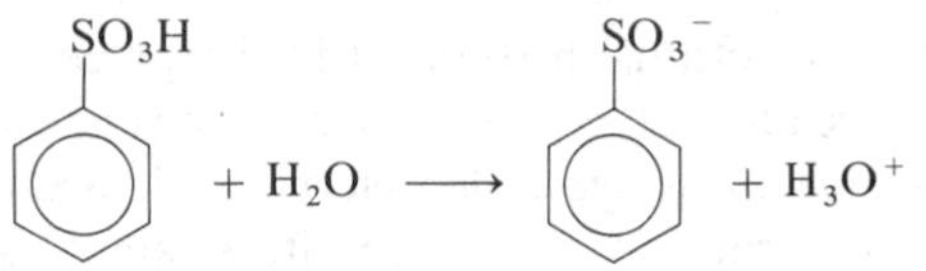

$$SO_3H \qquad + H_2O \longrightarrow \qquad SO_3^- \qquad + H_3O^+$$

Reaction of benzene with an alkyl halide in the presence of a catalyst introduces an alkyl group (**alkylation**).

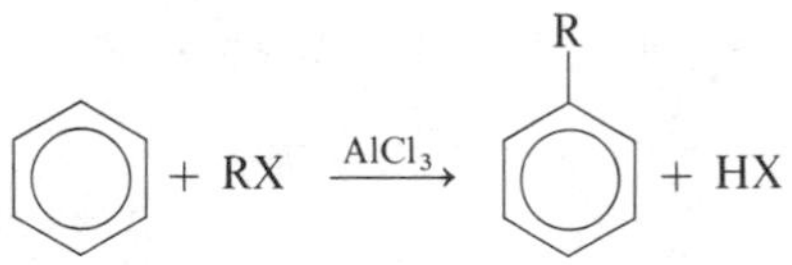

$$\text{benzene} + RX \xrightarrow{\text{AlCl}_3} R\text{-benzene} + HX$$

X is usually chlorine or bromine; aluminum chloride is the most common catalyst. Here are some specific examples of this reaction.

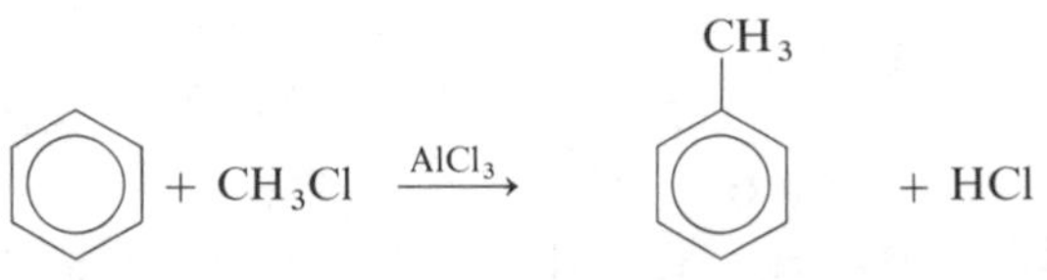

$$\text{benzene} + CH_3Cl \xrightarrow{\text{AlCl}_3} CH_3\text{-benzene} + HCl$$

Toluene
(methylbenzene)

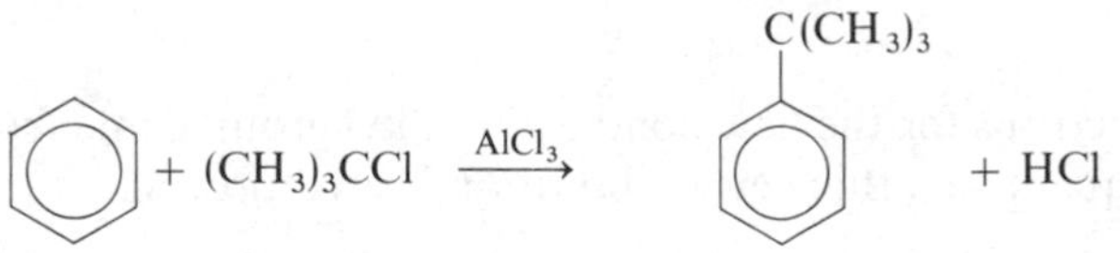

$$\text{benzene} + (CH_3)_3CCl \xrightarrow{\text{AlCl}_3} C(CH_3)_3\text{-benzene} + HCl$$

tert-Butylbenzene

An alkyl group can also be introduced by the reaction of benzene with an alkene in the presence of an acidic catalyst.

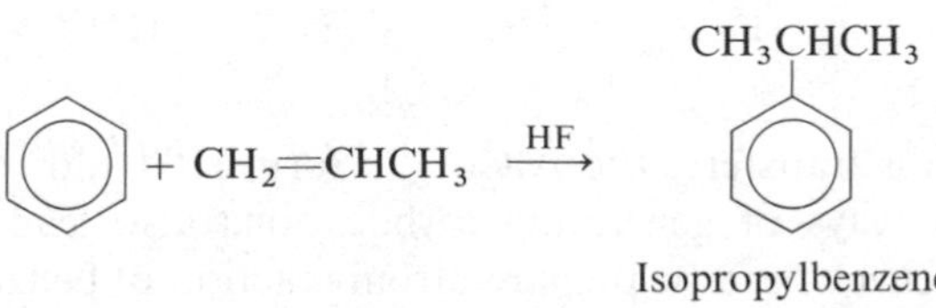

$$\text{benzene} + CH_2=CHCH_3 \xrightarrow{\text{HF}} CH_3CHCH_3\text{-benzene}$$

Isopropylbenzene

This is a substitution reaction of benzene but an addition reaction of the alkene. Note that the addition to the alkene follows Markovnikov's rule.

4.5
MECHANISM OF ELECTROPHILIC
AROMATIC SUBSTITUTION

The substitution reactions introduced in the preceding section all occur by similar mechanisms. Because attack on the aromatic ring by some electrophilic species is a key step, these reactions are called **electrophilic aromatic substitutions**.

Consider the reaction of benzene with an alkyl halide. The role of the aluminum chloride catalyst is to convert the alkyl halide to a carbocation.

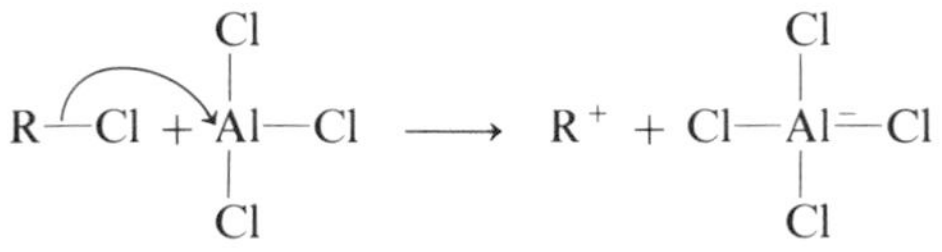

$$R\!-\!Cl + Al\!-\!Cl \longrightarrow R^+ + Cl\!-\!Al\!=\!Cl$$

The aluminum in aluminum chloride is surrounded by only six valence electrons and reacts with an alkyl halide to remove the halogen along with both electrons it shared with carbon. The carbocation that results is a potent electrophilic reagent that can attack the aromatic ring to form another carbocation.

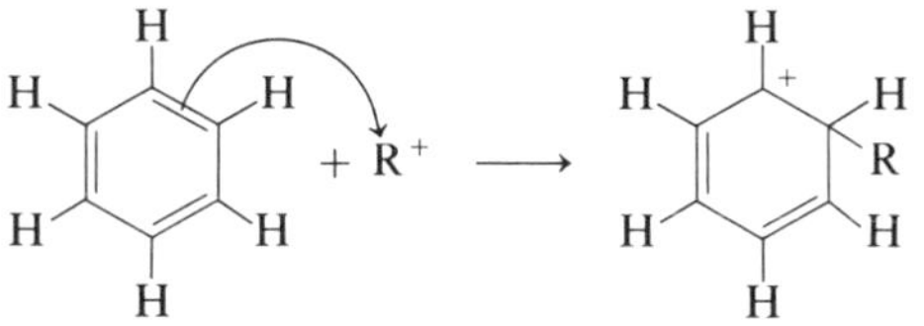

$$+ R^+ \longrightarrow$$

Both electrons for the new bond to the alkyl group come from the aromatic ring. In a subsequent step this new carbocation loses a proton.

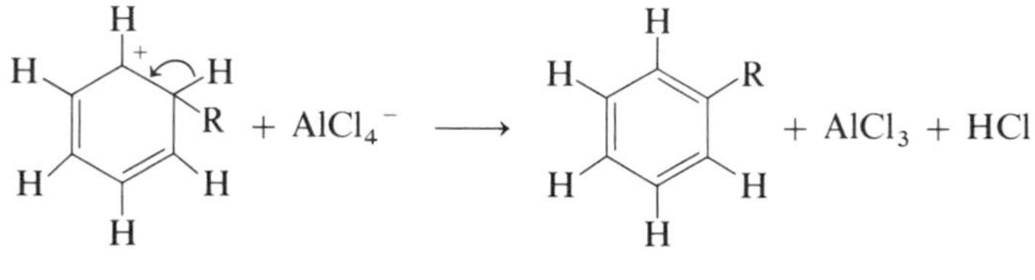

$$+ AlCl_4^- \longrightarrow + AlCl_3 + HCl$$

The proton is transferred to $AlCl_4^-$ to form HCl and regenerate the $AlCl_3$.

Other ways of generating carbocations also lead to alkylation of benzene. *tert*-Butylbenzene can be prepared from reaction of benzene with *tert*-butyl chloride and aluminum chloride. It is also formed, however, from reactions of benzene with 2-methyl-1-propene and acid or with *tert*-butyl alcohol and acid.

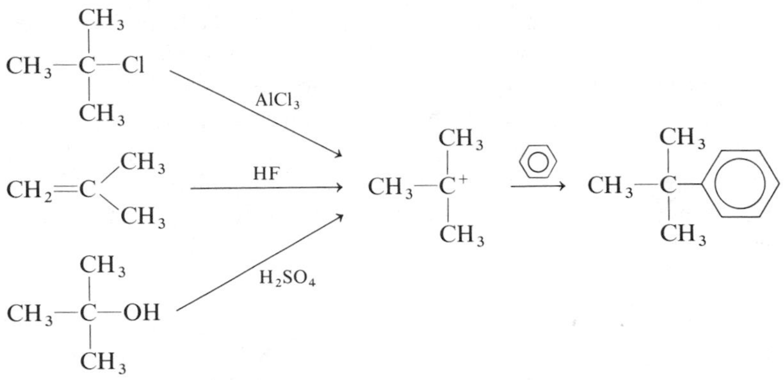

We already know that treating alkenes and alcohols with acids generates carbocation intermediates.

In halogenations the catalyst functions by combining with a halide ion, thereby making the halogen more electrophilic.

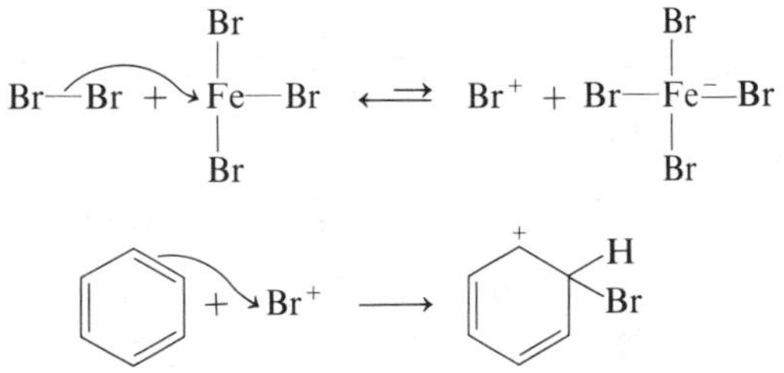

Although, for simplicity, the formation of Br^+ and its reaction with benzene are written as discrete steps, they may actually occur simultaneously. In a subsequent step a proton is lost from the carbocation intermediate.

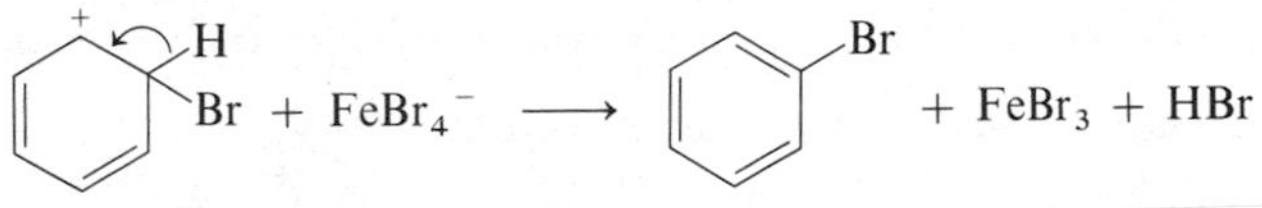

Recall that the first step in the reaction of an alkene with bromine also involves an electrophilic attack that generates a carbocation.

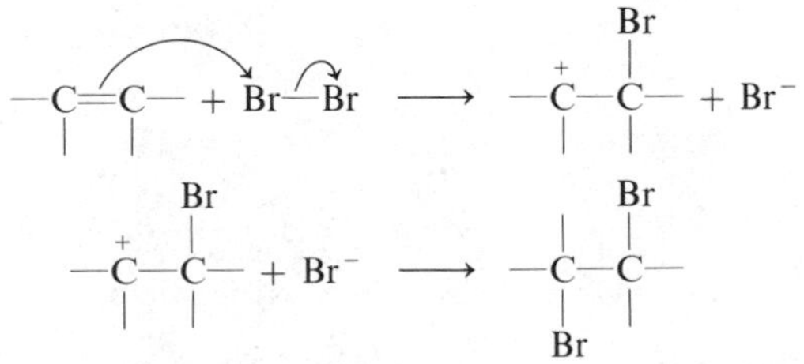

Because of the high stability (low reactivity) of benzene, a catalyst to increase the reactivity of the halogen and a higher reaction temperature are needed.

The stability of an aromatic ring also is responsible for the different fates of the carbocation intermediates formed from benzene and alkenes. Loss of a proton from the intermediate in the reaction of benzene is favored because it regenerates the particularly stable aromatic ring system. Attachment of a halide ion, the step that occurs with an alkene, would furnish an addition product lacking the resonance energy of the aromatic ring.

In nitration the electrophilic species that attacks the aromatic ring is NO_2^+, generated from the reaction of sulfuric and nitric acids.

$$H\text{—}O\text{—}NO_2 \underset{}{\overset{H_2SO_4}{\rightleftharpoons}} \underset{+}{H\text{—}O\text{—}NO_2} \overset{-H_2O}{\rightleftharpoons} NO_2^+$$

One electrophilic species in sulfonation is SO_3H^+.*

4.6
NOMENCLATURE OF AROMATIC COMPOUNDS

Substituted benzenes are given substitutive names, often based on benzene as the parent compound.

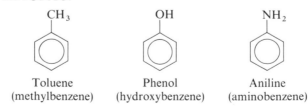

Nitrobenzene   Chlorobenzene   Ethylbenzene   Benzenesulfonic acid

Benzenesulfonic acid is a substitutive name in which the functional group name is placed after that of the parent.

Methylbenzene is the systematic name for the methyl-substituted benzene that we already encountered. The trivial name toluene, however, is in common use and is accepted by the IUPAC.

Toluene   Phenol   Aniline
(methylbenzene)   (hydroxybenzene)   (aminobenzene)

* Even more important is SO_3 ($H_2SO_4 \rightleftharpoons SO_3 + H_2O$), which although neutral is extremely electrophilic.

Trivial names are also often used for the benzenes substituted with a hydroxyl group, —OH (considered in the next chapter), and with an amino group, —NH$_2$ (considered in Chapter 8).

Two substituents can be arranged on a benzene ring in the three ways illustrated for the dichlorobenzenes.

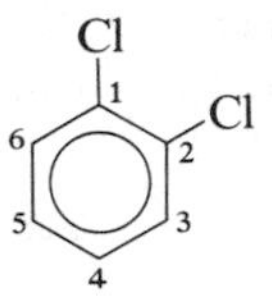

1,2-Dichlorobenzene
ortho-Dichlorobenzene
o-Dichlorobenzene

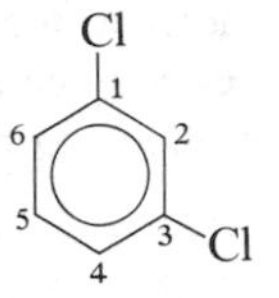

1,3-Dichlorobenzene
meta-Dichlorobenzene
m-Dichlorobenzene

1,4-Dichlorobenzene
para-Dichlorobenzene
p-Dichlorobenzene

Such constitutional isomers can be differentiated by using numbers to indicate the positions of the substituents. Alternatively, the prefixes *ortho* (for a 1,2-isomer), *meta* (for a 1,3-isomer), and *para* (for a 1,4-isomer) are frequently used. These prefixes are often abbreviated *o*, *m*, and *p*.

When the substituents of a disubstituted benzene are different, they are arranged alphabetically and the lowest number is assigned to the group named first.

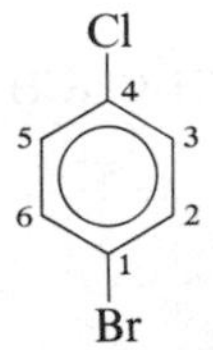

1-Bromo-4-chlorobenzene
p-Bromochlorobenzene

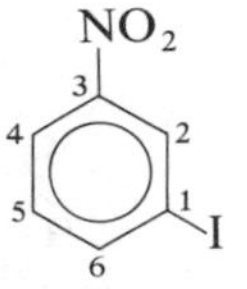

1-Iodo-3-nitrobenzene
o-Iodonitrobenzene

The special names for certain monosubstituted benzenes are often used as the parents for the names of more highly substituted benzenes.

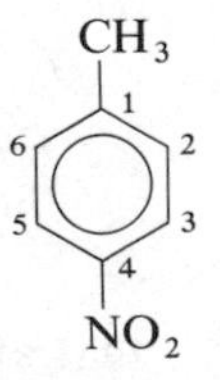

4-Nitrotoluene   3-Chlorophenol
p-Nitrotoluene   *m*-Chlorophenol

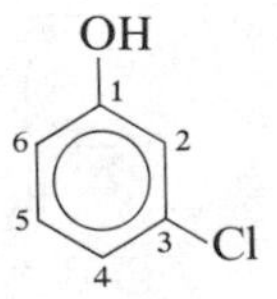

If one of the substituents is included in a special name for the aromatic ring (e.g., the methyl group is included in the name toluene), then the ring carbon bearing that substituent is assumed to be number one.

---

exercise with models 4.1 By constructing models, convince yourself that there is only one chlorobenzene. Construct models of the *ortho*-, *meta*-, and *para*-bromochlorobenzenes. Note the obvious differences between these constitutional isomers and convince yourself that there are only three bromochlorobenzenes.

If more than two substituents are present, only numbers can be used to indicate substituent positions.

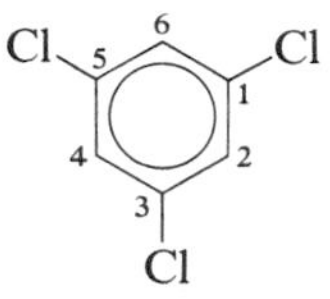

1,3,5-Trichlorobenzene 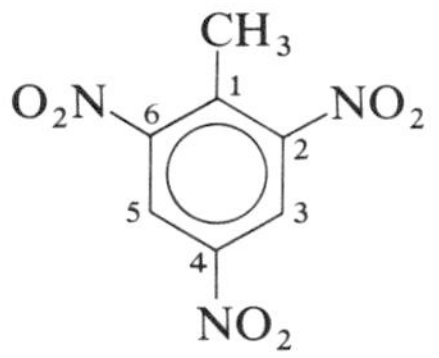   2,4,6-Trinitrotoluene

The group that corresponds to a benzene minus one hydrogen is called phenyl.

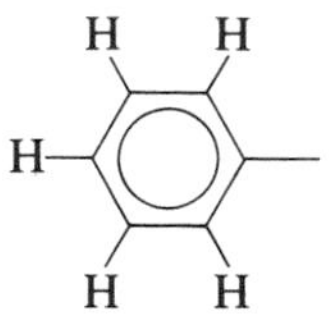

$CH_3CHCH_2CH_2CH_2CH_3$

Phenyl 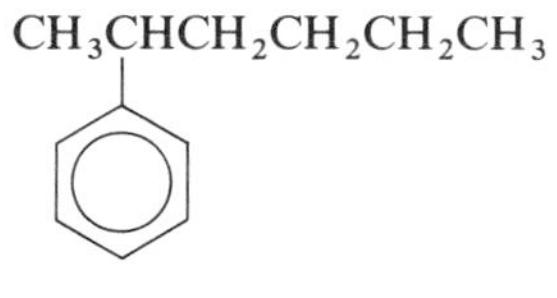   2-Phenylhexane

The group that corresponds to toluene minus one of the methyl hydrogens also occurs frequently and is called benzyl.

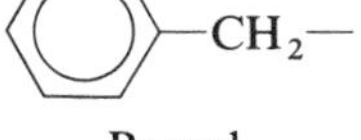

—CH_2—   —CH_2—Cl

Benzyl 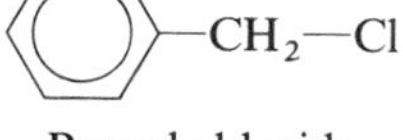   Benzyl chloride

Aromatic groups, such as phenyl, that are linked to something are given the general name **aryl groups**. Just as we have found it useful to let R stand for any alkyl group, so we will use Ar to represent any aryl group. Ar—Cl represents any compounds in which chlorine is attached directly to a carbon of an aromatic ring, regardless of any other substituents present.

---

problem 4.1   Name the following aromatic compounds.

(a)

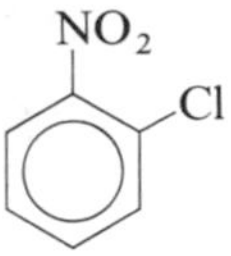

(b)

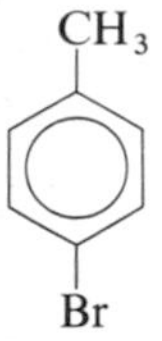

(c)

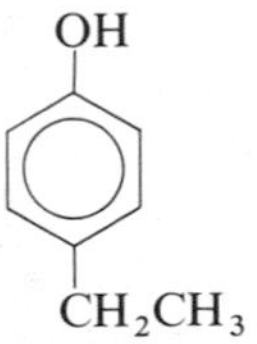

(d)

problem 4.2   Draw structures for the following compounds.
(a) 1-bromo-4-nitrobenzene
(c) 1,3-dichloro-5-nitrobenzene
(b) *o*-chloroaniline
(d) triphenylmethane

4.7
EFFECT OF SUBSTITUENTS
ON FURTHER SUBSTITUTION

Substituted benzenes also contain a stable aromatic ring and like benzene undergo substitution reactions. A new substituent could be introduced at a position ortho, meta, or para to a substituent already present. Substitutions of toluene, however, give mainly ortho and para products.

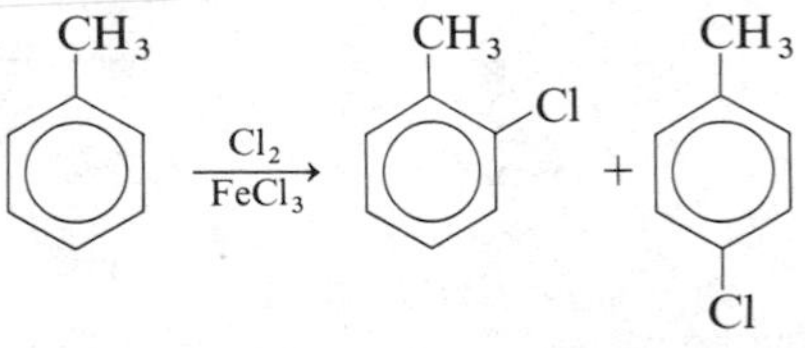

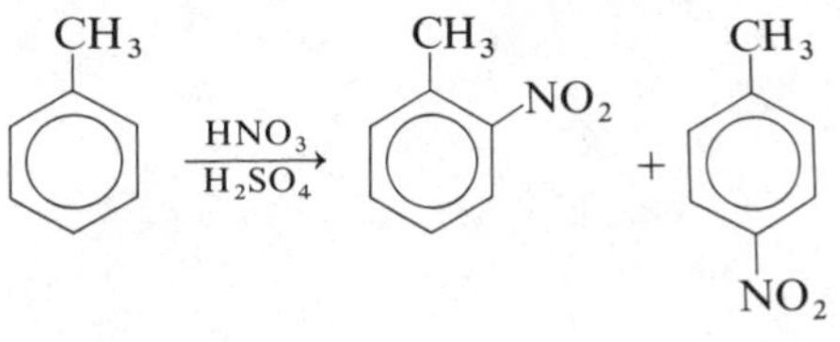

No more than a few percent of the meta isomers are formed. In contrast, reactions of nitrobenzene give principally meta products.

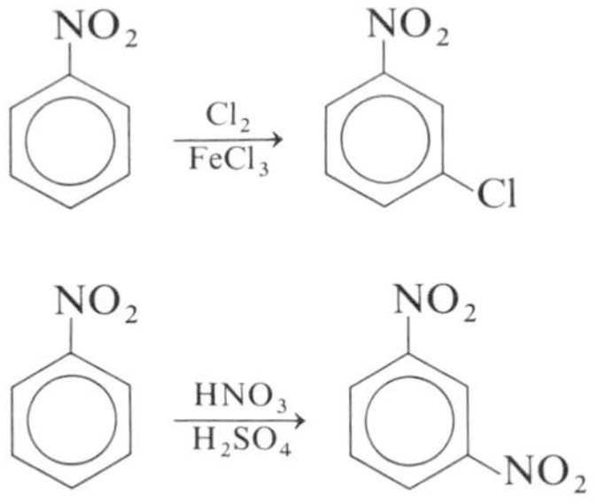

Only a few percent of the ortho and para isomers are formed.

As is evident in these examples, a substituent already present influences a further substitution reaction. A given substituent is either **ortho-para directing** (leading to formation mainly of ortho and para products) or **meta directing** (leading to formation mainly of meta products), regardless of which electrophilic substitution is taking place. The substituent also affects the ease of further substitution. Toluene is nitrated approximately 1 billion times more rapidly than is nitrobenzene. For nitration of nitrobenzene to occur in a reasonable length of time, more severe conditions (a higher concentration of sulfuric acid and a higher temperature) are needed than for the nitration of toluene. Toluene is also brominated, sulfonated, and alkylated more rapidly than is nitrobenzene.

The effects of different substituents on reactivity fall in the order given in Table 4.1. The most activating substituents are also ortho-para directing. The least activating substituents are meta directing.

---

problem 4.3 Draw structures for the major organic products of the following reactions.
(a) sulfonation of bromobenzene (b) bromination of benzenesulfonic acid

---

Table 4.1 Directing and activating effects of substituents

↑ increasing activating power	—OH, —OR $—NH_2$ —R (alkylbenzenes) (—H) (benzene) —F, —Cl, —Br, —I	ortho-para directing
	$-\overset{O}{\overset{\|}{C}}R$, $-\overset{O}{\overset{\|}{C}}H$, $-\overset{O}{\overset{\|}{C}}OH$ $—SO_3H$ $—NO_2$	meta directing

Aromatic substitution reactions are used to produce many familiar compounds. Alkylation of benzene with ethylene followed by high-temperature cracking is carried out on an enormous scale to prepare styrene, the monomer needed to prepare polystyrene.

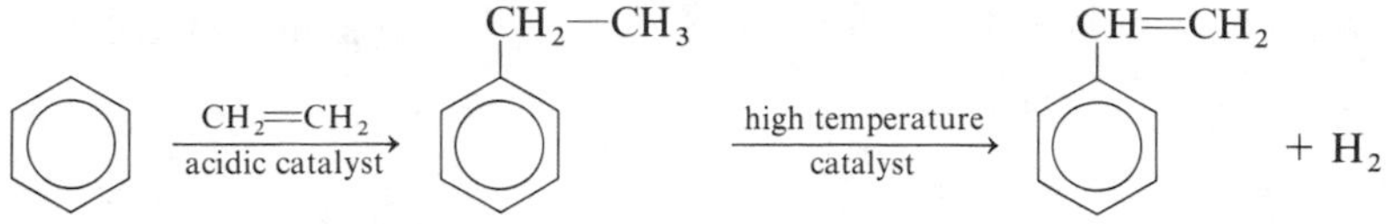

Styrene

Chlorination of chlorobenzene leads to a mixture of *o*- and *p*-dichlorobenzene.

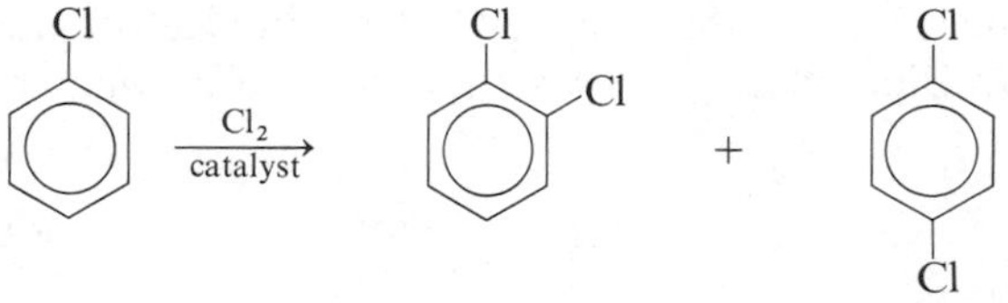

o-Dichlorobenzene   *p*-Dichlorobenzene

The isomers are separated and *p*-dichlorobenzene used as an insecticidal fumigant (e.g., in "moth crystals"). The ortho isomer is used as a solvent and in other applications.

Biphenyl has two benzene rings linked by a single bond. Chlorination of biphenyl produces liquid mixtures of compounds, known as polychlorobiphenyls (PCB's), containing up to 10 chlorine atoms.

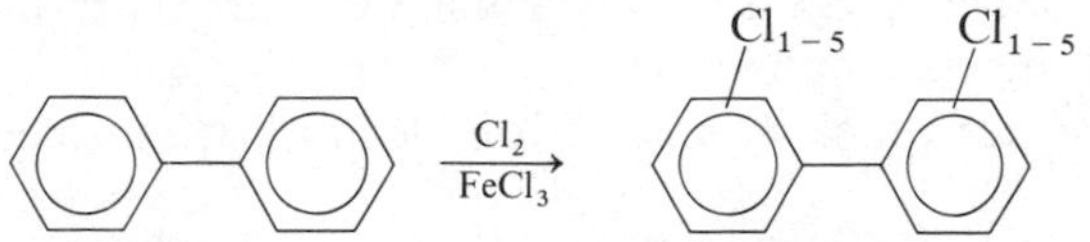

Biphenyl   A mixture of
                        chlorinated compounds

Because they have great thermal stability, these mixtures have been used extensively as cooling fluids in industrial heat exchangers and large transformers. However, because they have been found to be toxic and to disappear only slowly when released into the environment, their use in such applications is being phased out.

The high explosive TNT is prepared by nitration of toluene.

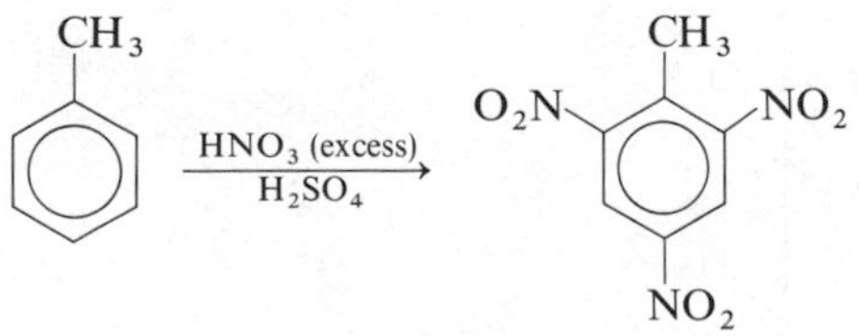

TNT
2,4,6-Trinitrotoluene

Because of the directive effects of the substituent groups, significant amounts of other isomers are not formed.

---

problem 4.4   How can the following compounds be prepared from benzene?
(a) *m*-bromonitrobenzene   (b) *p*-nitrotoluene

4.8
HOW SUBSTITUENTS AFFECT FURTHER SUBSTITUTION

Why does a substituent on an aromatic ring influence the position and ease of further substitution of that ring? Attack by the electrophile (E^+) to form the carbocation intermediate determines how rapidly and at what position a substitution will occur.

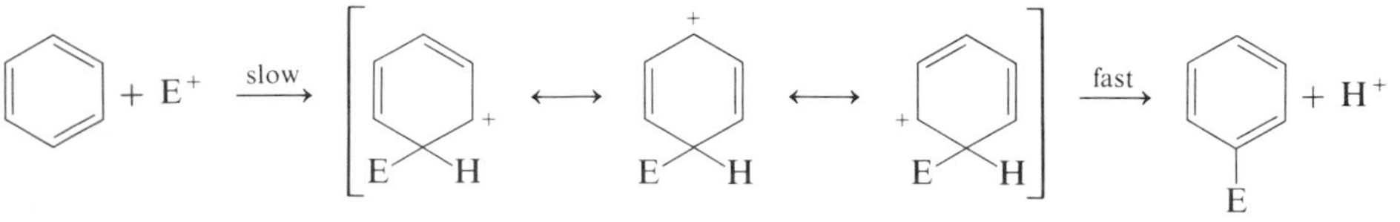

$$E^+ = R^+, NO_2{}^+, Br^+, \text{etc.}$$

By moving electron pairs, as we did with the structure of the allyl cation (Sec. 3.18), we can draw three reasonable resonance structures for the intermediate carbocation. This ion is best represented as a resonance hybrid of these structures, and we can see that its positive charge is divided among the three carbons that are ortho and para to the attacking electrophile.

Consider attack by an electrophile at a para or meta position of toluene.

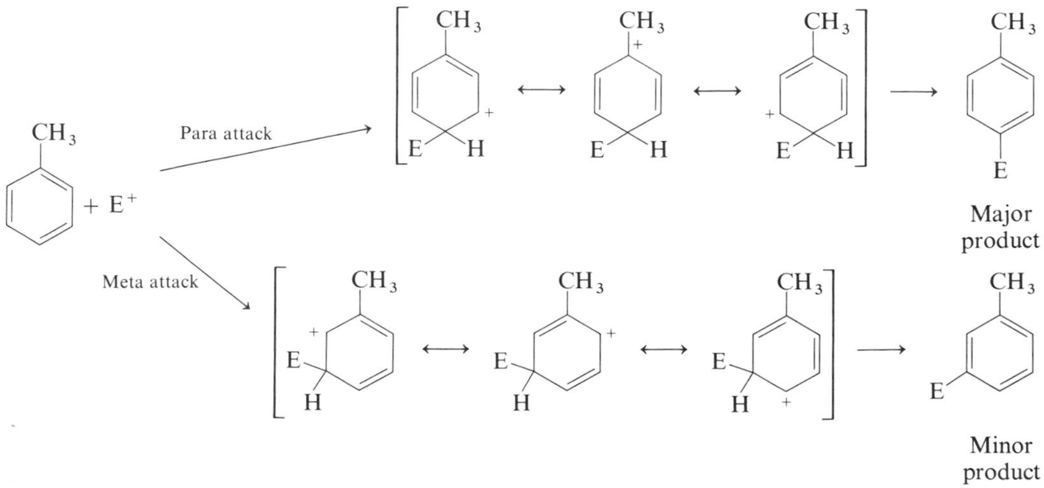

Compared to hydrogen, an alkyl group is slightly electron donating when attached to a positive carbon and hence stabilizes a carbocation. This was the explanation for the $3° > 2° > 1°$ order of carbocation stability. Because of stabilization by the methyl group, the ions formed by attack on any position of toluene are more stable and thus formed more rapidly than the ion formed by attack on benzene. The effect of the substituent is largest, however, when it is attached directly to a positive carbon. Note that in one resonance structure for the intermediate formed by para attack, the methyl group is attached directly to a positive carbon atom. None of the resonance structures for the intermediate formed by meta attack has this feature. Therefore the intermediate for para substitution is more stable and para attack is more rapid than meta attack.

---

problem 4.5   Draw resonance structures for the intermediate that results from attack by an electrophile at an ortho position of toluene. Note that the methyl group is attached to one of the partially positive carbons, just as observed in the intermediate for para substitution.

Other substituents more activating than a hydrogen also release electrons to a positive carbon and therefore *stabilize* carbocations and favor ortho and para substitution.

It is not immediately obvious why —OH and —NH$_2$ groups resemble a methyl group in their action. Oxygen and nitrogen are considerably more electronegative than carbon (Sec. 1.3) and might be expected to withdraw electrons shared with a ring carbon and thereby deactivate a carbocation intermediate. This effect, however, is less important than the partial donation of an unshared electron pair of nitrogen or oxygen to an adjacent positive carbon. This donation can be represented by additional resonance structures in which unshared electrons of the —OH and —NH$_2$ groups are used to form an additional bond to a positive carbon.

We are already familiar with the ability of the oxygen of water or the nitrogen of ammonia to use unshared electrons to form a bond to a proton.

The additional stability of the intermediate that results from part of its charge being on oxygen or nitrogen is sufficient to make ortho and para attack on —OH and —NH₂ substituted benzenes faster than on benzene.

Consider attack by an electrophile on the para and meta positions of nitrobenzene.

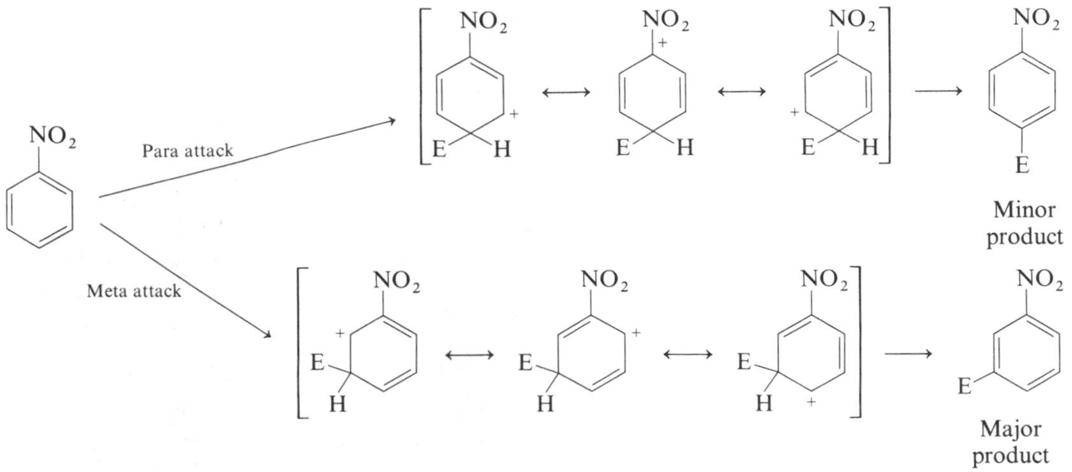

A nitro group *withdraws* electrons and *destabilizes* carbocations. Therefore substitution at any position of nitrobenzene is more difficult than substitution of benzene. Destabilization of the carbocation intermediates for para and ortho substitution is greatest because each of these ions has one resonance structure in which the nitro group is attached directly to a positive carbon. The nitro group is meta directing because it slows meta attack less than it does para and ortho attack.

Other substituents less activating than hydrogen also withdraw electrons from a positive carbon. Except for halogens, these substituents favor meta substitution. As do —OH and —NH₂ substituents, halogens withdraw electrons shared with a ring carbon but partially donate an unshared electron pair.

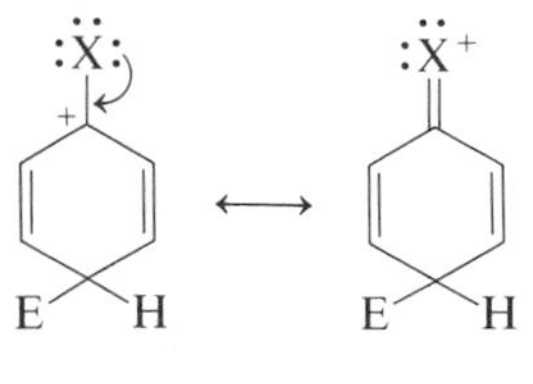

Because the withdrawing effect of halogen is largest, the carbocations are destabilized and the halobenzenes less readily substituted than benzene. Donation of the unshared electrons, however, is sufficient to make ortho and para substitution more favorable than meta substitution.

OTHER REACTIONS OF AROMATIC COMPOUNDS

Compounds that contain both a carbon-carbon double bond and an aromatic ring take part in typical alkene reactions. The double bond of styrene is readily hydrogenated.

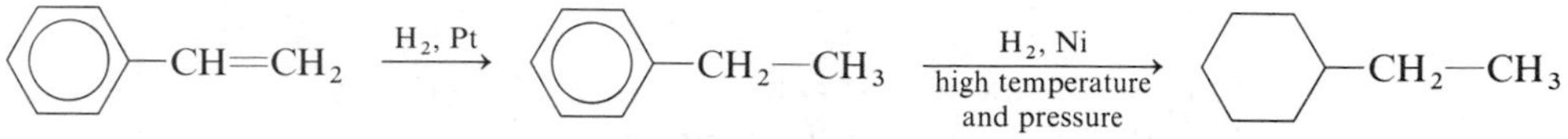

Far more stringent conditions are required to hydrogenate the aromatic ring. The double bond of styrene also reacts readily with bromine.

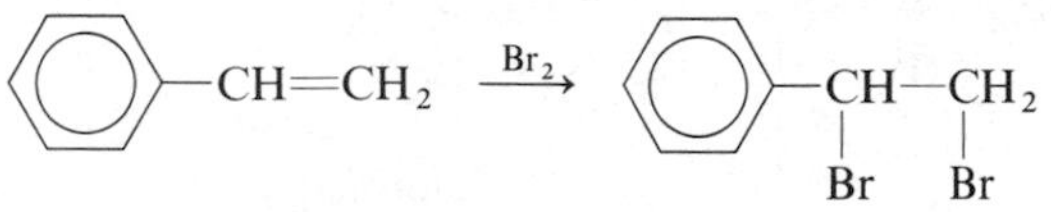

Alkylbenzenes undergo reactions characteristic of alkanes as well as aromatic compounds. The methyl group of toluene is chlorinated by using conditions that favor free-radical substitution.

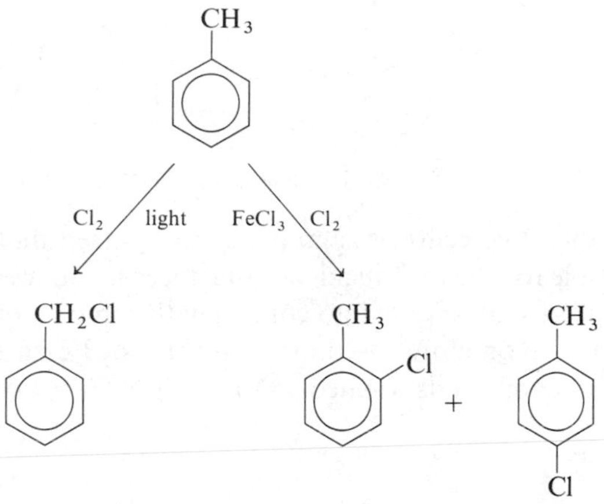

Chlorination of the aromatic ring requires conditions that favor electrophilic substitution.

4.10

POLYCYCLIC AROMATIC HYDROCARBONS

In many aromatic compounds a pair of carbon atoms is shared by two fused benzene rings. Naphthalene, the simplest of these compounds, can be represented in ways used for benzene.

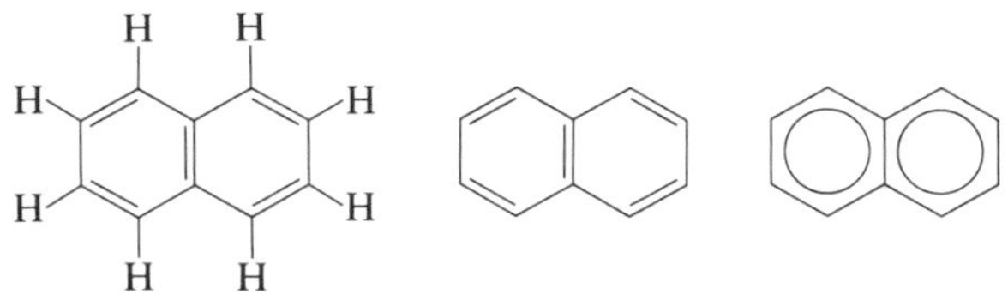

Naphthalene

Naphthalene is considered aromatic because like benzene it is considerably less reactive than alkenes and gives substitution products in many of its reactions.

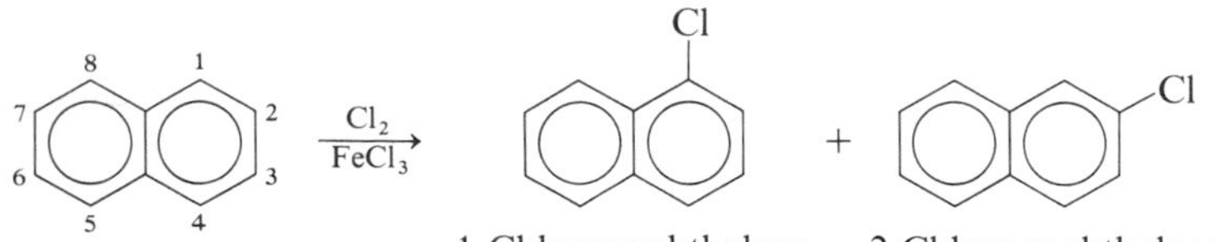

1-Chloronaphthalene   2-Chloronaphthalene

A numbering system is used to indicate the positions of substituents.

Anthracene and phenanthrene are isomeric aromatic hydrocarbons that contain three fused rings.

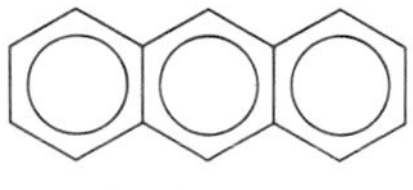

Anthracene 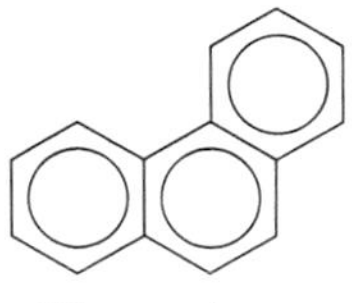   Phenanthrene

More than two centuries ago it was recognized that exposure to chimney soot was responsible for the high incidence of cancer in chimney sweeps. It is now known that these unfortunates came into contact with some highly carcinogenic compounds that contain four or more fused aromatic rings. For example, minute amounts of the following compounds applied to the skin of mice consistently produce tumors.

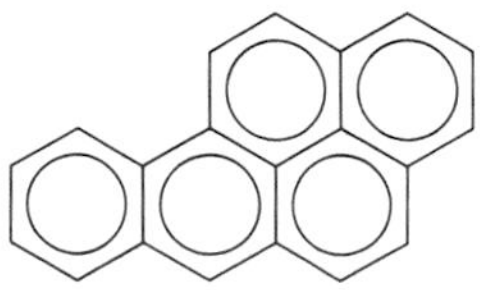

Benzo[a]pyrene 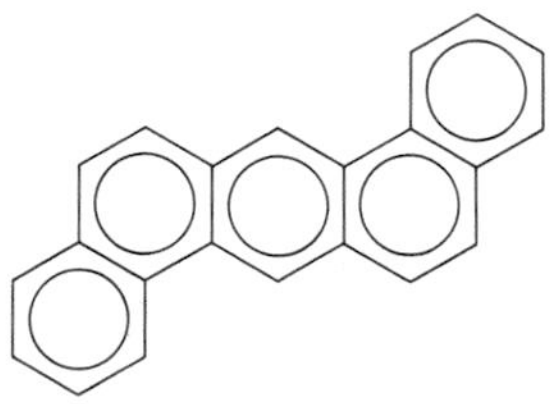   Dibenz[a,h]anthracene

Such compounds are found in the soot and tar that result from incomplete combustion of coal. They also occur in cigarette smoke.

Graphite, a form of carbon more common than diamond, consists of parallel layers of carbon atoms.

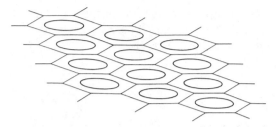

A small portion of one layer of graphite

Each carbon is bonded equally to three others to form a layer that can be regarded as a vast array of fused benzene rings. An individual layer of graphite is not easily deformed because its carbons are bonded together by strong covalent bonds. The forces between layers are relatively weak, however. Graphite's lubricating properties are due to the ease with which layers slide against one another. The "lead" in a pencil is graphite mixed with a binding agent.

4.11
SOURCES OF AROMATIC HYDROCARBONS

Aromatic hydrocarbons were once obtained principally from coal. When coal is heated at high temperatures in the absence of air, considerable amounts of gas and of a dark, viscous material known as coal tar are evolved.

$$\text{coal} \xrightarrow{\text{heat}} \text{coke} + \text{coal gas} + \text{coal tar}$$

The residue is coke, a solid composed principally of carbon. More than 10% of the world's output of coal is converted to coke for use in blast furnaces to reduce the iron oxides in iron ore to metallic iron.

Methane and hydrogen are the major constituents of coal gas. Coal tar is composed mainly of aromatic compounds, of which the most abundant generally are benzene, toluene, the dimethylbenzenes, naphthalene, and phenanthrene.

In spite of large-scale coal tar production, industrial demand for some aromatic hydrocarbons now far exceeds the supply available from this source. Therefore, petroleum has become the major source of some aromatic compounds. Crude petroleum generally contains only small amounts of aromatic compounds, but such compounds can be produced by dehydrogenation and cyclization reactions of petroleum fractions. Benzene, for example, is obtained from hexane.

$$CH_3CH_2CH_2CH_2CH_2CH_3 \xrightarrow[\text{catalyst}]{500°} \bighexagon + 4\,H_2$$

Except for ethylene and propene, benzene is the organic compound produced in largest amounts. Its annual production in the United States is about 27 kg per person. It is now known that leukemia and other blood disorders can arise from exposure to benzene, a compound to which as many as 2 million workers in the

United States potentially come into contact. As a result, increasingly stringent rules regulate exposure to benzene. Toluene is considerably less hazardous and, when feasible, is used in place of benzene.

4.12
COAL

As petroleum and natural gas become scarcer, interest is increasing in the earth's enormous coal deposits for use not only as a fuel but also as a source of the basic organic molecules needed by the chemical industry.

Coal had its origin in plant material that was deposited in swamps and removed from contact with oxygen by a covering of silt and debris before aerobic (oxygen-using) bacteria converted it completely to carbon dioxide and water. Partial decomposition by anaerobic bacteria then led to the material known as peat. Slow transformation of peat aided by the action of high pressure and moderately high temperature produced in sequence the materials known as lignites, bituminous coals, and anthracite coals. In the course of these transformations carbon content increased markedly. Carbon constitutes 92 to 95% of the weight (excluding water and mineral matter) of anthracite coals. Hydrogen and oxygen are generally the next most abundant elements, followed by nitrogen and sulfur.

Bituminous and anthracite coals are thought to consist of complex mixtures of large organic molecules that are rich in ring structures. The following structure has been proposed for a typical molecule of a bituminous coal.

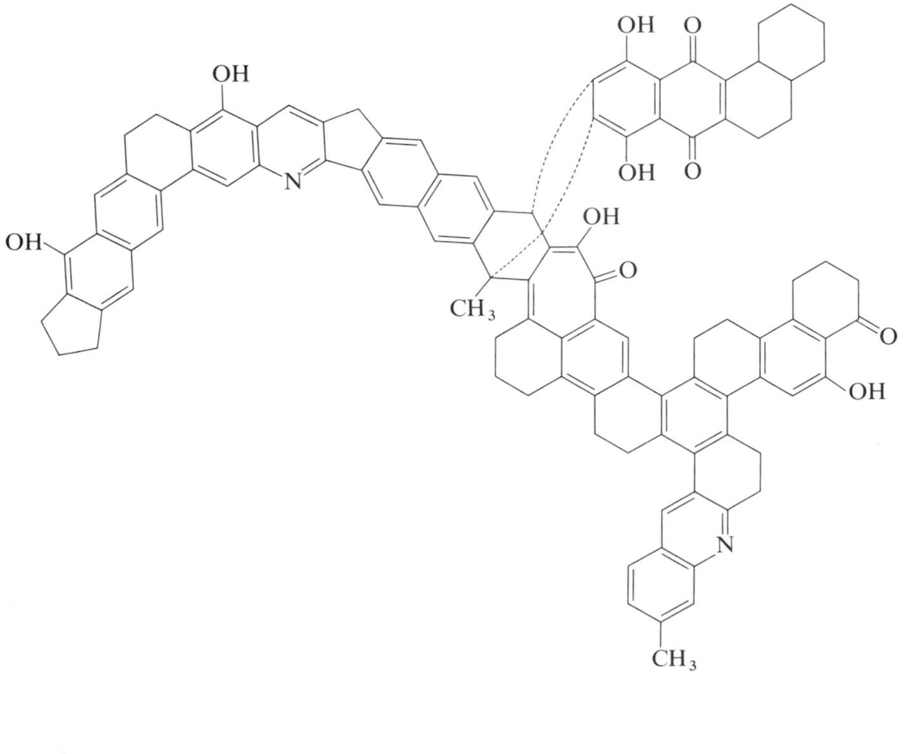

Coal can be used to produce alkanes, although at prices not presently competitive with petroleum and natural gas. When a mixture of coal and water is heated at high temperatures, the following reactions that convert coal to methane occur simultaneously.

1. Coal $\longrightarrow$ C + H$_2$ + CH$_4$
2. C + H$_2$O $\longrightarrow$ CO + H$_2$
3. C + 2 H$_2$ $\longrightarrow$ CH$_4$
4. CO + H$_2$O $\longrightarrow$ H$_2$ + CO$_2$

Reaction 1 is similar to the formation of coke. The extent to which reaction 4 is permitted to proceed adjusts the amounts of carbon monoxide and hydrogen to the 1:3 ratio that is optimal for a reaction that converts the remaining carbon monoxide and hydrogen to methane.

$$CO + 3 H_2 \xrightarrow{\text{Ni}} CH_4 + H_2O$$

A mixture of carbon monoxide and hydrogen obtained from the reaction of coal and water can also be converted to a hydrocarbon mixture that contains mainly linear alkanes and 1-alkenes.

$$CO + H_2 \xrightarrow{\text{Fe or Co catalyst}} \text{hydrocarbons}$$

During World War II this process was used on a large scale to produce gasoline in Germany, a country poor in petroleum resources.

4.13
MORE ABOUT COVALENT BONDING

The simple description of covalent bonds outlined in Chapter 1 is remarkably successful in rationalizing the structures of many organic molecules. We have encountered some limitations, however. The descriptions of benzene and the allyl cation are inadequate, and we had to introduce the resonance theory (Sect. 3.18) to deal with the structures of such species. Moreover, the simple description of bonding gives us no understanding of the geometries of molecules. Why are bond angles 109° in methane, 120° in ethylene, and 180° in acetylene? Why is rotation easy around the carbon-carbon bond of ethane but difficult around the carbon-carbon bond of ethylene? We will now learn a more detailed way of describing covalent bonds that affords a better understanding of some properties of organic molecules.

4.14
ATOMIC ORBITALS

Electron arrangements in atoms are probably familiar to you from an introductory course in chemistry. We will, however, briefly review those features needed to consider the electrons in the covalent bonds of organic molecules.

1 H							2 He
3 Li	4 Be	5 B	6 C	7 N	8 O	9 F	10 Ne
11 Na	12 Mg	13 Al	14 Si	15 P	16 S	17 Cl	18 Ar

Figure 4.2 The first 18 elements of the periodic table.

All atoms of a particular element have the same number of protons in the nucleus—the **atomic number** of that element. In a neutral atom the number of electrons surrounding the nucleus also equals the atomic number. Except for bromine and iodine, the elements found most commonly in organic compounds (C, H, N, O, F, P, S, and Cl) are included in the first 18 elements, shown in the abbreviated periodic table in Figure 4.2.

Electrons in an atom can have only certain discrete energies. Electrons closest to the nucleus have the lowest energy (greatest stability) because of the large electrostatic attraction by the positive nucleus. Electrons more distant from the nucleus are of higher energy (lower stability). The electrons of an atom are divided into **shells** that are given the designations 1, 2, 3, 4 . . . . Number 1 indicates the shell of lowest energy, 2 the next shell, and so on. The electrons within one of these shells are further divided into subshells called **atomic orbitals**. An orbital can hold no more than two electrons.

The first shell has only a single orbital, designated 1s. The second shell has four orbitals: one orbital that is designated 2s and three orbitals of identical energy that are designated 2p. The third shell has one 3s orbital, three 3p orbitals, and five 3d orbitals. The energies of electrons in these atomic orbitals increase in the following order:

$$1s < 2s < 2p < 3s < 3p < 3d$$
$$\xrightarrow{\text{increasing energy}}$$

In the most stable state of an atom, called the **ground state**, its electrons have the lowest possible energies. Therefore the single electron of a hydrogen atom and the two electrons of a helium atom are in a 1s orbital. Because a 1s orbital has room for only two electrons, one of the three electrons of a beryllium atom is in the 2s orbital, the most stable of the other orbitals. The electron arrangements of the ground states of the first 18 elements are shown in Table 4.2.

Helium (He), neon (Ne), and argon (Ar) are particularly stable. In fact, these elements are sometimes called the inert gases because they do not form stable compounds. The gain or loss of electrons by atoms in forming the ionic and covalent bonds of most compounds can be ascribed to a tendency to attain the stable electron arrangements of the inert gases. Consider sodium chloride ($Na^+ Cl^-$) as an example of an ionic compound: Na^+ has one electron less than a sodium atom and hence

Table 4.2 The most stable electron arrangements of the first 18 elements. The two electrons in an orbital differ in the property called spin and so are represented by arrows pointed in opposing directions.

Atom	Atomic number	First shell 1s	Second shell 2s	2p			Third shell 3s	3p		
H	1	↑								
He	2	↑↓								
Li	3	↑↓	↑							
Be	4	↑↓	↑↓							
B	5	↑↓	↑↓	↑						
C	6	↑↓	↑↓	↑	↑					
N	7	↑↓	↑↓	↑	↑	↑				
O	8	↑↓	↑↓	↑↓	↑	↑				
F	9	↑↓	↑↓	↑↓	↑↓	↑				
Ne	10	↑↓	↑↓	↑↓	↑↓	↑↓				
Na	11	↑↓	↑↓	↑↓	↑↓	↑↓	↑			
Mg	12	↑↓	↑↓	↑↓	↑↓	↑↓	↑↓			
Al	13	↑↓	↑↓	↑↓	↑↓	↑↓	↑↓	↑		
Si	14	↑↓	↑↓	↑↓	↑↓	↑↓	↑↓	↑	↑	
P	15	↑↓	↑↓	↑↓	↑↓	↑↓	↑↓	↑	↑	↑
S	16	↑↓	↑↓	↑↓	↑↓	↑↓	↑↓	↑↓	↑	↑
Cl	17	↑↓	↑↓	↑↓	↑↓	↑↓	↑↓	↑↓	↑↓	↑
Ar	18	↑↓	↑↓	↑↓	↑↓	↑↓	↑↓	↑↓	↑↓	↑↓

the electron arrangement of neon; Cl^- has one electron more than a chlorine atom and thus the electron arrangement of argon. If shared electrons are counted, carbon, nitrogen, oxygen, and fluorine in covalent compounds are usually surrounded by eight second-shell electrons, the number found in neon. In covalent compounds hydrogen is surrounded by two first-shell electrons, the number in helium. Phosphorus, sulfur, and chlorine are often surrounded by eight third-shell electrons, as is argon. Because of the $3d$ orbitals in the third shell, however, these atoms are

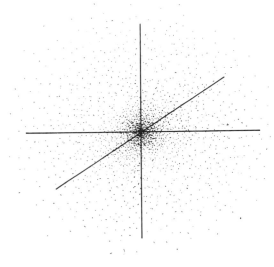

Figure 4.3 Electron-density distribution of the hydrogen atom. The nucleus is at the origin of the three coordinate axes.

sometimes found in compounds (H_3PO_4 and H_2SO_4 are familiar examples) in which they are surrounded by larger numbers of third-shell electrons.

Because the energy of an electron in a given orbital is fixed, the uncertainty principle tells us that the position at any given time of this small particle cannot be precisely defined. Using **quantum mechanics**, however, the branch of physics that describes the energy and motion of small particles, such as electrons, it is possible to describe the probability of finding an electron in different regions around the nucleus. Consider the electron of a ground state hydrogen atom. If it were possible to take a large number of instantaneous snapshots of this electron, we would observe a pattern like that in Figure 4.3. The probability of finding the electron in a given region is greatest near the nucleus and diminishes with distance. Electrons in $1s$ orbitals of other atoms exhibit a similar, spherically symmetrical pattern. Because drawing such a picture is tedious, an orbital is often represented more simply by drawing a boundary that encloses a substantial fraction (say 90%) of the total electron density. Such representations of $1s$ and $2s$ orbitals are shown in Figure 4.4. The $2s$ orbital is larger because the average distance of a $2s$ electron from the nucleus is greater.

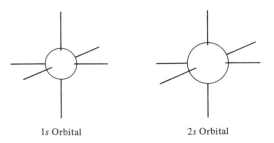

1s Orbital   2s Orbital

Figure 4.4 Representations of the $1s$ and $2s$ orbitals by surfaces that enclose about 90% of the electron density. The nucleus is at the center.

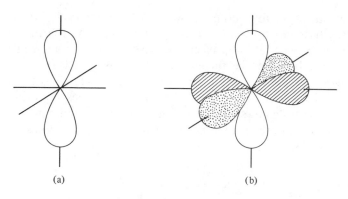

(a)   (b)

Figure 4.5   (a) Representation of a $2p$ orbital by a surface that encloses about 90% of the electron density. The nucleus is at the origin of the coordinate axes. (b) The three $2p$ orbitals surrounding a nucleus. The orbitals are shaded differently to show which lobes belong together.

A $2p$ orbital has a more complex, dumbbell-like shape. As shown in Figure 4.5(a), the electron density is concentrated in two lobes. The three $2p$ orbitals of an atom are identical in energy and shape but, as pictured in Figure 4.5(b), are at right angles to each other. The simplified drawings of orbitals that indicate only boundaries enclosing most (but not all) of the electron density can be somewhat misleading. If the three $2p$ orbitals all contain the same number of electrons (one or two), then the electron density due to all the $2p$ electrons is exactly spherical.

4.15
COVALENT BONDS DESCRIBED BY OVERLAP
OF ATOMIC ORBITALS

Covalent bonds result from the attraction of electrons by two or more nuclei. Electrons of a molecule occupy orbitals, which can hold no more than two electrons. These are not the atomic orbitals of the constituent atoms but rather new orbitals, called **molecular orbitals**, that are characteristic of the molecule.

In principle, the properties of the molecular orbitals of any compound can be calculated by using quantum mechanics. In practice, such calculations are too complicated to be carried out exactly except for the molecule H_2 and a few other simple chemical species. Simplifications are needed to apply quantum mechanics to even the smallest organic molecules. We will examine the results of an approach, particularly useful to organic chemists, that relates the distribution of electrons in molecules to shapes of atomic orbitals of individual atoms.

Using this approach, we consider a covalent bond to form from **overlapping** of atomic orbitals. Formation of the hydrogen molecule can be pictured as shown in Figure 4.6. As two hydrogen atoms approach one another, their $1s$ orbitals,

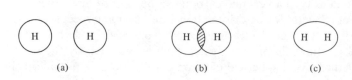

(a)   (b)   (c)

Figure 4.6.   (a) Two isolated hydrogen atoms, each containing one electron in a $1s$ orbital. (b) Two hydrogen atoms close enough for their $1s$ orbitals to overlap significantly. (c) The molecular orbital occupied by the two electrons of H_2.

each containing one electron, begin to overlap. Some of the electron density is placed in a region where attraction by both nuclei is significant. As the nuclei approach even closer, further concentration of electron density between them leads to increasing attractive forces between the nuclei and electrons. Note that the electron density in the molecular orbital of the hydrogen molecule [Fig. 4.6(c)] is heavily concentrated between the nuclei. Of course, as approach continues, repulsion between the positive nuclei also becomes significant. Two hydrogen atoms do not approach more closely than in the hydrogen molecule because additional repulsion would exceed additional attraction.

4.16
sp^3 HYBRID ORBITALS

How can the bonds in methane, the compound with which we began our study of organic chemistry, be described in terms of overlap of orbitals of carbon and of hydrogen? Consider the valence electrons (Table 4.2) of a ground state carbon atom.

The 2s orbital is filled and one 2p orbital is empty. Therefore only the two 2p orbitals that are occupied by single electrons seem available for effective overlap with the 1s orbitals of hydrogen atoms. We might expect carbon to form only two bonds.

For a carbon atom to have the capacity to form four bonds, one of its 2s electrons must be excited (raised in energy) and placed into the empty 2p orbital.

Now there are four orbitals with which 1s orbitals of hydrogen can overlap to form bonds. The extra energy given off in forming two additional covalent bonds is greater than that necessary to excite the electron.

Electron density in 2s and 2p atomic orbitals of carbon is not arranged in a way that would lead to efficient overlap with the orbitals of hydrogen atoms. The 1s orbital of hydrogen could overlap with just one of the two lobes of a 2p orbital and with only one side of the nondirectional 2s orbital. Moreover, overlap with these 2s and 2p orbitals would not lead to a tetrahedral geometry for methane, because the three hydrogens bonded by overlap with the 2p orbitals should be at angles of 90°.

The valence electron density of an excited carbon can be divided into other sets of four orbitals that are more useful for describing bonds in organic molecules. The electron density due to all the valence electrons is exactly spherical. A 2s orbital is inherently spherical, and the three 2p orbitals (when each is occupied by the same number of electrons) together give a spherical electron distribution. This spherical

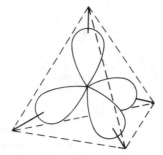

Figure 4.7 *sp³* Hybrid orbitals. The nucleus is at the point where the four orbitals meet.

cloud of valence electron density can be divided into four identical orbitals, shown in Figure 4.7, that point to the corners of a regular tetrahedron. One-fourth of the electron density of each new orbital corresponds to that of an atomic 2*s* orbital and three-fourths to that of atomic 2*p* orbitals.

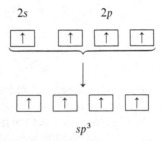

To indicate this 1:3 ratio, the new orbitals are called **sp³ hybrid orbitals.**

Carbon *sp³* hybrid orbitals are useful for describing methane. Each bond is considered to result from overlap of a hydrogen 1*s* orbital with an *sp³* hybrid orbital [Fig. 4.8(a)]. Tetrahedral bond angles result, for overlap is greatest when each hydrogen is located at the end of an *sp³* orbital. The resulting molecular orbitals of methane are shown in Figure 4.8(b). Because its electron density is concentrated in a particular direction from the nucleus, an *sp³* hybrid orbital overlaps more effectively with the 1*s* orbital of hydrogen than does a 2*s* or 2*p* orbital.

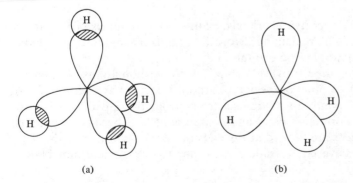

(a)   (b)

Figure 4.8 (a) Overlap of hydrogen 1*s* orbitals with four *sp³* hybrid orbitals of a carbon to form methane. (b) The molecular orbitals of methane.

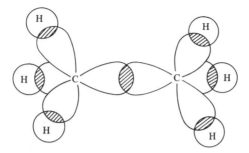

Figure 4.9   Overlap of orbitals to form ethane.

sp^3 Hybrid orbitals are also used to describe bonds in other compounds in which carbons are bonded to four groups. The carbon-hydrogen bonds in ethane (Fig. 4.9), for example, result from overlap of hydrogen $1s$ orbitals with carbon sp^3 hybrid orbitals. The carbon-carbon bond results from the overlap of two sp^3 hybrid orbitals.

4.17
sp^2 HYBRID ORBITALS

The electron density of a carbon atom in which one $2s$ electron has been excited to the empty $2p$ orbital can be divided into another set of orbitals that is useful for describing ethylene. We consider the electron density of the $2s$ orbital and two of the $2p$ orbitals to be divided into three equivalent orbitals.

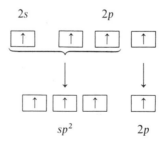

The resulting orbitals, called sp^2 **hybrid orbitals**, lie in a plane at angles of $120°$ from each other [Fig. 4.10(a)]. The remaining $2p$ orbital retains its characteristic shape and is perpendicular to the sp^2 hybrid orbitals [Fig. 4.10(b)].

The overlap involving the sp^2 hybrid orbitals of the carbon atoms of ethylene is shown in Figure 4.11(a). A bond between the carbons is formed by overlap of two sp^2 hybrid orbitals. The bonds to hydrogens result from overlap of sp^2 hybrid orbitals with hydrogen $1s$ orbitals. Bond angles are about $120°$ because overlap is greatest when the hydrogens are located at the ends of the sp^2 hybrid orbitals. The $2p$ orbitals on each carbon overlap in a sideways fashion to form an additional bond [Fig. 4.11(b)].

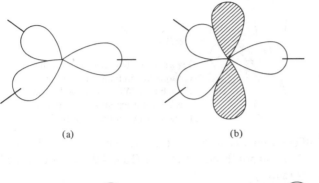

(a)   (b)

Figure 4.10   (a) *sp*² Hybrid orbitals. The nucleus is at the point where the three orbitals meet. (b) The three *sp*² hybrid orbitals and the 2*p* orbital (shaded) of an *sp*² hybridized carbon.

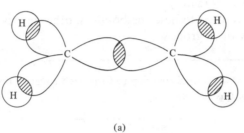
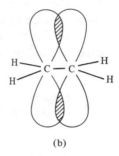

(a)   (b)

Figure 4.11   (a) Overlap in ethylene involving the *sp*² hybrid orbitals. (b) Overlap in ethylene involving the 2*p* orbitals.

The resulting molecular orbitals of ethylene are shown in Figure 4.12. The electron density in the bond that results from overlap of the 2*p* orbitals lies partly above and partly below the plane of the other atoms. Molecular orbitals divided in this manner are called **π (pi) orbitals**. The electron density in the other molecular orbitals of ethylene (and in the molecular orbitals of methane or ethane) is symmetrical about a line joining the nuclei. Such molecular orbitals are called **σ (sigma) orbitals**. The double bond of ethylene consists of one σ and one π bond. Because the sideways overlapping of 2*p* orbitals is not as great as the end-on overlapping of *sp*² or *sp*³ hybrid orbitals, the π bond of ethylene is somewhat weaker than the σ carbon-carbon bonds of ethylene and ethane.

As shown in Figure 4.13, 2*p* orbitals on adjoining carbons do not overlap at all if they are at right angles. Rotation around a carbon-carbon double bond is so

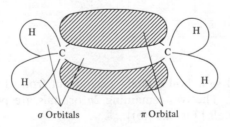

σ Orbitals   π Orbital

Figure 4.12   The σ and π (shaded) molecular orbitals of ethylene.

Figure 4.13 Rotation around the carbon-carbon bond of an alkene breaks the π bond. When rotation has proceeded to the extent of 90°, no overlap remains between the 2p orbitals.

difficult because it requires sufficient energy to break the covalent π bond. In contrast, the geometry of the σ carbon-carbon bond of ethane (Fig. 4.9) is not changed by rotation, and rotation occurs readily.

sp^2 Hybrid orbitals are used to describe bonds in other compounds in which carbons are bonded to only three groups.

---

problem 4.6 Indicate which orbitals overlap to form each covalent bond of propylene. Describe the geometry of this alkene.

Alkenes are more reactive than alkanes. Reactions of alkenes often result from attack by electrophiles on their π-electron clouds. Electrons in alkanes are concentrated between the nuclei in σ orbitals. The π electrons in alkenes, farther from the nuclei and not as strongly held as σ electrons, are more available for attack.

4.18
sp HYBRID ORBITALS

The electron density of an excited carbon atom can be divided into another set of orbitals useful for describing acetylene and other compounds in which carbons are bonded to only two groups. We consider the electron density of the 2s and one 2p orbital to be divided into two equivalent orbitals.

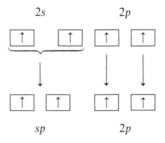

The resulting orbitals, called ***sp* hybrid orbitals**, extend out from the nucleus at an angle of 180° [Fig. 4.14(a)]. The two remaining 2p orbitals are perpendicular to the axis of the *sp* hybrid orbitals [Fig. 4.14(b)].

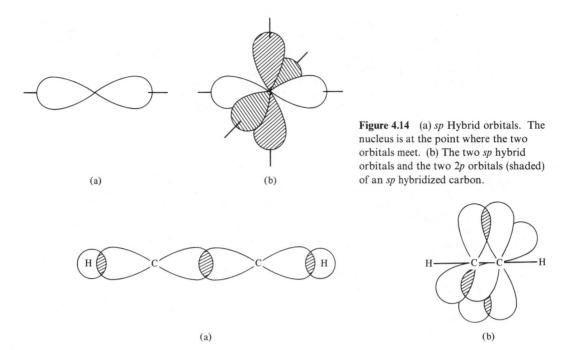

(a)   (b)

Figure 4.14 (a) *sp* Hybrid orbitals. The nucleus is at the point where the two orbitals meet. (b) The two *sp* hybrid orbitals and the two 2*p* orbitals (shaded) of an *sp* hybridized carbon.

(a)   (b)

Figure 4.15 (a) The σ molecular orbitals of acetylene result from overlap involving *sp* hybrid orbitals. (b) The π oribitals of acetylene result from overlap of 2*p* orbitals.

A σ bond between the carbons of acetylene results from overlap of two *sp* hybrid orbitals [Fig. 4.15(a)]. Carbon-hydrogen σ bonds result from overlap of *sp* hybrid orbitals with hydrogen 1*s* orbitals. The linear geometry of acetylene is a result of the directionality of the *sp* orbitals. Sideways overlap of the 2*p* orbitals forms two π bonds [Fig. 4.15(b)]. The triple bond consists of one σ bond and two somewhat weaker π bonds.

4.19
DESCRIPTION OF ORGANIC MOLECULES BY USING HYBRID ORBITALS

Considering bonds of organic molecules to result from overlap with carbon sp^3, sp^2, and *sp* hybrid orbitals has provided useful descriptions of methane, ethylene, and acetylene. It may have seemed that a new type of hybrid orbital was needed for each molecule considered. In fact, the three types of hybrid orbitals introduced are sufficient to describe the bonding in most other organic molecules.

Let us describe the bonds of benzene by using hybrid orbitals. Each carbon of benzene is attached to three other atoms at angles of 120° and hence is sp^2 hybridized. The carbon-carbon and carbon-hydrogen σ bonds shown in Figure 4.16(a) are formed by overlap involving the three sp^2 orbitals of each carbon.

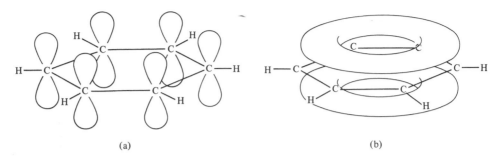

(a)   (b)

Figure 4.16   (a) The σ bonds (depicted by lines) in benzene result from overlap
involving sp^2 hybrid orbitals. A $2p$ orbital remains on each carbon (for clarity,
the $2p$ orbitals are drawn at a reduced size). (b) The π-electron cloud that results
from overlap of the $2p$ orbitals.

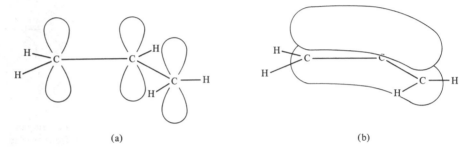

(a)   (b)

Figure 4.17   (a) The σ bonds (depicted by lines) in the allyl cation result from
overlap involving sp^2 hybrid orbitals. A $2p$ orbital remains on each carbon.
(b) The π-electron cloud that results from overlap of the $2p$ orbitals.

The $2p$ orbital that remains on each carbon overlaps equally with the $2p$
orbitals on the two adjoining carbons. The six electrons contributed by the $2p$
orbitals form a cloud of π electrons [Fig. 4.16(b)] above and below the plane of the
ring of carbons.* Attraction of each electron of this cloud by more than two nuclei
is responsible for the added stability (resonance energy) of benzene.

The simple description of bonding (using lines to represent electron pairs)
provides a structure with alternating single and double bonds for benzene. The
orbital description of benzene makes evident the equivalence of the carbon-carbon
bonds.

The allyl cation was not described adequately by the simple bonding theory.

$$\begin{array}{c} \text{H} \\ \text{H} \qquad | \qquad \text{H} \\ \diagdown \qquad\qquad \diagup \\ \text{C}=\text{C}-\text{C}^+ \\ \diagup \qquad\qquad \diagdown \\ \text{H} \qquad\qquad \text{H} \end{array}$$

Its description using overlap of orbitals is more successful. Each carbon is bonded
to only three other atoms and so has sp^2 hybridization. Carbon-carbon and carbon-

---

* Because an orbital can hold only two electrons, the six electrons that constitute the π-electron
cloud of benzene occupy three molecular orbitals.

hydrogen σ bonds result from overlap involving the sp^2 hybrid orbitals [Fig. 4.17(a)]. The $2p$ orbitals that remain on each carbon overlap to form a π cloud [Fig. 4.17(b)] containing two electrons. Clearly the central carbon is bonded equally to the two terminal carbons.

4.20
AROMATIC COMPOUNDS

The term **aromatic** is used to refer to unsaturated compounds that are considerably more stable than analogous alkenes. This stability is associated with a low reactivity toward reagents that readily attack alkenes and a tendency to undergo substitution rather than addition reactions. The aromatic compounds considered previously have all contained single or fused benzene rings. Do compounds having rings of other sizes exhibit aromatic properties?

We might expect cyclobutadiene and cyclooctetraene to be aromatic.

Cyclobutadiene   Benzene   Cyclooctatetraene

Cyclobutadiene, however, is highly reactive; only at exceedingly low temperatures does it survive long enough to be observed. Cyclooctatetraene is stable at room temperature but lacks a significant resonance energy and reacts as do alkenes. Cyclooctatetraene would need to be planar for the maximum overlap of $2p$ orbitals that is necessary to form a continuous π cloud of electrons. In fact, it is nonplanar and contains alternating single and double bonds.

Theory and experiment both indicate that cyclic hydrocarbons are aromatic only if they are fully conjugated (a $2p$ orbital at each ring carbon) and have a total of $(4n + 2)$ π electrons, where n is 0, 1, 2, 3, 4 . . . . Compounds with 2, 6, 10, 14, 18 . . . π electrons exhibit aromatic properties; those with 4, 8, 12, 16 . . . do not. Just as the numbers of electrons of inert gas atoms are associated with an unusual stability, so are certain numbers of π electrons. Conjugated systems that have $(4n + 2)$ π electrons can be considered to have filled π-electron shells.

Cyclobutadiene (four π electrons) and cyclooctatetraene (eight π electrons) do not have numbers of π electrons associated with aromatic systems. The following hydrocarbon with an 18-membered ring and 18 π electrons has aromatic properties.

All its carbon-carbon bonds have the same length. The vast majority of aromatic compounds, however, have six π electrons.

SUMMARY

Definitions and Ideas: Aromatic Compounds

Cyclic compounds, such as benzene, that are unsaturated but unusually stable and unreactive are called **aromatic**.

Benzene is planar and has bond angles of 120°. Its carbon-carbon bonds are all identical and intermediate in length (1.39 Å) between typical carbon-carbon single and double bonds.

Benzene can be represented as a hybrid of two resonance structures that differ only in placement of electrons of carbon-carbon bonds. As predicted by the resonance theory, benzene has a large **resonance energy** (an unusual stability).

Substituted benzenes are given substitutive names, using benzene or a monosubstituted benzene (such as toluene, phenol, and aniline) with an accepted trivial name as the parent. The positions of substituents are usually indicated by numbers. If only two substituents are present, **ortho** (for a 1,2-isomer), **meta** (for a 1,3-isomer), and **para** (for a 1,4-isomer) can also be used.

Substitutions in which a hydrogen is replaced by some other group but the stable aromatic system is preserved are characteristic reactions of aromatic compounds.

A substituent already present on an aromatic ring influences both the position and ease of further substitution (see Table 4.1). A given substituent is either **ortho-para directing** or **meta directing**, regardless of which substitution is taking place.

The reactions called **electrophilic aromatic substitutions** proceed by a two-step mechanism. In the first step an electrophile attacks the aromatic ring to form a carbocation intermediate. In the second step the carbocation rapidly loses a proton, regenerating the aromatic ring system. Catalysts are often needed to help generate a reactive electrophile. The effects of a substituent are due to stabilization or destabilization of the carbocation intermediates that lead to products of ortho, meta, and para substitution.

Important Reactions of Aromatic Compounds

1. *Halogenation* $ArH \xrightarrow[FeX_3]{X_2} Ar{-}X$
   Ar = an aryl group   X = Cl or Br

2. *Nitration* $ArH \xrightarrow[H_2SO_4]{HNO_3} Ar{-}NO_2$

3. *Sulfonation* $ArH \xrightarrow{H_2SO_4} Ar{-}SO_3H$

4. *Alkylation* $ArH \xrightarrow[AlCl_3]{RX} Ar{-}R$
   R = an alkyl group   X = Cl or Br

Definitions and Ideas: Covalent Bonding

The electrons of an atom are divided into shells (1, 2, 3 . . .) and within those shells into subshells called **atomic orbitals**, which can hold no more than two electrons. The first shell has one orbital, designated 1s. The second shell has one orbital designated 2s and three orbitals designated $2p$. The third shell has one $3s$, three $3p$, and five $3d$ orbitals. The energies of electrons in these orbitals increase in the following order: $1s < 2s < 2p < 3s < 3p < 3d$. The electrons of the most stable state of an atom, called the **ground state**, have the lowest possible energies (see Table 4.2).

An s orbital is spherical about the nucleus. A $2p$ orbital is dumbbell shaped, with the lobes on either side of the nucleus. The three $2p$ orbitals are at right angles to one another.

The gain or loss of electrons by an atom in forming ionic or covalent bonds can often be ascribed to a tendency to attain the stable electron arrangement of an inert gas (He, Ne, Ar . . .).

Electrons in a molecule occupy orbitals, called **molecular orbitals**, that are characteristic of the particular molecule. It is useful to describe molecular orbitals as the result of overlapping of atomic orbitals.

For carbon to form four bonds, one of its $2s$ electrons must be excited to a $2p$ orbital so that it has one $2s$ and three $2p$ valence electrons. The valence electron density of such an excited carbon can be divided into different sets of four orbitals.

One manner of division provides a set of four equivalent orbitals, called **sp^3 hybrid orbitals**, at angles of 109° to each other. These orbitals are useful for describing covalent bonds in methane and other compounds in which carbons are bonded to four groups.

Another manner of division provides a set of three **sp^2 hybrid orbitals**, which lie in a plane at angles of 120° to each other, and a $2p$ orbital, which is perpendicular to this plane. This set of orbitals is useful for describing bonding in ethylene and other compounds in which carbons are bonded to only three groups. Three bonds to such a carbon result from overlap with the sp^2 orbitals. The $2p$ orbital can overlap in a sideways fashion with a $2p$ orbital on an adjoining atom to form an additional bond.

Another mode of division provides a set of two **sp hybrid orbitals**, which are at an angle of 180° from each other, and two $2p$ orbitals, which are perpendicular to the axis of the sp hybrid orbitals and to each other. This set of orbitals is useful for describing covalent bonds in acetylene and other compounds in which carbons are bonded to only two groups. Two bonds to such a carbon result from overlap with the sp orbitals. The $2p$ orbitals can overlap with $2p$ orbitals on adjoining atoms to form two additional bonds.

Molecular orbitals that are symmetrical about a line joining two atoms bonded together are called **σ (sigma) orbitals**. Molecular orbitals that lie partly above and partly below two or more atoms bonded together are called **π (pi) orbitals**.

The $2p$ orbital on each carbon of benzene overlaps equally with the $2p$ orbitals on the two adjoining carbons, resulting in a continuous cloud of π electrons. The added stability of benzene is due to the attraction of each π electron by more than two nuclei.

To be aromatic, a cyclic hydrocarbon must be fully conjugated (have a $2p$ orbital at each ring carbon) and have a total of $(4n + 2)$ π electrons, where $n = 0, 1, 2, 3 \ldots$.

PROBLEMS

1. Draw structures for the following compounds.

(a) chlorobenzene

(b) isopropylbenzene

(c) *o*-dinitrobenzene

(d) *m*-chloronitrobenzene

(e) 1,3,5-trinitrobenzene

(f) 2-chlorotoluene

(g) *p*-nitrobenzenesulfonic acid

(h) 4-bromophenol

(i) 3-bromo-1-nitrobenzene

(j) aniline

(k) triphenylmethane

(l) benzylcyclohexane

(m) 2-phenyloctane

(n) 1-nitronaphthalene

2. Name the following compounds.

(a) 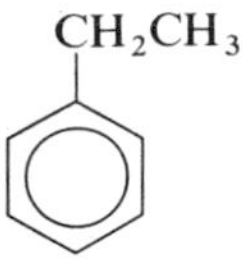 CH_2CH_3

(b) 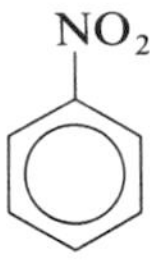 NO_2

(c) 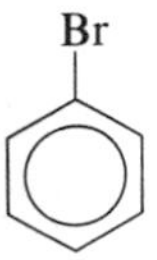 Br

(d) 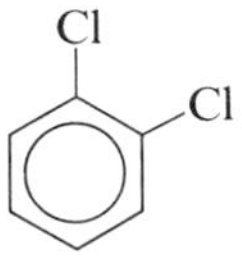 Cl — Cl

(e) 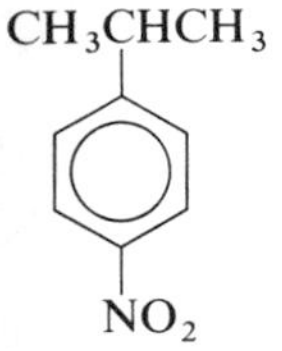 CH_3CHCH_3 ... NO_2

(f) 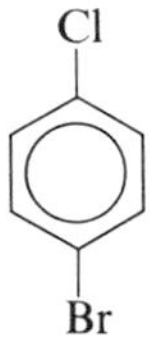 Cl ... Br

(g) 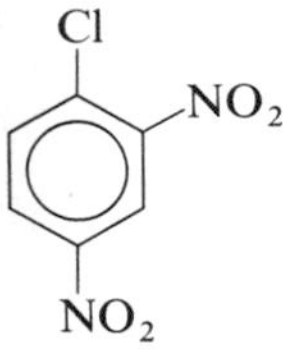 Cl ... NO_2 ... NO_2

(h) 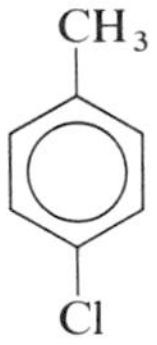 CH_3 ... Cl

(i) 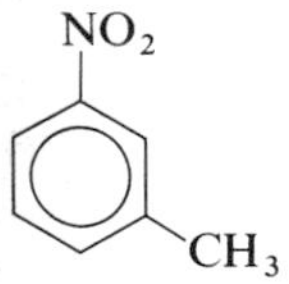 NO_2 ... CH_3

(j) 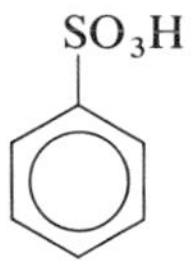 SO_3H

(k)  SO_3H ... Br

(l) 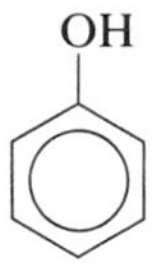 OH

(m) 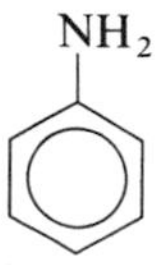 NH_2

(n) 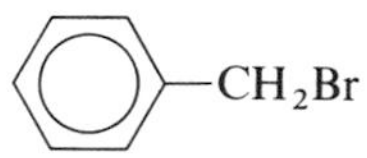 ⬡—CH_2Br

(o) 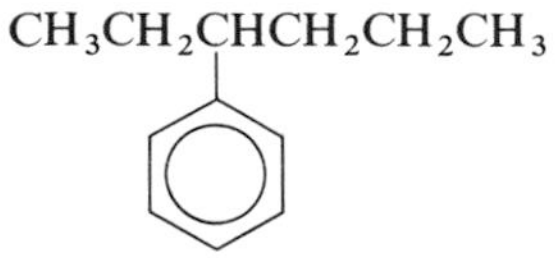 $CH_3CH_2CHCH_2CH_2CH_3$

(p) 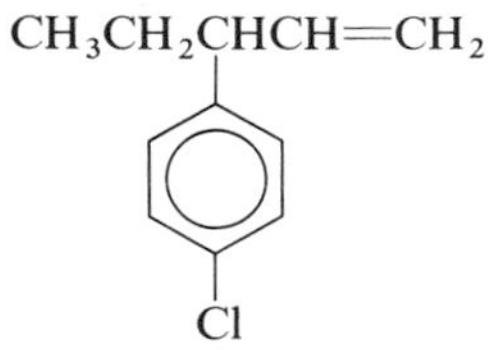 $CH_3CH_2CHCH=CH_2$ ... Cl

3. Draw structures of all compounds that fit the following descriptions.
   (a) benzene derivatives with the molecular formula C_7H_7Br
   (b) chloronitrobenzenes
   (c) dibromobenzenes
   (d) trichlorobenzenes
   (e) chloroanthracenes
   (f) dinitronaphthalenes

4. There are three isomeric dibromobenzenes. When treated with a mixture of nitric and sulfuric acids, isomer A (mp 87°) gives one mononitro derivative ($C_6H_3Br_2NO_2$); not even traces of a second derivative are formed. Treated in the same way, isomer B (mp 7°) gives two mononitro derivatives and isomer C (mp −7°) three mononitro derivatives. Assign structures to A, B, and C. This procedure of finding how many monosubstitution products form was an early method for determining the arrangement of substituents in a disubstituted benzene.

5. Draw structures for the following compounds. (Count as products even those that might form in only small amounts.)
   (a) any compounds of molecular formula C_8H_{10} that form three monobromo substitution products
   (b) a compound of molecular formula C_8H_{10} that forms only one mononitro substitution product
   (c) a compound of molecular formula C_9H_{12} that forms four mononitro substitution products
   (d) a compound of molecular formula $C_6H_3Br_3$ that forms only one monochloro substitution product

6. Why don't two compounds exist that have the isomeric structures shown here?

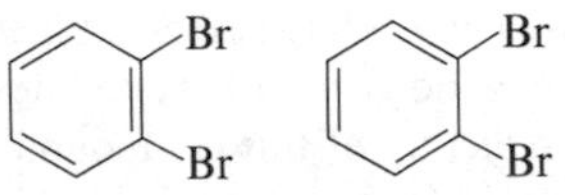

7. (a) Draw structures of the monosubstitution products that result from reactions of benzene with the following reagents.
   1. H_2SO_4 2. HNO_3 (H_2SO_4)
   3. Cl_2 ($FeCl_3$)   4. CH_3CH_2Cl ($AlCl_3$)
   (b) Do the same for reactions of toluene with the same reagents.
   (c) Do the same for reactions of nitrobenzene with the same reagents.

8. Draw structures for the major organic products of the following reactions.

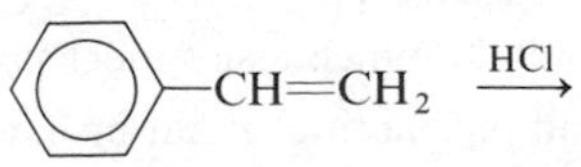

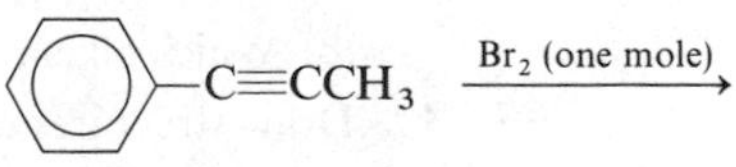

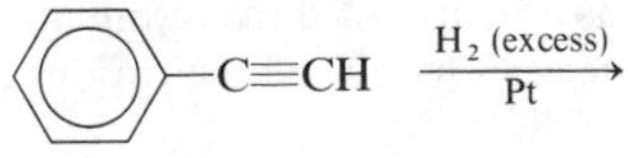

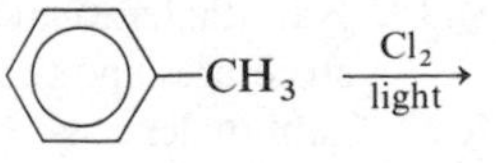

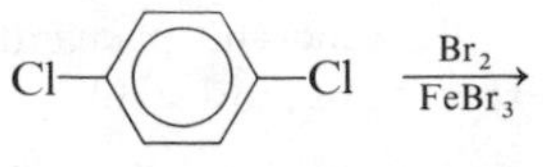

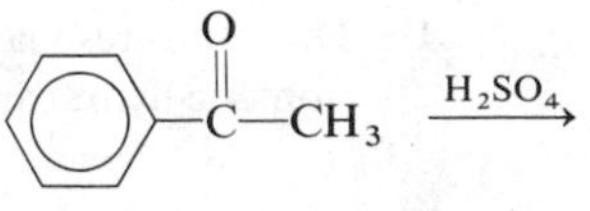

(g)

(h)

(i)

9. Write equations to show how each of the following compounds can be synthesized from benzene. Most of the syntheses will require at least two reactions. Assume that a mixture of ortho and para isomers can be separated into the pure compounds.
   (a) isopropylbenzene
   (b) methylcyclohexane
   (c) benzyl chloride
   (d) diphenylmethane
   (e) 1-chloro-3-nitrobenzene
   (f) 1-chloro-4-nitrobenzene
   (g) *p*-ethylbenzenesulfonic acid
   (h) *m*-chlorobenzenesulfonic acid

10. Why does bromination of *tert*-butylbenzene produce *para*-bromo-*tert*-butylbenzene but very little of the ortho isomer?

11. Alkenes react with sulfuric acid to give addition products. The first step in these reactions is addition of a proton to the alkene. In contrast, aromatic compounds react with sulfuric acid to give substitution products (arylsulfonic acids). The first step in these reactions is addition to the aromatic compound of an electrophilic species, SO_3 or SO_3H^+. Actually, attack by a proton on aromatic compounds is far more rapid than is sulfonation. Why isn't this attack by a proton ordinarily noted? How could you show that it is occurring?

12. When treated with a large excess of methyl chloride and aluminum chloride, benzene is converted in good yield to a crystalline compound of molecular formula $C_{13}H_{21}AlCl_4$. Draw the structure of this compound and explain why it is stable.

13. To convert benzene to a monoalkylbenzene, it is customary to use considerably more benzene than alkyl halide. What is the result of using equimolar amounts of benzene and an alkyl halide?

14. Which one of the isomeric dimethylbenzenes reacts most rapidly with sulfuric acid?

15. It is difficult to convert benzene to ethylbenzene without forming substantial amounts of diethylbenzenes. Benzene, however, can be converted to nitrobenzene without significant formation of dinitrobenzenes. Explain this difference in behavior.

16. (a) Draw structures of the products of addition of HCl to the following alkenes.
       1. $CH_2 = CH - OCH_3$ 2. $CH_2 = CH - NO_2$ 3. $CH_2 = CH - CH_3$
   (b) Which alkene reacts most rapidly? Which reacts most slowly?

17. Draw structures of anthracene and phenanthrene that include every hydrogen.

18. Naphthalene can be described as a hybrid of three resonance structures. Draw these structures and then predict the relative lengths of the various carbon-carbon bonds of naphthalene.

19. Draw the resonance structures of anthracene and phenanthrene. Which of these isomeric hydrocarbons would you expect to be most stable?

20. What gain or loss of valence electrons is necessary for each of the following atoms to attain an inert gas electron configuration.

(a) Na   (b) O   (c) Mg
(d) N   (e) Cl   (f) Li
(g) F   (h) C

21. Indicate the hybridization (sp^3, sp^2, or sp) of each carbon in the following molecules.

(a)

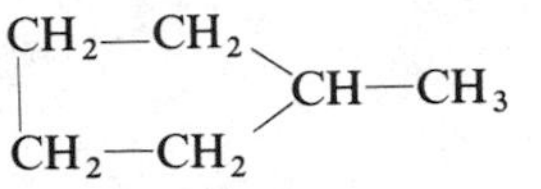

(b)

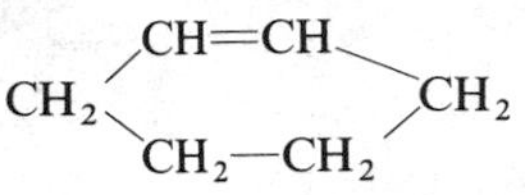

(c)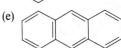

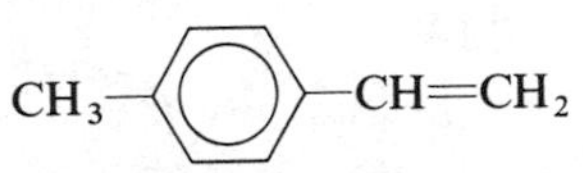

(d) CH_3—$C\equiv C$—CH=CH_2

22. Predict the geometry of CH_2=C=CH_2 (allene or 1,2-propadiene).

23. Which of the following exhibit aromatic properties?

(a)

(b) CH_2=CH—CH=CH_2

(c)

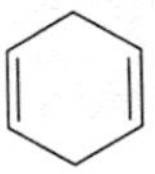

(d)   —CH_3

(e)

(f)

24. Why doesn't the following molecule exhibit aromatic properties? (Constructing a model may be helpful.)

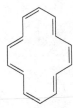

5

Alcohols, Ethers, and Phenols

The **hydroxyl group**, —OH, is a commonly occurring and important functional group, particularly in the compounds of living systems. Compounds (R—OH) in which the hydroxyl group is bonded to an alkyl group are called **alcohols**. Compounds (Ar—OH) in which a hydroxyl group is bonded directly to an aryl group differ significantly from alcohols in chemical behavior. For this reason, they are given a different name—**phenols**—and will be considered separately.

So far we have discussed several families of hydrocarbons—alkanes, alkenes, alkynes, and aromatic hydrocarbons. Although the hydroxyl group in alcohols or phenols is an adornment to a hydrocarbon structure, it can dominate the properties of a molecule. Alcohols and phenols undergo a variety of reactions that involve the hydroxyl group. In addition, the hydroxyl group profoundly affects some physical properties, including solubility. The importance of hydroxyl groups in biological compounds is sometimes due as much to their effect on solubility as to their chemical reactivity.

Alcohols and phenols can be regarded as derivatives of water and some of their properties are related to those of water. Compounds in which both hydrogens of water are replaced by organic groups are called **ethers**. Ethers are also considered in this chapter.

5.1
NOMENCLATURE OF ALCOHOLS AND ETHERS

Simple alcohols are often given radicofunctional names, with the word **alcohol** used to indicate the function.

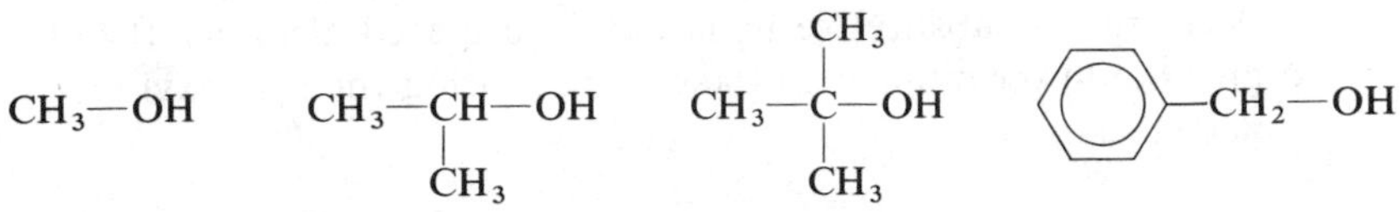

$$CH_3-OH \qquad CH_3-\overset{\displaystyle |}{\underset{\displaystyle CH_3}{CH}}-OH \qquad CH_3-\overset{\displaystyle CH_3}{\underset{\displaystyle CH_3}{\overset{\displaystyle |}{\underset{\displaystyle |}{C}}}}-OH \qquad \langle\!\!\bigcirc\!\!\rangle-CH_2-OH$$

Methyl alcohol   Isopropyl alcohol   *tert*-Butyl alcohol   Benzyl alcohol

Simple as well as more complex alcohols can be given substitutive names. The longest chain of carbons that includes the carbon attached to the hydroxyl group is selected, and the name of the corresponding alkane with the ending -*e* changed to -*ol* is used as the parent name.

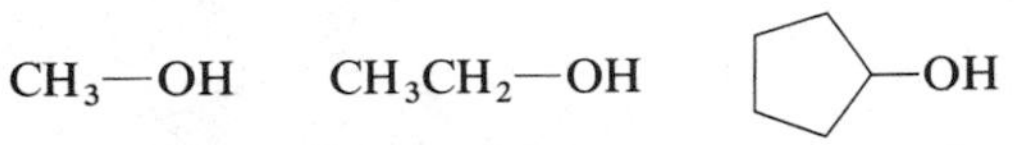

$$CH_3-OH \qquad CH_3CH_2-OH \qquad \bigcirc\!\!-OH$$

Methanol   Ethanol   Cyclopentanol

When isomers are possible, the position of the hydroxyl group must be indicated by the number of the carbon to which it is attached. Carbons of the chain are numbered, starting from the end nearer the hydroxyl group. Of course, any other substituents must be named and assigned numbers.

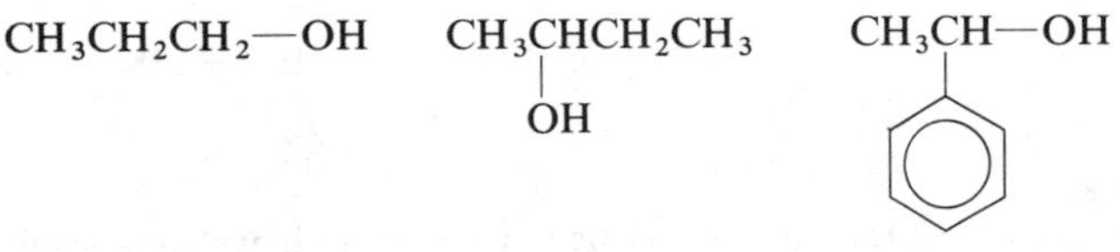

$$CH_3CH_2CH_2-OH \qquad CH_3\overset{\displaystyle |}{\underset{\displaystyle OH}{CH}}CH_2CH_3 \qquad CH_3\overset{\displaystyle |}{CH}-OH$$

1-Propanol   2-Butanol   1-Phenylethanol

Consider the following alcohol.

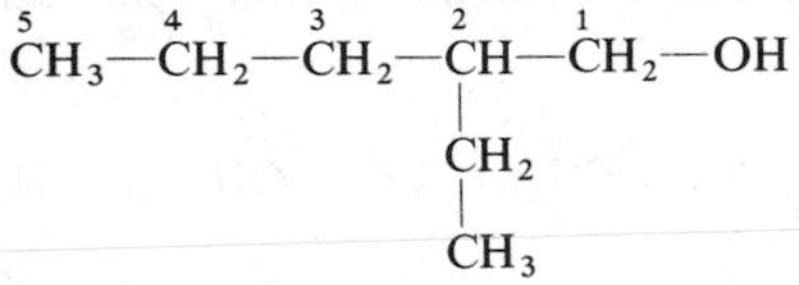

$$\overset{5}{C}H_3-\overset{4}{C}H_2-\overset{3}{C}H_2-\overset{2}{C}H-\overset{1}{C}H_2-OH$$
$$\underset{\displaystyle CH_3}{\overset{\displaystyle |}{\underset{\displaystyle |}{CH_2}}}$$

The longest chain containing the carbon attached to the hydroxyl group has five carbons; hence this alcohol is named as a pentanol. Note that the molecule has a six-carbon chain, but that chain does not include the carbon linked to the hydroxyl group. Numbering the chain from the right-hand end gives the lowest possible number to the carbon bearing the hydroxyl group. The full name is 2-ethyl-1-pentanol.

Compounds containing two hydroxyl groups are named as **diols**.

$$HO-CH_2CH_2CH_2CH_2-OH$$

1,4-Butanediol

In deriving the substitutive names of unsaturated alcohols, the chain is numbered to give the lower number to the carbon bearing the hydroxyl group and its ending is placed last.

$$CH_2{=}CHCH_2{-}OH$$

2-Propen-1-ol (not 1-propen-3-ol)
Allyl alcohol

In naming cyclic alcohols, the hydroxyl group is ordinarily assumed to be at carbon number 1.

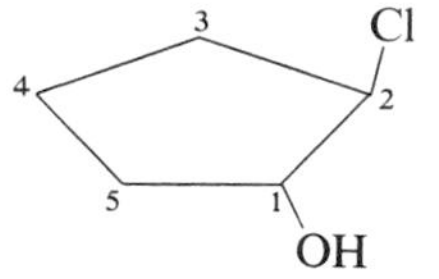

trans-2-Chlorocyclopentanol

The prefix **hydroxy** is used when an —OH group must be named as a substituent.

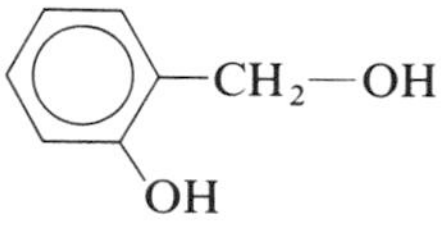

o-Hydroxybenzyl alcohol 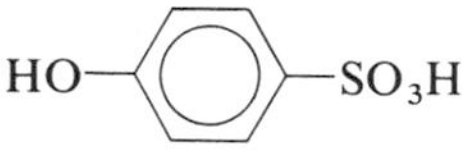   *p*-Hydroxybenzenesulfonic acid

The chemical behavior of alcohols sometimes depends on whether the hydroxyl group is attached to a primary, secondary, or tertiary carbon. As illustrated below for the four butyl alcohols, an alcohol can be designated primary (1°), secondary (2°), or tertiary (3°).

$$CH_3CH_2CH_2CH_2{-}OH$$

A primary alcohol
1-Butanol
Butyl alcohol

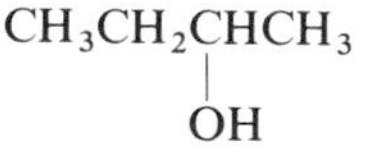

$$CH_3CH_2CHCH_3$$
$$|$$
$$OH$$

A secondary alcohol
2-Butanol
sec-Butyl alcohol

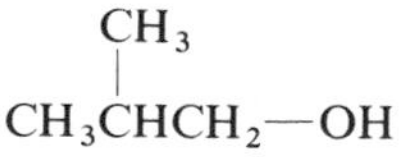

$$CH_3$$
$$|$$
$$CH_3CHCH_2{-}OH$$

A primary alcohol
2-Methyl-1-propanol
Isobutyl alcohol

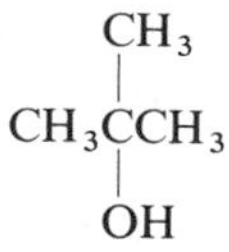

$$CH_3$$
$$|$$
$$CH_3CCH_3$$
$$|$$
$$OH$$

A tertiary alcohol
2-Methyl-2-propanol
tert-Butyl alcohol

Radicofunctional names are usually used for simple ethers. In forming these names, both organic groups must be named (in alphabetical order), followed by the word **ether**.

$$CH_3CH_2-O-CH_3 \qquad CH_3CH_2-O-CH_2CH_3$$

Ethyl methyl ether   Diethyl ether

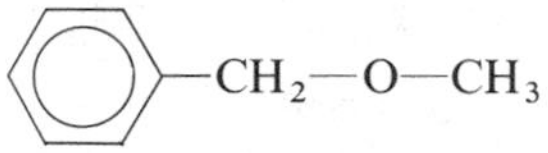

Benzyl methyl ether 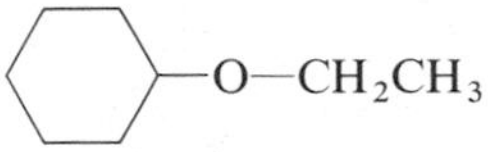   Cyclohexyl ethyl ether

To form its substitutive name, an ether is considered a derivative of the hydrocarbon corresponding to the larger organic group. The smaller alk*yl* group and the oxygen attached to it are indicated by the name alk*oxy*.

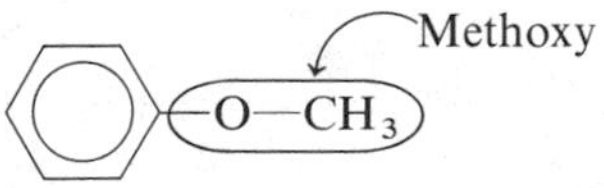

Methoxybenzene
Methyl phenyl ether
Anisole

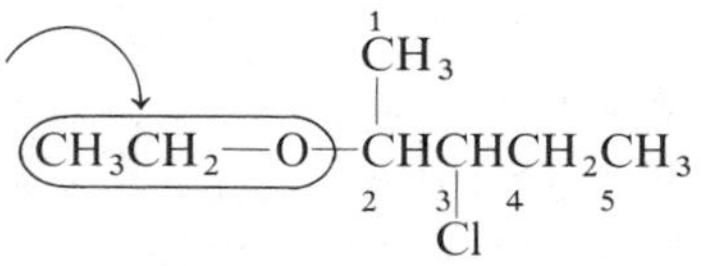

3-Chloro-2-ethoxypentane

Anisole is a trivial name accepted by the IUPAC.

---

problem 5.1   Name the following compounds.

(a) 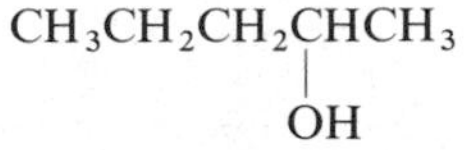 $CH_3CH_2CH_2CHCH_3$
        $\underset{OH}{|}$

(b)

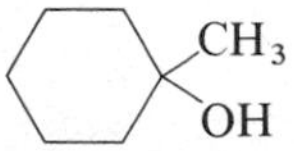

(c) 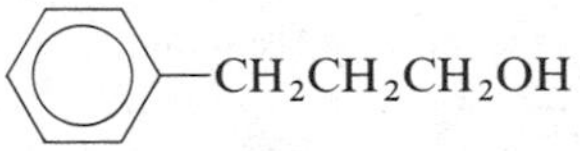
    $-CH_2CH_2CH_2OH$

(d) 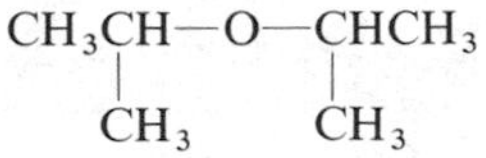 $CH_3CH-O-CHCH_3$
        $\underset{CH_3}{|}$ $\underset{CH_3}{|}$

problem 5.2   Draw structures for the following compounds and indicate for each alcohol whether it is primary, secondary, or tertiary.
(a) isopropyl alcohol   (b) 2-bromo-1-propanol
(c) 3-methyl-3-pentanol   (d) allyl methyl ether

5.2
PHYSICAL PROPERTIES OF ALCOHOLS AND ETHERS

Some physical properties of alcohols are very different from those of hydrocarbons. Examine the boiling points in Table 5.1 of a group of organic compounds that have similar weights. Ethanol boils much higher than the hydrocarbons. Unusually

Table 5.1 Boiling points and molecular weights of some simple organic compounds

Structural formula	Name	Molecular weight	Boiling point (°C)
$CH_3CH_2CH_3$	propane	44	−42
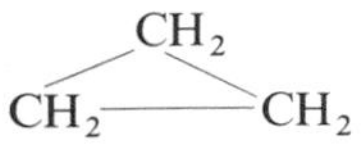 CH_2 (cyclopropane ring) CH_2——CH_2	cyclopropane	42	−33
$CH_3CH{=}CH_2$	propene	42	−47
$CH_3C{\equiv}CH$	propyne	40	−23
CH_3CH_2—OH	ethanol	46	78
CH_3—O—CH_3	dimethyl ether	46	−23
CH_3CH_2—NH_2	ethylamine	45	17
CH_3CH_2—F	ethyl fluoride	48	−38

high boiling points are characteristic of alcohols. The high boiling points are not due just to the presence of an oxygen, for dimethyl ether, an isomer of ethanol, boils in the same range as the hydrocarbons. Ethylamine also has a boiling point higher than the corresponding hydrocarbons, but ethyl fluoride does not.

Molecules must break away from the liquid during boiling. Attractive forces between ethanol molecules to be overcome when liquid ethanol boils must be considerably greater than the forces between dimethyl ether or propane molecules.

The degree of solubility of compounds in water is particularly important in biological systems. Compounds that must move within a cell or between cells must be soluble in the aqueous biological environment. Conversely, compounds that

Table 5.2 Solubility of alcohols in water

Structural formula	Name	Solubility (g/100 g of H_2O at 20°C)
CH_3—OH	methanol	∞
CH_3CH_2—OH	ethanol	∞
$CH_3CH_2CH_2$—OH	1-propanol	∞
CH_3CHCH_3 | OH	2-propanol	∞
$CH_3CH_2CH_2CH_2$—OH	1-butanol	8.0
$CH_3CHCH_2CH_3$ | OH	2-butanol	12.5
$CH_3(CH_2)_3CH_2$—OH	1-pentanol	2.5
$CH_3(CH_2)_4CH_2$—OH	1-hexanol	0.7
$CH_3(CH_2)_6CH_2$—OH	1-octanol	0.05
$CH_3(CH_2)_8CH_2$—OH	1-decanol	0.004

belong to such components as cell walls, membranes, tendons, and skin must have low water solubilities.

In contrast to hydrocarbons, alcohols tend to be water soluble. As shown in Table 5.2, alcohols with only a few carbons are infinitely soluble in water—any amount of one of these alcohols will dissolve in a fixed amount of water. As the number of carbons increases, solubility of alcohols decreases and ultimately approaches the low values characteristic of alkanes. Compounds with more carbons, however, are often quite water soluble if they have several hydroxyl groups. $HO(CH_2)_5OH$ (1,5-pentanediol), for instance, is infinitely soluble in water.

Solubilities of ethers and amines (compounds of structure $R—NH_2$) are comparable to those of alcohols. Diethyl ether, an isomer of 1-butanol, has a solubility of 7.9 g in 100 g of water. Note that ethers resemble alkanes in boiling point but alcohols in solubility. Alkyl halides, however, resemble alkanes both in solubility and in boiling point. The explanation for these observations is related to the attractive forces between molecules, the subject of the next section.

5.3
ATTRACTIVE FORCES BETWEEN MOLECULES; THE HYDROGEN BOND

Attractive forces between molecules are important in determining boiling points and solubilities as well as some other properties. We will consider two important kinds of forces. The first, called a **dispersion force** (or a van der Waals force), arises from the constant motion of electrons. This motion results in momentary imbalances between positive and negative charges in a molecule. Such imbalances slightly affect the motions of electrons in a nearby molecule in a way that produces a small net electrostatic attraction between the molecules.

Because dispersion forces operate only at short distances, the attraction between two molecules is related to the area of close contact between them. Boiling points of alkanes increase with increasing number of carbons, for large alkanes have larger areas of close contact than small alkanes and hence are attracted together by larger dispersion forces. Contact between pentane molecules (boiling point 36°) is greater than between molecules of its more branched and compact isomer, neo-pentane (boiling point 9°). Because of reduced contact between molecules, boiling points decrease somewhat with increasing chain branching.

The second kind of attractive force is called a **hydrogen bond**. You probably encountered the hydrogen bond in an earlier chemistry course when learning about the properties of water. A hydrogen bond requires two features:

1. a hydrogen attached to one of the most electronegative atoms, usually oxygen or nitrogen.*

2. an atom with available, unshared electron pairs, generally also an oxygen or nitrogen.

---

* Fluorine, chlorine, and bromine also are very electronegative but are attached to hydrogen only in the inorganic compounds HF, HCl, and HBr.

The hydrogen bond, often represented by a dashed line, can be regarded as an electrostatic attraction between the relatively positive hydrogen and the available electron pair.

$$\overset{\delta^-}{O}-\overset{\delta^+}{H}-----\overset{\delta^-}{\ddot{O}}- \qquad \underset{/}{\overset{\backslash}{N}}-H-----\ddot{O}-$$

$$-O-H-----\ddot{N}- \qquad \underset{/}{\overset{\backslash}{N}}-H-----\ddot{N}-$$

Hydrogen bonds ordinarily are 5 to 10% as strong as typical covalent bonds.

Alcohols contain both features needed for a hydrogen bond. The hydrogen of the hydroxyl group of one alcohol molecule can form a hydrogen bond to the oxygen of another.

$$\ddot{O}-R$$
$$|$$
$$H$$
$$\vdots$$
$$:\ddot{O}-H-----:\ddot{O}-H-----:\ddot{O}-H$$
$$|\qquad\qquad|\qquad\qquad|$$
$$R\qquad\qquad R\qquad\qquad R$$

Alcohol molecules are held together by hydrogen bonds in addition to dispersion forces. The extra energy needed to break the hydrogen bonds is responsible for the high boiling points of alcohols. Amines ($R-NH_2$) also have both features needed for formation of a hydrogen bond. Of course, the simple inorganic molecules water and ammonia also form hydrogen bonds and have unusually high boiling points. An ether molecule, however, has no hydrogen attached to oxygen and cannot hydrogen bond to another ether molecule. Therefore ethers have boiling points comparable to those of hydrocarbons.*

Molecules other than alcohols and amines—dimethyl ether is an example—also have relatively positive atoms and electron-donor atoms that can lead to additional intermolecular attractions.

$$CH_3-\overset{\overset{\displaystyle H}{|}}{\underset{\underset{\displaystyle H}{|}}{\overset{\delta^-}{O}-\overset{\delta^+}{C}}}-----\overset{\overset{\displaystyle CH_3}{}}{\underset{\underset{\displaystyle CH_3}{}}{\overset{\delta^-}{\ddot{O}}}} \qquad \text{A very weak interaction}$$

* Changes in molecular size, molecular shape, and hydrogen bonding affect melting points less regularly than boiling points. Melting involves the destruction of the regular, repeating structure of a solid. Relatively small changes in molecular structure sometimes greatly alter the efficiency with which molecules pack together in a crystal and therefore significantly change the attractive forces that maintain the crystal structure.

Attractions are much smaller when a hydrogen is not involved, however. Because of its uniquely small radius, hydrogen can approach more closely to an electron-donor atom than can larger atoms, such as carbon.

Hydrogen bonds affect properties other than boiling points and solubilities. In contrast to butane, $HOCH_2CH_2OH$ (1,2-ethanediol or ethylene glycol), the common antifreeze in automobile cooling systems, is more stable in a gauche than anti conformation.

H   H
CH_3 H   less stable than
CH_3 H   Butane

H   H
CH_3 CH_3 (Sec. 1.8)
H   H

H   H
O   H   more stable than
H   O   H
H   1,2-Ethanediol

H   H
OH   OH
H   H

Formation of an intramolecular (internal) hydrogen bond, possible in the gauche but not in the anti conformation, is responsible.

Hydrogen bonding is critically important in determining essential properties of the important polymeric molecules of living systems: carbohydrates (Chapter 10), proteins (Chapter 14), and nucleic acids (Chapter 15).

5.4
SOLUBILITY

Intermolecular forces also determine the extent to which a compound dissolves in a particular solvent. Molecules are in constant motion. Other things being equal, molecules of a compound placed in contact with a solvent tend to dissolve—to move into and become dispersed in the solvent. For a molecule to dissolve and move into a sea of solvent molecules, however, the attractive forces that held it to molecules of its own kind must be broken. To provide space for the dissolved molecule, some solvent molecules must be separated from each other and the attractions between them lost or reduced. Of course, new attractive forces form between the dissolved molecule and solvent molecules. The extent to which a compound dissolves in a solvent is related to the relative magnitudes of these various intermolecular forces.

Consider an alkane dissolving in water. Dispersion attractions are broken between the alkane molecules and are formed between water molecules and dissolved alkane molecules. These attractions, however, are much weaker than the hydrogen

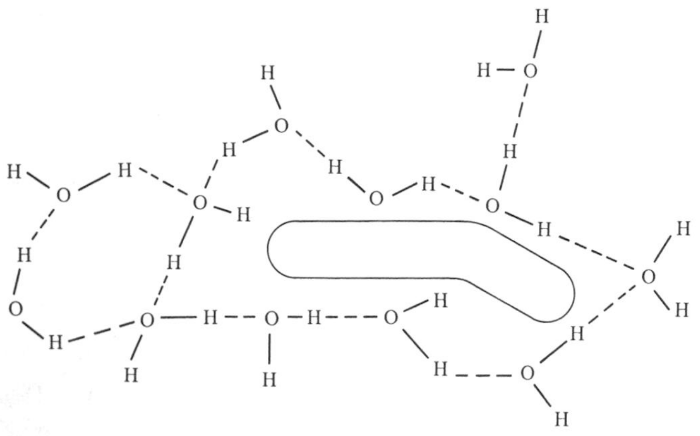

Figure 5.1 When an alkane molecule dissolves in water, some water molecules adjacent to the alkane participate in fewer hydrogen bonds.

bonds between water molecules. To accommodate an alkane molecule, some hydrogen bonds between water molecules must be broken. As shown in Figure 5.1, some hydrogens and oxygens of water molecules next to the alkane cannot participate fully in hydrogen bonding. As a result, alkanes are very insoluble in water.

Consider methanol dissolving in water. As shown in Figure 5.2, the hydrogen bonds that must be broken between water molecules are to a considerable extent replaced by hydrogen bonds between methanol and water molecules. As a result, methanol and other small alcohols are very water soluble. Of course, as the organic group becomes larger, the hydroxyl group constitutes a smaller portion of the molecule and solubility decreases. An ether oxygen bears no hydrogen to contribute to a hydrogen bond, but it can form hydrogen bonds to the hydrogens of water

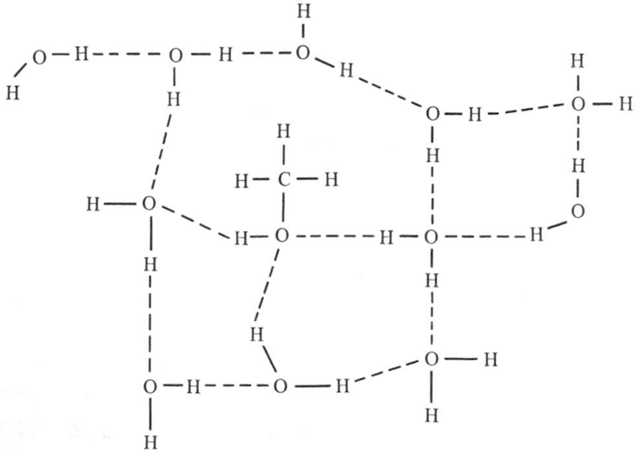

Figure 5.2 Methanol molecules dissolved in water are extensively hydrogen bonded to water molecules.

molecules. The resulting hydrogen bonding is sufficient to give ethers solubilities comparable to those of isomeric alcohols.

This is a convenient place to review what was learned in an earlier course about the solubilities of salts in water. Even though the attractive forces between positive and negative ions must be overcome in order for a salt to dissolve, many salts are water soluble. Water is an excellent solvent for salts because attractive forces are formed between the ions and partially positive or negative portions of water molecules. As shown for sodium chloride, positive ions are surrounded by oxygens of water and negative ions by hydrogens.

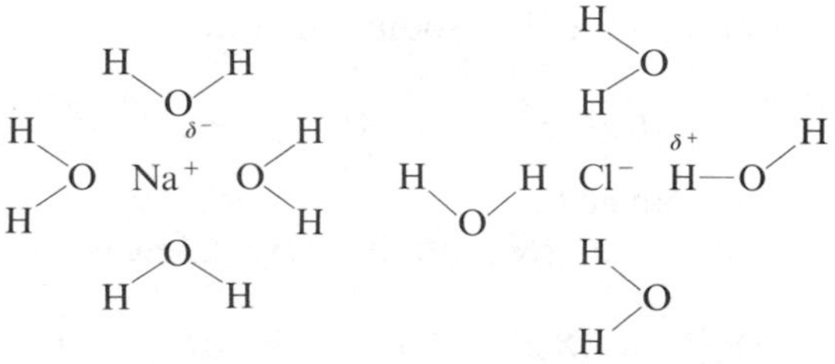

In addition, water has a particularly high value of a solvent property, called the dielectric constant, that reduces the attractive forces between oppositely charged ions separated by solvent molecules.

Biological molecules that must be water soluble contain hydrogen bonding groups or charged groups or both.

5.5
REACTIONS OF ALCOHOLS AND ETHERS

Alcohols undergo a variety of reactions. One important reaction, acid-catalyzed dehydration, was discussed in Section 3.14.

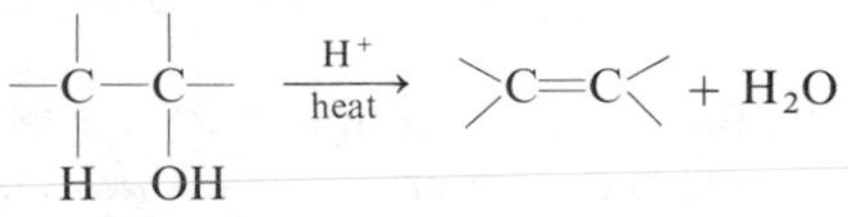

Several other significant reactions are introduced in the following sections. Because ethers have no hydrogen attached to oxygen, we will see that they do not undergo reactions comparable to some characteristic alcohol reactions. In fact, ethers undergo relatively few reactions and in that respect resemble alkanes rather than alcohols.

5.6
REACTIONS AS ACIDS AND BASES

An alcohol is both a weak acid and a weak base, just as is water. We know that small concentrations of hydronium and hydroxide ions (10^{-7} M) exist in pure water as a result of protonation of one water molecule by another.

$$2\,H\!-\!O\!-\!H \;\underset{\longleftarrow}{\longrightarrow}\; H\!-\!\overset{+}{\underset{\underset{H}{|}}{O}}\!-\!H \;+\; {}^{-}O\!-\!H$$

Strong acids, such as hydrochloric, nitric, and sulfuric, react almost completely with water to produce hydronium ions.

$$H\!-\!O\!-\!H + HX \;\longrightarrow\; H\!-\!\overset{+}{\underset{\underset{H}{|}}{O}}\!-\!H \;\; X^{-} \qquad \text{(HX is a strong acid.)}$$

In such reactions water acts as a base. Water is converted to hydroxide salts by reaction with strong bases like sodium hydride.

$$H\!-\!O\!-\!H + Na^{+}H^{-} \;\longrightarrow\; H\!-\!O^{-}\,Na^{+} + H_{2}$$

Here water acts as an acid.

An alcohol is similarly protonated by strong acids

$$R\!-\!O\!-\!H + HX \;\longrightarrow\; R\!-\!\overset{+}{\underset{\underset{H}{|}}{O}}\!-\!H \;\; X^{-} \qquad \text{(HX is a strong acid.)}$$

and loses a proton to form a salt on treatment with strong bases.

$$CH_{3}\!-\!O\!-\!H + Na^{+}\,H^{-} \;\longrightarrow\; CH_{3}\!-\!O^{-}\,Na^{+} + H_{2}$$
$$\text{Sodium methoxide}$$

Salts of alcohols are also formed by treating alcohols with active metals, such as sodium or potassium.

$$(CH_{3})_{3}C\!-\!O\!-\!H + 2\,K \;\longrightarrow\; 2\,(CH_{3})_{3}C\!-\!O^{-}\,K^{+} + H_{2}$$
$$\text{Potassium } \textit{tert}\text{-butoxide}$$

The negative ions formed from simple alcohols can be named by changing the -*yl* ending of the name of the alkyl group to -*oxide*. As a class, they are called **alkoxide ions**. These ions are even somewhat stronger bases than hydroxide ions, and alkoxide salts are often used when strongly basic reagents are needed in organic reactions. Because alkoxide ions are more basic than hydroxide ion, they cannot be generated in significant concentrations by treating alcohols with hydroxide salts.

$$R\!-\!O\!-\!H + Na^{+}\,{}^{-}O\!-\!H \;\underset{\longleftarrow}{\longrightarrow}\; R\!-\!O^{-}\,Na^{+} + H\!-\!O\!-\!H$$

Ethers are also protonated by strong acids.

$$R\!-\!O\!-\!R + HX \;\longrightarrow\; R\!-\!\overset{+}{\underset{\underset{R}{|}}{O}}\!-\!R \;\; X^{-}$$

Lacking a hydrogen attached to oxygen, however, ethers do not react with bases.

$$R—O—R + Na^+ H^- \not\longrightarrow$$

5.7
CONVERSION OF ALCOHOLS TO ESTERS

An **ester** is a compound formed directly or indirectly by the linkage together of an alcohol and an acid with loss of water. As illustrated for glycerol, many alcohols react with cold, concentrated nitric acid to form nitrate esters.

$$
\begin{array}{l}
CH_2—OH \\
| \\
CH—OH \\
| \\
CH_2—OH
\end{array}
\quad + 3\,HONO_2 \longrightarrow
\begin{array}{l}
CH_2—ONO_2 \\
| \\
CH—ONO_2 \\
| \\
CH_2—ONO_2
\end{array}
\quad + 3\,H_2O
$$

1,2,3-Propanetriol   Glyceryl trinitrate
Glycerol   Nitroglycerin

Like many other organic compounds that are high explosives, nitroglycerin contains approximately the amount of oxygen needed to convert each carbon to carbon dioxide and each hydrogen to water.

$$4\,C_3H_5N_3O_9 \longrightarrow 12\,CO_2 + 10\,H_2O + 6\,N_2 + O_2$$

Nitroglycerin

Nitroglycerin is very sensitive to shock. It is, however, much more stable when absorbed on a material like wood pulp to produce the substance known as dynamite. Income from the fortune that Nobel amassed from his invention of dynamite established and maintains the Nobel prizes. Nitroglycerin also dilates blood vessels and is used in the treatment of hypertension.

Depending on the amounts of reactants used, monoesters or diesters can be produced from the reactions of some alcohols with cold, concentrated sulfuric acid.

$$
CH_3—OH + HO{-}\!\!\overset{\displaystyle O}{\underset{\displaystyle O}{\overset{\|}{\underset{\|}{S}}}}\!\!{-}OH \longrightarrow CH_3—O{-}\!\!\overset{\displaystyle O}{\underset{\displaystyle O}{\overset{\|}{\underset{\|}{S}}}}\!\!{-}OH + H_2O
$$

Methyl hydrogen sulfate

$$
2\,CH_3—OH + HO{-}\!\!\overset{\displaystyle O}{\underset{\displaystyle O}{\overset{\|}{\underset{\|}{S}}}}\!\!{-}OH \longrightarrow CH_3—O{-}\!\!\overset{\displaystyle O}{\underset{\displaystyle O}{\overset{\|}{\underset{\|}{S}}}}\!\!{-}O—CH_3 + 2\,H_2O
$$

Dimethyl sulfate

Alkyl hydrogen sulfates formed in this manner from alcohols with long alkyl chains are used to produce one class of detergents (Sec. 13.7).

Esters of phosphoric, diphosphoric acid, and triphosphoric acids are important in biological systems.

$$HO-\overset{\overset{\textstyle O}{\|}}{\underset{\underset{\textstyle OH}{|}}{P}}-OH \qquad HO-\overset{\overset{\textstyle O}{\|}}{\underset{\underset{\textstyle OH}{|}}{P}}-O-\overset{\overset{\textstyle O}{\|}}{\underset{\underset{\textstyle OH}{|}}{P}}-OH \qquad HO-\overset{\overset{\textstyle O}{\|}}{\underset{\underset{\textstyle OH}{|}}{P}}-O-\overset{\overset{\textstyle O}{\|}}{\underset{\underset{\textstyle OH}{|}}{P}}-O-\overset{\overset{\textstyle O}{\|}}{\underset{\underset{\textstyle OH}{|}}{P}}-OH$$

Phosphoric acid   Diphosphoric acid   Triphosphoric acid
                        (pyrophosphoric acid)   (tripolyphosphoric acid)

For example, 3-methyl-3-butenyl diphosphate is the fundamental unit from which terpenes are constructed (Secs. 3.21 and 7.9).

$$CH_2{=}CCH_2CH_2-O-\overset{\overset{\textstyle O}{\|}}{\underset{\underset{\textstyle OH}{|}}{P}}-O-\overset{\overset{\textstyle O}{\|}}{\underset{\underset{\textstyle OH}{|}}{P}}-OH$$
$$\underset{\textstyle CH_3}{|}$$

3-Methyl-3-butenyl diphosphate

Note that sulfur and phosphorus, because their valence electrons are in the third shell (Sec. 4.14), can be surrounded by more than eight electrons.

Alcohols also form esters with carboxylic acids, a family of organic acids. This important group of esters is discussed in Chapter 12.

5.8
OXIDATION OF ALCOHOLS

As do most organic compounds containing carbon-hydrogen bonds, alcohols burn in air to form carbon dioxide and water. Under milder conditions, however, alcohols undergo selective oxidations that specifically involve the hydroxyl function.

What do we mean by oxidation? We consider an organic molecule to be oxidized if it gains carbon-oxygen bonds or loses carbon-hydrogen bonds. All the following high-temperature reactions are oxidations of the carbon-containing species; the reverse of each reaction is a reduction.

$$CH_4 + 2\,O_2 \longrightarrow CO_2 + 2\,H_2O$$
$$C\ + O_2 \longrightarrow CO_2$$
$$2\,CH_3OH + 3\,O_2 \longrightarrow 2\,CO_2 + 4\,H_2O$$
$$CH_3{-}CH_3 \longrightarrow CH_2{=}CH_2 + H_2$$

Consider the degrees of oxidation possible for a single carbon atom.

		Compound	Number of carbon-oxygen bonds
	H | H—C—H | H	methane	0
	OH | H—C—H | H	methanol	1
Oxidation Reduction	O || H—C—H	formaldehyde	2
	O || H—C—OH	formic acid	3
	O=C=O	carbon dioxide	4

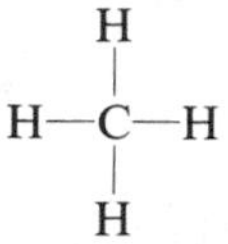

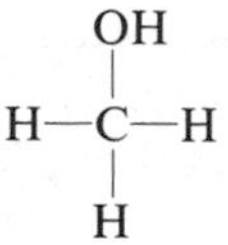

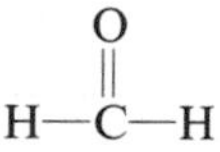

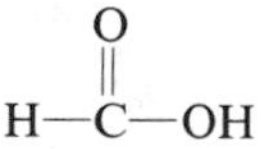

Because all its bonds are to hydrogen, the carbon of methane is completely reduced. Methanol has one carbon-oxygen bond. Formaldehyde, the simplest member of the family of compounds known as aldehydes, has two bonds between carbon and oxygen. **Aldehydes** (Chapter 9) contain a $-\overset{\text{O}}{\underset{\|}{\text{C}}}$H group bonded to an organic group $(\text{R}-\overset{\text{O}}{\underset{\|}{\text{C}}}\text{H})$ or (only in formaldehyde) to a hydrogen. A compound with two hydroxyl groups attached to a carbon has the same degree of oxidation. Such compounds, however, tend to lose water and form compounds with carbon-oxygen double bonds.

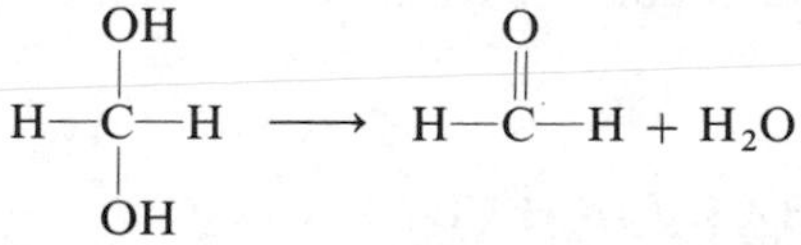

$$\text{H}-\underset{\overset{\|}{\text{OH}}}{\overset{\overset{\text{OH}}{\|}}{\text{C}}}-\text{H} \longrightarrow \text{H}-\overset{\overset{\text{O}}{\|}}{\text{C}}-\text{H} + \text{H}_2\text{O}$$

Formic acid, the simplest member of the important family of organic compounds known as carboxylic acids, has three bonds between carbon and oxygen. **Carboxylic acids** (Chapter 11) contain a $-\overset{\overset{\text{O}}{\|}}{\text{C}}-\text{OH}$ group bonded to an organic group $(\text{R}-\overset{\overset{\text{O}}{\|}}{\text{C}}-\text{OH})$ or (only in formic acid) to a hydrogen. Because all four of its bonds are to oxygen, the carbon in carbon dioxide is completely oxidized.

The course of oxidation of an alcohol depends on whether it is primary, secondary, or tertiary. A primary alcohol can be oxidized to an aldehyde.

$$CH_3CH_2CH_2CH_2OH + CrO_3\text{-pyridine} \longrightarrow CH_3CH_2CH_2\overset{\overset{\displaystyle O}{\|}}{C}H$$

1-Butanol

Butanal
Butyraldehyde

A complex of chromium trioxide (CrO_3) with the organic molecule pyridine (Sec. 8.9) is one reagent used for this purpose. Aldehydes are quite readily oxidized. Consequently, oxidations of primary alcohols with most reagents do not stop at the aldehyde stage but instead produce carboxylic acids.

$$CH_3CH_2OH + K_2Cr_2O_7 \longrightarrow CH_3-\overset{\overset{\displaystyle O}{\|}}{C}-OH$$

Ethanol

Acetic acid

$$\text{\Large$\pentagon$}-CH_2OH + KMnO_4 \longrightarrow \text{\Large$\pentagon$}-\overset{\overset{\displaystyle O}{\|}}{C}-OH$$

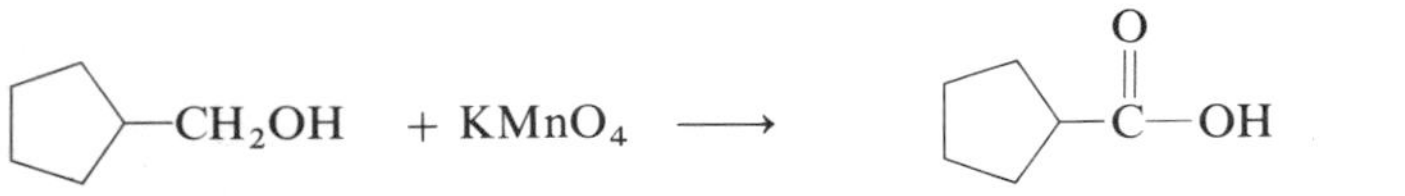

Cyclopentylmethanol

Cyclopentanecarboxylic acid

Aqueous solutions of potassium dichromate ($K_2Cr_2O_7$) plus an acid or of potassium permanganate ($KMnO_4$) plus a base are common oxidants. Oxidation of a carboxylic acid requires cleaving a carbon-carbon single bond and does not occur readily with such reagents.*

Dichromate salts are orange but, in oxidizing an alcohol, are reduced to green Cr^{3+} salts. Some instruments that test the ethanol content of suspected drunken drivers use this color change. A given volume of the suspect's breath is passed through a tube containing a dichromate salt and the development of a green color is monitored.

Secondary alcohols are oxidized to ketones by the same reagents.

$$\overset{\overset{\displaystyle OH}{|}}{CH_3CH_2CHCH_2CH_3} + K_2Cr_2O_7 \longrightarrow CH_3CH_2\overset{\overset{\displaystyle O}{\|}}{C}CH_2CH_3$$

3-Pentanol

3-Pentanone
Diethyl ketone

$$\text{\Large$\hexagon$}-OH + KMnO_4 \longrightarrow \text{\Large$\hexagon$}=O$$

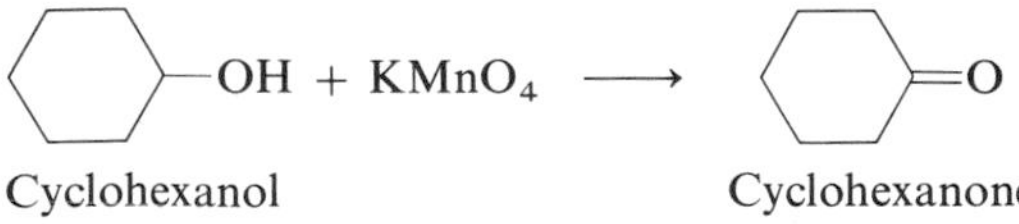

Cyclohexanol

Cyclohexanone

---

* Formic acid, unique among carboxylic acids in having a hydrogen attached to the carbon that is bonded to oxygens, is easily oxidized to carbon dioxide.

Ketones (Chapter 9) are compounds that have two organic groups attached to a

$$\overset{\displaystyle O}{\underset{\displaystyle \|}{}}$$

—C— (carbonyl) group. Oxidation of a ketone requires cleaving a carbon-carbon single bond and does not occur readily with such reagents.

Tertiary alcohols are not easily oxidized.

$$R\overset{\displaystyle R}{\underset{\displaystyle R}{-\overset{|}{\underset{|}{C}}-}}OH + K_2Cr_2O_7 \text{ or } KMnO_4 \;\; \xrightarrow{\;\;/\!\!/\;\;}$$

Note that their oxidation would require cleaving a carbon-carbon bond. Ethers are also inert to many oxidizing agents.

$$R—O—R + K_2Cr_2O_7 \text{ or } KMnO_4 \;\; \xrightarrow{\;\;/\!\!/\;\;}$$

5.9
CONVERSION OF ALCOHOLS AND ETHERS TO ALKYL HALIDES

Alcohols react with hydrochloric, hydrobromic, or hydriodic acids to form alkyl halides.

$$CH_3CH_2CH_2CH_2OH + HBr \longrightarrow CH_3CH_2CH_2CH_2Br + H_2O$$

$$\overset{\displaystyle OH}{\underset{\displaystyle |}{CH_3CHCH_3}} + HI \longrightarrow \overset{\displaystyle I}{\underset{\displaystyle |}{CH_3CHCH_3}} + H_2O$$

$$\underset{\displaystyle \underset{|}{CH_3}}{\overset{\displaystyle \overset{OH}{|}}{CH_3CH_2CCH_3}} + HCl \longrightarrow \underset{\displaystyle \underset{|}{CH_3}}{\overset{\displaystyle \overset{Cl}{|}}{CH_3CH_2CCH_3}} + H_2O$$

The necessary conditions depend on the nature of the alcohol. Tertiary alcohols react most readily, often if simply shaken with an aqueous solution of a hydrogen halide. Primary alcohols are the least reactive; heating and the presence of sulfuric acid or some other catalyst generally are necessary for their reactions with hydrogen halides to proceed at useful rates.

Ethers also react with hydrogen halides to form alkyl halides.

$$CH_3CH_2—O—CH_2CH_3 + HX \longrightarrow CH_3CH_2X + CH_3CH_2—OH$$
$$X = Cl, Br, \text{ or } I \qquad\qquad\qquad \Big\downarrow {\scriptstyle HX}$$
$$CH_3CH_2X$$

The other product of ether cleavage is an alcohol. In the presence of an excess of hydrogen halide this alcohol also is converted to an alkyl halide.

Aryl-oxygen bonds are not cleaved by hydrogen halides. Therefore reaction of an alkyl aryl ether with a hydrogen halide (even in excess) gives an alkyl halide and a phenol (but not an aryl halide).

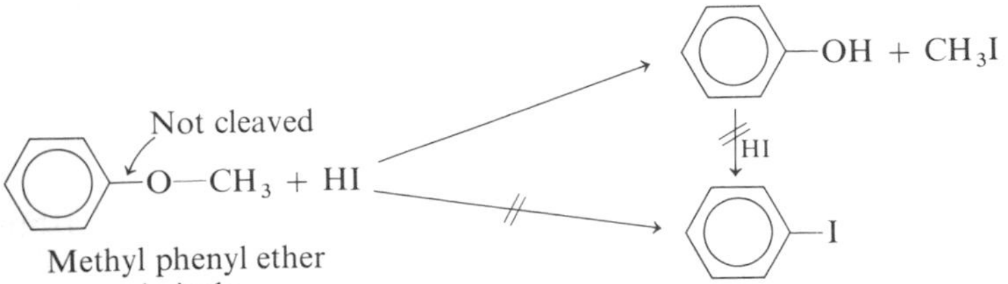

Alcohols, but not ethers, are also converted to alkyl halides by some other inorganic, halogen-containing compounds. Phosphorous trihalides (PCl_3, PBr_3, and PI_3) and thionyl chloride ($SOCl_2$) are frequently used.

$$3\ CH_3CHCH_2CH_2OH + PBr_3 \longrightarrow 3\ CH_3CHCH_2CH_2Br + H_3PO_3$$
$$\underset{CH_3}{|} \qquad\qquad\qquad\qquad \underset{CH_3}{|}$$

3-Methyl-1-butanol   1-Bromo-3-methylbutane

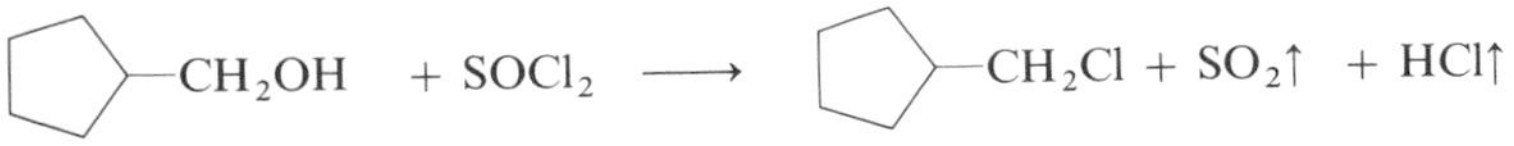

Cyclopentylmethanol   Cyclopentylmethyl chloride

Thionyl chloride is a particularly convenient reagent; evolution of gaseous SO_2 and HCl drives the reaction to completion and facilitates purification of the alkyl chloride.

---

problem 5.3 Draw structures of the organic products of the following reactions.

(a) $(CH_3)_2CHOH \xrightarrow{\text{NaH}}$

(b)
$$\overset{\displaystyle OH}{\underset{\displaystyle |}{(CH_3)_2CHCHCH_3}} \xrightarrow{\text{KMnO}_4}$$

(c) 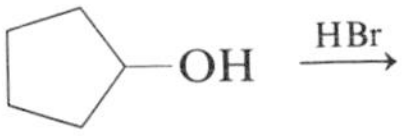 $\xrightarrow{\text{HBr}}$

(d) $(CH_3)_2CHCH_2OH \xrightarrow{\text{K}_2\text{Cr}_2\text{O}_7}$

5.10
MECHANISMS OF REACTIONS OF ALCOHOLS
WITH HYDROGEN HALIDES

Substitution reactions in which hydroxyl is replaced by a halogen occur by two different mechanisms. The initial step of either mechanism is protonation of the alcohol by the hydrogen halide (or by another acid added as a catalyst).

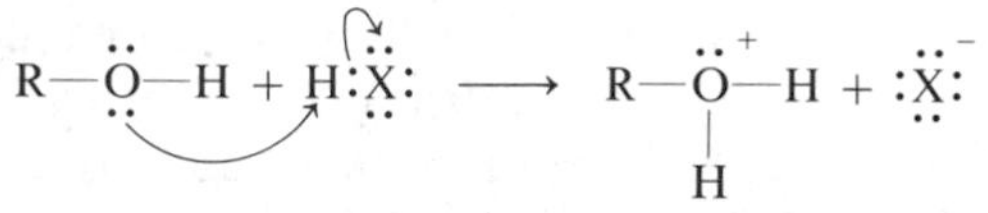

$$R-\ddot{O}-H + H\!:\!\ddot{X}\!: \longrightarrow R-\overset{+}{\ddot{O}}-H + :\ddot{X}\!:^{-}$$
$$\qquad\qquad\qquad\qquad\quad |$$
$$\qquad\qquad\qquad\qquad\quad H$$

One mechanism involves loss of water from the protonated alcohol to form a carbocation, which in a subsequent step reacts with halide ion.

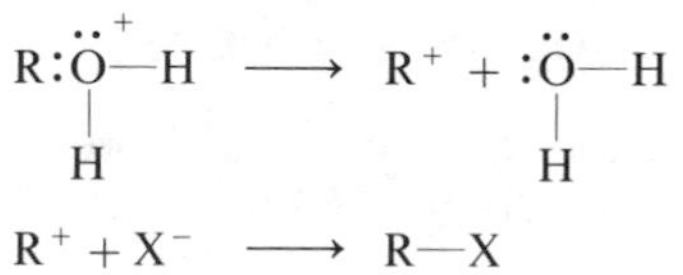

$$R\!:\!\overset{+}{\ddot{O}}-H \longrightarrow R^+ + :\ddot{O}-H$$
$$\quad\ |\qquad\qquad\qquad\quad |$$
$$\quad\ H\qquad\qquad\qquad\quad H$$

$$R^+ + X^- \longrightarrow R-X$$

Alcohols, such as tertiary and allylic, that form relatively stable carbocations react in this manner. Benzyl alcohol, although nominally a primary alcohol, also tends to react this way. The benzyl cation is stabilized by resonance structures that place a portion of the positive charge on carbons of the aromatic ring.

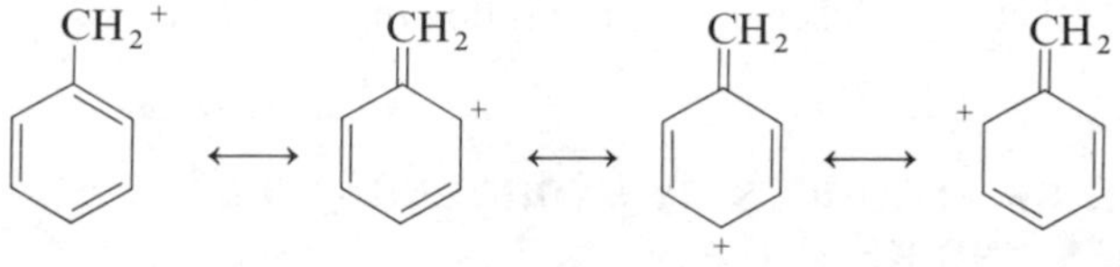

Protonation of the alcohol is necessary for formation of the carbocation. In principle, an alcohol might ionize to form a carbocation (R—OH → R$^+$ + OH$^-$). This cleavage, however, requires overcoming the electrostatic attraction between positive and negative ions and does not occur readily. In contrast, cleavage of a protonated alcohol (R—OH$_2^+$ → R$^+$ + H$_2$O) produces a positive and a neutral fragment and there is no electrostatic attraction to overcome.

The other mechanism involves direct attack by a halide ion on the protonated alcohol. An unshared electron pair of the halide ions forms a bond to carbon at the same time that the oxygen departs, taking with it the electron pair that it had shared with carbon.

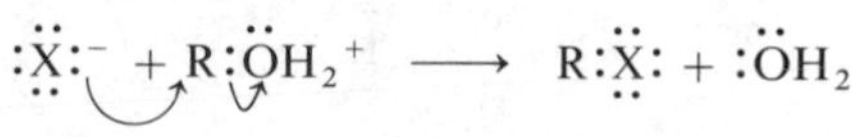

$$:\ddot{X}\!:^{-} + R\!:\!\overset{+}{\ddot{O}}H_2 \longrightarrow R\!:\!\ddot{X}\!: + :\ddot{O}H_2$$

If we could halt such a reaction near the middle of the reaction pathway, we would find halide and water both partially bonded to carbon (X----R----$\overset{\delta+}{OH_2}$) where the X is $\overset{\delta-}{}$. Methanol and primary alcohols react in this fashion, as generally do secondary alcohols. Protonation of the hydroxyl group is also essential for reaction in this manner. Protonation makes the carbon attached to the hydroxyl group relatively more positive, facilitating attack by the halide anion. Protonation also make it easier for oxygen to attract and depart with the electron pair that had bonded it to carbon.

In the reactions of alcohols with hydrogen halides, an electron pair furnished by a halogen replaces one furnished by oxygen. Such reactions in which one nucleophilic group replaces another are members of a large and important family of reactions called **nucleophilic substitutions**. We will learn more about such reactions in Chapter 7.

5.11
PREPARATIONS OF ALCOHOLS AND ETHERS

We learned already that alcohols can be prepared from the acid-catalyzed addition of water to alkenes (Secs. 3.8 and 3.9).

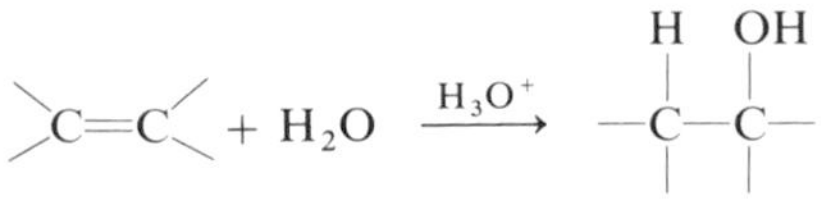

$$\underset{}{C=C} + H_2O \xrightarrow{H_3O^+} \overset{\overset{H \quad OH}{\mid \quad \mid}}{-C-C-}$$

Conditions must be used that favor hydration over the reverse reaction, alcohol dehydration (Sec. 3.14). Two additional preparations of alcohols and one preparation of ethers are described in the following sections.

5.12
PREPARATION OF ALCOHOLS BY HYDROGENATION
OF ALDEHYDES AND KETONES

Alcohols can be prepared by hydrogenating aldehydes and ketones. These additions of H_2 to $C=O$ are similar to additions of H_2 to $C=C$ and also require a catalyst, usually platinum or nickel.

$$\overset{\overset{O}{\parallel}}{CH_3CH_2CCH_2CH_3} \xrightarrow[Ni]{H_2} \overset{\overset{OH}{\mid}}{CH_3CH_2CHCH_2CH_3}$$

3-Pentanone
Diethyl ketone

3-Pentanol

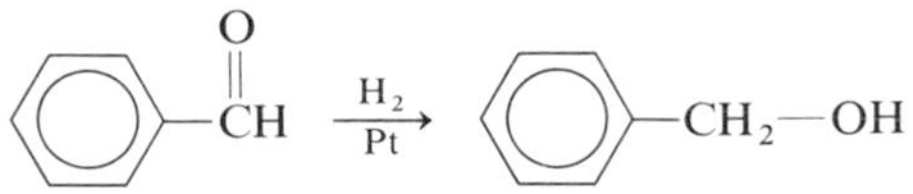

Benzaldehyde

Benzyl alcohol

Note that the carbonyl group is reduced more readily than is an aromatic ring. Lithium aluminum hydride ($LiAlH_4$), another type of reagent that reduces aldehydes and ketones to alcohols, is described in Section 9.9.

5.13
PREPARATION OF ALCOHOLS AND ETHERS
FROM ALKYL HALIDES

Alcohols can be prepared by treating alkyl halides with hydroxide salts dissolved in water.

$$CH_3CH_2CH_2CH_2Br + Na^+ OH^- \xrightarrow{H_2O} CH_3CH_2CH_2CH_2{-}OH + Na^+ Br^-$$

$$CH_2{=}CHCH_2Cl + Na^+ OH^- \xrightarrow{H_2O} CH_2{=}CHCH_2{-}OH + Na^+ Cl^-$$

Dialky ethers are prepared in a similar fashion by treating alkyl halides with alkoxide salts.

$$CH_3CH_2Br + Na^+ \ {}^-O{-}\underset{\underset{CH_3}{|}}{CH}CH_3 \longrightarrow CH_3CH_2{-}O{-}\underset{\underset{CH_3}{|}}{CH}CH_3 + Na^+ Br^-$$

These reactions in which hydroxide or alkoxide ions replace halide ions are nucleophilic substitution reactions. Additional examples and mechanisms of nucleophilic substitution reactions of alkyl halides are discussed in Sections 7.5 and 7.6.

We learned earlier (Sec. 3.13) that alkyl halides can be dehydrohalogenated to alkenes by treatment with strong bases. In fact, reactions of tertiary halides with hydroxide or alkoxide salts furnish predominantly alkenes. Secondary and primary halides give mixtures of substitution products (alcohols or ethers) and alkenes. By appropriate choice of conditions, however, either can often be made to predominate. In reactions with hydroxide, alcohol formation is often favored by using water as the solvent and alkene formation by using a less polar solvent, such as an alcohol.

---

problem 5.4   How can *tert*-butyl ethyl ether be prepared from an alkyl halide and an alcohol?

5.14
IMPORTANT ALCOHOLS

Methanol, a liquid boiling at 65°, is an important solvent and chemical intermediate. This simplest alcohol is sometimes called "wood alcohol" because it was once obtained from the liquid that distills from wood heated to a high temperature in the absence of air. Use of methanol is extensive (about 16 kg per person each year in the United States) and most is now synthesized by hydrogenation of carbon monoxide.

$$CO + 2H_2 \xrightarrow[\substack{300 \text{ atmospheres} \\ 350°}]{\text{catalyst}} CH_3OH$$

The carbon monoxide and hydrogen are generally produced by heating petroleum fractions or natural gas with water.

$$C_nH_{2n+2} + nH_2O \longrightarrow nCO + (2n+1)H_2$$

As we already saw (Sec. 4.12), carbon monoxide and hydrogen are also available from heating coal with water. Because of the possibility of its large-scale production from coal, methanol has been suggested as an alternative to petroleum or natural gas as a fuel. Only slight modification of an internal combustion engine permits using methanol in place of gasoline.

Ethanol is certainly the best known of the alcohols. Many a person has met misfortune by not realizing that the word "alcohol" on a label might refer to something other than this familiar compound. Ethanol is used in large amounts in the chemical industry as a solvent and a raw material for synthesis of other organic compounds. Most of this ethanol is produced in a prosaic fashion from the acid-catalyzed hydration of ethylene.

Ethanol can also be produced by fermentation of aqueous solutions of simple carbohydrates. Particularly important are sucrose (common table sugar), obtained from molasses and grapes, and glucose, obtained from hydrolysis of the starch in potatoes and grains.

$$C_{12}H_{22}O_{11} + H_2O \xrightarrow{\text{yeast}} 4CH_3CH_2OH + 4CO_2$$
Sucrose

$$C_6H_{12}O_6 \xrightarrow{\text{yeast}} 2CH_3CH_2OH + 2CO_2$$
Glucose

Carbohydrates are discussed in Chapter 10. The ability to produce ethanol from renewable plant sources has led to interest in gasohol, a blend of gasoline and ethanol, for use in internal combustion engines.

Wine is produced by fermentation of the juice of grapes or other fruits. Table wines typically contain 10 to 13% of ethanol. Sparkling wines also contain carbon dioxide, either produced naturally by fermentation or added artificially. Fermentation cannot produce ethanol concentrations much greater than 15% because the yeast becomes lethargic above that concentration. Therefore wines like port, sherry, or vermouth that typically contain 17 to 21% of ethanol have been "fortified" by addition of ethanol from other sources.

Beer is produced from barley, sometimes mixed with other grains. Before fermentation, the barley is soaked in water and then slowly dried by heating to produce malt. This treatment plus the addition of hops leads to the characteristic taste of beer.

Gin, vodka, and the various whiskeys contain 40 to 50% of ethanol. They result from distillation of less concentrated ethanol solutions obtained from fermentation of carbohydrates from mixtures of grains. Rum is obtained in the same manner from molasses. Brandy is obtained from distillation of wines.

Alcohol concentration in distilled beverages is often expressed in terms of **proof**. The original "proof" of concentration was to pour the distilled liquor over gunpowder and touch the mixture with a flame. Below a certain concentration of ethanol, the gunpowder would not burn. In the United States, proof is now defined as twice the percentage of ethanol content by volume—an 80-proof beverage has an ethanol content equal to 40% of the volume.

Distilled beverages are heavily taxed. In the United States the federal tax alone is about 20 times the cost of industrially produced ethanol. Because of the widespread use of ethanol in industry, this discrepancy in price creates an obvious enforcement problem. To prevent its diversion to beverage use, most nontaxed ethanol is **denatured** by addition of poisonous or unpalatable substances. Common additives include methanol (which in contrast to ethanol is highly toxic), gasoline, nicotine, and kerosene.

Hydroxyl groups are found in many naturally occurring compounds, including the major components of the sex attractants of the female silkworm moth and the male boll weevil.

Bombykol
(secreted by the female silkworm moth)

Grandisol
(one component of a mixture
of four compounds secreted
by the male boll weevil)

Synthetic grandisol mixed with the other compounds secreted by the male boll weevil has been used in traps placed in cotton fields in an effort to control this insect.

5.15
IMPORTANT ETHERS

Because diethyl ether dissolves most organic compounds, is unreactive toward many reagents, and is volatile (boiling point 35°) and hence easily removed, it is used extensively as a solvent. Diethyl ether was once an important inhalation anaesthetic, but other compounds are now usually used for that purpose.

An ether oxygen can be part of a ring, as in the common solvent tetrahydro-furan (boiling point 66°).

Tetrahydrofuran

Cyclic compounds, such as tetrahydrofuran, that have at least one atom other than carbon as a member of a ring are called **heterocyclic**. Most heterocyclic ethers have properties similar to those of acyclic ethers. Tetrahydrofuran, for example, is inert to most reagents but is cleaved by heating with hydrogen halides.

$$\text{(ring)} \xrightarrow{\text{HBr}} BrCH_2CH_2CH_2CH_2OH \xrightarrow{\text{HBr}} BrCH_2CH_2CH_2CH_2Br$$

In contrast to most other ethers, cyclic ethers with three-membered rings are highly reactive. Oxirane is used both as a general name for this group of heterocyclic ethers and as the specific name of its simplest and most extensively used member. The reactivity of oxiranes is due to considerable angle strain that makes their carbon-oxygen bonds unusually weak. Ethylene glycol is produced from oxirane (sometimes called ethylene oxide) and water in the presence of a small amount of an acid catalyst.

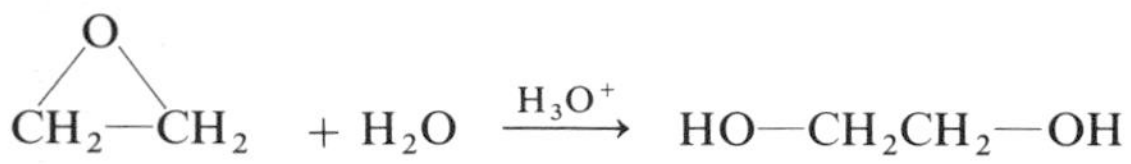

$$CH_2\text{—}CH_2 + H_2O \xrightarrow{H_3O^+} HO\text{—}CH_2CH_2\text{—}OH$$

Oxirane   1,2-Ethanediol
(ethylene oxide)   (ethylene glycol)

Oxirane also reacts readily with other nucleophilic reagents.

$$CH_2\text{—}CH_2 + CH_3OH \longrightarrow CH_3O\text{—}CH_2CH_2\text{—}OH$$

2-Methoxyethanol

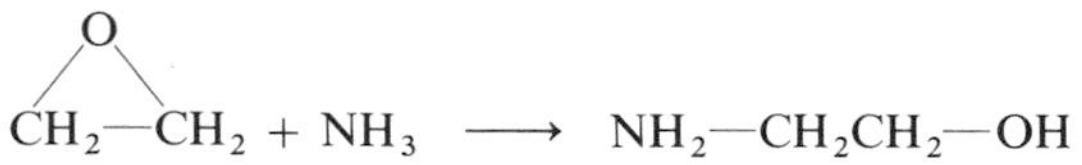

$$CH_2\text{—}CH_2 + NH_3 \longrightarrow NH_2\text{—}CH_2CH_2\text{—}OH$$

2-Aminoethanol

Disparlure, the sex attractant emitted by egg-bearing female gypsy moths, is an oxirane.

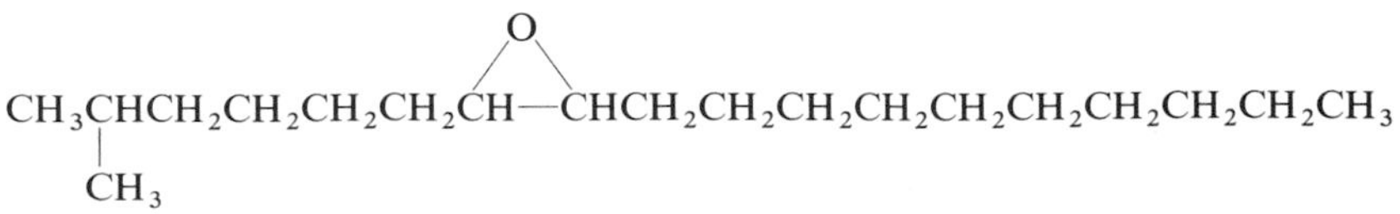

$$CH_3CHCH_2CH_2CH_2CH_2CH\text{—}CHCH_2CH_2CH_2CH_2CH_2CH_2CH_2CH_2CH_3$$
$$\mid$$
$$CH_3$$

The carcinogenic activity in some organisms of benzo[a]pyrene (Sec. 4.10) is thought to be due to its conversion to an oxirane that reacts readily with oxygen- and nitrogen-containing functional groups of proteins (Chapter 14) or nucleic acids (Chapter 15).

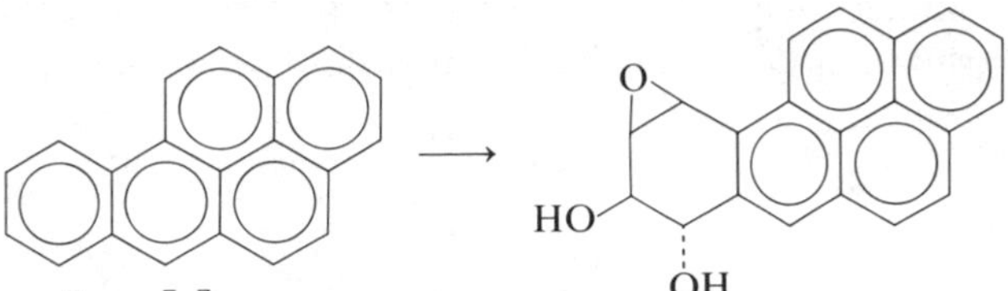

Benzo[a]pyrene

Some cyclic compounds with several ether functions interact strongly with cations. The six ether oxygens in the 18-membered ring of the following "crown ether," for instance, surround a potassium ion, just as do the oxygens of water molecules.

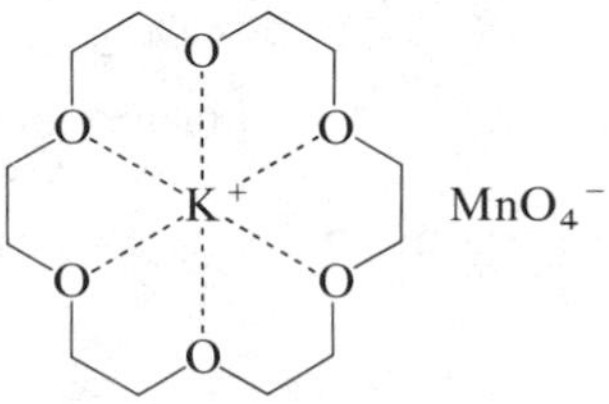

Salts in which the positive ion is enveloped in this manner by a sphere of organic material often have solubilities more like those of hydrocarbons than of inorganic salts. The foregoing potassium permanganate complex is soluble in benzene.

Oxygen atoms in nonactin, an antibiotic isolated from several species of **streptomyces**, interact strongly with K^+.

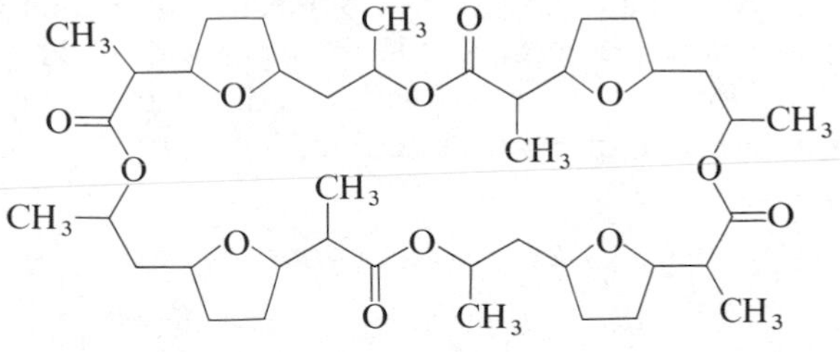

Nonactin

An **antibiotic** is a compound produced by a living organism and having the capacity to inhibit the growth of a microorganism. Nonactin's antibacterial action is due to abnormal leakage through membranes of K^+ enveloped by nonactin. The normal transport of biologically important ions, such as K^+ and Ca^{2+}, across membranes is thought to involve interactions of the ions with oxygens and nitrogens of large protein "carrier molecules" located in the membranes (Sec. 13.11).

THIOLS AND SULFIDES, SULFUR ANALOGS
OF ALCOHOLS AND ETHERS

Sulfur is immediately below oxygen in the periodic table and so it is not surprising that it forms compounds analogous to alcohols and ethers.

$$R-SH \qquad R-S-R$$

A thiol   A sulfide

Thiol and **sulfide** functions appear in some biologically important molecules.

Thiols and sulfides can be regarded as derivatives of hydrogen sulfide. Hydrogen sulfide smells like rotten eggs and thiols and sulfides also have abominable odors. Small amounts of ethanethiol are sometimes added to natural gas, otherwise almost odorless, to make leaks easily detectable.

$$CH_3CH_2-SH$$

Ethanethiol

The odor of the striped skunk is due to a mixture of compounds that include the following.

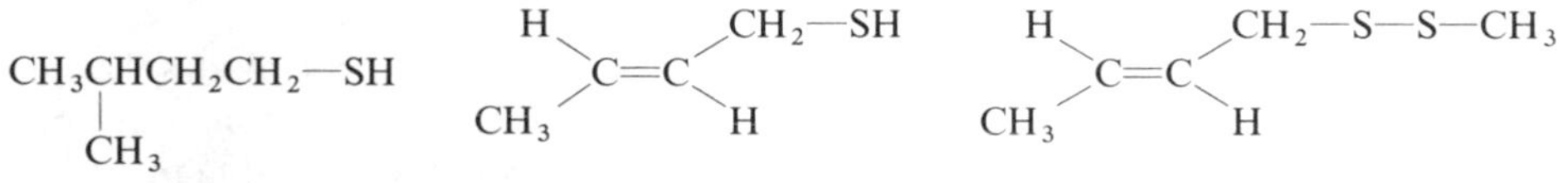

$$CH_3CHCH_2CH_2-SH$$
$$|$$
$$CH_3$$

3-Methyl-1-
butanethiol

$$\underset{CH_3}{\overset{H}{\diagdown}}C=C\underset{H}{\overset{CH_2-SH}{\diagup}}$$

trans-2-Butene-
1-thiol

$$\underset{CH_3}{\overset{H}{\diagdown}}C=C\underset{H}{\overset{CH_2-S-S-CH_3}{\diagup}}$$

trans-2-Butenyl
methyl disulfide

The last compound shown is a disulfide, the sulfur analog of a peroxide (ROOR). Disulfides generally are more stable than peroxides. They are readily formed by oxidation of thiols—for example, with O_2, dilute H_2O_2 (hydrogen peroxide), or I_2. Disulfides are easily reduced back to thiols.

$$2\ R-SH \underset{\text{reduction}}{\overset{\text{oxidation}}{\rightleftarrows}} R-S-S-R$$

Thiol   Disulfide

This readily reversible oxidation–reduction plays an important role in protein chemistry (Chapter 14).

Thiols can be prepared from alkyl halides in a manner similar to that used for alcohols.

$$CH_3CH_3CH_3Br + Na^+\ {}^-SH \longrightarrow CH_3CH_2CH_2-SH + Na^+\ Br^-$$

1-Propanethiol

In contrast to alcohols, thiols are sufficiently acidic to be converted to salts by sodium hydroxide. As do salts of alcohols, these salts react with alkyl halides, providing a synthesis of sulfides.

$$CH_3CH_2CH_2\!-\!SH \xrightarrow{\ Na^+\,OH^-\ } CH_3CH_2CH_2\!-\!S^-\,Na^+ \xrightarrow{\ CH_3I\ } CH_3CH_2CH_2\!-\!S\!-\!CH_3$$

Methyl propyl sulfide

Oxidation of dimethyl sulfide leads to dimethyl sulfoxide and dimethyl sulfone.

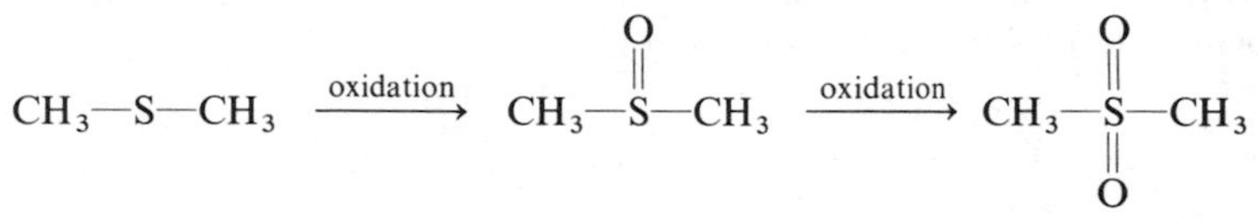

Dimethyl sulfide   Dimethyl sulfoxide   Dimethyl sulfone

These oxidation products are examples of the ability of sulfur to form more bonds (be surrounded by more groups) than can an oxygen. Because it dissolves many inorganic salts and also most organic compounds, dimethyl sulfoxide is an important solvent.

5.17
NOMENCLATURE OF PHENOLS

Phenol is used both as a general name for the group of compounds with one or more hydroxyl groups attached to an aromatic ring and as the specific name for its simplest member.

Phenol

More complicated phenols are often named as derivatives of phenol.

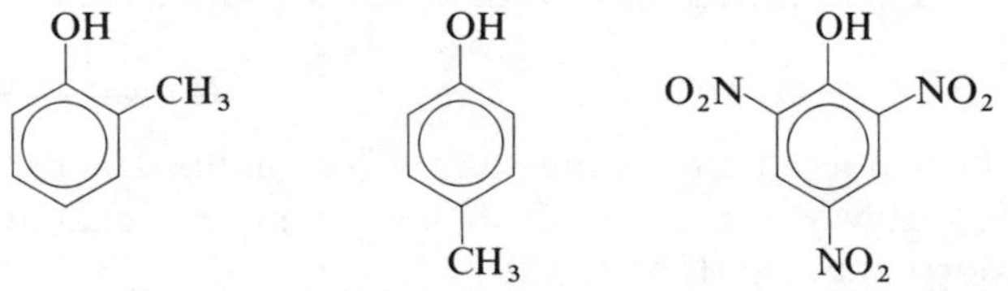

o-Methylphenol   p-Methylphenol   2,4,6-Trinitrophenol
(o-cresol)   (p-cresol)   (picric acid)

Trivial names (indicated in parentheses) are often used for some frequently encountered phenols.

5.18
REACTIONS OF PHENOLS

In much of their chemical behavior phenols differ significantly from alcohols. Of course, phenols undergo aromatic substitution reactions. We learned that the —OH substituent attached to an aromatic ring is ortho-para directing and strongly activating. Treatment of phenol with bromine, even without a catalyst, leads to rapid formation of 2,4,6-tribromophenol.

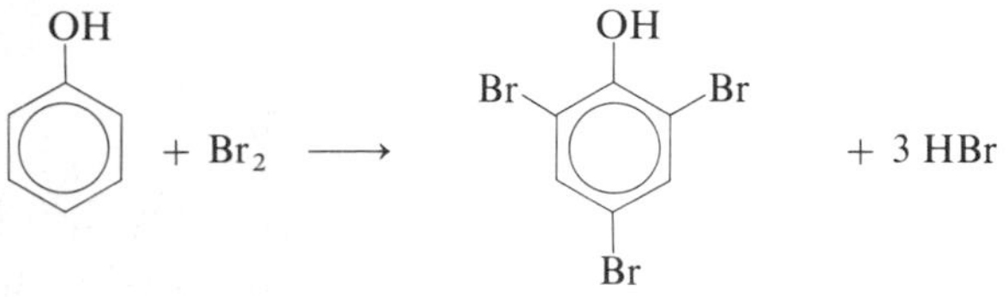

2,4,6-Tribromophenol

Phenols cannot be dehydrated and do not react with hydrogen halides (Sec. 5.9) to form aryl halides.

$$\text{Ar—OH} + \text{HX} \nrightarrow \text{Ar—X} + \text{H}_2\text{O}$$

Esters of phenols with inorganic acids are more reactive than the corresponding esters of alcohols and are not generally prepared by direct reaction of a phenol and an inorganic acid. As described in the following sections, phenols also differ significantly from alcohols in acidity and in behavior toward oxidizing agents.

5.19
REACTIONS OF PHENOLS AS ACIDS

Phenols are much more acidic than alcohols. They are converted almost completely to anions by treatment with aqueous solutions of hydroxide salts.

$$\text{ArOH} + \text{Na}^+ \text{OH}^- \longrightarrow \text{ArO}^- \text{Na}^+ + \text{H}_2\text{O}$$

The salts that result are usually more water soluble than the neutral phenols.

Probably you are already familiar with the quantitative treatment of the ionization of an acid (HA) in water.

$$\text{HA} + \text{H}_2\text{O} \rightleftharpoons \text{H}_3\text{O}^+ + \text{A}^-$$

The extent to which ionization occurs is indicated by the numerical value of K_a, called the **acidity constant**. K_a is related to the concentrations of the species involved in the equilibrium by the following equation.*

$$K_a = \frac{[H_3O^+][A^-]}{[HA]}$$

The brackets [ ] indicate concentrations of the species that they enclose. K_a is characteristic of the acid and increases with increasing acidity.

K_a's of typical alcohols are about 10^{-18}, but the K_a of phenol is 1.0×10^{-10}

$$ROH + H_2O \longleftrightarrow RO^- + H_3O^+ \qquad K_a \simeq 10^{-18}$$

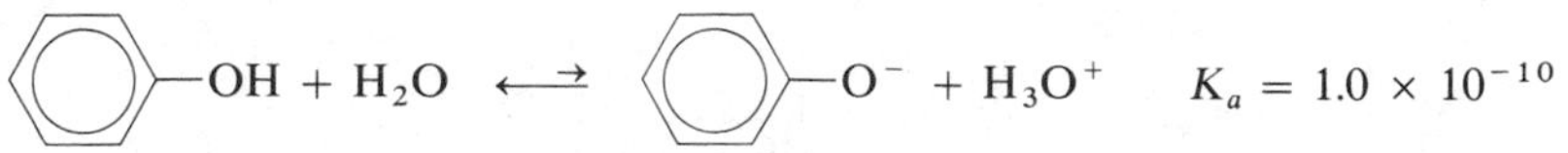

$$\text{C}_6\text{H}_5-OH + H_2O \longleftrightarrow \text{C}_6\text{H}_5-O^- + H_3O^+ \qquad K_a = 1.0 \times 10^{-10}$$

Phenoxide ion

Because it reacts only incompletely with water to form H_3O^+, phenol is a weak acid. Nevertheless, it is 100 million times more acidic than a typical alcohol.

Resonance stabilization of the phenoxide anion formed on ionization is responsible for the greater acidity of phenol.

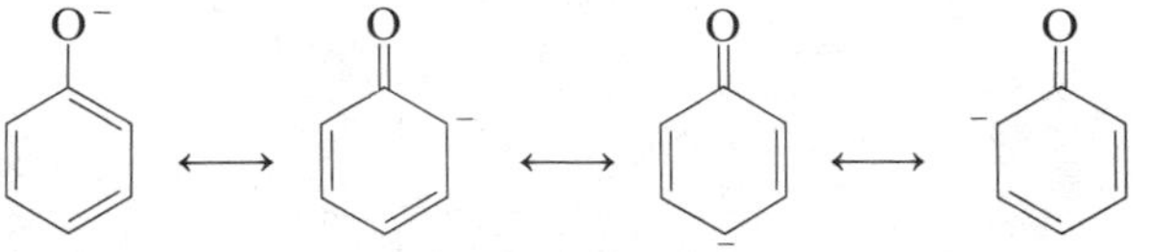

As indicated by the resonance structures, a portion of the negative charge is distributed over carbons of the aromatic ring. In contrast, an alkoxide ion has no resonance stabilization. The position of an equilibrium depends on the relative stabilities of the chemical species to the right and left of the arrows. Phenol is more acidic than an alcohol because phenoxide ion is more stable relative to phenol than is an alkoxide ion relative to an alcohol.

Phenol acidity is influenced by substituents. Because the electron-withdrawing effect of the nitro group further disperses the charge of the phenoxide ion, p-nitrophenol is more acidic than phenol.

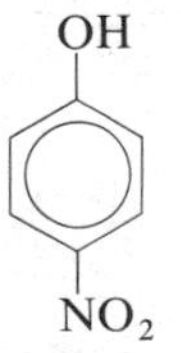

$$K_a = 7 \times 10^{-8}$$

p-Nitrophenol

---

* The concentration of water (the solvent) is so large that it does not change significantly in this reaction. The essentially constant value of $[H_2O]$ is customarily incorporated into the value of K_a.

5.20
OXIDATION OF PHENOLS; QUINONES

A variety of oxidizing agents readily oxidize phenols to products that are often complex mixtures. Phenols are frequently used as antioxidants in organic materials, such as foodstuffs, soaps, and paints. Small amounts of BHT or similar compounds, for instance, are commonly added to foods containing fats or oils.

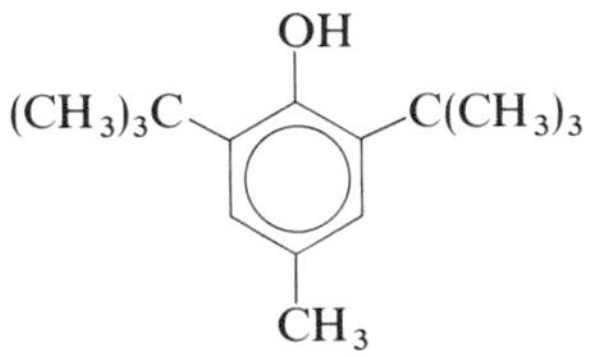

2,6-Di-*tert*-butyl-4-methylphenol
"Butylated hydroxy toluene" (BHT)

These additives act as inhibitors of free-radical chain reactions with oxygen that lead to development of rancidity. It is suspected that one function of vitamin E (note its terpene structure) is as an antioxidant to protect highly unsaturated lipids (Chapter 13) from oxidation.

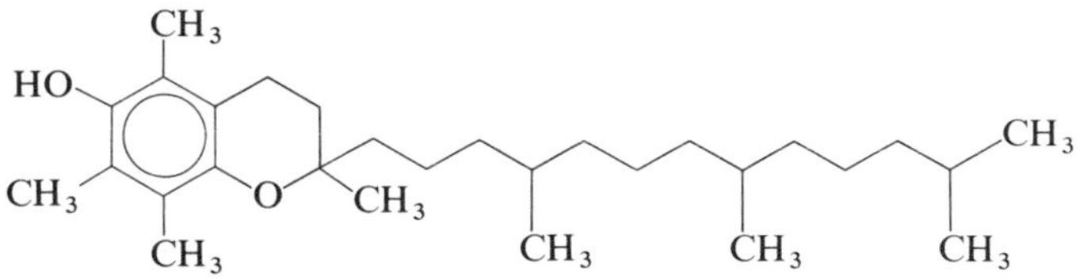

Vitamin E
(α-tocopherol)

Hydroquinone is readily oxidized to *p*-benzoquinone by a variety of oxidizing agents.

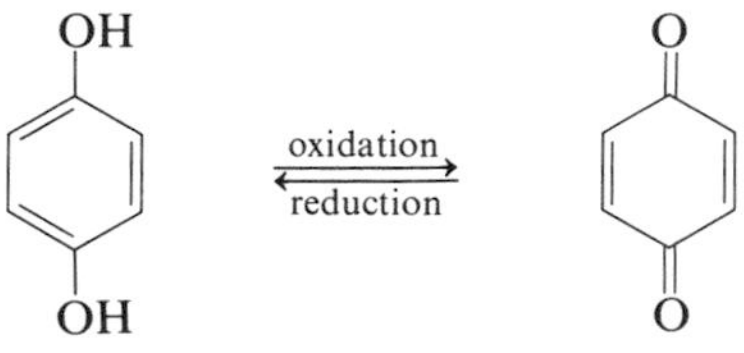

Hydroquinone *p*-Benzoquinone

Such oxidations are characteristic of 1,4-dihydroxy-substituted aromatic rings.

A bombardier beetle utilizes the oxidation of hydroquinone in a remarkable way. When attacked, this beetle adds a stored mixture of hydroquinone and hydrogen peroxide to two enzymes. One enzyme catalyzes the exothermic oxidation of hydroquinone by hydrogen peroxide to form *p*-benzoquinone and water. The other enzyme catalyzes the decomposition of hydrogen peroxide to O_2 and water. The pressure of the O_2 expels the hot mixture (temperature about $100°C$) explosively toward the attacker.

The ready oxidation of hydroquinone to *p*-benzoquinone is also responsible for its use as a photographic developer. Hydroquinone is oxidized most rapidly by those silver bromide crystals in the portion of a film that has been exposed to light. The metallic silver that results produces the photographic image.

p-Benzoquinone is easily reduced to hydroquinone. Such reductions are characteristic of **quinones**, compounds that contain two carbonyl groups and two alkene functions in a six-membered ring.

Terpene quinones known as coenzyme Q or (because of its ubiquitous occurrence in microorganisms, plants, and animals) ubiquinone are involved in the respiratory oxidation-reduction chain of reactions in the mitochondria of cells.

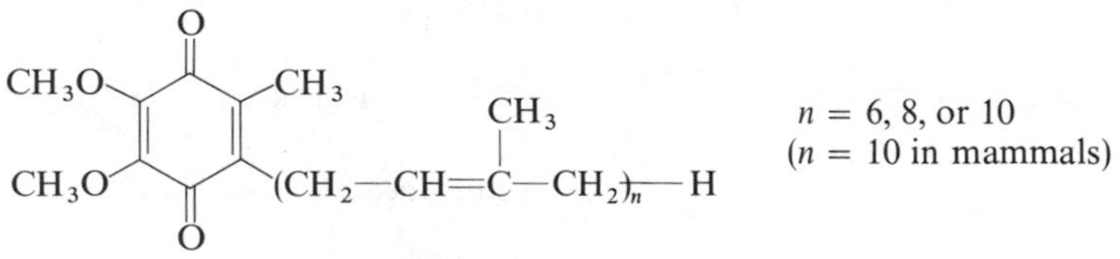

$n = 6, 8,$ or 10
$(n = 10$ in mammals$)$

Coenzyme Q

All quinones are colored and many natural pigments are quinones.

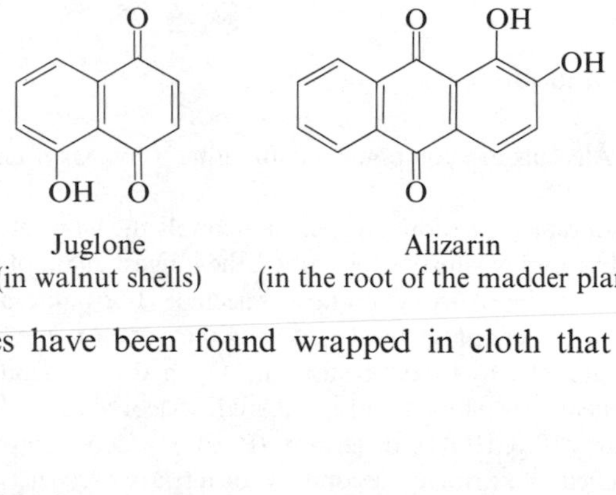

Juglone
(in walnut shells)

Alizarin
(in the root of the madder plant)

Egyptian mummies have been found wrapped in cloth that was dyed red with alizarin.

5.21
SOME IMPORTANT PHENOLS

Most phenols have antibacterial properties. A dilute solution of phenol ("carbolic acid") was the first generally used antiseptic. Because it is particularly corrosive to tissue and highly poisonous when taken internally, phenol is now only rarely used for this purpose. Nevertheless, some substituted phenols, including *p*-chlorophenol and a mixture of *o*-, *m*-, and *p*-methylphenols (cresols), are in current use as antiseptics and disinfectants.

The irritants in poison ivy, poison oak, and poison sumac are the following phenol and others with the same skeleton but having one, two, or three double bonds in the alkyl chain.

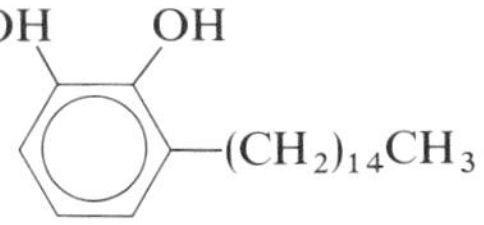

The phenol Δ^1-tetrahydrocannabinol has probably been responsible for more arrests than any other organic compound except ethanol.

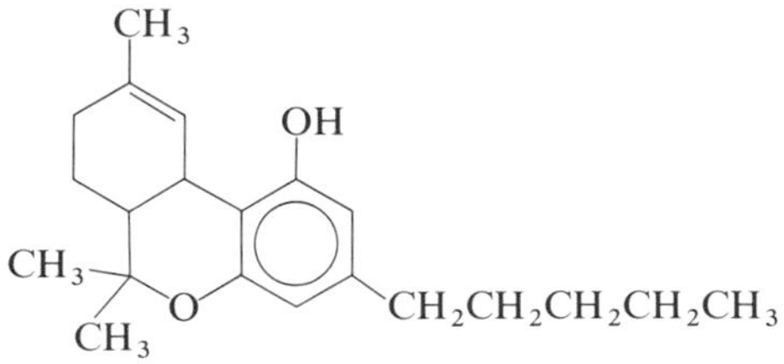

Δ^1-Tetrahydrocannabinol

Commonly known as THC, it is the principal active constituent of marihuana.

SUMMARY

Definitions and Ideas

Alcohols are compounds containing a hydroxyl group (—OH) bonded to an alkyl group.

In radicofunctional names of alcohols the word **alcohol** is used to indicate the function. In forming substitutive names, the longest chain of carbons that includes the carbon attached to the hydroxyl group is selected. The name of the corresponding alkane with the ending *-e* changed to *-ol* is used as the parent name. The chain is numbered from the end nearer the hydroxyl group, and the hydroxyl group is assigned a number. Other substituents are named and numbered. Alcohols are classified as primary (RCH_2OH), secondary (R_2CHOH), or tertiary (R_3COH), depending on whether the hydroxyl group is attached to a primary, secondary, or tertiary carbon.

Attractive forces called **dispersion forces** (or van der Waals forces) exist between all molecules. Dispersion forces are weak and operate only at short distances.

A **hydrogen bond**, another kind of attractive force between molecules, can be formed between a hydrogen attached to an oxygen or nitrogen and an oxygen or nitrogen with an unshared electron pair.

Because alcohol molecules can hydrogen bond to one another, attractive forces between them are unusually great. Consequently, alcohols have unusually high boiling points.

Because they cannot form hydrogen bonds to water molecules but disrupt the hydrogen bonds between water molecules, hydrocarbons have little solubility in water. Alcohols, ethers, and other molecules that can form hydrogen bonds to water are considerably more water soluble.

In their reactions as acids and bases, alcohols resemble water. They are converted by strong bases (such as NaH) to ions (RO^-), which are themselves strong bases. As a group, these negative ions are called **alkoxides**. A specific alkoxide ion can be named by changing the -*yl* ending of the name of the alkyl group R to -*oxide*.

An **ester** is a compound formed directly or indirectly by the linkage together of an alcohol and an acid with loss of water.

An organic molecule is **oxidized** if it gains carbon-oxygen bonds or loses carbon-hydrogen bonds.

Reactions in which one nucleophilic group is replaced by another are called **nucleophilic substitutions**. Reactions of alcohols with hydrogen halides to produce alkyl halides are nucleophilic substitutions and proceed by two mechanisms.

1. The protonated alcohol cleaves to form a carbocation intermediate ($R—OH_2^+ \rightarrow R^+ + OH_2$) that in a second step reacts with a halide ion ($R^+ + X^- \rightarrow RX$). Alcohols that can form relatively stable carbocations usually react in this manner.

2. In a one-step process a halide ion attacks the carbon bonded to the protonated hydroxyl group ($X^- + R—OH_2^+ \rightarrow RX + OH_2$). Methanol, primary alcohols, and often secondary alcohols react in this fashion.

Ethers are compounds containing two organic groups bonded to an oxygen.

Radicofunctional names of ethers are formed by naming both organic groups attached to the oxygen, followed by the word **ether**. In substitutive names —OR can be named alk*oxy*, derived from the name of the alk*yl* group R.

In contrast to alcohols, ethers undergo relatively few reactions.

Compounds that have at least one atom other than carbon as a member of a ring are called **heterocyclic**. Heterocyclic compounds ordinarily have properties similar to those of related acyclic compounds.

Phenols are compounds containing a hydroxyl group bonded to an aryl group.

Substituted phenols are often named as derivatives of phenol.

Due to resonance stabilization of their anions (ArO^-), phenols are much more acidic than alcohols.

Important Reactions Involving Alcohols

1. *Dehydration* $\overset{\displaystyle |\quad\ |}{\underset{\displaystyle H\ \ OH}{-C-C-}} \xrightarrow[\text{heat}]{H^+} \ \diagdown\!\!{C}\!\!=\!\!{C}\diagup$

   (discussed in Chapter 3)

2. *Reactions as acids* $ROH \xrightarrow{\text{NaH}} RO^-\ Na^+$

   NaH and other strong bases convert alcohols to salts.

3. *Conversion to esters*   ROH $\xrightarrow{\text{H}_2\text{SO}_4}$ ROSO$_3$H

Alcohols react with some inorganic acids, including H_2SO_4 and HNO_3, to form esters.

4. *Oxidation*   Primary alcohols are usually oxidized to carboxylic acids.

$$RCH_2OH \xrightarrow{\text{K}_2\text{Cr}_2\text{O}_7 \text{ or KMnO}_4} \overset{\displaystyle O}{\underset{\displaystyle \|}{R}}COH$$

With an appropriate reagent, however, it is possible to stop at the aldehyde stage.

$$RCH_2OH \xrightarrow{\text{CrO}_3\text{-pyridine}} \overset{\displaystyle O}{\underset{\displaystyle \|}{R}}CH$$

Secondary alcohols are oxidized to ketones.

$$\overset{\displaystyle OH}{\underset{\displaystyle |}{R}}CHR \xrightarrow{\text{K}_2\text{Cr}_2\text{O}_7 \text{ or KMnO}_4} \overset{\displaystyle O}{\underset{\displaystyle \|}{R}}CR$$

Tertiary alcohols are not readily oxidized.

5. *Conversion to alkyl halides*   ROH $\xrightarrow{\text{HX}}$ RX

PX$_3$ and SOCl$_2$ are alternative reagents.   X = Cl, Br, or I

6. *Preparation by hydration of alkenes* $\underset{/}{\overset{\backslash}{C}}=\overset{/}{\underset{\backslash}{C}}$ $\xrightarrow[\text{H}_3\text{O}^+]{\text{H}_2\text{O}}$ $-\underset{\underset{H}{|}}{C}-\underset{\underset{OH}{|}}{C}-$

(discussed in Chapter 3)

7. *Preparation by hydrogenation of aldehydes and ketones*

$$-\overset{\displaystyle O}{\underset{\displaystyle \|}{C}}- \xrightarrow[\text{Pt or Ni}]{\text{H}_2} -\overset{\displaystyle OH}{\underset{\displaystyle \underset{H}{|}}{C}}-$$

8. *Preparation from alkyl halides*   RX $\xrightarrow[\text{H}_2\text{O}]{\text{NaOH}}$ ROH

Dehydrohalogenation is usually the major reaction with tertiary halides.

Important Reactions Involving Ethers

1. *Cleavage by hydrogen halides*   ROR′ $\xrightarrow{\text{HX}}$ RX + R′OH

$\downarrow$HX

R′X

X = Cl, Br, or I

2. *Preparation from alkyl halides* $\quad$ RX $\xrightarrow{\text{NaOR'}}$ ROR'

$\quad$ Dehydrohalogenation is usually the major reaction with tertiary halides.

Important Reactions Involving Phenols

1. *Aromatic substitution* $\quad$ Phenols undergo the usual electrophilic aromatic substitution reactions, such as halogenation and sulfonation (discussed in Chapter 4). The —OH group is ortho-para directing and activating.

2. *Reactions as acids* $\quad$ ArOH $\xrightarrow{\text{NaOH}}$ ArO$^-$ Na$^+$

PROBLEMS

1. Draw structures for the following compounds.
   - (a) 1-butanol
   - (b) cyclopentanol
   - (c) isopropyl alcohol
   - (d) 3-buten-2-ol
   - (e) 3-chloro-1-propanol
   - (f) *sec*-butyl alcohol
   - (g) 1-phenyl-1-propanol
   - (h) 2,3-butanediol
   - (i) *trans*-2-methyl-3-penten-2-ol
   - (j) diphenyl ether
   - (k) dimethyl ether
   - (l) methyl vinyl ether
   - (m) cyclohexyl methyl ether
   - (n) 2-methoxyhexane
   - (o) ethyl *p*-nitrophenyl ether
   - (p) anisole
   - (q) *p*-chlorophenol
   - (r) 2,4-dinitrophenol
   - (s) hydroquinone
   - (t) sodium ethoxide
   - (u) ethanethiol
   - (v) dimethyl sulfide

2. Name the following compounds.

   (a) $(CH_3)_2CHCH_2OH$

   (b)

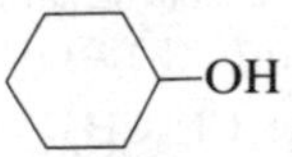

   (c) $CH_3CH_2CH_2OH$

   (d) $CH_3CH_2CH_2CH_2CH_2CH_2OH$

   (e)

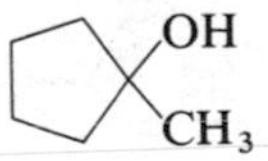

   (f)

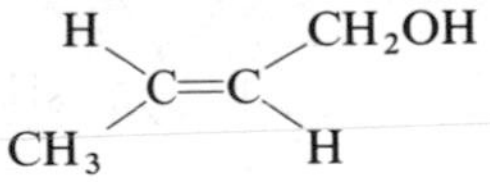

   (g)

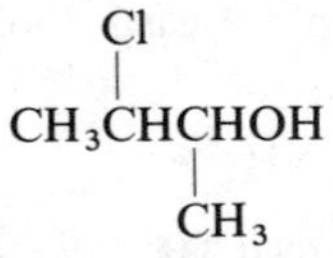

   (h)

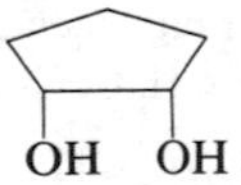

   (i) $CH_3CH_2OCH_2CH_3$

   (j) $CH_3CH_2OCH_2CH_2CH_2CH_3$

   (k)

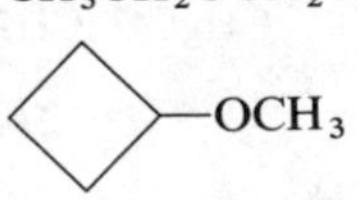

   (l)

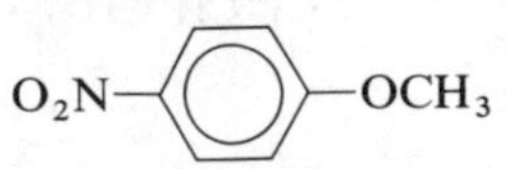

   (m) $CH_2{=}CHCH_2OCH_3$

   (n) 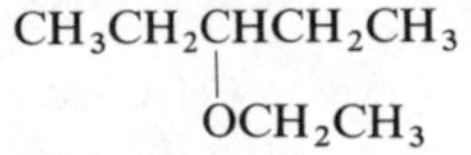 $CH_3CH_2CHCH_2CH_3$
   $\qquad\qquad\qquad |$
   $\qquad\qquad\quad OCH_2CH_3$

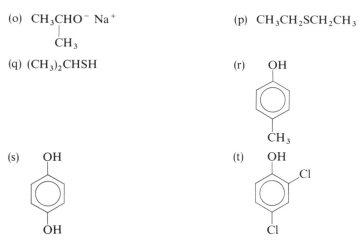

(o) $CH_3CHO^-\ Na^+$
       $|$
       CH_3

(p) $CH_3CH_2SCH_2CH_3$

(q) $(CH_3)_2CHSH$

(r)   OH
       [benzene ring]
       CH$_3$

(s)   OH
       [benzene ring]
       OH

(t)   OH
       [benzene ring with Cl]   Cl
       Cl

3. Give another name for each of the following compounds.
   (a) *tert*-butyl alcohol
   (b) 2-propanol
   (c) 2-methyl-1-propanol
   (d) methyl phenyl ether
   (e) 2-methoxypropane
   (f) *p*-cresol

4. Draw structures of all compounds that fit the following descriptions.
   (a) compounds of molecular formula $C_4H_{10}O$
   (b) alcohols of molecular formula $C_5H_{12}O$
   (c) ethers of molecular formula $C_5H_{12}O$

5. Classify each of the following alcohols as primary, secondary, or tertiary.
   (a) compounds (a) to (g) in problem 2
   (b) compounds that were the answers to problem 4(b).

6. Which of the following molecules are capable of forming hydrogen bonds to molecules of the same kind?
   (a) $(CH_3)_3COH$
   (b) $CH_3OCH_2CH_3$
   (c) CH_3CH_2Cl
   (d) $CH_3CH_2NCH_3$
       $|$
       CH_3

   (e) [cyclopentane ring]—NH_2

   (f) [benzene ring]—OH

7. Arrange each group of compounds in order of increasing boiling point.
   (a) $CH_3CH_2CH_2OH,\qquad CH_3CH_2OCH_3$

   (b)   OH   F   NH$_2$
          $|$ $|$ $|$
          $CH_3CHCH_3,$ $CH_3CHCH_3,$ CH_3CHCH_3

   (c) $CH_3CH_2NCH_3,\qquad CH_3CH_2CHOH,\qquad CH_3CH_2CHNH_2$
          $|$ $|$ $|$
          CH_3 CH_3 CH_3

   (d)   CH$_3$
                                     $|$
          $CH_3CH_2CH_2CH_2CH_2CH_3,$ $CH_3CCH_2CH_3,$ $CH_3CH_2CH_2CHCH_3$
                                     $|$ $|$
                                     CH_3 CH_3

8. Arrange each group of compounds in order of increasing solubility in water.

(a) $CH_3CH_2OH,$ $CH_3CH_2OCH_2CH_3,$ $CH_3CH_2CH_2CH_2CH_3$

(b) $CH_3CH_2OH,$ $CH_3CH_2CH_2CH_2OH,$ CH_3CH_2Cl

(c) $CH_3CH_2CH_2CH_2CH_3,$ $CH_3OCH_2CH_3,$ $CH_3CH_2CH_2CH_2CH_2CH_2OH$

(d) $CH_3CH_2CH_2CH_2CH_2CH_2OH,$ $HOCH_2CH_2CH_2CH_2CH_2CH_2OH,$

$HOCH_2CH_2CHCH_2CH_2OH$
$\qquad\qquad\;\;|$
$\qquad\qquad\;\;OH$

(e) $CH_3OCH\text{-}CH_2OH,$ $CH_3CH_2CH_2CH_2OH$

(f)

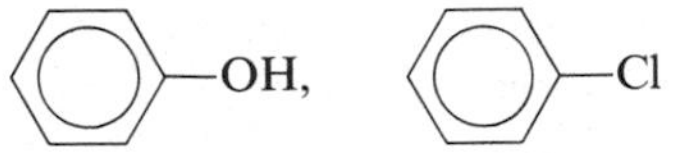

—OH,   —Cl

9. Draw structures for any organic products that result from the following reactions.

(a) $CH_3CH_2CH_2OH \xrightarrow{\;HCl\;}$

(b) $CH_3CH_2OH \xrightarrow{\;NaH\;}$

(c)
$$\qquad\qquad OH$$
$$\qquad\qquad |$$
$$CH_3CH_2CHCH_2CH_3 \xrightarrow{\;K_2Cr_2O_7\;}$$

(d) $(CH_3)_2CHCH_2Cl \xrightarrow[\;H_2O\;]{\;NaOH\;}$

(e) —$CH_2OH \xrightarrow{\;PBr_3\;}$

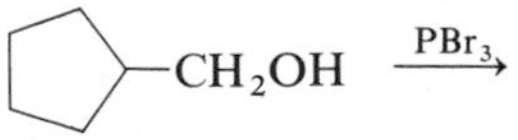

(f) $(CH_3)_3COH \xrightarrow{\;KMnO_4\;}$

(g) $(CH_3)_3CO^- K^+ \xrightarrow{\;HCl\;}$

(h) —$CH_2OH \xrightarrow{\;CrO_3\text{-pyridine}\;}$

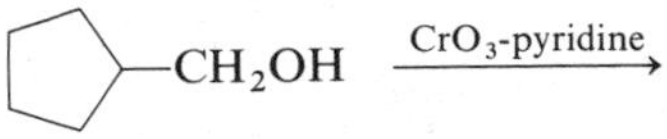

(i) $CH_3CHCH_2CH_2OH \xrightarrow{\;K_2Cr_2O_7\;}$
$\quad\;\;\;|$
$\quad\;\;\;CH_3$

(j) —$CHCH_2CH_3 \xrightarrow{\;SOCl_2\;}$
$\qquad\qquad|$
$\qquad\qquad OH$

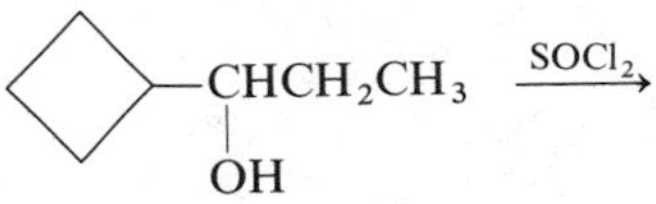

(k) $CH_3CH_2CH_2CH_2OH \xrightarrow{\;cold\ conc.\ H_2SO_4\;}$

(l) $CH_3CH_2OCH_2CH_3 \xrightarrow{\;NaH\;}$

(m) $(CH_3)_2CHOCH(CH_3)_2 \xrightarrow{\;HBr\ (excess)\;}$

(n) $CH_3CH_2OCH_3 \xrightarrow{\;K_2Cr_2O_7\;}$

(o) $CH_3CH_2CH_2OCH_2CH_2CH_3 \xrightarrow{\;NaOH\;}$

(p) —$O^- Na^+ \xrightarrow{\;CH_3I\;}$

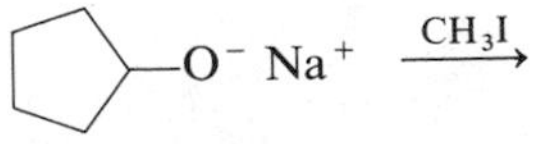

(q) $\xrightarrow{\;HCl\ (excess)\;}$

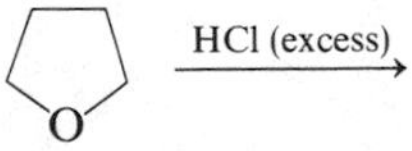

(r)
$$CH_2\overset{O}{\underset{}{-}}CH_2 \xrightarrow[\;H_3O^+\;]{\;H_2O\;}$$

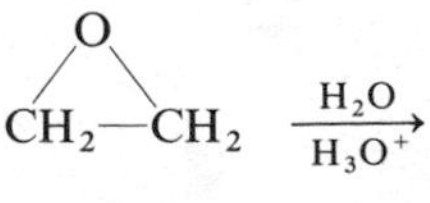

(s)
$$CH_2\overset{O}{\underset{}{-}}CH_2 \xrightarrow{\;HCl\;}$$

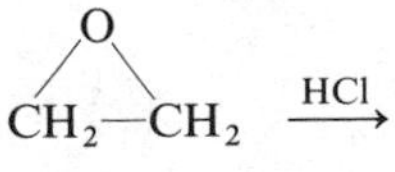

(t) —$O^- Na^+ \xrightarrow{\;HCl\;}$

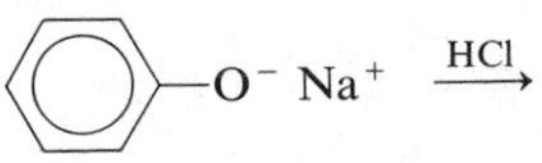

(u) —$OH \xrightarrow{\;Br_2\ (excess)\;}$

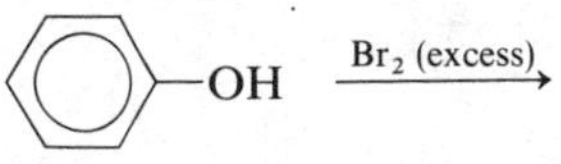

(v) CH_3——$OH \xrightarrow{\;NaOH\;}$

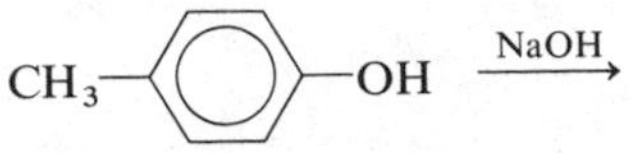

10. The following compound contains several of the functional groups discussed so far.

$$CH_2\text{=}CCH_2CH_2OCH_3$$
$$|$$
$$CH_2OH$$

What organic reaction products would result from treating it with each of the following reagents?

(a) Br_2 (b) NaOH

(c) HBr (excess)   (d) NaH

11. What are the major and minor organic products expected from the following reactions?

(a) NaOH in water with the following halides

     1. $CH_3CH_2CH_2Br$ 2.

(b) $NaOCH_2CH_3$ with the following halides

     1. $(CH_3)_3CBr$ 2. $(CH_3)_2CHCH_2Br$

12. Write equations to show how isopropyl alcohol can be converted to each of the following compounds. Some of the conversions require more than one reaction.

(a)   Cl
           |
     CH_3CHCH_3 (b) OCH_3
                                             |
                                      CH_3CHCH_3

(c) $CH_2\text{=}CHCH_3$ (d)   O
                                                ‖
                                      CH_3CCH_3

(e) 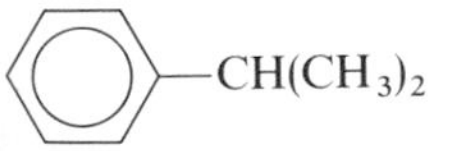 $\bigcirc\text{—}CH(CH_3)_2$ (f) 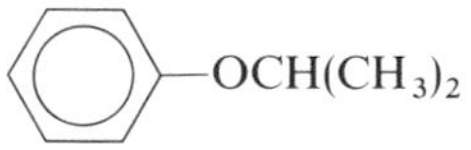 $\bigcirc\text{—}OCH(CH_3)_2$

13. Write equations to show how each of the following conversions can be carried out. Many will require more than one reaction.

(a)  $\bigcirc\text{—}OH \longrightarrow \bigcirc\text{—}Cl$

(b)   OH   O
         | ‖
     $CH_3CHCH_2CH_2CH_3 \longrightarrow CH_3CCH_2CH_2CH_3$

(c) 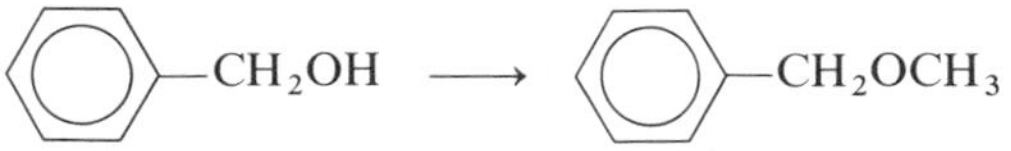 $\bigcirc\text{—}CH_2OH \longrightarrow \bigcirc\text{—}CH_2OCH_3$

(d) CH_3
                                   |
     $CH_3OH \longrightarrow CH_3OCCH_3$
                                   |
                                   CH_3

(e) $(CH_3)_2CHCH_2Cl \longrightarrow (CH_3)_2CHCH_2ONO_2$

(f)

$$CH_3CH_2CH_2CH{=}CH_2 \longrightarrow CH_3CH_2CH_2\overset{\overset{\displaystyle O}{\displaystyle \|}}{C}CH_3$$

(g)

⬠—OH ⟶ ⬠

(h) $CH_3CH_2OH \longrightarrow CH_3CH_2ONa$

(i)

$$CH_3CH_2Cl \longrightarrow CH_3CH_2O{-}\!\!\bigcirc\!\!\!\!\hexagon$$

(j)

$$CH_3CH_2CH_2CH_2OH \longrightarrow CH_3CH_2\overset{\overset{\displaystyle Cl}{\displaystyle |}}{C}HCH_3$$

(k)

$$(CH_3)_2CHCH_2OH \longrightarrow (CH_3)_2CH\overset{\overset{\displaystyle O}{\displaystyle \|}}{C}OH$$

(l)

⬡—$CH_2Cl \longrightarrow$ ⬡—$\overset{\overset{\displaystyle O}{\displaystyle \|}}{C}H$

(m) $CH_3CH_2CH_2CH_2Br \longrightarrow CH_3CH_2CH_2CH_2SH$

(n)

⬡—$CH_3 \longrightarrow$ ⬡—CH_2OH

14. Draw the structure of a compound of molecular formula $C_4H_{10}O_3$ that reacts with NaH to give H_2 and with an excess of HBr to give 1,2-dibromoethane as the only organic product.

15. Draw the structure of a compound of molecular formula C_7H_8O—only slightly soluble in water but very soluble in aqueous base—that reacts rapidly with an excess of bromine to give a compound of molecular formula $C_7H_5Br_3O$.

16. What is the structure of a compound of molecular formula C_5H_8O formed by treating cyclopentene with chlorine in aqueous base followed by very strong base. Draw structures of the intermediates involved in this reaction sequence.

17. When 3-methyl-2-buten-1-ol is treated with HCl, two isomeric unsaturated chlorides (C_5H_9Cl) are formed. What are the structures of the chlorides and why are two isomers formed?

18. One of the compounds obtained in the acid-catalyzed dehydration of ethanol is diethyl ether. In fact, the reaction conditions can be modified to make diethyl ether the major product. Show by equations the mechanism of formation of diethyl ether.

19. Cleavage of *tert*-butyl methyl ether with a minimal amount of HCl produces *tert*-butyl chloride and methanol. What is the mechanism of this reaction?

20. Describe a procedure for separating phenol and cyclohexanol without distilling or crystallizing.

21. Arrange the following phenols in order of increasing acidity: phenol, *p*-methylphenol, *p*-nitrophenol, 2,4-dinitrophenol.

22. The phenoxide anion is stabilized considerably by resonance interactions in which an electron pair is transferred from oxygen to a carbon of the ring. Why is an equivalent transfer of an electron pair from the oxygen far less important in phenol itself?

23. Describe a test, simple to carry out and with easily observable results, that can be used to distinguish between the members of each of the following groups of compounds.
(a) pentane and ethanol
(b) propyl alcohol and allyl alcohol
(c) phenol and cyclohexanol
(d) dipropyl ether and hexanol
(e) cyclohexanol and 1-methylcyclopentanol
(f) ethylene glycol and 1-butanol

Chiral Compounds

In this chapter we encounter another type of stereoisomerism. Some organic molecules differ the same way as left and right shoes—they are mirror images of each other. This subtle type of isomerism was not introduced earlier and so you may consider it a curiosity, found only in obscure or esoteric molecules. Actually, the structural features needed for such isomerism are found in the majority of organic molecules, including most biological molecules. Reactions in a living organism usually involve only one of a pair of such isomers. Therefore part of the quest for evidence of extraterrestrial life has been a search for a similar preference in the organic materials obtained from the moon and from meteorites. In a living system isomers that differ in this seemingly trivial way often differ more in behavior than compounds that are constitutional isomers or not even isomers at all.

6.1
ENANTIOMERS

Consider the simple organic molecule bromochlorofluoromethane.

$$\begin{array}{c} \text{Br} \\ | \\ \text{F}-\text{C}-\text{Cl} \\ | \\ \text{H} \end{array}$$

A three-dimensional model of this molecule can be constructed by placing four balls of different color or size in a tetrahedral fashion around a central ball. As

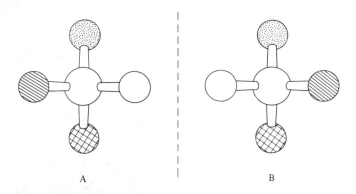

A   B

Figure 6.1 Two models that can represent a bromochlorofluoromethane molecule.

shown in Figure 6.1, *two such models can be made.* Models A and B are not identical. They cannot be superimposed (placed so that the same atoms of each occupy the same spaces). Imagine, for example, sliding A to the right until the balls at the top and bottom of A and B coincide. The balls at the left and right will not match. Two kinds of bromochlorofluoromethane molecules, corresponding to models A and B, actually exist.

How do these models differ? They are **mirror images** of one another. If you place one of the models in front of a mirror, its reflection would look like the other model. If the dashed line in Figure 6.1 is a mirror that is perpendicular to the plane of the paper, then each model is the reflection in that mirror of the other model. The two bromochlorofluoromethanes differ just as do other familiar objects that are mirror images—for instance, right and left hands or right and left gloves.

---

exercise with models 6.1 Construct two models of bromochlorofluoromethane that correspond to A and B in Figure 6.1. By trying to superimpose these models, convince yourself that they are not identical and therefore represent isomeric molecules. Note the mirror image relationship between the two models—the reflection of one model in a mirror is identical to the other model. Convince yourself that any additional model of bromochlorofluoromethane that you make is identical to one of the two models that you have already constructed.

Any object—a foot, a glove, or a bromochlorofluoromethane molecule—that is not identical with its mirror image is called **chiral**. Two chiral isomers, such as the two bromochlorofluoromethane molecules, that are mirror images of one another are called **enantiomers**. Enantiomers differ only in the way that atoms are arranged in space and hence are stereoisomers. Earlier we encountered two examples of stereoisomers: cis and trans alkenes and cis and trans disubstituted cycloalkanes. Such cis-trans isomers are not mirror images of each other; they actually differ in shape. Stereoisomers that are not mirror images are called **diastereomers** (or diastereoisomers). The different types of isomers are classified in Table 6.1. Note that

Table 6.1 Different classes of isomers

Class of isomers	How members of the class differ	Examples	
		Names	Representations
1. Constitutional isomers	differ in which atoms are bonded to which	butane and isobutane (both C_4H_{10})	$CH_3-CH_2-CH_2-CH_3$ $\quad$ $CH_3-CH-CH_3$ with CH_3 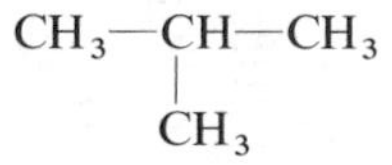
		ethanol and dimethyl ether (both C_2H_6O)	CH_3-CH_2-OH $\qquad$ CH_3-O-CH_3
2. Stereoisomers	differ only in the spatial arrangements of atoms		
a. Enantiomers	stereoisomers that are related as mirror images	the two bromo-chlorofluoro-methanes	
b. Diastereomers	stereoisomers that are not related as mirror images	*cis-* and *trans*-2-butene 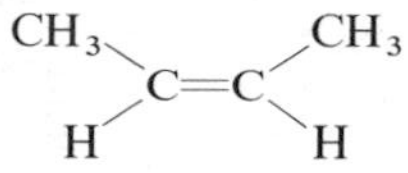 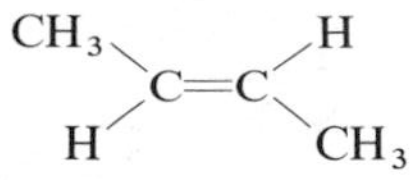	
		cis- and *trans*-1,3-dichloro-cyclobutane 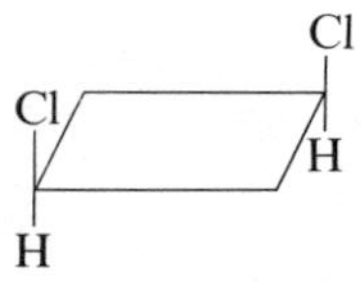 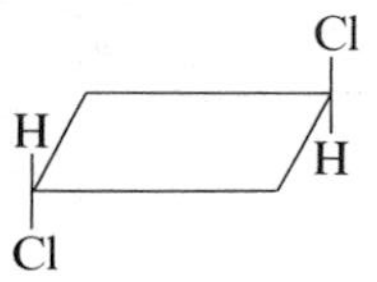	

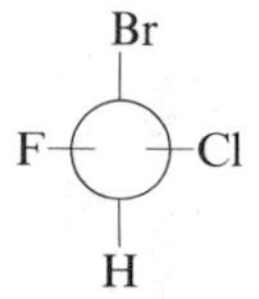

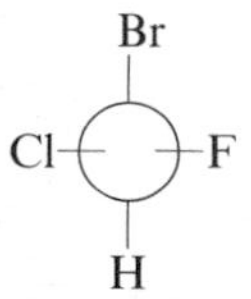

isomers are either constitutional isomers or stereoisomers and that stereoisomers are related either as enantiomers or as diastereomers.

Which of the following molecules are not identical with their mirror images and hence chiral?

$$
\begin{array}{cc}
\overset{\displaystyle Br}{\underset{\displaystyle H}{CH_3-\overset{|}{\underset{|}{C^*}}-Cl}}
&
\overset{\displaystyle OH}{\underset{\displaystyle H}{CH_3-CH_2-CH_2-CH_2-\overset{|}{\underset{|}{C^*}}-CH_2-CH_3}}
\\[2em]
\text{1-Bromo-1-chloroethane} & \text{3-Heptanol}
\end{array}
$$

$$
\begin{array}{cc}
\overset{\displaystyle Br}{\underset{\displaystyle H}{Cl-\overset{|}{\underset{|}{C}}-Cl}}
&
\overset{\displaystyle OH}{\underset{\displaystyle H}{CH_3-\overset{|}{\underset{|}{C}}-CH_3}}
\\[2em]
\text{Bromodichloromethane} & \begin{array}{c}\text{2-Propanol}\\\text{Isopropyl alcohol}\end{array}
\end{array}
$$

We can determine whether any structure is chiral by constructing a model, then constructing a mirror image model, and finally comparing the two models to see if they are different. We need not make this effort for 1-bromo-1-chloromethane or 3-heptanol, however. Each structure has one carbon (starred in the structures) to which four different groups are attached. Note that the two alkyl groups attached to the starred carbon in 3-heptanol are different even though both are linked by —CH_2—. Therefore both 1-bromo-1-chloroethane and 3-heptanol can be represented by the same pair of models (A and B in Fig. 6.1) that we already used. In fact, there are two isomeric 1-bromo-1-chloroethanes and two isomeric 3-heptanols. A carbon that has four different groups attached to it is often called an **asymmetric carbon** (or a chiral carbon). Any structure containing one asymmetric carbon can be represented by the models that we used and corresponds to a pair of enantiomers.*

Is bromodichloromethane chiral? Figure 6.2 shows two models of this molecule that are mirror images (imagine the dashed line to be a mirror perpendicular to the plane of the paper). The two models are identical. A bromodichloromethane molecule is **achiral** (not chiral) and has no enantiomer. Only one kind of bromodichloromethane molecule exists. Of course, many everyday objects are identical with their mirror images and hence achiral; examples include caps, pencils, tin cans, hammers, cups, and sticks of chewing gum. Isopropyl alcohol and other compounds in which two of the four groups attached to a tetrahedral carbon are the same can also be represented by the models shown in Figure 6.2 and are achiral. Compounds with only two different groups attached to a tetrahedral carbon (e.g., CH_2Cl_2) also are identical to their mirror images and hence achiral.

---

* Molecules that contain more than one asymmetric carbon are considered in Sections 6.5 to 6.7.

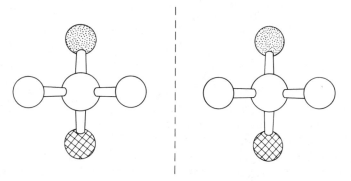

Figure 6.2   Two models that can represent bromodichloromethane (the identical balls representing chlorine atoms). The models are mirror images but also are identical.

exercise with models 6.2   Construct two models of bromodichloromethane as shown in Figure 6.2. Convince yourself that these models are mirror images but also are superimposable (identical).

problem 6.1   Which of the following structures represent chiral molecules?

(a) $\bigcirc$ —CH—CH$_3$
              |
             OH

(b) CH$_3$—CH—CH$_2$—CH$_3$
            |
           CH$_3$

(c) CH$_3$—CH—CH$_2$OH
            |
            Cl

(d)   OH
      $\bigcirc$

6.2
REPRESENTING ENANTIOMERS

Enantiomers differ in the three-dimensional arrangement of atoms. The difference between enantiomers is easily seen when they are represented by three-dimensional models. Ordinarily, however, we want to represent molecules simply and quickly on the two-dimensional surface of a sheet of paper. The structural drawings that we generally use, such as the following one of bromochlorofluoromethane, do not indicate geometry and can stand for either enantiomer.

$$
\begin{array}{c}
\text{Br} \\
| \\
\text{F—C—Cl} \\
| \\
\text{H}
\end{array}
$$

How can we represent a specific enantiomer on a flat sheet of paper? Several conventions for doing so are shown in Figure 6.3. In A, the three-dimensional arrangement around an asymmetric carbon is shown by a perspective drawing. B uses a convention that we already know. We assume the asymmetric carbon to be in the plane of the paper, groups attached to it by wedge-shaped lines (◄) to be in front of the paper, and groups attached by dashed lines to be behind the paper. Of course, the same pair of enantiomers could be represented in different orientations.

The even simpler representations in C are most commonly used. Such drawings, called **projections** (or Fischer projections), look exactly like structural drawings. When understood to be projections, however, these drawings have additional

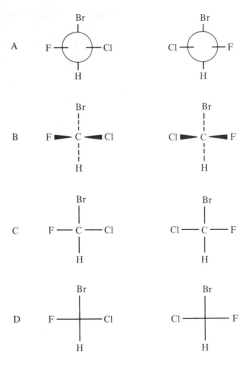

Figure 6.3 Different representations of the pair of enantiomeric bromochlorofluoromethanes.

meaning. We understand that the groups at the right and left of the chiral carbon are above the plane of the paper and the groups at the top and bottom are below the plane. Each projection in C is equivalent to the drawings above it. Because projection formulas look identical to the usual structural drawings, it is the responsibility of a writer to make it clear when a drawing is to be considered a projection. As shown in D, projections are sometimes simplified by omitting the asymmetric carbon.

6.3
SPECIFYING ABSOLUTE CONFIGURATION

The word configuration has been used to refer to the arrangement of atoms that characterizes a particular stereoisomer. Of the two possible mirror image configurations, the one that a particular enantiomer has is called its **absolute configuration**.

How can we specify a particular absolute configuration? Imagine that you need to tell a distant doctor to give a patient the following enantiomer of methamphetamine.

Methamphetamine

This enantiomer is used in some inhalers to relieve nasal congestion. Its enantiomer, commonly called "speed," is a central nervous system stimulant that, because of its harmful physiological action, is carefully regulated by federal law. You could communicate the information by sending a model or a suitable drawing of the appropriate enantiomer. It is obviously more convenient, however, to have a simple designation to specify the absolute configuration.

This situation resembles familiar problems that we face every day. If you have an abdominal pain, you don't need to visit the doctor's office to point to the side that hurts—you can report over the phone that the pain is on the right or left side. Or imagine that you are in the control room of a nitroglycerine plant and suddenly realize that a critical valve in the farthest reaches of the plant must be shut immediately—opening it even the slightest bit more will cause the plant to explode. How do you tell the newly employed technician which way to turn the valve? You shout to turn the valve clockwise or counterclockwise. We have all learned arbitrary but universally understood standards, such as right and left or clockwise and counterclockwise, that enable us to describe directional properties.

To specify the configurations of chiral compounds, we need convenient standards for molecules that correspond to those used for familiar objects and operations. In the most widely used method for specifying configuration, the four groups attached

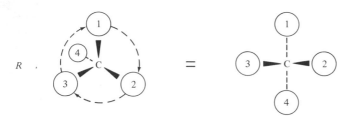

R

S

Figure 6.4 Two views of the R and S standards used for specifying the configuration of an asymmetric carbon. Numbers 1 to 4 are rankings assigned to the four different groups attached to an asymmetric carbon. Each drawing on the right can be converted to the "steering wheel" drawing on the left by tilting it so that group 4 is placed farther behind the paper and group 1 brought forward.

to an asymmetric carbon are ranked in order—1, 2, 3, 4—according to a set of rules.* The two arbitrary standards for specifying configuration are called R (from the Latin *rectus*, right) and S (from the Latin *sinister*, left). As shown in Figure 6.4, these standards correspond to the two possible ways of arranging the four groups around an asymmetric carbon. It is convenient to think of R and S standards in terms of the first ("steering wheel") representation of each in Figure 6.4. In these representations we sight down the bond from carbon to group 4 (with group 4 away from us). Groups 1, 2, and 3 are arranged like the spokes of a wheel. The configuration of the carbon is R when the sequence $1 \rightarrow 2 \rightarrow 3$ is clockwise but S when the same sequence is counterclockwise. Of course, the same models can be drawn in other orientations.

The four groups attached to an asymmetric carbon are ranked by a set of rules.

1. Rank the groups in order of decreasing atomic number of the atoms attached directly to the asymmetric carbon.

Consider the bromochlorofluoromethanes. The atomic numbers of the atoms attached to the asymmetric carbon decrease in the order Br (35) > Cl (17) > F (9) > H (1), so the groups are ranked in that order.

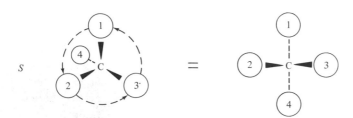

(R)-Bromochlorofluoromethane   (S)-Bromochlorofluoromethane

* Another procedure often used with carbohydrates and some other important biological compounds is described in Section 10.1.

The placement of groups in the first drawing corresponds to the *R* standard and so this enantiomer has the *R* configuration.

Consider 2-chlorobutane.

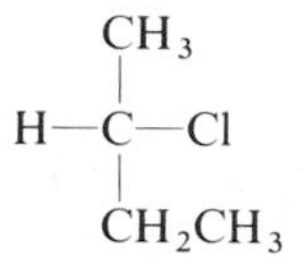

$$
\begin{array}{c}
CH_3 \\
| \\
H-C-Cl \\
| \\
CH_2CH_3
\end{array}
$$

On the basis of rule 1, the chlorine is ranked 1 and the hydrogen ranked 4. Both the methyl and ethyl groups, however, are attached to the asymmetric carbon by a carbon atom. To rank such groups, the following rule is needed.

2.   When two identical atoms are attached to the asymmetric carbon, assign the higher ranking to the atom that is attached to an atom of higher atomic number. If a decision cannot be made at this stage, continue along the chain one atom at a time until you find a difference in atomic number.

The carbon of the methyl group is attached only to hydrogen atoms but that of the ethyl group to another carbon. Therefore ethyl outranks methyl and the enantiomer of 2-chlorobutane shown has the *S* configuration.*

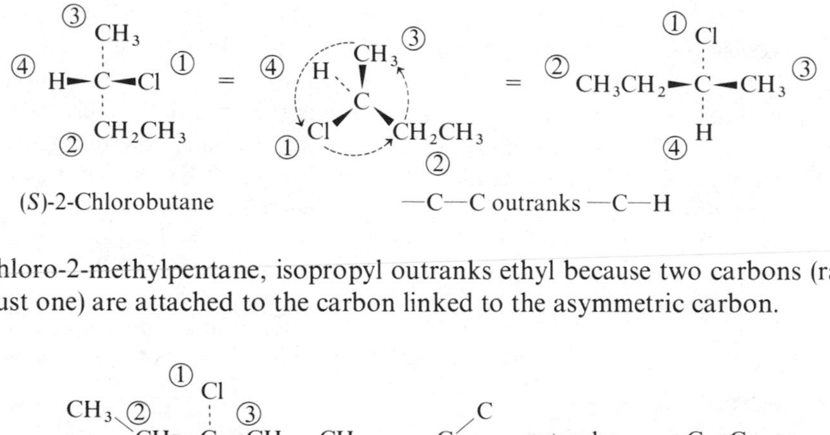

(S)-2-Chlorobutane   —C—C outranks —C—H

In 3-chloro-2-methylpentane, isopropyl outranks ethyl because two carbons (rather than just one) are attached to the carbon linked to the asymmetric carbon.

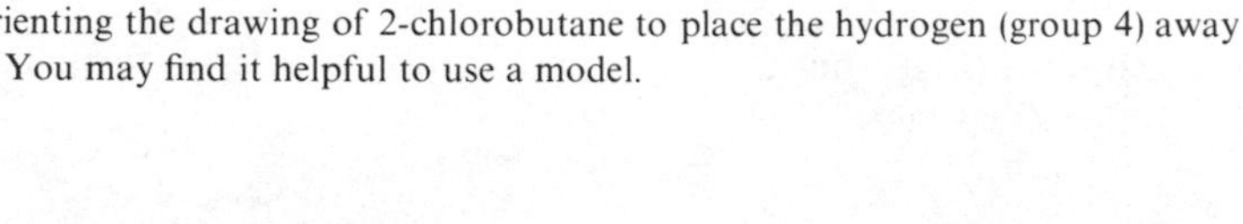

(S)-3-Chloro-2-methylpentane

---

* Reorienting the drawing of 2-chlorobutane to place the hydrogen (group 4) away from you can be confusing. You may find it helpful to use a model.

An additional rule is needed to distinguish between the ethyl and vinyl groups of 3-chloro-1-pentene.

3.   Consider each atom of a double (or triple) bond to be doubled (or tripled).

$$\overset{(C)\quad(C)}{}$$

Therefore $-CH=CH_2$ is considered to be $-\overset{|}{C}H-\overset{|}{C}H_2$ and outranks $-CH_2-CH_3$.

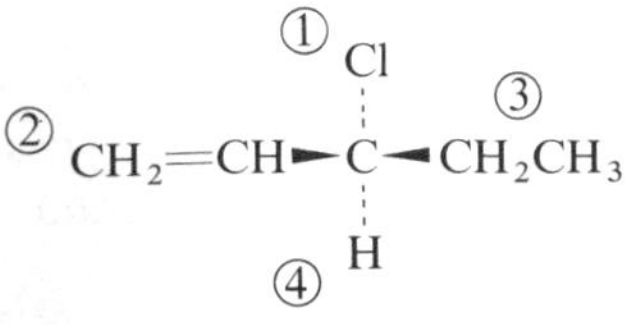

①
Cl

② ③
CH_2=CH➤C◄CH_2CH_3 $-C=C =$ $-\overset{|}{C}-\overset{|}{C}$ outranks $-C-C$

④ H

(S)-3-Chloro-1-pentene

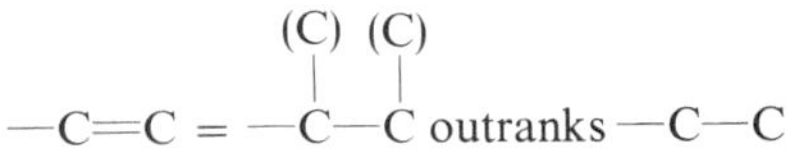

problem 6.2   Specify the configurations of the following compounds.

(a)   Br

CH_3CH_2➤C◄Cl

CH_3

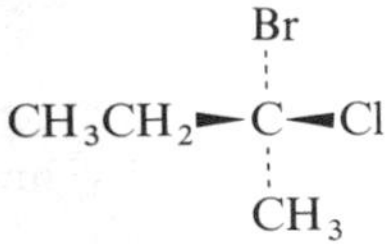

(b) CH_2OH

H➤C◄Cl

CH_2CH_3

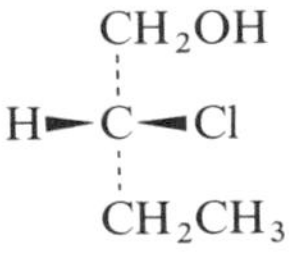

When we discussed cis and trans alkenes, you probably wondered how to specify the configuration of an alkene having four different groups attached to the carbons of the double bond. The configuration of any alkene can be specified unambiguously by the "*EZ*" system. Using the preceding rules, the two substituents at each carbon of the double bond are ranked 1 and 2. The configuration is called *E* (from the German *entgegen*, opposite) or *Z* (from the German *zusammen*, together), depending on which of the following arrangements results.

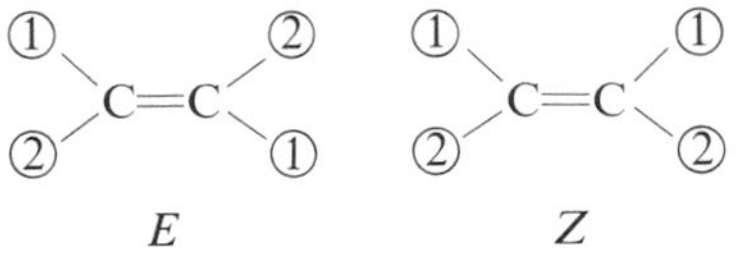

① ②   ① ①
C=C   C=C
② ①   ② ②

E   *Z*

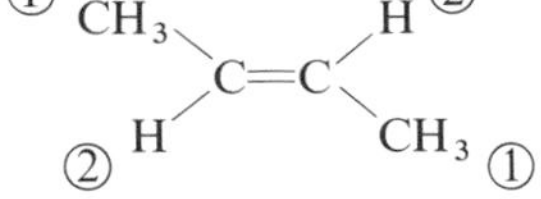

① CH_3 H ②   ① Br CH_2CH_3 ①
C=C   C=C
② H CH_3 ①   ② Cl CH_3 ②

(E)-2-Butene   (Z)-1-Bromo-1-chloro-
trans-2-Butene   2-methyl-1-butene

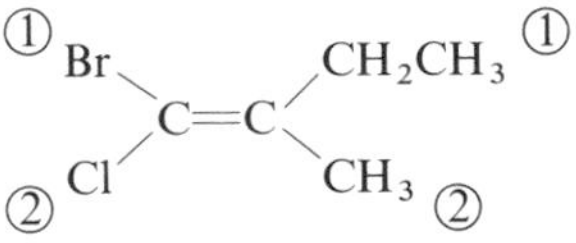

The configuration of the second alkene would be difficult to specify using cis or trans.

6.4
PHYSICAL PROPERTIES OF ENANTIOMERS; OPTICAL ACTIVITY

Most physical properties of a pair of enantiomers are identical. They have the same melting point, boiling point, density, and solubility in water. For example, (R)- and (S)-2-butanol both boil at 99.5° and have a density of 0.807 g/ml. The similarity in properties is hardly surprising for molecules that differ only in being mirror images. Think of the properties of familiar chiral objects—a right and a left shoe have the same length, color, weight, and density.

Enantiomers differ in one easily measured physical property—their interaction with plane-polarized light. In some of its properties light behaves as waves vibrating at right angles to the direction in which the light is moving. The vibrations in a beam of ordinary light are oriented in all possible planes. If such a beam is passed through a polarizing filter, however, only the portion of the light whose vibrations are in a particular plane is transmitted.* The filter may be a Polaroid sheet (a sheet of plastic containing specially oriented crystals) or a specially prepared prism constructed from certain crystals. Light whose vibrations occur in only one plane is called **plane-polarized** light.

| A beam of | A beam of |
| ordinary light | plane-polarized light |

An enantiomer rotates the plane of plane-polarized light. This property can be observed with an instrument called a **polarimeter** (Fig. 6.5). In a polarimeter a beam of plane-polarized light is prepared by passing ordinary light through a polarizing filter. The beam of plane-polarized light then passes through a solution of the compound to be investigated. A second polarizing filter is used to determine the plane of the plane-polarized light that emerges from the solution. The maximum amount of light is transmitted if the orientation of the second filter corresponds to the plane of polarization of the beam of light. If the compound is not an enantiomer, then the orientation of the plane is unchanged; the maximum amount of light is transmitted when the two polarizing filters have the same orientation. If the compound is an enantiomer, however, then the plane of the plane-polarized light is altered; the second polarizing filter must be rotated to maximize light transmission. If the rotation is clockwise (from the viewpoint of the observer), then the enantiomer is called **plus** or **dextrorotatory**. If the rotation is counterclockwise, then the enantiomer is **minus** or **levorotatory**.

The amount by which the light is rotated is indicated by an angle (α). The magnitude of α depends on the nature of the compound and is proportional to the

* More accurately, only the component of vibrations in a particular plane is transmitted.

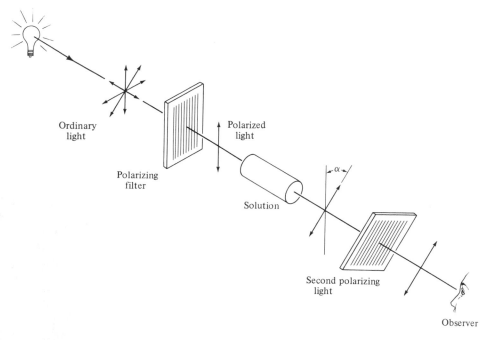

Ordinary
light

Polarized
light

Polarizing
filter

Solution

Second polarizing
light

Observer

Figure 6.5 A schematic drawing of a polarimeter. A beam of plane-polarized light is passed through a solution of the compound to be investigated. If the compound is an enantiomer, then the plane of polarization is rotated. The second polarizing filter must be rotated (by the angle α) to maximize the transmission of light to the observer. In this drawing the observer had to turn the second polarizing filter in a clockwise fashion, so the compound was plus or dextrorotatory.

number of its molecules that the light encounters in passing through the sample tube. To remove variations due to concentration of the solution and length of the sample tube, the observed rotation, α, is converted to a **specific rotation**. The specific rotation, $[\alpha]$, is defined as the rotation caused by a concentration of 1 g/ml in a sample tube 1 decimeter (10 cm) long.

$$\text{specific rotation} = [\alpha]_{\text{wavelength}}^{\text{temperature}} = \frac{\text{observed rotation (in degrees)}}{\text{length (in decimeters)} \times \text{concentration (in g/ml)}}$$

Since the specific rotation also depends on the wavelength of the light and the temperature, they must be indicated.

The specific rotations of (R)- and (S)-lactic acids have the same magnitude. They differ, however, in direction. (R)-Lactic acid is minus and its enantiomer is plus. The members of any pair of enantiomers also exhibit rotations of identical magnitude but opposite direction. Unlike physical properties, such as melting point or solubility, rotation of plane-polarized light is directional. It is reasonable that mirror image molecules rotate plane-polarized light in opposite directions.

$$
\begin{array}{cc}
\text{CO}_2\text{H} & \text{CO}_2\text{H} \\
| & | \\
\text{H}\!-\!\text{C}\!-\!\text{OH} & \text{HO}\!-\!\text{C}\!-\!\text{H} \\
| & | \\
\text{CH}_3 & \text{CH}_3
\end{array}
$$

(R)-$(-)$-Lactic acid $\qquad$ (S)-$(+)$-Lactic acid

$[\alpha]_{546\,\text{nm}}^{22^\circ} = -2.6^\circ$ $\qquad$ $[\alpha]_{546\,\text{nm}}^{22^\circ} = +2.6^\circ$

In the names of the lactic acids, prefixes (R) or (S) and $(+)$ or $(-)$ refer to very different things. R or S indicates the configuration of an enantiomer specified by using an arbitrary set of rules. Plus or minus indicates the experimentally determined direction in which an enantiomer rotates plane-polarized light. *There is no simple relationship between these designations*; some molecules that have R configurations are plus and some are minus.

Because of their effect on plane-polarized light, enantiomers are said to be *optically active*. Measurement of optical activity is useful in recognizing and in working experimentally with enantiomers. If enantiomers differed only in interaction with plane-polarized light, however, their stereoisomerism would be merely a curiosity, probably not worth mentioning in this book. In fact, enantiomers differ in a far more significant way (see Secs. 6.8 and 6.9). First, however, we consider compounds containing two asymmetric carbons because understanding their stereoisomeric relationships provides the key to understanding when two enantiomers may behave very differently.

6.5
COMPOUNDS CONTAINING TWO NONIDENTICAL
ASYMMETRIC CARBONS

Consider 2-bromo-3-chlorobutane, a simple structure that contains two asymmetric carbons (C-2 and C-3).

$$
\overset{1}{\text{CH}_3}\!-\!\overset{2}{\text{CH}}\!-\!\overset{3}{\text{CH}}\!-\!\overset{4}{\text{CH}_3}
$$
$$
\quad\;\;\; | \quad\;\; |
$$
$$
\quad\;\;\; \text{Br} \quad \text{Cl}
$$

2-Bromo-3-chlorobutane

The asymmetric carbons are not identical; one of the four different groups at C-2 is a bromine but at C-3 is a chlorine. As shown in Figure 6.6, four stereoisomers have this structure. To describe the configurations of these stereoisomers, the configuration (R or S) of *each* asymmetric carbon must be specified.

Four isomers are possible because each asymmetric carbon has two possible configurations. Isomers A and D (or B and C) have different configurations at C-2 but the same configuration at C-3. Isomers A and C (or B and D) have the same configuration at C-2 but different configurations at C-3. Isomers A and B (or C

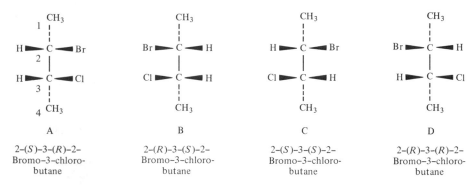

A

2-(S)-3-(R)-2-
Bromo-3-chloro-
butane

B

2-(R)-3-(S)-2-
Bromo-3-chloro-
butane

C

2-(S)-3-(S)-2-
Bromo-3-chloro-
butane

D

2-(R)-3-(R)-2-
Bromo-3-chloro-
butane

Figure 6.6 The four stereoisomeric 2-bromo-3-chlorobutanes. A and B are enantiomers; so are C and D. Other combinations are diastereomers.

and D) differ in configuration at both asymmetric carbons. Compare to the possible combinations of hands and gloves. Hands are left and right and so are gloves, resulting in four combinations:

A′ left hand + right glove
B′ right hand + left glove
C′ left hand + left glove
D′ right hand + right glove

How are the four stereoisomers related? A and B are enantiomers. Both C-2 and C-3 in B have a mirror image relationship to these asymmetric carbons in A. Isomers C and D also are enantiomers. What, however, is the relationship between A and C? Because they have the same configuration at C-2, they are not enantiomers but belong instead to the other class of stereoisomers called **diastereomers** (classes of isomers are reviewed in Table 6.1). A and D also are diastereomers. So are B and C or B and D. A particular stereoisomer has no more than one enantiomer. Any other stereoisomers must be related to it as diastereomers. The analogy with hands and gloves is useful. Combination D′ (right hand + right glove) is the mirror image of combination C′ (left hand + left glove). Yet A′ and B′ (each with one right and one left) are not mirror images of C′ nor are they identical with C′.

All the 2-bromo-3-chlorobutanes are optically active. Because A and B are enantiomers, their rotations have the same magnitude but opposite signs. Other properties—boiling point, melting point, solubility—are identical. The same is true of enantiomers C and D. However, just as do the diastereomers (such as *cis*- and *trans*-2-butene) discussed in earlier chapters, A and C *differ in all properties*. Because diastereomers are not related as mirror images, a difference in properties is to be expected. All other combinations of diastereomers (A and D, B and C, B and D) also differ in properties. The hand-glove analogy again is useful. Combinations C′ and D′—properly gloved left and right hands—have identical properties. Combinations C′ and A′ are very different, however; a left hand tucked snugly into a

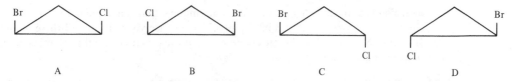

Figure 6.7   The four stereoisomeric 1-bromo-2-chlorocyclopropanes. A and B are enantiomers; so are C and D. Other combinations are diastereomers.

left glove on a cold winter's day differs noticeably from a left hand mismatched with a right glove.

Each 2-bromo-3-chlorobutane is shown in Figure 6.6 in a particular eclipsed conformation. By rotation around the central carbon-carbon bond, each can assume a variety of conformations. Such rotation, however, does not convert one stereoisomer into another. The number of possible stereoisomers does not depend on the particular conformations in which they are depicted or on conformational changes. Consider A and C again. These stereoisomers differ because the configurations at C-3 are different (*R* in A but *S* in C). Rotation around carbon-carbon bonds does not alter the configurations at C-3 that make these isomers different, just as no turning of the chiral bromochlorofluoromethanes pictured in Figure 6.1 makes them equivalent.

1-Bromo-2-chlorocyclopropane

The 1-bromo-2-chlorocyclopropane structure also contains two nonidentical asymmetric carbons. Figure 6.7 shows the four stereoisomers that have this simple cyclic structure. The cis isomers A and B are one pair of enantiomers and the trans isomers C and D another. These isomers correspond to the four 2-bromo-3-chlorobutanes, although the differences between isomers may be easier to visualize with the cyclic compounds. Since rotation around carbon-carbon bonds is impossible in these cyclic compounds, it is obvious that these stereoisomers cannot be interconverted without breaking and re-forming carbon-carbon bonds.

---

exercise with models 6.3   Construct four models that correspond to the 2-bromo-3-chlorobutanes pictured in Figure 6.6. Convince yourself that each asymmetric carbon of A has a configuration opposite to that of the same carbon of B. Do the same with isomers C and D. Convince yourself that one asymmetric carbon has the same configuration in A and C (or in A and D) but that the other asymmetric carbon has opposite configurations. Note that rotation around the central carbon-carbon bond never converts one of these isomers into another.

exercise with models 6.4 Construct four models that correspond to the 1-bromo-2-chlorocyclopropanes pictured in Figure 6.7. Convince yourself that all these models are different. Note that A and B are enantiomers, as are C and D, and that all other combinations are diastereomers.

6.6
COMPOUNDS CONTAINING TWO IDENTICAL
ASYMMETRIC CARBONS

In considering the examples in the preceding section, we stressed the nonidentity of the two asymmetric carbons. When the two asymmetric carbons in a structure are identical (have the same four substituents), then there are only three instead of four stereoisomers. Consider the 2,3-dichlorobutanes.

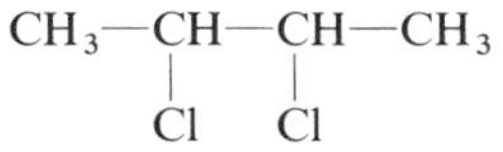

$$CH_3—CH—CH—CH_3$$
$$\underset{Cl}{|} \quad \underset{Cl}{|}$$

2,3-Dichlorobutane

Note that the same four groups (—H, —Cl, —CH$_3$, —CHCH$_3$) surround each of
$$\underset{Cl}{|}$$
the asymmetric carbons. We can quickly draw structures of these compounds by replacing the bromine in each of the 2-bromo-3-chlorobutanes in Figure 6.6 by a chlorine. C and D in Figure 6.8 represent a pair of enantiomers. A and B, however, although mirror images, are identical; they can be superimposed if either is rotated one-half turn in the plane of the paper. Compounds that are achiral even though they contain asymmetric carbons are called **meso compounds**. Because meso compounds are not chiral, they are optically inactive.

The identity of a meso compound with its mirror image may be easier to see if we consider the 1,2-dichlorocyclopropanes. The representations in Figure 6.9 correspond to those for the 1-bromo-2-chlorocyclopropanes in Figure 6.7 with each bromine replaced by a chlorine. A and B are mirror images but obviously identical

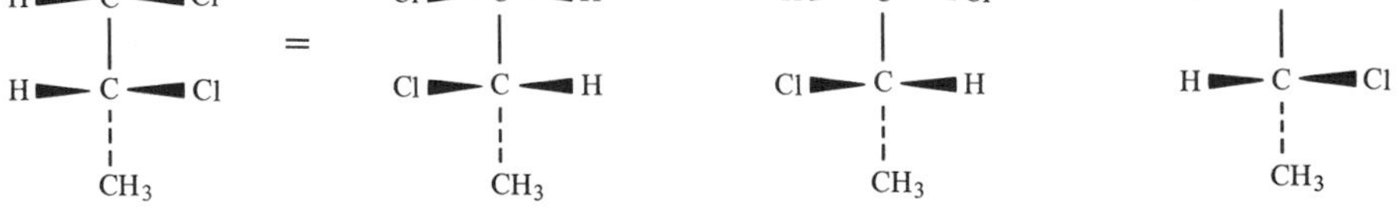

A meso–2, 3–Dichlorobutane
B
C 2–(S)–3–(S)–2, 3–Dichlorobutane
D 2–(R)–3–(R)–2, 3–Dichlorobutane

Figure 6.8 The three 2,3-dichlorobutanes. A and B are two representations of the same meso compound. C and D are enantiomers.

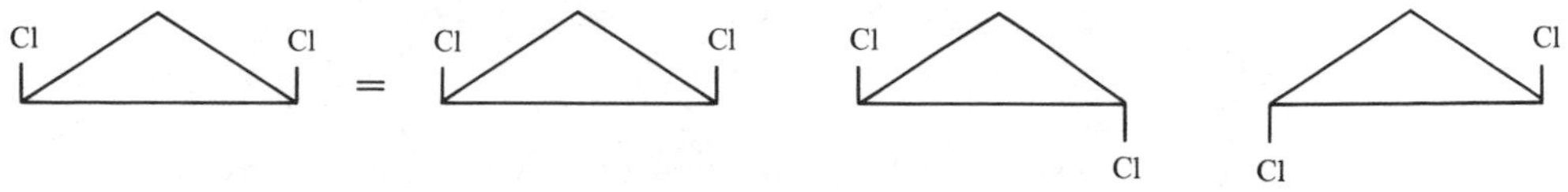

Figure 6.9   The three 1,2-dichlorocyclopropanes. A and B are identical representations of a meso compound. C and D are enantiomers.

and hence represent one meso compound. C and D are nonidentical mirror images and hence represent a pair of enantiomers.

The analogy with familiar objects is again useful. Although there are four possible glove–hand combinations, there are only three hand–hand combinations:

A′ (= B′)   left hand + right hand (= right hand + left hand)
C′   left hand + left hand
D′   right hand + right hand

exercise with models 6.5   Construct four models that correspond to the 2,3-dichlorobutanes shown in Figure 6.8. Convince yourself that A and B are mirror images but identical and hence represent a single meso compound. Note that C and D are nonidentical mirror images and hence represent enantiomers.

Do the same with the 1,2-dichlorocyclopropanes in Figure 6.9.

problem 6.3   How many stereoisomers correspond to each of the following structures?

(a)

(b) $CH_3—CH_2—CH—CH_2—CH—CH_2—CH_3$
             | |
             CH_3 CH_3

(c) $CH_3—CH—CH—CH_3$
            | |
            OH CH_3

(d)

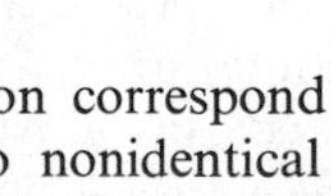

6.7
COMPOUNDS CONTAINING MORE THAN TWO
ASYMMETRIC CARBONS

Structures with one asymmetric carbon correspond to two isomers (one pair of enantiomers) and structures with two nonidentical asymmetric carbons to four stereoisomers (two pairs of enantiomers). Each additional nonidentical asymmetric carbon doubles the number of isomers. Imagine adding a third asymmetric carbon (—CH—CH₃) to the 2-bromo-3-chlorobutane structure.
   |
   OH

$$CH_3-CH-CH-CH-CH_3$$

$$\underset{Br}{|} \quad \underset{Cl}{|} \quad \underset{OH}{|}$$

Either configuration (R or S) of the new asymmetric carbon can be added to each stereoisomer shown in Figure 6.6 to give two new isomers, resulting in a total of eight isomers. The general rule is that a structure with n nonidentical asymmetric carbons corresponds to 2^n isomers [$2^{(n-1)}$ pairs of enantiomers].*

The structure of the steroid cholesterol has eight nonidentical chiral carbons (circled in the drawing) and hence $2^8 = 256$ stereoisomers.

Cholesterol

As is ordinarily the case with biological compounds, a living organism contains only one of these stereoisomers.

6.8
RESOLUTION OF RACEMIC MIXTURES

Only one of a pair of enantiomers is usually found in a living system. Laboratory reactions (see Sec. 6.9), however, ordinarily give both enantiomers and in equal amounts. Because an exactly equal mixture of enantiomers is frequently encountered, it is assigned a special name: **racemic mixture**. In addition to bottles of (+)-lactic acid and (−)-lactic acid, a chemical stockroom may have a bottle labeled racemic lactic acid or (±)-lactic acid. The effects of its (+) and (−) enantiomers on a beam of plane-polarized light exactly cancel, so a racemic mixture is optically inactive.

Because laboratory syntheses often give racemic mixtures, it is important to separate such mixtures into the individual enantiomers, a process called **resolution**. A pair of enantiomers cannot be separated by the procedures ordinarily used to separate organic compounds. Having identical boiling points, enantiomers cannot be separated by distillation. Having identical solubilities, they cannot be separated by ordinary crystallization procedures.

A general procedure for separating enantiomers involves their chemical conversion to diastereomers. It is worth examining this procedure carefully, for it gives us some insight into the nature of situations in which enantiomers behave differently.

---

* If some asymmetric carbons are identical, then (as we saw for the case of two identical asymmetric carbons) the number of isomers will be less.

Consider the specific problem of resolving a racemic mixture of a chiral amine. To show how to do so, we need to know that amines react rapidly with carboxylic acids to form salts (Eq. 1) and that the amines can be regenerated by treating the salts with base (Eq. 2).

$$RNH_2 + RCO_2H \longrightarrow RNH_3^+ \; RCO_2^- \tag{1}$$

$$RNH_3^+ \; RCO_2^- + OH^- \longrightarrow RNH_2 + RCO_2^- + H_2O \tag{2}$$

For the present, the details of these reactions (considered in Chapters 8 and 11) are unimportant; we need recognize only that they occur.

Consider the reaction of a racemic mixture of 1-phenylethanamine with the *S* enantiomer of malic acid, a compound isolated from apples and many other fruits.

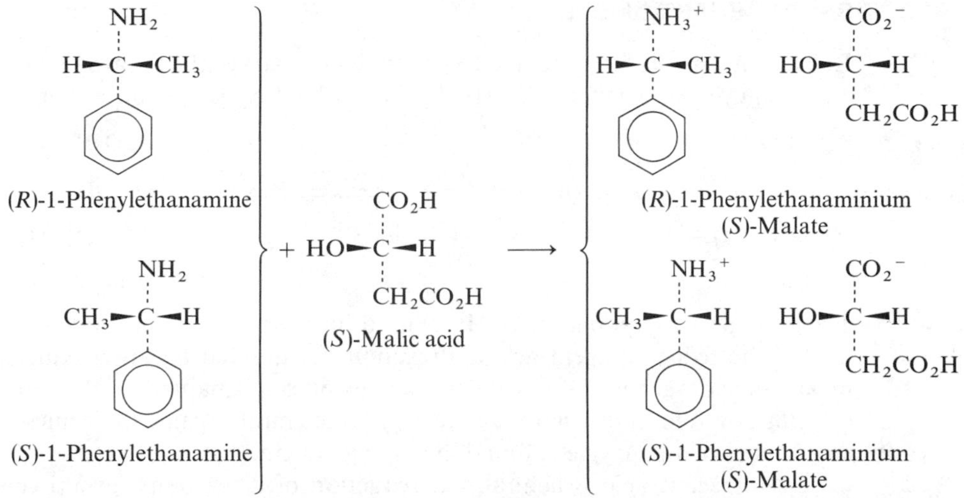

Both the (*R*) and (*S*) enantiomers of the amine react with (*S*)-malic acid and so two salts form. The two salts *are diastereomers.* Because their properties are different, they can be separated by conventional methods. In this particular example advantage is taken of solubility properties. One salt crystallizes rapidly from an aqueous solution.

Once the salts have been separated, each can be converted to one of the enantiomeric amines. Treating (*R*)-1-phenylethanaminium (*S*)-malate with base, for example, forms the *R* enantiomer of the amine.

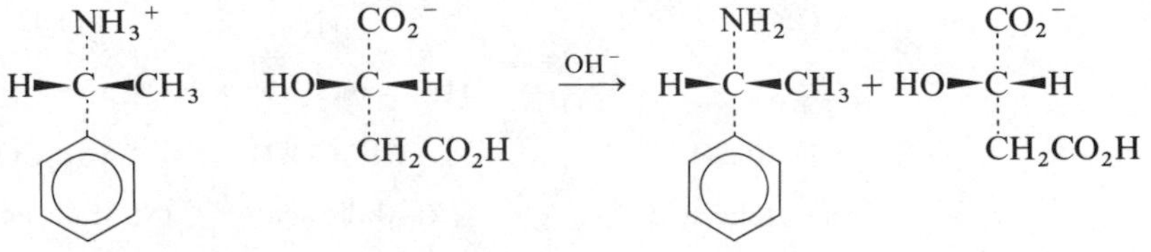

(*R*)-1-Phenylethanaminium (*S*)-malate

This resolution took advantage of the chemical process in which enantiomers differ—their reaction with a chiral compound. An analogy with feet and shoes may be useful. How can you separate a bin filled with right and left shoes? You cannot do it on the basis of such physical properties of the shoes as weight or length; they are identical for right and left shoes. You can do it by jamming your left foot repeatedly into the bin, however. Occasionally your foot will plunge into a left shoe (but presumably not into a right shoe). You can then withdraw your foot, remove the shoe, and repeat the process until all left shoes have been removed from the bin. Two objects, whether molecules or shoes, that are enantiomers of each other *differ in their interactions with another chiral object.*

6.9
REACTIONS OF ENANTIOMERS

Reduction of oxaloacetic acid to (*S*)-malic acid is part of the tricarboxylic acid cycle, a critical cycle of reactions in virtually all aerobic (oxygen-requiring) organisms.

$$
\begin{array}{ccc}
CO_2H & & CO_2H \\
| & & \vdots \\
C{=}O \quad + NADH + H^+ \xrightleftharpoons{(S)\text{-malate dehydrogenase}} & HO{\blacktriangleright}C{\blacktriangleleft}H \quad + NAD^+ \\
| & & \vdots \\
CH_2CO_2H & & CH_2CO_2H
\end{array}
$$

Oxaloacetic acid .... (*S*)-Malic acid

The biological compound NADH (reduced form of nicotinamide adenine dinucleotide) is the reducing agent and the reaction is catalyzed by an enzyme called (*S*)-malate dehydrogenase. We will consider enzymes in Chapter 14. For the moment, it is important to know that they are very large chiral organic molecules—only one enantiomer of an enzyme is found in a living system.

Oxaloacetic acid is achiral, but reduction of its carbonyl group generates an asymmetric carbon. Only (*S*)-malic acid is produced in this reduction. (*R*)-malic acid is not detected. The reaction is reversible and, in the presence of the enzyme, malic acid can be oxidized to oxaloacetic acid. If racemic malic acid is added to a solution of the enzyme and NAD$^+$, only (*S*)-malic acid molecules are oxidized—(*R*)-malic acid molecules do not react. Such specificity for a single enantiomer is typical of biological reactions.

Consider the same oxidations and reductions carried out with ordinary laboratory reagents. Catalytic hydrogenation of oxaloacetic acid produces equal amounts (a racemic mixture) of (*R*)-malic acid and (*S*)-malic acid.

$$
\begin{array}{cccc}
CO_2H & & CO_2H & CO_2H \\
| & & \vdots & \vdots \\
C{=}O & \xrightleftharpoons[KMnO_4]{H_2(Pt)} & H{\blacktriangleright}C{\blacktriangleleft}OH \quad + & HO{\blacktriangleright}C{\blacktriangleleft}H \\
| & & \vdots & \vdots \\
CH_2CO_2H & & CH_2CO_2H & CH_2CO_2H
\end{array}
$$

Oxaloacetic acid .... (*R*)-Malic acid ...... (*S*)-Malic acid

Malic acid can be oxidized to oxaloacetic acid by reaction with potassium permanganate. The *R* and *S* isomers, however, are oxidized at the same rate.

It is easy to understand why the laboratory reactions show no preference for one enantiomer. (R)-Malic acid is formed by attachment of hydrogen (adsorbed on the catalyst) to one of the two sides of the carbonyl group and (S)-malic acid by addition of hydrogen to the other side. The probability of the two sides of the carbonyl group of oxaloacetic acid approaching the catalyst surface are equal. In the reaction of a large number of molecules (even 1 mg contains more than 10^{18} molecules), essentially equal numbers of the enantiomeric molecules are formed. Similarly, there is an equal probability of potassium permanganate encountering and reacting with each malic acid enantiomer.

What is different in the biological reactions? It is the presence of the chiral enzyme. Enzymes show a remarkable selectivity for their **substrates**, the compounds whose reactions they catalyze. A substrate fits into a cavity in an enzyme. The shape of the cavity and the arrangement around it of the functional groups that interact with the substrate lead to the remarkable specificity. Because the cavity is chiral, it fits just one enantiomer of a substrate (just as a left glove fits a left but not a right hand). The substrate, any other molecules involved in the reaction (NADH in the oxaloacetic acid reduction), and functional groups of the enzyme that take part in the reaction are brought together in the cavity. Figure 6.10 illustrates one hypothetical way in which the enzyme could be specific for (S)-malic acid.

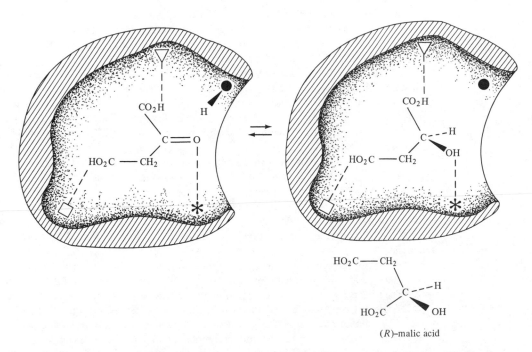

Figure 6.10   One hypothetical way in which an enzyme could be specific for oxidation and reduction involving (S)-malic acid. Around the cavity are sites (□ and ▽) that interact (perhaps by hydrogen bonding) with the carboxyl groups, a site (∗) that interacts with the carbonyl or hydroxyl group, and a site (●) that can donate or accept a hydrogen. These sites do not accommodate (R)-malic acid.

Consider more generally the interaction of an enzyme ((*R*)-ENZ) with the enantiomers of some substrate (SUB).

$$
(R)\text{-ENZ} +
\begin{cases}
(R)\text{-SUB} \\
(S)\text{-SUB}
\end{cases}
\longrightarrow
\begin{array}{c}
(R)\text{-ENZ}\text{-}\text{-}\text{-}\text{-}(R)\text{-SUB} \\
+ \\
(R\text{-})\text{ENZ}\text{-}\text{-}\text{-}\text{-}(S)\text{-SUB}
\end{array}
$$

The two species that can potentially form are diastereomers and hence have different properties. This is analogous to the formation of diastereomers when racemic 1-phenylethanamine is treated with a single enantiomer of a chiral carboxylic acid (Sec. 6.8). The stabilities of diastereomeric complexes involving enzymes are so different that one complex generally does not form.

Compounds that are enantiomers of each other behave differently toward another chiral molecule. It is for this reason that learning about chiral molecules is so important.

SUMMARY

Definitions and Ideas

An object is called **chiral** if it is not identical with its mirror image and **achiral** if identical with its mirror image. Stereoisomers are either enantiomers or diastereomers. **Enantiomers** are stereoisomers that are mirror images of one another. **Diastereomers** are stereoisomers that are not mirror images.

A carbon that has four different groups attached to it is called an **asymmetric carbon** (or a chiral carbon). A structure that contains one asymmetric carbon corresponds to a pair of enantiomers. A structure that contains two different asymmetric carbons corresponds to four stereoisomers (two pairs of enantiomers). The general rule is that a structure having n different asymmetric carbons corresponds to 2^n stereoisomers [$2^{(n-1)}$ pairs of enantiomers].

A structure with two identical asymmetric carbons corresponds to three isomers: a pair of enantiomers and a meso compound. A **meso compound** is a compound that is achiral even though it contains asymmetric carbons.

Enantiomers are often represented by **projections** (or Fischer projections). It is understood that groups drawn to the left and right of an asymmetric carbon are above the plane of the paper and groups drawn to the top and bottom of that carbon are below the plane.

Of the two possible mirror image configurations, the one that a particular enantiomer has is called its **absolute configuration**. Absolute configuration can be specified by using R or S. The four groups attached to an asymmetric carbon are ranked 1, 2, 3, and 4, using a set of rules. Group 4 is placed away from the observer and the sequence $1 \rightarrow 2 \rightarrow 3$ traced. The configuration is R if this sequence is clockwise but S if counterclockwise.

The configuration about the double bond of an alkene can be specified as E or Z. The two groups at each double bond carbon are ranked 1 and 2. The configuration is Z if the groups numbered 1 are on the same side of the double bond but E if they are on opposite sides.

Physical properties of diastereomers are different. Most physical properties of a pair of enantiomers are identical, although they differ in their interaction with plane-polarized light. An enantiomer rotates the plane of plane-polarized light and is said to be *optically active*. Two compounds that are enantiomers of each other rotate plane-polarized light by the same amount but in opposite directions. Rotation is called **plus** or **dextro-rotatory** if clockwise and **minus** or **levorotatory** if counterclockwise.

An equal mixture of a pair of enantiomers is called a **racemic mixture**. Separation of a racemic mixture into the individual enantiomers is called **resolution**. A racemic mixture can be resolved by using a reaction with a single enantiomer of a chiral compound to form two products that are diastereomers. The diastereomers can be separated and then each converted to one of the enantiomers.

Two compounds that are enantiomers of each other behave differently in a reaction with a chiral compound. This is the significant difference between enantiomers and is particularly important in biological reactions.

PROBLEMS

1. Which of the following articles are chiral?

(a) a cup   (b) a pair of glasses
(c) a wood screw   (d) a screwdriver
(e) a bicycle   (f) a chair
(g) a pair of scissors   (h) a skateboard
(i) a pin   (j) a clock

2. Which of the following structures or names represent chiral molecules?

(a) $CH_3CH_2CH_2CH_2CHCH_3$
$\qquad\qquad\qquad\quad |$
$\qquad\qquad\qquad\;\; Br$

(b) $CH_3CH_2CHCH_2CH_3$
$\qquad\qquad\qquad |$
$\qquad\qquad\quad OH$

(c) $CH_3CH{=}CHBr$

(d) $CH_3CHCH_2CHCH_3$
$\qquad\qquad | \qquad\;\; |$
$\qquad\quad CH_3 \quad\; CH_3$

(e) 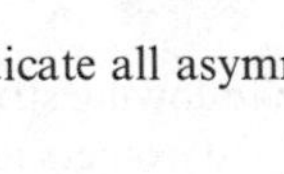 —OH

(f) 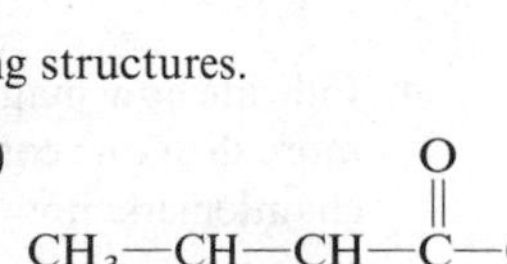 —$CHCH_2CH_3$
$\qquad\qquad\qquad\qquad\qquad |$
$\qquad\qquad\qquad\qquad CH_3$

(g)$\quad$ Cl
$\qquad\quad |$
$\;\; CH_3CCH_2CH_3$
$\qquad\quad |$
$\qquad CH_2CH_3$

(h) *trans*-1,2-dichlorocyclobutane

3. Indicate all asymmetric carbons in the following structures.

(a)

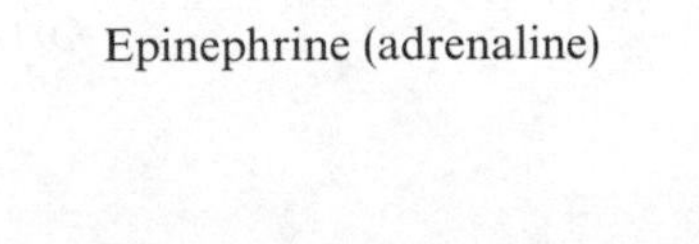

HO——CH—CH$_2$
$\qquad\qquad\qquad\qquad |$
HO
$\qquad\qquad\qquad$ OH
$\qquad\qquad\qquad\qquad\quad$ NH—CH$_3$

Epinephrine (adrenaline)

(b)

$\qquad\qquad\qquad\qquad\qquad O$
$\qquad\qquad\qquad\qquad\qquad \|$
$CH_3—CH—CH—C—OH$
$\qquad\quad | \qquad\; |$
$\qquad\;\; OH \;\; NH_2$

Threonine (an amino acid)

(c) limonene (a terpene)

(structure in
Sec. 3.21)

(d) 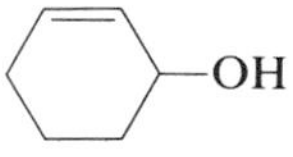 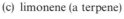—OH

(e)

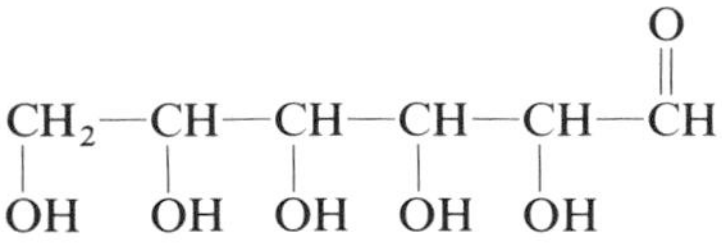

$$CH_2—CH—CH—CH—CH—\overset{\overset{\displaystyle O}{\|}}{CH}$$
$$\quad|\quad\quad|\quad\quad|\quad\quad|\quad\quad|$$
$$OH\quad OH\quad OH\quad OH\quad OH$$

Glucose (a carbohydrate)

(f)

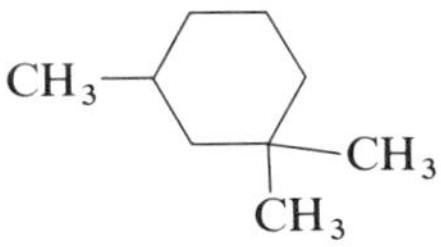

CH_3— cyclohexane —CH_3
CH_3

(g)

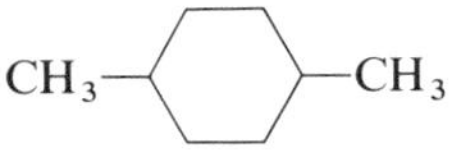

CH_3— cyclohexane —CH_3

(h)

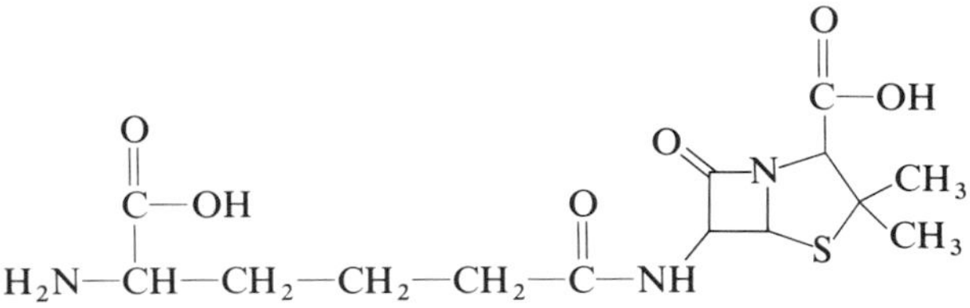

$$\overset{\overset{\displaystyle O}{\|}}{C}—OH$$

$$H_2N—\overset{|}{CH}—CH_2—CH_2—CH_2—\overset{\overset{\displaystyle O}{\|}}{C}—NH$$

Penicillin N

(i)

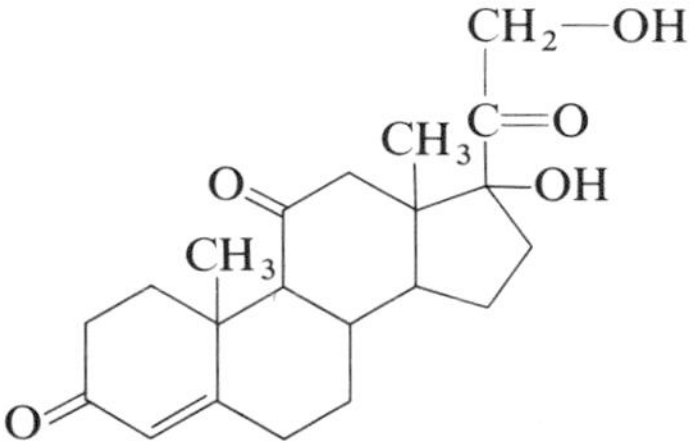

$CH_2—OH$

$CH_3\overset{\displaystyle C=O}{}$

—OH

CH_3

Cortisone (a steroid)

(j)

$$OH \quad O \quad OH \quad O \qquad\qquad \overset{\overset{\displaystyle O}{\|}}{C}—NH_2$$
$$OH$$
$$OH$$
$$HO \quad CH_3 \qquad N—CH_3$$
$$CH_3$$

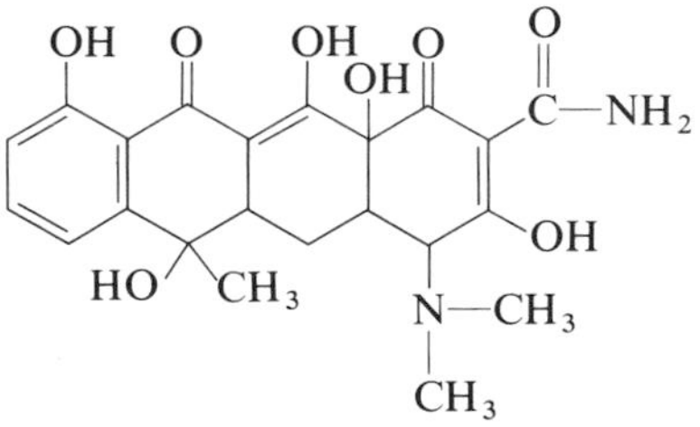

Tetracycline (an antibiotic)

4. Indicate how many compounds have each of the following structures. When there is
more than one compound, indicate the nature of the isomers (e.g., how many pairs of
enantiomers, how many meso compounds, etc.).

(a) 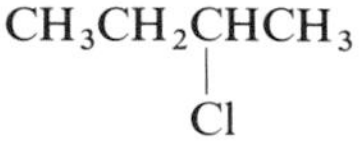 $CH_3CH_2\overset{|}{C}HCH_3$
$\quad\quad\quad Cl$

(b) 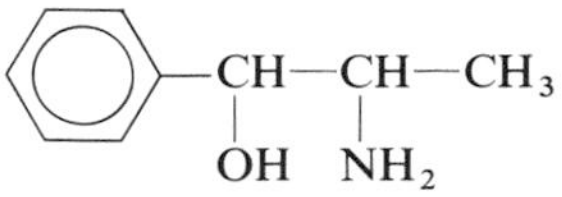 phenyl—$\overset{|}{C}H—\overset{|}{C}H—CH_3$
$\quad\quad\quad OH \quad NH_2$

(c) 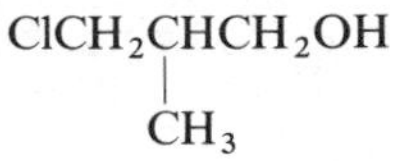 $ClCH_2CHCH_2OH$
　　　　　|
　　　　CH_3

(d) 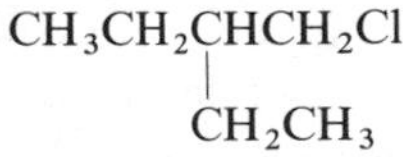 $CH_3CH_2CHCH_2Cl$
　　　　　　　|
　　　　　　CH_2CH_3

(e)

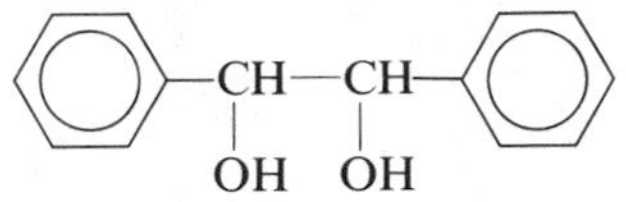

$-CH-CH-$
　|　　|
OH　OH

(f) 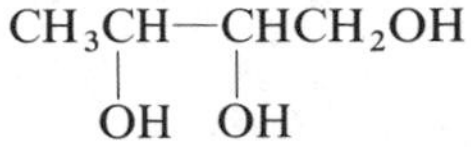 $CH_3CH-CHCH_2OH$
　　　　　|　　|
　　　　OH　OH

(g) 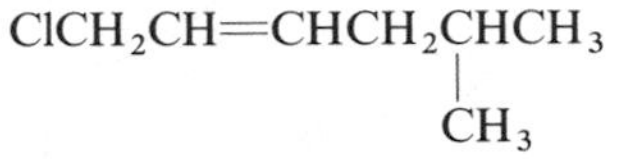 $ClCH_2CH=CHCH_2CHCH_3$
　　　　　　　　　　　|
　　　　　　　　　　CH_3

(h) 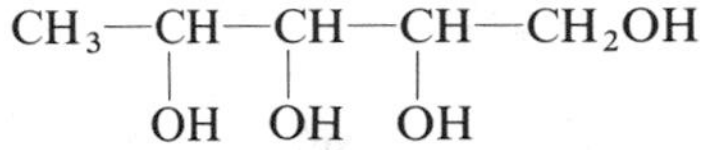 $CH_3-CH-CH-CH-CH_2OH$
　　　　　　|　　|　　|
　　　　　OH　OH　OH

(i) 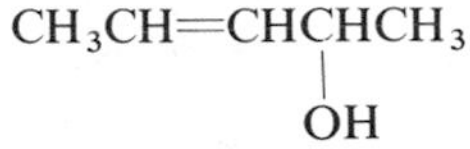 $CH_3CH=CHCHCH_3$
　　　　　　　　|
　　　　　　　OH

(j)

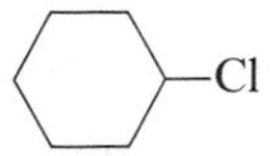

$-Cl$

(k)

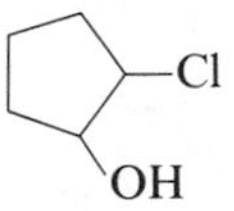

$-Cl$
OH

(l)

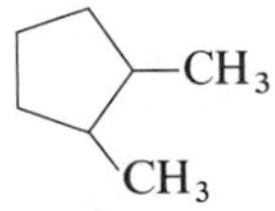

$-CH_3$
CH_3

(m)

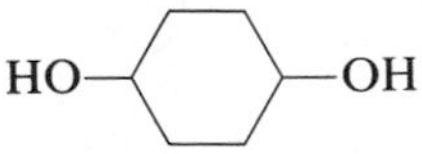

$HO-$ $-OH$

(n)

$-Cl$

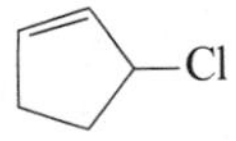

5. Draw all chiral compounds that satisfy the following descriptions.
　(a) have the molecular formula C_4H_9Cl
　(b) are named as methyloctanes
　(c) have molecular formula $C_4H_{10}O$
　(d) have molecular formula $C_5H_{12}O$
　(e) have molecular formula C_6H_{12} and give an optically inactive alkane on hydrogenation

6. Draw representations of the following.
　(a) the smallest acyclic alkane that is chiral
　(b) the smallest cycloalkane that is chiral

7. Assign an *R* or *S* configuration to each compound.

(a) 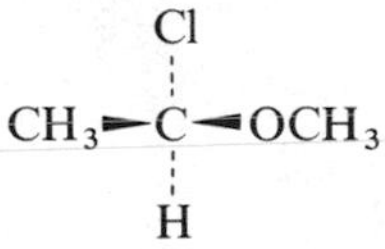　　　　Cl
　　　　　|
　$CH_3-C-OCH_3$
　　　　　|
　　　　　H

(b)　　　　NH_2
　　　　　|
　$CH_3-C-CH_2CH_3$
　　　　　|
　　　　　H

(c)　　CH_3
　　　　|
　$H-C-OH$
　　　　|
　　CH_2Cl

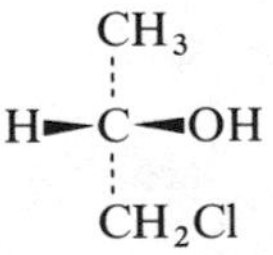

(d) 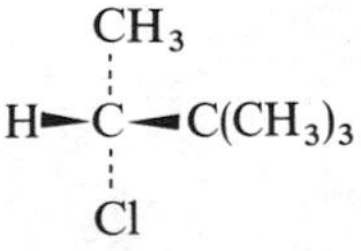　　CH_3
　　　　|
　$H-C-C(CH_3)_3$
　　　　|
　　　Cl

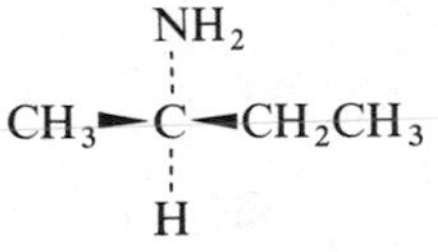

(e)

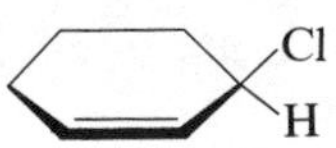

Cl
H

(f) 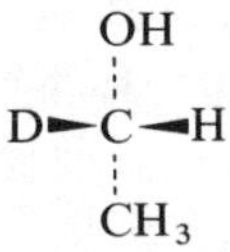　　OH
　　　　|
　$D-C-H$
　　　　|
　　CH_3

(D is a heavier isotope of H.)

8. Which of the following drawings represent (*R*)-2-chlorobutane and which (*S*)-2-chlorobutane?

(a)

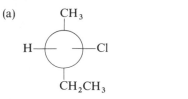

(b)

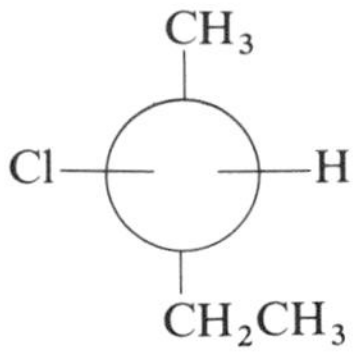

(c)

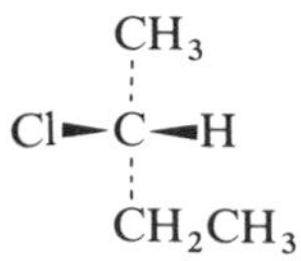

CH₃
Cl►C◄H
CH₂CH₃

(d)

CH₂CH₃
Cl►C◄H
CH₃

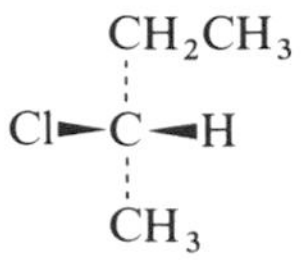

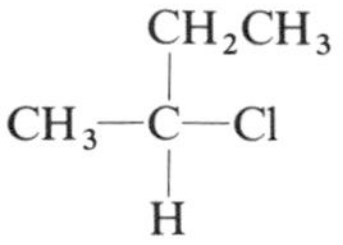

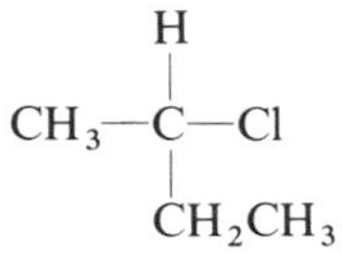

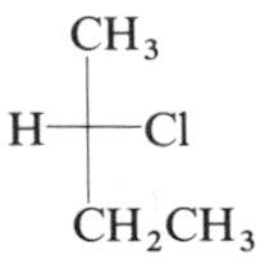

(e)

CH₂CH₃
CH₃─C─Cl
H

(f)

H
CH₃─C─Cl
CH₂CH₃

(g)

CH₃
H┼Cl
CH₂CH₃

(h)

Cl
H┼CH₃
CH₂CH₃

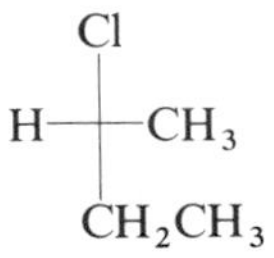

9. Draw suitable representation of the following compounds.
(a) (*R*)-1-bromo-1-chloroethane   (b) (*S*)-2-bromo-1-propanol
(c) (*R*)-2-chloropentane   (d) *meso*-1,2-cyclopentanediol

10. Assign an *E* or *Z* configuration to each compound.

(a) 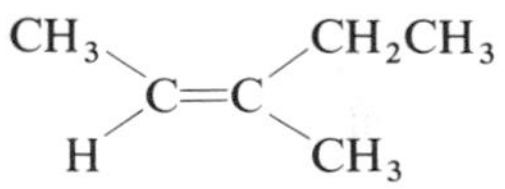 CH₃   CH₂CH₃
       C=C
    H   CH₃

(b)

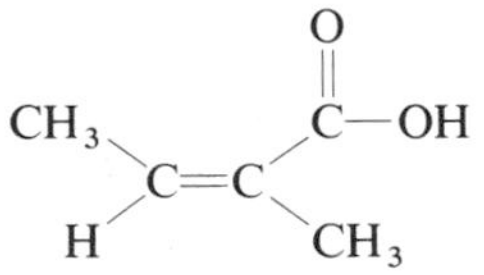

         O
         ‖
CH₃   C─OH
    C=C
H   CH₃

Angelic acid

(c)

CH₃(CH₂)₇   (CH₂)₇─C─OH   (with O above double bonded)
          C=C
       H   H

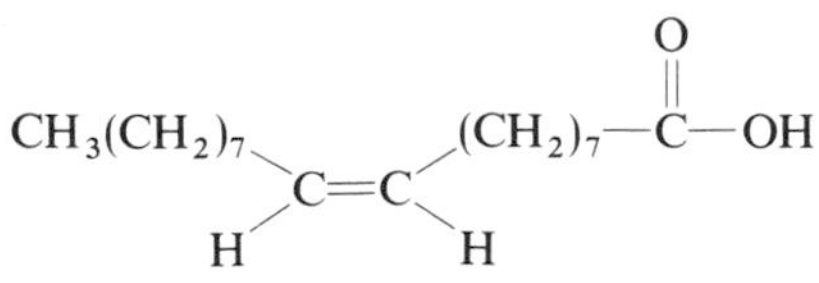

Oleic acid
(obtained from fats and oils)

(d) cyclohexene

11. Consider the following physical properties: melting point, boiling point, density, solubility in water, and specific rotation.
(a) In which of these properties do (*R*)-2-chlorobutane and (*S*)-2-chlorobutane differ?
(b) In which of these properties do (+)-tartaric acid and *meso*-tartaric acid differ?
(c) In which of these properties do *cis*-3-hexene and *trans*-3-hexene differ?

12. Calculate $[\alpha]_{589\,nm}^{20°}$ for (R)-glyceraldehyde. A solution of 2.00 g dissolved in sufficient water to give 100 ml of solution gave an observed rotation (α) of $+0.174°$ in a 10-cm polarimeter sample tube at 20°, using light with a wavelength of 589 nm.

13. A solution of a compound gives an observed rotation of 20° in a polarimeter. How can you tell whether the plane of polarized light is really rotated 20°, 380° (20° + 360°), $-340°$, etc.?

14. Indicate the configurations of the organic products of the following reactions.

(a)
$$CH_3$$
$$H{-}C{-}CH_2CH_2OH \xrightarrow{HBr}$$
$$CH_2CH_3$$

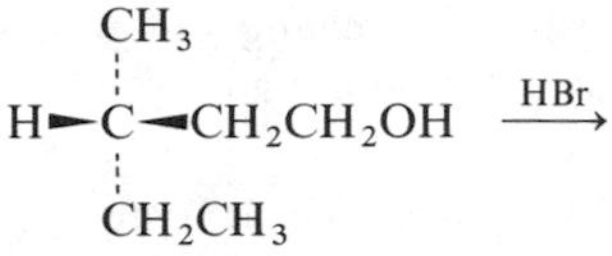

(b) $CH_3CH_2CH{=}CH_2 \xrightarrow{HBr}$

(c)
$$CH_2CH_3$$
$$H{-}C{-}OH \xrightarrow[Pt]{H_2}$$
$$CH{=}CH_2$$

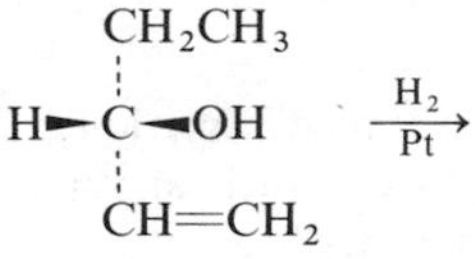

(d)

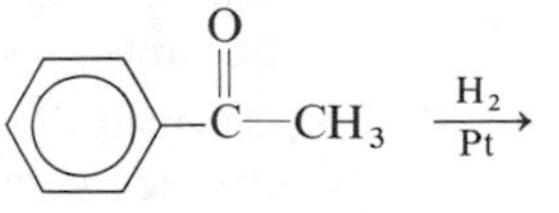

(e)

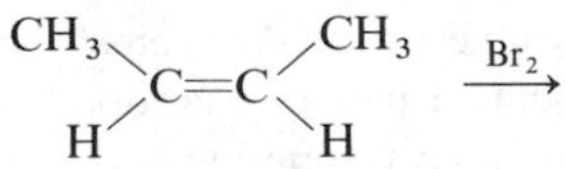

(f)

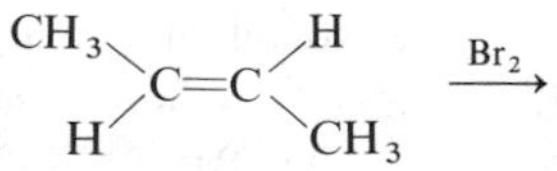

15. Reduction of the R ketone that is shown leads to generation of a new chiral carbon and therefore to two isomeric 3-phenyl-2-butanols.

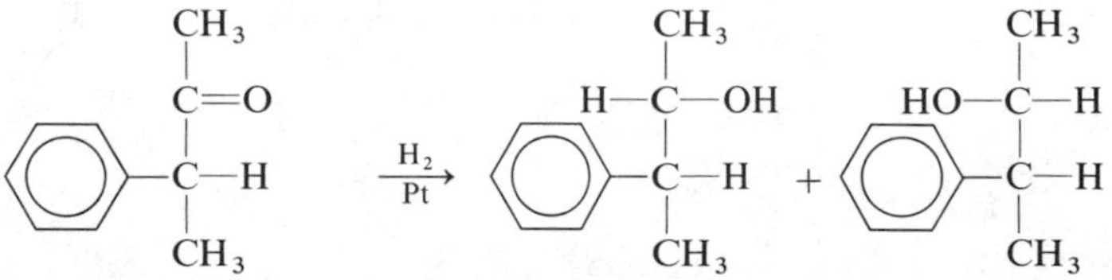

(R)-3-Phenyl-2-butanone   Major product   Minor product

In contrast to the reduction of oxaloacetic acid discussed in Section 6.9, however, the reduction products are formed in unequal amounts. Why is that reasonable in this case?

16. The resolution of racemic 1-phenylethanamine described in Section 6.8 was carried out by using (S)-malic acid. Why couldn't the more readily available racemic malic acid have been used instead?

17. (a) In projection formulas all bonds appear to be the same. Yet they don't all have the same meaning. Projection B, obtained by a 90° rotation of projection A, appears to be the same as A. Show by making models that A and B represent enantiomers.

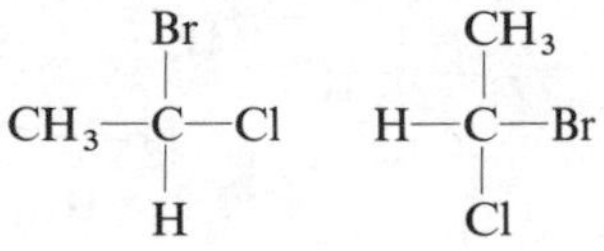

The moral is that projection drawings cannot be moved about at random.

(b) Which of the following projections represent the same enantiomer?

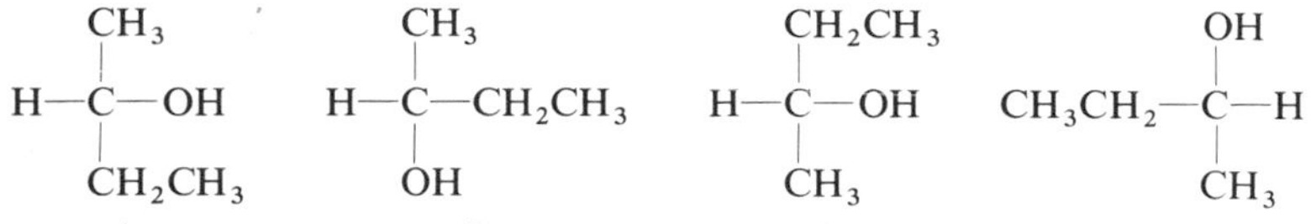

$$
\begin{array}{cccc}
CH_3 & CH_3 & CH_2CH_3 & OH \\
| & | & | & | \\
H-C-OH & H-C-CH_2CH_3 & H-C-OH & CH_3CH_2-C-H \\
| & | & | & | \\
CH_2CH_3 & OH & CH_3 & CH_3 \\
A & B & C & D
\end{array}
$$

(c) Determine a simple rule for which interchanges of groups in a projection lead to another representation of the same isomer and which to a representation of its enantiomer.

18. The two **gauche** conformations of butane (Sec. 1.8) are mirror images that are not identical. If in doubt about this, construct models. Why don't we consider the two gauche forms to be enantiomers?

19. The insecticide lindane (Sec. 7.10) is one of nine isomeric 1,2,3,4,5,6-hexachlorocyclohexanes. Draw representations of the nine isomers and indicate which are optically active.

20. Not all chiral compounds have asymmetric carbon atoms. Try to figure out why each of the following structures represents a pair of enantiomers. Make suitable representations of the two enantiomers that correspond to each structure. If you get stuck, constructing models may help.

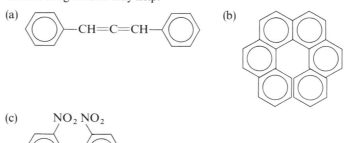

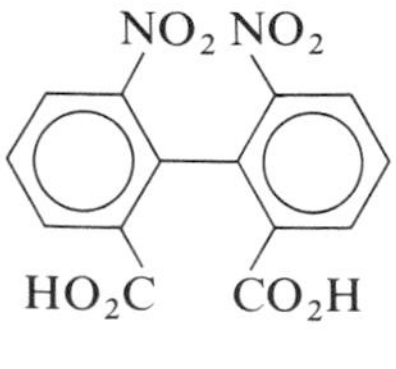

ADDITIONAL EXERCISE WITH MODELS

Construct models of the chair conformation of *cis*-1-bromo-2-chlorocyclohexane shown in A and of its mirror image in B.

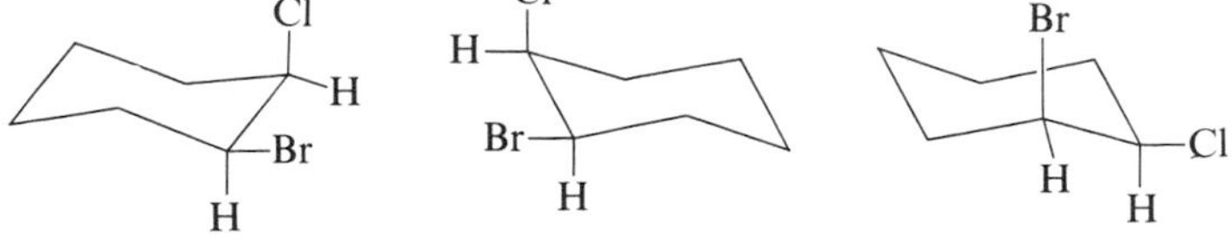

Note that A and B are enantiomers (mirror images but not superimposable). Each may assume other conformations. For instance, A may be flipped to chair conformation A′. Note, however, that flipping never makes A and B identical.

Construct models of the chair conformation of *trans*-1-bromo-2-chlorocyclohexane shown in C and of its mirror image in D.

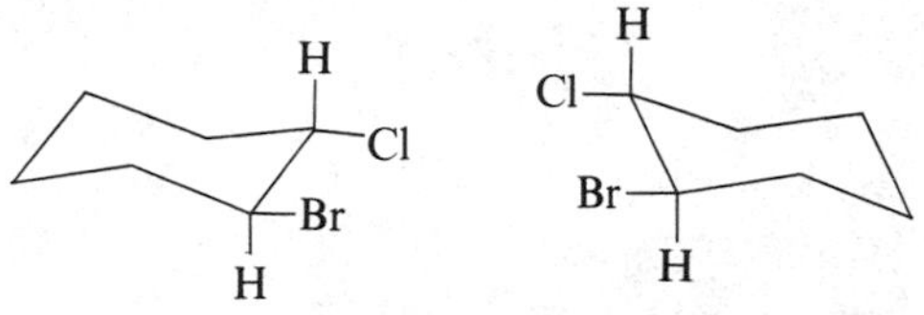

Note that C and D are enantiomers and that flipping never makes them identical (nor identical to A or B).

Therefore there are four stereoisomeric 1-bromo-2-chlorocyclohexanes (two pairs of enantiomers). Although each stereoisomer can assume more than one conformation, the number of stereoisomers is the same as for acyclic compounds with two chiral carbons.

Organic Halides

Relatively few of the organic compounds in living organisms contain halogens. Nevertheless, halogen-containing compounds play an important role in organic chemistry, particularly in the synthesis of other groups of compounds. Moreover, many commercially important organic compounds contain halogens. In this chapter we review what we already learned about organic halides and learn more about their properties and reactions.

7.1
INTERCONVERSIONS OF IMPORTANT GROUPS OF COMPOUNDS

It is easy to be overwhelmed by the seemingly endless number of organic reactions. It may be useful to stop and survey some of the fundamental reactions already discussed or soon to be encountered.

The relation between alkenes, alcohols, and alkyl halides is particularly important in organic chemistry. Earlier we encountered reactions that interconvert these species.

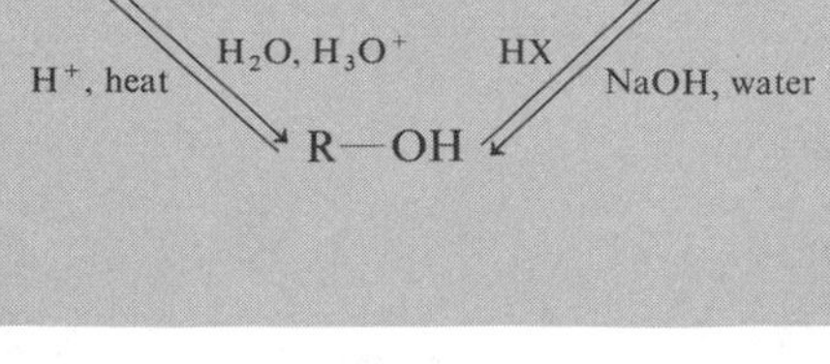

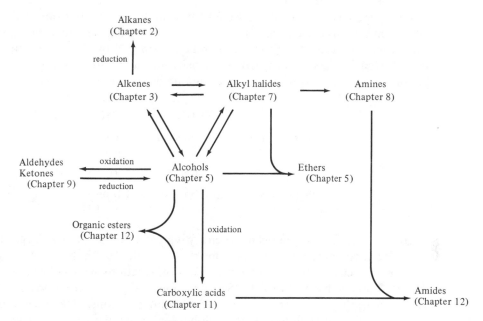

Figure 7.1   Some important interconversions of major families of organic compounds.

Additional reactions, summarized in Figure 7.1, link alkenes, alcohols, and alkyl halides to the other important classes of compounds discussed in this book. We have already encountered nucleophilic substitution reactions of alkyl halides with salts of alcohols to produce ethers. Nucleophilic substitution reactions of alkyl halides with ammonia produce amines. We learned that reactions of alcohols with inorganic acids lead to inorganic esters; reactions with carboxylic acids (the principal organic acids) lead to organic esters. Other important groups of compounds are related to alcohols and alkenes by oxidation-reduction reactions. Reduction of alkenes produces alkanes. Oxidation of alcohols leads to aldehydes, ketones, and carboxylic acids. Earlier we learned that reduction of aldehydes and ketones is an important preparation of alcohols.

7.2
ORGANIC HALIDES: NOMENCLATURE, PHYSICAL PROPERTIES, AND PREPARATIONS

We have already learned about the nomenclature for alkyl and aryl halides. We also know that organic halides do not participate significantly in hydrogen bonding. Therefore their solubilities in water are low and their boiling points are similar to those of hydrocarbons of comparable molecular weight. Density is one property in which halides differ from the other compounds considered. Because the density of atoms increases markedly as we go from the top to the bottom of the periodic table, compounds containing Cl, Br, or I are denser than hydrocarbons or organic compounds with functional groups containing only N or O. As evident from the values given for the following halides, density increases with increasing atomic number of the halogen and with an increasing number of halogen atoms.

$CHCl_3$	$CHBr_3$	CHI_3
Chloroform	Bromoform	Iodoform
1.48	2.89	4.01

CH_3Br	CH_3CH_2Br	$CH_3(CH_2)_5Br$
Methyl bromide	Ethyl bromide	1-Bromohexane
1.68	1.46	1.17

Of course, the effect of a halogen on density is greater in small than in large molecules.

We encountered several reactions that furnish alkyl halides. Alkyl chlorides or bromides can be prepared from alkanes by reactions with chlorine or bromine that are initiated by heat or light (Sec. 2.8). Generally these reactions give mixtures of products that result from attack at different positions. Halogenation is most useful for alkanes, such as ethane, cyclohexane, or neopentane, that can form only one monosubstitution product.

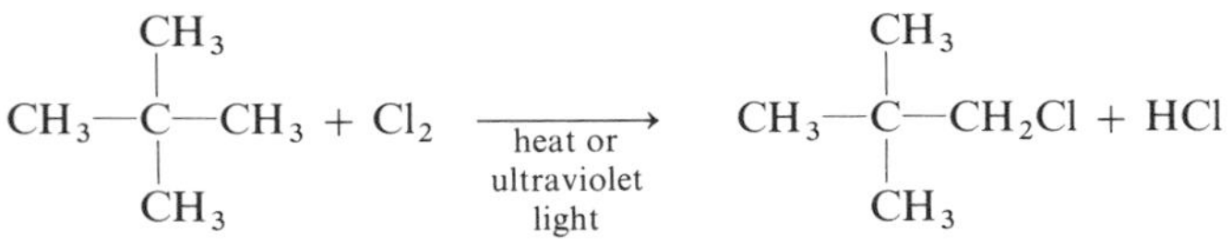

$$CH_3-\underset{\underset{CH_3}{|}}{\overset{\overset{CH_3}{|}}{C}}-CH_3 + Cl_2 \xrightarrow[\substack{\text{heat or} \\ \text{ultraviolet} \\ \text{light}}]{} CH_3-\underset{\underset{CH_3}{|}}{\overset{\overset{CH_3}{|}}{C}}-CH_2Cl + HCl$$

Neopentane   1-Chloro-2,2-dimethylpropane

Alkyl halides can also be prepared from reactions of alcohols (Sec. 5.9) or alkenes (Sec. 3.7) with hydrogen halides.

Unlike the halogenations of alkanes, these reactions usually introduce halogens at specific locations.

Aryl chlorides and bromides are most commonly prepared from reactions of halogens with aromatic compounds (Secs. 4.4 and 4.7).

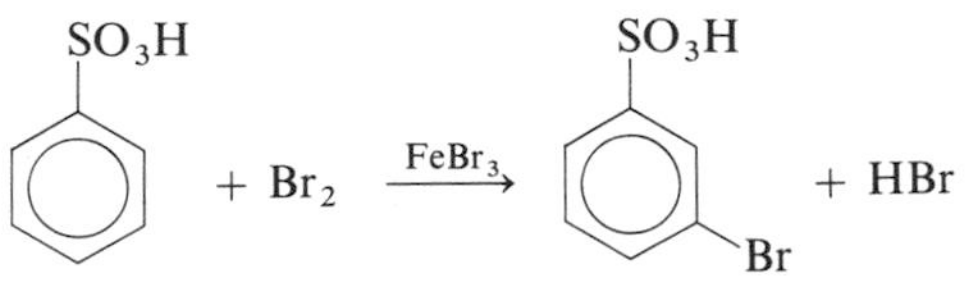

These reactions are electrophilic substitutions, very different in nature from the free-radical reactions of alkanes with halogens. Except with the most reactive aromatic compounds, catalysts like $FeBr_3$ must be used. As noted, the ease and position of substitution are determined by any substituents already present in the aromatic ring.

7.3
REACTIONS OF ORGANIC HALIDES

Nucleophilic substitutions are the most characteristic and versatile reactions of alkyl halides. Earlier we saw examples of these reactions and also of the dehydrohalogenations to alkenes that may accompany or supersede the substitutions when alkyl halides are treated with nucleophilies. In the following sections we will learn more about these important reactions. We will also encounter one new reaction of both alkyl and aryl halides: their conversion to organometallic compounds.

7.4
CONVERSION OF ORGANIC HALIDES
TO GRIGNARD AND ORGANOLITHIUM REAGENTS;
ORGANOMETALLIC COMPOUNDS

Compounds that contain carbon-metal bonds are called **organometallic**. Because of their role in reactions for forming carbon-carbon bonds (see Sec. 9.8 for one example), organomagnesium and organolithium compounds are particularly important. They are prepared by reaction of an alkyl or aryl halide with the metal.

$$CH_3CHCH_2{-}Cl + Mg \xrightarrow{\text{diethyl ether}} CH_3CHCH_2{-}Mg{-}Cl$$
$$\underset{CH_3}{|} \qquad\qquad\qquad\qquad \underset{CH_3}{|}$$

Isobutyl chloride   Isobutylmagnesium chloride

$$\bigcirc{-}Br + Mg \xrightarrow{\text{diethyl ether}} \bigcirc{-}Mg{-}Br$$

Bromobenzene   Phenylmagnesium bromide

$$CH_3CH_2CH_2CH_2{-}Cl + 2\,Li \xrightarrow{\text{pentane}} CH_3CH_2CH_2CH_2{-}Li + LiCl$$

Butyl chloride   Butyllithium

Organomagnesium compounds are usually called **Grignard reagents** after Victor Grignard, the French chemist who in 1912 won a Nobel Prize for their synthesis and study. Grignard reagents are most frequently prepared in diethyl ether but organolithium compounds in pentane or hexane. In these reactions the alkyl halide in solution must react at the surface of the solid metal. Therefore the metals are often used in finely divided form to provide a large surface area that maximizes contact between the reactants.

Carbon (electronegativity 2.5) has a considerably greater tendency than magnesium (1.3) or lithium (1.0) to attract shared electron pairs. Therefore the C—Mg and C—Li bonds in Grignard reagents and organolithium compounds are very polar.

$$-\overset{\displaystyle |\delta^-}{\underset{\displaystyle |}{C}}-\overset{\delta^+}{MgX} \qquad -\overset{\displaystyle |\delta^+}{\underset{\displaystyle |}{C}}-\overset{\delta^-}{Li}$$

Grignard reagents and organolithium compounds are extremely basic. They readily abstract a proton from water to give a hydrocarbon and a hydroxide salt.

$$CH_3\underset{\underset{\displaystyle CH_3}{|}}{CH}CH_2{-}MgCl + H_2O \longrightarrow CH_3\underset{\underset{\displaystyle CH_3}{|}}{CH}CH_3 + Mg(OH)Cl$$

They also abstract protons from organic compounds containing O—H or N—H bonds.

⬡—MgBr + CH₃CH₂OH ⟶ ⬡ + CH₃CH₂O⁻ ⁺MgBr

$$CH_3CH_2CH_2CH_2{-}Li + (CH_3)_2NH \longrightarrow CH_3CH_2CH_2CH_3 \;+\; (CH_3)_2N^- \; Li^+$$

Lithium diisopropylamide

The basicity of anions increases in going from right to left across the periodic table.

$$\overset{increasing\ basicity}{\longleftarrow}$$

$$\delta^+Li{-}R^{\delta^-} \quad Li^+\ ^-NR_2 \quad Li^+\ ^-OR \quad Li^+\ ^-X$$

We know that halide ions are not too basic but (Sec. 5.6) that alkoxide ions (⁻OR) are strong bases. Amide ions (⁻NR₂) are even more basic and are often used when very strong bases are needed. Although generally covalent rather than ionic, Grignard reagents and organolithium compounds are highly polar and even more basic than alkoxide and amide ions. Because of reacting so avidly with water, they must be prepared and handled under rigorously dry conditions.

A reaction that converts an organometallic compound to a hydrocarbon can be used to introduce an isotope of hydrogen at a particular site.

$$CH_3O{-}⬡{-}Br \xrightarrow{Mg} CH_3O{-}⬡{-}MgBr \qquad CH_3O{-}⬡{-}D$$

Naturally occurring hydrogen is mainly ^1H but contains 0.016% of ^2H, a stable isotope used so frequently by chemists that it is given a special name, deuterium (D). Because of the slightly different physical properties of D_2O and H_2O, essentially pure D_2O can be isolated from a sample of water.

Many organometallic compounds of other metals are also known. Earlier we discussed tetraethyllead (Sec. 2.17). Ethylmercuric chloride is one of a number of simple organomercury compounds used as fungicides—for example, in preserving seeds.

$$CH_3CH_2-Hg-Cl$$

Ethylmercuric chloride

The electronegativities of Pb (2.3) and Hg (2.0) are considerably greater than those of Li and Mg. As a result, the carbon-metal bonds in organolead and organomercury compounds are less polar than in organolithium and organomagnesium compounds and do not react readily with water.

A health hazard associated with mercury pollution arises from the capacity of some bacteria to convert inorganic mercury compounds to methylmercuric chloride ($CH_3-Hg-Cl$) and dimethylmercury ($CH_3-Hg-CH_3$). These organomercury compounds, which are highly toxic to the nervous system, are considerably more soluble in organic materials than are inorganic mercury compounds and tend to accumulate in living systems. Algae have higher concentrations of mercury than does the water in which they grow. Small crustaceans feeding on the algae have yet higher concentrations and fish feeding on the crustaceans even higher concentrations. Increases in mercury concentration of 100,000-fold have been observed along such a food chain.

7.5
NUCLEOPHILIC SUBSTITUTIONS
AND DEHYDROHALOGENATIONS OF ALKYL HALIDES

Alkyl halides react with a wide variety of nucleophiles to give products in which the halogen has been replaced by another group.

$$R\!:\!\ddot{X}\!: + :\!\ddot{N}u\!:^- \longrightarrow R\!:\!\ddot{N}u\!: + :\!\ddot{X}\!:^-$$

Such reactions are members of the very extended family of reactions called **nucleophilic substitutions** (S_N reactions) in which one nucleophile replaces another. The group that departs with the electron pair that it shared with carbon is called the **leaving group**. In reactions of alkyl halides the leaving group is a halogen. Some important nucleophilic substitution reactions of alkyl halides are listed in Table 7.1.

Earlier we saw examples of the use of reactions 1 and 2 to prepare alcohols and ethers and of reactions 3 and 4 to prepare their sulfur analogs, thiols and sulfides. Reaction 5, with cyanide salts, produces the compounds called nitriles (cyanides).

$$\langle\bigcirc\rangle\!-\!CH_2Br + Na^{+\,-}C\!\equiv\!N \longrightarrow \langle\bigcirc\rangle\!-\!CH_2-C\!\equiv\!N + Na^+\,Br^-$$

Phenylacetonitrile
(benzyl cyanide)

Table 7.1 Some nucleophilic substitution reactions of alkyl halides (X = Cl, Br, or I)

Reaction	Product	Reference
1. $R{-}X + {}^{-}OH \rightarrow R{-}OH$	alcohol	Section 5.13
2. $R{-}X + {}^{-}O{-}R' \rightarrow R{-}O{-}R'$	ether	Section 5.13
3. $R{-}X + {}^{-}SH \rightarrow R{-}SH$	thiol	Section 5.16
4. $R{-}X + {}^{-}S{-}R' \rightarrow R{-}S{-}R'$	sulfide	Section 5.16
5. $R{-}X + {}^{-}C{\equiv}N \rightarrow R{-}C{\equiv}N$	nitrile (cyanide)	Section 7.5
6. $R{-}X + {}^{-}X' \rightarrow R{-}X'$	halide	Section 7.5
7. $R{-}X + NH_3 \rightarrow R{-}NH_3{}^{+}$	alkylammonium ion	Sections 7.5 and 8.5

Alkyl halides even undergo substitution by other halide ions (reaction 6). Such reactions tend to be reversible. A solution of sodium iodide in the solvent acetone, however, is effective in preparing high yields of iodides from chlorides or bromides.

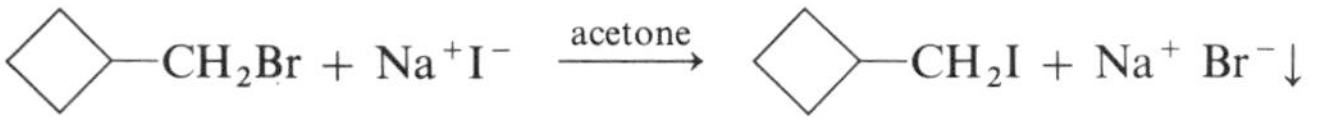
CH2Br + Na+I−
acetone
CH2I + Na+ Br− ↓

Insolubility of NaCl and NaBr in acetone displaces the equilibrium toward formation of the iodide.

Nucleophilic reagents are usually negative but can be neutral if an unshared electron pair is available that readily attacks carbon. We are familiar with the ability of the unshared electron pair of ammonia to form a bond to a proton.

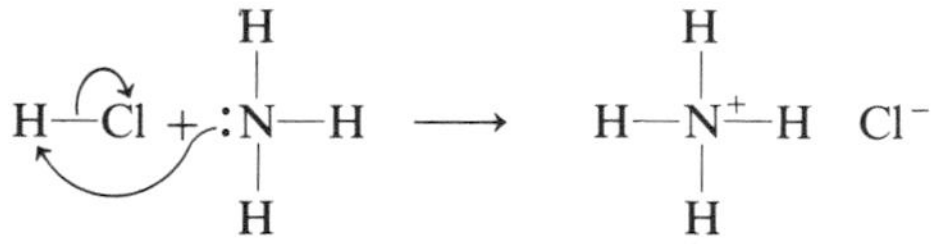
H—Cl + :NH3
H—N+—H Cl−

Ammonium chloride

Reaction 7 of ammonia and alkyl halides produces alkylammonium ions.

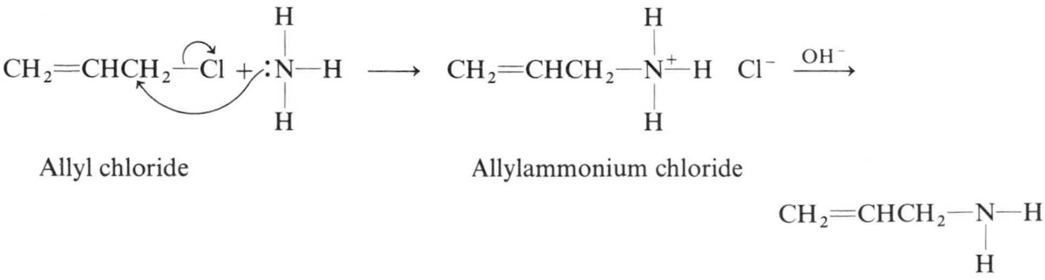
CH2=CHCH2—Cl + :N—H
CH2=CHCH2—N+—H Cl−
OH−
CH2=CHCH2—N—H

Allyl chloride

Allylammonium chloride

Allylamine

This reaction provides a synthesis of amines, the subject of the next chapter.

Nucleophilic substitutions are ordinarily useful only with primary or secondary halides. We learned (Sec. 5.13) that when an alkyl halide is treated with hydroxide or alkoxide ions, elimination of hydrogen halide (dehydrohalogenation) to form an alkene competes with substitution. Elimination usually predominates in reactions of tertiary halides and often can be made the major reaction with primary or secondary halides. Dehydrohalogenation can also accompany nucleophilic substitution when other nucleophiles are used.

7.6
MECHANISMS OF NUCLEOPHILIC SUBSTITUTIONS
OF ALKYL HALIDES

Nucleophilic substitutions are so numerous and important that chemists have intensively studied their mechanisms. Many of these reactions fall into two mechanistic categories, often labeled bimolecular nucleophilic substitution or S_N2 and unimolecular nucleophilic substitution or S_N1.

As illustrated for the reaction of one enantiomer of 2-bromooctane with hydroxide ion, an S_N2 reaction occurs in one step.

$$
HO^- \quad \overset{H}{\underset{CH_3}{\underset{\diagdown}{C}}}{\overset{\diagup R}{-}}Br \quad \longrightarrow \quad HO-\overset{R}{\underset{CH_3}{\underset{\diagup}{C}}}{\diagdown}H \quad + \; Br^-
$$

$$R{-} = CH_3CH_2CH_2CH_2CH_2CH_2{-}$$

Hydroxide ion (the nucleophilic reagent) attacks the carbon (from the side opposite bromide (the leaving group). A carbon-oxygen bond forms as the carbon-bromine bond breaks. If we could stop the reaction near the middle of the reaction pathway, we would find hydroxide and bromide both partially bonded to carbon.

$$
\overset{H \quad R}{HO\overset{\delta^-}{----}\underset{\underset{CH_3}{|}}{C}\overset{\delta^-}{----}Br}
$$

Two types of evidence are particularly significant in establishing the mechanisms of S_N2 and S_N1 reactions. The first is provided by **kinetics**, the area of chemistry concerned with the rates at which chemical reactions occur. The rate at which 2-octanol forms is proportional to the concentrations of both the alkyl bromide and hydroxide ion. This dependence is found because the two reactants must collide for this single-step reaction to occur, and the frequency of such collisions is proportional to the concentration of each reactant. Reactions whose rates show such a dependence are called **bimolecular** (or second order). The rates of other S_N2

reactions are also proportional to the concentrations of both the alkyl halide and the nucleophile.

The second important type of evidence is stereochemical. In an S_N2 reaction the nucleophilic reagent attacks carbon from the side opposite the leaving group. During the course of the reaction the other three groups attached to the carbon move from one side to the other, much as do the ribs of an umbrella that is blown inside out. A single enantiomer of the substrate is converted to a single enantiomer of the product. The reactant and product have a mirror image relationship (if we consider Br and OH to be equivalent groups). Such **inversion of configuration** is a characteristic of S_N2 reactions. Attack from the side opposite to the leaving group is favored because it keeps the electron-rich attacking and leaving groups as far apart as possible.

As illustrated for the reaction of one enantiomer of 3-chloro-3,7-dimethyloctane with water, an S_N1 reaction occurs in two steps.

$$R\!\!-\!\!- = (CH_3)_2CHCH_2CH_2CH_2-$$

The first step is a slow ionization to form a carbocation intermediate. In the fast second step the carbocation reacts with water ($R^+ + OH_2 \rightarrow R-OH_2^+$); rapid loss of the proton gives R—OH.

The rate at which the product forms is independent of the concentration or nature of the nucleophile. Although hydroxide is a much more reactive nucleophile than water, the reaction proceeds at the same rate in neutral water as in water to which varying amounts of hydroxide ion are added. The rate of this two-step reaction is identical to the rate of the slower (rate-determining) step—ionization to form the carbocation. The unstable carbocation reacts rapidly with any available nucleophile. Addition of the more nucleophilic hydroxide ion does not change the rate of the slow step.

Formation of product enantiomers in equal amounts is a characteristic of S_N1 reactions. A racemic product arises because carbocations are planar. One enantiomer results from attack of water at one face of the carbocation, the other enantiomer from attack at the other face. Because attack at the two faces of the planar carbocation is equally probable, the enantiomers form in equal amounts.

Which halides react by S_N1 and which by S_N2 mechanisms? The ease of S_N2 reactions decreases with increasing substitution at the carbon bonded to the halogen.

decreasing ease of S_N2 reactions
$$CH_3 > 1° > 2° > 3°$$

Methyl halides react readily but tertiary halides hardly at all by this mechanism. The order of reactivity is due to the bulkiness of alkyl groups. A nucleophile must closely approach a carbon already bonded to four groups. This approach becomes increasingly difficult as fewer hydrogens and more alkyl groups are attached to the carbon. The ease of S_N1 reactions follows the reverse order.

$$\xrightarrow{\text{increasing ease of } S_N1 \text{ reactions}}$$
$$CH_3 < 1° < 2° < 3°$$

The rate of an S_N1 reaction is identical to the rate of carbocation formation, and that is related to carbocation stability.

As a result of their low reactivity in S_N2 processes and the relatively great stabilities of the carbocations they form, tertiary halides generally react by S_N1 mechanisms. The mechanisms of reactions of secondary halides depend on the reaction conditions. A solvent like water that interacts strongly with ions promotes ionization to a carbocation and favors an S_N1 process. A nucleophile that is particularly effective at attacking carbon favors an S_N2 process. Although nominally primary, benzyl and allyl halides form particularly stable carbocations and often react by an S_N1 mechanism.

7.7
MECHANISMS OF DEHYDROHALOGENATIONS
OF ALKYL HALIDES

Dehydrohalogenation generally accompanies reactions of halides with nucleophiles and sometimes is the dominant reaction. One mechanism involves attack by the nucleophile on a hydrogen of the halide.

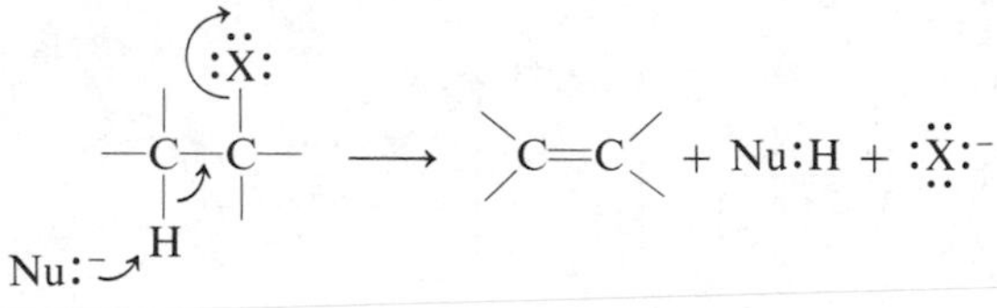

This mode of attack by the nucleophile competes with (S_N2) attack at carbon that leads to substitution.

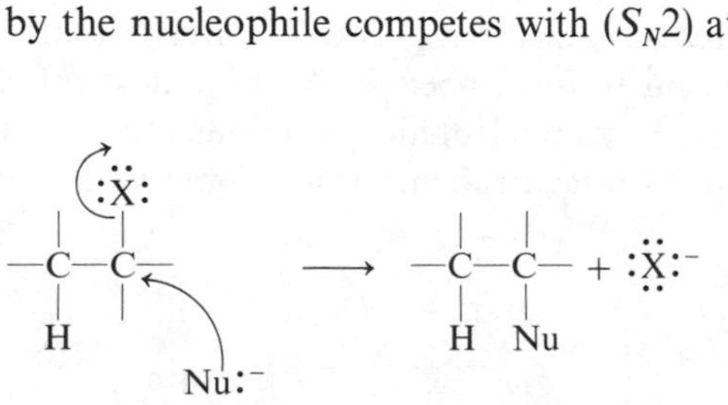

Tertiary halides react with strong bases to give mainly elimination products, since attack at carbon is particularly difficult. Choice of conditions sometimes determines whether elimination or substitution dominates in reactions of primary or secondary halides. As noted (Sec. 5.13), the use of a hydroxide salt in an alcohol favors elimination. Bulkier bases, such as $(CH_3)_3CO^- K^+$, have greater difficulty in approaching carbon and increase the tendency toward elimination.

The first step in another mechanism of dehydrohalogenation is the same as in an S_N1 reaction: ionization to a carbocation.

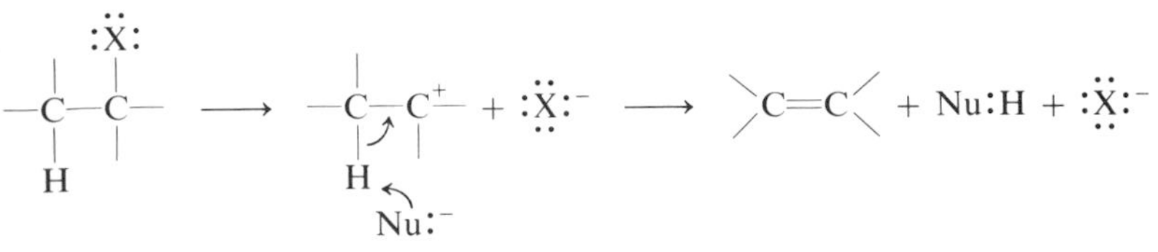

Attack by a nucleophile on the hydrogen to give an elimination product competes with attack at carbon to give a substitution product. With strongly basic nucleophiles, such as OH^-, elimination predominates. With weakly basic nucleophiles, such as H_2O or Cl^-, substitution is sometimes significant.

7.8
NUCLEOPHILIC SUBSTITUTIONS OF ARYL HALIDES

Most aryl halides undergo nucleophilic substitution reactions only with great difficulty. The reaction of chlorobenzene and hydroxide, one procedure for the commercial production of phenol, requires concentrated aqueous sodium hydroxide and high temperature and pressure.

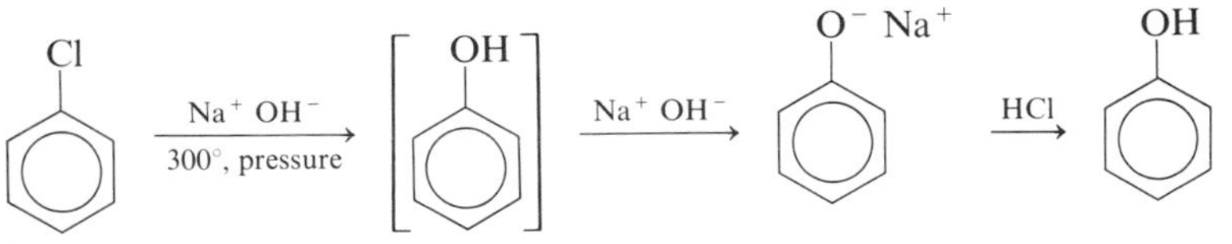

In the basic reaction medium phenol is converted into its sodium salt, which must be treated with acid to form phenol. Aryl halides with strongly electron-withdrawing substituents undergo nucleophilic reactions more readily. Hydroxide reacts rapidly with 1-chloro-2,4,6-trinitrobenzene at room temperature.

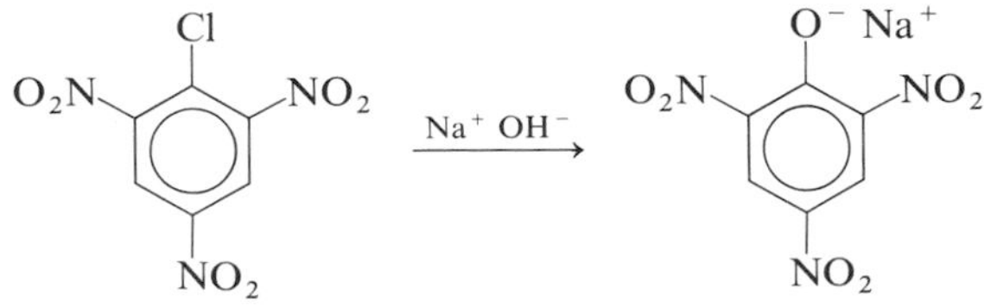

For a number of reasons, aryl halides do not react by S_N1 or S_N2 pathways. For instance, the aromatic ring blocks approach of the nucleophile in the manner characteristic of S_N2 reactions—from the side of the carbon opposite to the halide. Ionization to a carbocation, necessary for an S_N1 reaction, is difficult because aryl cations are relatively unstable.

Aromatic nucleophilic substitution generally takes place by an addition-elimination pathway resembling that for electrophilic aromatic substitution.

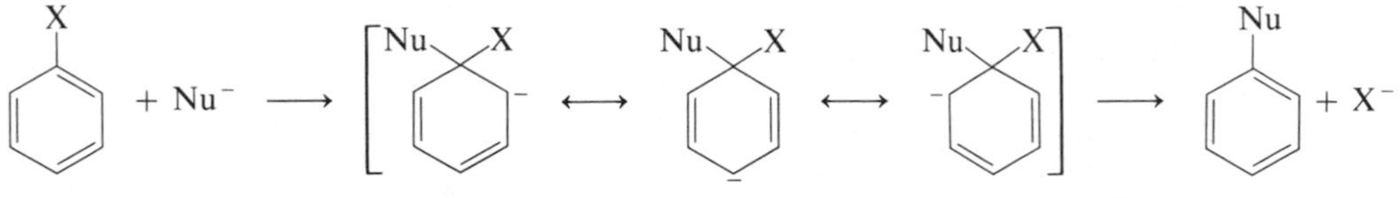

The resonance structures for the intermediate make it apparent why electron-withdrawing groups, particularly when ortho and para to the halogen, facilitate this type of aromatic substitution.

7.9
OTHER COMPOUNDS THAT UNDERGO
NUCLEOPHILIC SUBSTITUTIONS AND ELIMINATIONS;
SULFONATE AND PHOSPHATE ESTERS

In both nucleophilic substitutions and dehydrohalogenations of alkyl halides, halogen departs as a negative ion, taking with it the electron pair that it had shared with carbon. Other nucleophilic groups that are weakly basic like halogen also tend to be good leaving groups in such reactions.

We already saw one example. Alcohols form alkyl halides when treated with hydrogen halides and alkenes when heated with strong acids, such as sulfuric acid.

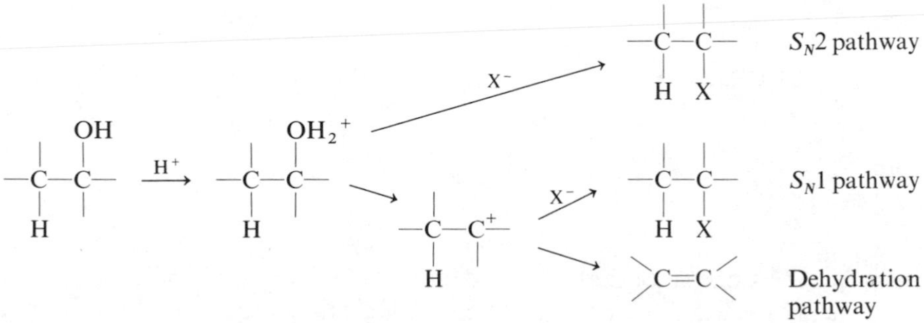

We learned (Sec. 5.10) that the reaction with hydrogen halides can occur by attack of a halide ion on the protonated alcohol. We now recognize this as an S_N2 pathway.

Alternatively, the protonated alcohol can lose water to form a carbocation, which in a second step reacts with halide. This is an S_N1 pathway. An alternative reaction of the carbocation—loss of a proton—forms an alkene. Protonation of the alcohol is essential for these reactions to occur. It converts a poor leaving group (OH^-) into a much less basic and much better leaving group (OH_2). The readier cleavage of $C-OH_2^+$ than of $C-OH$ parallels the easier cleavage of the strong acid $H-OH_2^+$ than of the weak acid $H-OH$.

The hydroxyl group of an alcohol can also be transformed into a better leaving group by conversion to an ester of a sulfonic acid ($R-OH \rightarrow R-O-\overset{\overset{\displaystyle O}{\|}}{\underset{\underset{\displaystyle O}{\|}}{S}}-R'$).

Sulfonic acids ($R'-\overset{\overset{\displaystyle O}{\|}}{\underset{\underset{\displaystyle O}{\|}}{S}}-OH$) are very strong acids, and the corresponding sulfonate ions ($R'-\overset{\overset{\displaystyle O}{\|}}{\underset{\underset{\displaystyle O}{\|}}{S}}-O^-$) are therefore very weak bases and excellent leaving groups. Substitution and elimination reactions of arylsulfonate esters can be achieved without the strongly acidic conditions necessary for similar reactions of alcohols.

$$H-\overset{\overset{\displaystyle CH_3}{|}}{\underset{\underset{\displaystyle R}{|}}{C}}-O-\overset{\overset{\displaystyle O}{\|}}{\underset{\underset{\displaystyle O}{\|}}{S}}-\langle\bigcirc\rangle-CH_3 \xrightarrow{CH_3CH_2O^- \, Na^+} CH_3CH_2O-\overset{\overset{\displaystyle CH_3}{|}}{\underset{\underset{\displaystyle R}{|}}{C}}-H \quad (S_N2)$$

$$R- = CH_3CH_2CH_2CH_2CH_2CH_2-$$

Complete inversion of configuration

$$\overset{\overset{\displaystyle CH_3}{|}}{CH_3CHCHCH_3} \\ \underset{\underset{\underset{\displaystyle O}{\|}}{O-\overset{\overset{\displaystyle O}{\|}}{S}-\langle\bigcirc\rangle-CH_3}}{|} \xrightarrow{(CH_3)_3CO^- \, K^+} \underset{\underset{\displaystyle 77\%}{}}{CH_3\overset{\overset{\displaystyle CH_3}{|}}{C}=CHCH_3} + \underset{\underset{\displaystyle 23\%}{}}{CH_3\overset{\overset{\displaystyle CH_3}{|}}{CH}CH=CH_2}$$

Esters of phosphoric acid and the related diphosphoric and triphosphoric acids (Sec. 5.7) are abundant in biological systems and sometimes take part in biological displacement reactions. One example occurs in the synthesis of terpenes. Some 3-methyl-3-butenyl diphosphate, the key compound in terpene biosynthesis, is isomerized to 3-methyl-2-butenyl diphosphate. These two isomers then react together in an enzyme-catalyzed reaction.

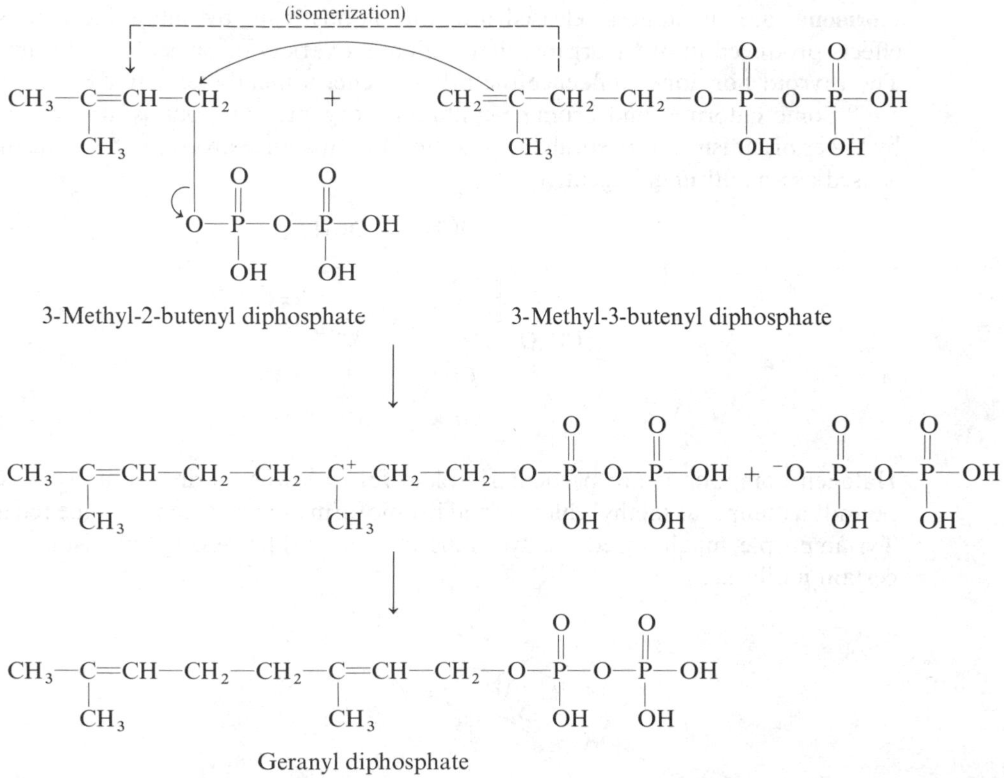

(isomerization)

$$CH_3-\underset{\underset{CH_3}{|}}{C}=CH-CH_2 \quad + \quad CH_2=\underset{\underset{CH_3}{|}}{C}-CH_2-CH_2-O-\underset{\underset{OH}{|}}{\overset{\overset{O}{\|}}{P}}-O-\underset{\underset{OH}{|}}{\overset{\overset{O}{\|}}{P}}-OH$$

$$^+O-\underset{\underset{OH}{|}}{\overset{\overset{O}{\|}}{P}}-O-\underset{\underset{OH}{|}}{\overset{\overset{O}{\|}}{P}}-OH$$

3-Methyl-2-butenyl diphosphate   3-Methyl-3-butenyl diphosphate

$$CH_3-\underset{\underset{CH_3}{|}}{C}=CH-CH_2-CH_2-\overset{+}{\underset{\underset{CH_3}{|}}{C}}CH_2-CH_2-O-\underset{\underset{OH}{|}}{\overset{\overset{O}{\|}}{P}}-O-\underset{\underset{OH}{|}}{\overset{\overset{O}{\|}}{P}}-OH \quad + \quad ^-O-\underset{\underset{OH}{|}}{\overset{\overset{O}{\|}}{P}}-O-\underset{\underset{OH}{|}}{\overset{\overset{O}{\|}}{P}}-OH$$

$$CH_3-\underset{\underset{CH_3}{|}}{C}=CH-CH_2-CH_2-\underset{\underset{CH_3}{|}}{C}=CH-CH_2-O-\underset{\underset{OH}{|}}{\overset{\overset{O}{\|}}{P}}-O-\underset{\underset{OH}{|}}{\overset{\overset{O}{\|}}{P}}-OH$$

Geranyl diphosphate

The double bond of one isomer acts as a nucleophilic reagent, furnishing the electron pair needed to form a new carbon-carbon bond. The allylic diphosphate of the other isomer is the leaving group. This reaction can also be regarded as an electrophilic attack on an alkene. Loss of a proton results in formation of the diphosphate ester of the terpene geraniol (Sec. 3.21).

7.10
SOME IMPORTANT ORGANIC HALOGEN COMPOUNDS

Even though chloride ion is found in large amounts in living systems, no chlorine- or bromine-containing organic compounds have been observed to result from the normal metabolism of higher animals. Iodide ion, however, is extracted from the blood and concentrated in the thyroid gland, where the hormones thyroxine and triiodothyronine are formed.

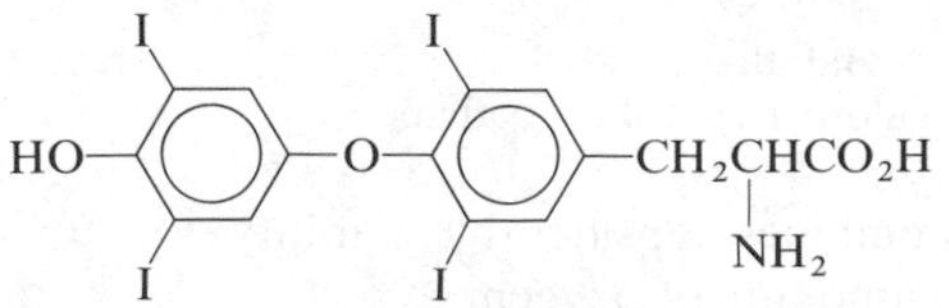

Thyroxine 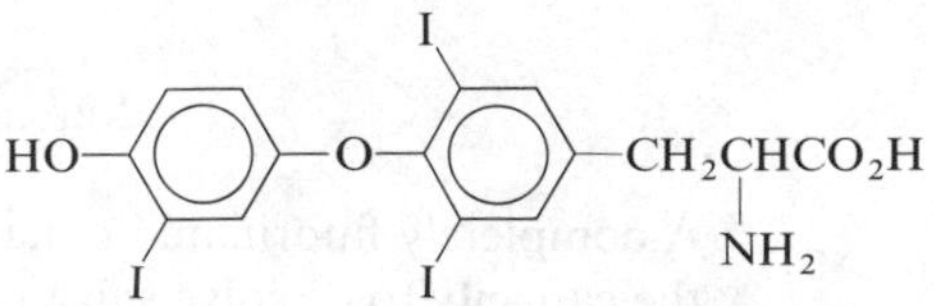   Triiodothyronine

Hormones are substances released into the bloodstream by endocrine glands. By effects produced in other organs, they regulate a variety of physiological functions. The thyroid hormones influence the rate of general metabolic activity.

Some chlorine- and bromine-containing organic compounds are synthesized by lower organisms. Griseofulvin, an antibiotic substance produced by some molds, is used as an antifungal agent.

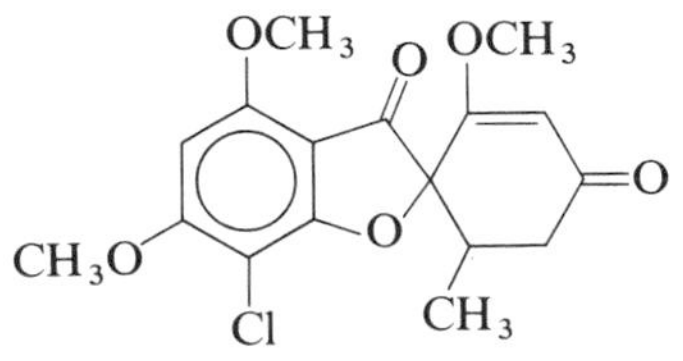

Griseofulvin

Halogen compounds are particularly plentiful in lower marine organisms. Compounds as simple as methyl chloride and bromoform are produced by some red algae. Tyrian purple, much prized as a dye in the ancient world, was originally isolated from certain mollusks.

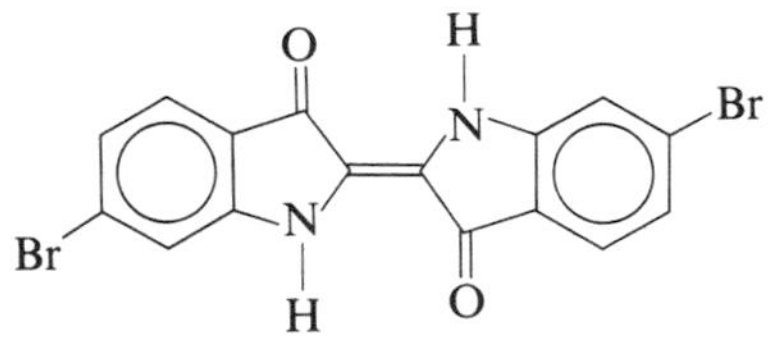

Tyrian purple

Many synthetic halogen-containing compounds have specific applications. Dichloromethane is used in paint removers, trichloroethylene for degreasing metals, and tetrachloroethylene as a dry-cleaning liquid and a drug active against hookworms. Ethyl chloride is used as a topical anesthetic for minor surgery. When liquid ethyl chloride (boiling point 12°) is sprayed from a pressurized container onto an area of skin, its rapid evaporation cools adjoining tissues and deadens nerve response. Halothane, a nonflammable and relatively nontoxic compound, is now the most widely used inhalation anesthetic.

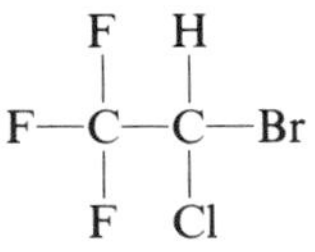

Halothane
2-Bromo-2-chloro-1,1,1-trifluoroethane

A completely fluorinated nonane combines considerable stability, low toxicity, and the capacity to dissolve substantial amounts of oxygen.

$$\begin{array}{cc}
CF_3 & CF_3 \\
| & | \\
CF_3CCF_2CCF_3 \\
| & | \\
CF_3 & CF_3
\end{array}$$

Rats immersed in this liquid can live and breathe for extended periods of time. Such compounds have been used experimentally in humans as emergency blood substitutes.

The two most common members of the group of compounds known as Freons are prepared from carbon tetrachloride by nucleophilic substitution reactions.

$$CCl_4 \xrightarrow[SbF_5]{HF} \quad CFCl_3 \quad \xrightarrow[SbF_5]{HF} \quad CF_2Cl_2$$

Trichlorofluoromethane   Dichlorodifluoromethane
Freon 11   Freon 12
bp 24°   bp −30°

In 1973 the United States produced 370×10^6 kg of Freons 11 and 12, nearly one-half of the world's production. Freon 11 is used particularly as a propellant in spray cans. A solution or emulsion in liquid Freon of the material to be sprayed is sealed in a pressurized can. When the valve is opened, rapid evaporation of the Freon produces a very fine spray (aerosol). Freon 12, used particularly as a refrigerant, is found in the majority of household and commercial refrigeration units and air conditioners. Because of their low toxicity, lack of odor, nonflammability, insolubility in water, and great chemical stability, Freons are used in place of other compounds having suitable volatilities for these applications.

Today it is recognized that the stability of the Freons may pose a serious environmental problem. There seems to be no process that rapidly removes these compounds from the troposphere (the lower atmosphere in which we live). A substantial portion of all the Freons ever released into the atmosphere may still be there. The Freons presumably diffuse slowly into the stratosphere (the next highest atmospheric layer), where they encounter high-intensity ultraviolet light. Little of this light reaches the lower atmosphere because it is mostly absorbed by the small amount of ozone (O_3) that exists in the stratosphere. Ultraviolet light cleaves C—Cl bonds to produce chlorine atoms.

$$CF_3Cl \xrightarrow{\text{ultraviolet light}} CF_3 \cdot + Cl \cdot$$

Chlorine atoms have been demonstrated to participate in a catalytic fashion in a chain reaction that destroys ozone $(2\,O_3 \rightarrow 3\,O_2)$. Therefore the Freons slowly working their way into the stratosphere may significantly reduce the stratospheric ozone layer. Ozone is the only natural stratospheric species that strongly absorbs the high-energy ultraviolet radiation that is of sufficient energy to break covalent bonds in a wide range of organic compounds. Until the stratospheric ozone layer developed, life on earth was probably restricted to habitats, such as under several feet of water, that provided protection from this high-energy radiation.

Many pesticides (substances that eliminate organisms that we consider pests) are chlorine-containing organic compounds. The inexpensive insecticide DDT is the best known.

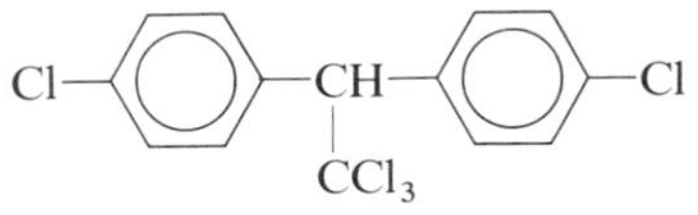

DDT
"*D*ichloro*d*iphenyl*t*richloroethane"
1,1,1,-Trichloro-2,2-di(4-chlorophenyl)ethane

It was first used on a large scale in World War II to control insect-borne diseases, such as malaria, typhus, and yellow fever. Its activity against a wide spectrum of insects is related to some interaction with insect nerve membranes that interferes with the transmission of nerve impulses.

In spite of DDT's amazing success in controlling disease, its use was banned in the United States in 1972 for all but a few applications. DDT is not rapidly degraded by biological systems. As you would guess from its structure, the solubility of DDT is low in water but high in nonpolar organic compounds. Because of its solubility properties and the lack of mechanisms to remove it rapidly, DDT ingested by an organism tends to remain and to be concentrated in fatty tissues. Moreover, DDT is concentrated along food chains. Considerable concern arose about the levels that developed in humans—about 10 parts per million in adults in the United States by the early 1960s. The results of the much higher levels found in some other organisms are evident. Large concentrations of DDT in birds, for example, interfere with calcium metabolism, resulting in eggs shells too weak to survive incubation.

Mirex and lindane are other chlorine-containing insecticides that have proved persistent in the environment.

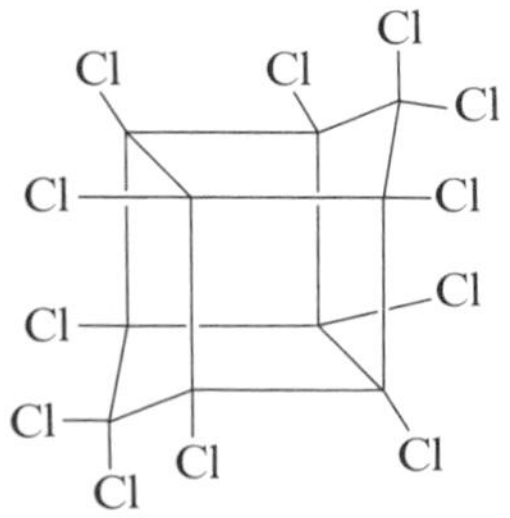

Mirex 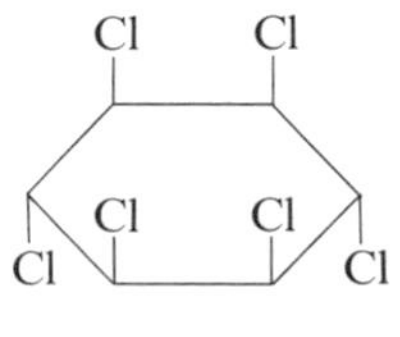   Lindane

Of the nine stereoisomers that have the 1,2,3,4,5,6-hexachlorocyclohexane structure, only lindane is an effective insecticide.

Definitions and Ideas

Organic halides have low solubilities in water and boiling points similar to those of hydrocarbons with comparable molecular weights. Their densities are greater than those of hydrocarbons.

Compounds that contain carbon-metal bonds are called **organometallic**. Grignard reagents (RMgX) and organolithium compounds (RLi) are very basic.

Nucleophilic substitutions are reactions in which one nucleophile replaces another. Many of the reactions of alkyl halides are nucleophilic substitutions. Dehydrohalogenation competes with substitution when an alkyl halide is treated with a nucleophile and usually predominates in reactions of tertiary halides.

Many nucleophilic substitutions of alkyl halides occur either by S_N2 or S_N1 mechanisms. An S_N2 reaction occurs by a one-step mechanism: the nucleophile bonds to the carbon at the same time that the bond to the **leaving group** breaks. S_N2 reactions are **bimolecular** and proceed with **inversion of configuration**. The ease of S_N2 reactions decreases with increasing substitution at the carbon attached to halogen: $CH_3 > 1° > 2° > 3°$. An S_N1 reaction occurs by a two-step mechanism: a slow ionization to form a carbocation intermediate, followed by a rapid reaction of the carbocation with the nucleophile. S_N1 reactions are **unimolecular** and proceed with racemization. The ease of S_N1 reactions parallels carbocation stabilities: $CH_3 < 1° < 2° < 3°$.

Dehydrohalogenations of alkyl halides occur either by direct attack of a nucleophile on a hydrogen of the halide or by ionization of the halide to a carbocation, followed by attack of a nucleophile on a hydrogen of the carbocation.

Important Reactions Involving Organic Halides

1. *Conversion to organometallic compounds*

$$RX \xrightarrow{\text{Mg}} RMgX \quad \text{(Grignard reagents)}$$
$$R = \text{alkyl or aryl} \quad X = \text{Cl, Br, or I}$$
$$RX \xrightarrow{\text{Li}} RLi \quad \text{(organolithium compounds)}$$

2. *Nucleophilic substitution reactions*   Summarized in Table 7.1. Dehydrohalogenation is often the predominant reaction with tertiary halides.

3. *Dehydrohalogenation* $\quad -\overset{|}{\underset{H}{C}}-\overset{|}{\underset{X}{C}}- \xrightarrow[\text{alcohol}]{\text{KOH}} \overset{}{\underset{}{C}}=\overset{}{\underset{}{C}}$

(discussed in Chapter 3)

4. *Preparation by halogenation of alkanes* $\quad RH \xrightarrow[\substack{\text{heat or} \\ \text{ultraviolet light}}]{X_2} RX$

(discussed in Chapter 2) $X = \text{Cl or Br}$

5. *Preparation from alcohols*   ROH $\xrightarrow{\text{HX, PX}_3, \text{ or SOCl}_2}$ RX

   (discussed in Chapter 5)   X = Cl, Br, or I

6. *Preparation by addition of hydrogen halides to alkenes*

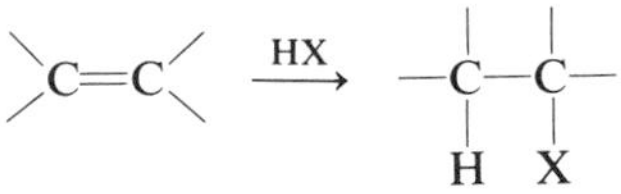

$$\overset{\diagdown}{\underset{\diagup}{C}}=\overset{\diagup}{\underset{\diagdown}{C}} \xrightarrow{\text{HX}} -\overset{|}{\underset{|}{C}}-\overset{|}{\underset{|}{C}}-$$
$$\qquad\qquad\qquad\quad \text{H} \quad \text{X}$$

   (discussed in Chapter 3)   X = Cl, Br, or I

7. *Preparation by halogenation of aromatic compounds*   ArH $\xrightarrow[\text{FeX}_3]{\text{X}_2}$ ArX

   (discussed in Chapter 4)   X = Cl or Br

PROBLEMS

1. Draw structures for the following compounds.
   (a) isopropyl chloride
   (b) 2-chloro-2-methylbutane
   (c) 3-chloro-1-butene
   (d) dichlorodifluoromethane
   (e) bromoform
   (f) methylene chloride
   (g) chlorocyclohexane
   (h) cyclopentyl bromide
   (i) benzyl bromide
   (j) allyl chloride
   (k) vinyl iodide
   (l) iodobenzene
   (m) *p*-chlorotoluene
   (n) 2,4-dichlorotoluene
   (o) *p*-chlorobenzyl chloride
   (p) diphenylmethyl bromide

2. Name the following compounds.

   (a) $CH_3CH_2CH_2CH_2CHCH_3$
   $$\qquad\qquad\qquad\qquad\quad |$$
   $$\qquad\qquad\qquad\qquad\; Br$$

   (b) ⬡—CH_2CH_2Br

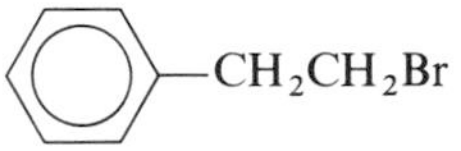

   (c) $\overset{CH_3}{\diagdown}C=C\overset{\diagup CH_2Cl}{\underset{\diagdown H}{}}$
   $$\;\; H \diagup$$

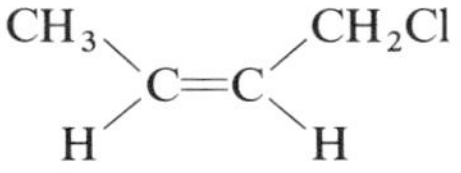

   (d) $(CH_3)_2C{=}CHI$

   (e) ⬡—$CHCH_2CH_2CH_3$
   $$\qquad\qquad |$$
   $$\qquad\quad\; Cl$$

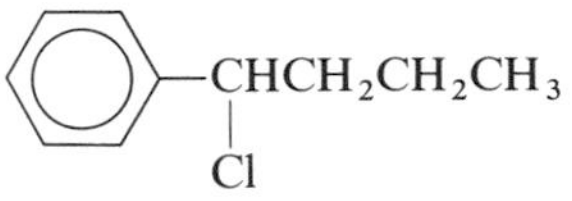

   (f) ▱—Cl

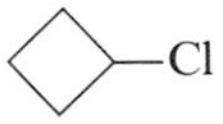

   (g) Cl⬡Cl

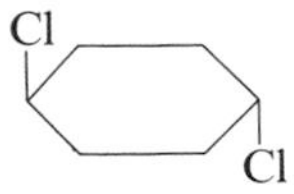

   (h) Cl⬡NO$_2$, Cl

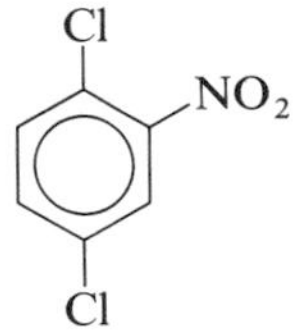

3. Give two names for each of the following compounds.

(a) $(CH_3)_3CCl$

(b) $CH_3CH_2CHCH_3$
  |
  Br

(c) $CH_2{=}CHCH_2Br$

(d) CHI_3

(e) CH_2Br_2

(f) 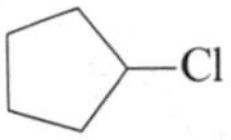 ⬠—Cl

4. Classify the following halides as primary, secondary, or tertiary.

(a) CH_3CHCH_3
  |
  Cl

(b) 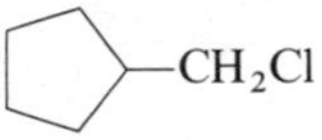 ⬠—CH_2Cl

(c)  ⬠ with CH_3 and Cl

(d) $\quad CH_2CH_3$
  |
  CH_3CCH_2Cl
  |
  CH_3

(e) $\qquad CH_3$
  |
  $CH_3CH_2CCH_2CH_3$
  |
  Cl

(f) 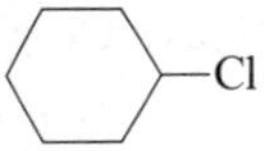 ⬡—Cl

5. Draw all compounds of molecular formula $C_5H_{11}Cl$ and classify each as a primary, secondary, or tertiary chloride.

6. Draw structures for any organic products that result from the following reactions.

(a) $(CH_3)_3COH \xrightarrow{HCl}$

(b) 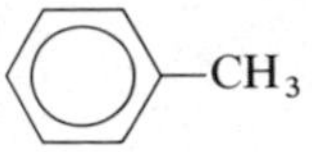 ⬡—$CH_3 \xrightarrow{Cl_2}{FeCl_3}$

(c) 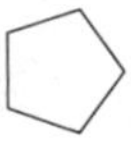 ⬠ $\xrightarrow[heat]{Cl_2}$

(d) 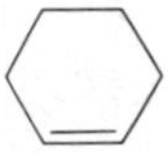 ⬡ $\xrightarrow{Br_2}$

(e) 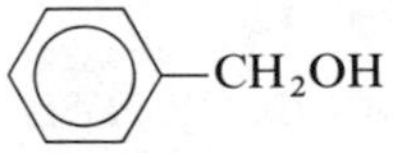 ⬡—$CH_2OH \xrightarrow{HBr}$

(f) 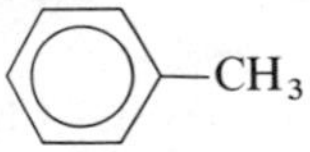 ⬡—$CH_3 \xrightarrow[light]{Br_2}$

(g) $CH_3CH_2CH{=}CH_2 \xrightarrow{HBr}$

(h) 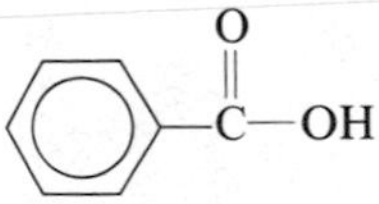 ⬡—$\overset{\displaystyle O}{\overset{\|}{C}}{-}OH \xrightarrow{Br_2}{FeBr_3}$

(i) $CH_3CH_2CH_2Br \xrightarrow{NaOCH_3}$

(j) $\quad Br$
  |
  $CH_3CCH_2CH_3 \xrightarrow[ethanol]{KOH}$
  |
  CH_3

(k) $CH_3CH_2CH_2Cl \xrightarrow[acetone]{NaI}$

(l) $CH_2{=}CHCH_2Cl \xrightarrow{KSH}$

(m) $(CH_3)_2CHCH_2CH_2Cl \xrightarrow[H_2O]{NaOH}$

(n) $CH_3CH_2CH_2CH_2Br \xrightarrow{NaCN}$

(o) 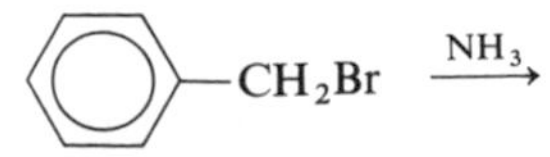 ⬡—CH_2Br $\xrightarrow{NH_3}$

(p) 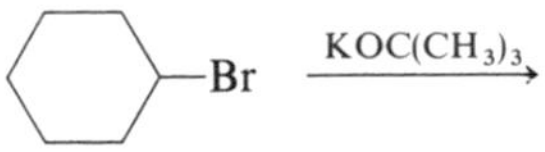 ⬡—Br $\xrightarrow{KOC(CH_3)_3}$

(q) CH_3CHCH_3 $\xrightarrow{Mg}$
    |
   Br

(r) 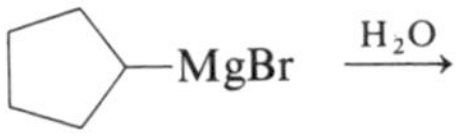 ⬠—MgBr $\xrightarrow{H_2O}$

(s) 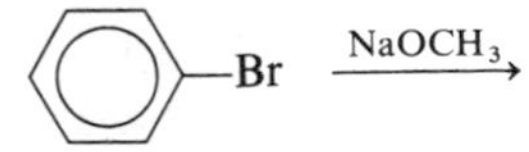 ⬡—Br $\xrightarrow{NaOCH_3}$

(t)

Cl
⬡ NO_2 $\xrightarrow{NH_3}$
  NO_2

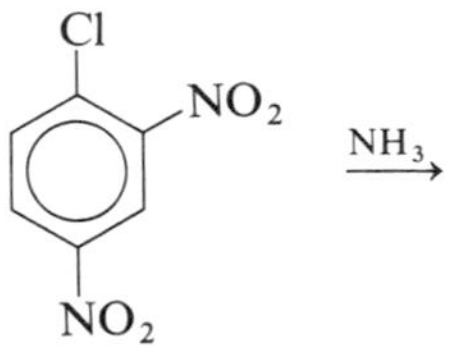

7. Write equations to show how the following conversions can be carried out. More than one step will sometimes be necessary.

(a)  ⬠=CH_2 $\longrightarrow$ ⬠ with Cl and CH_3

(b) $CH_3CH_2CH_2CH_2I$ $\longrightarrow$ $CH_3CH_2CH_2CH_2NH_3{}^+$

(c)

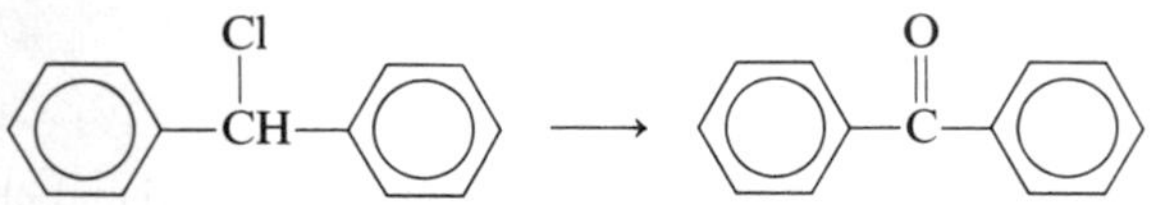

Cl
|
⬡—CH—⬡ $\longrightarrow$ ⬡—C(=O)—⬡

(d) 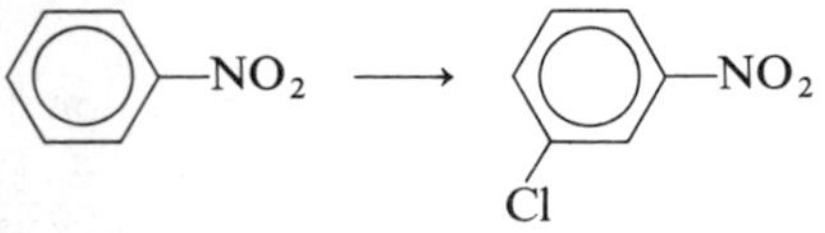 ⬡—NO_2 $\longrightarrow$ ⬡ with NO_2 and Cl

(e) 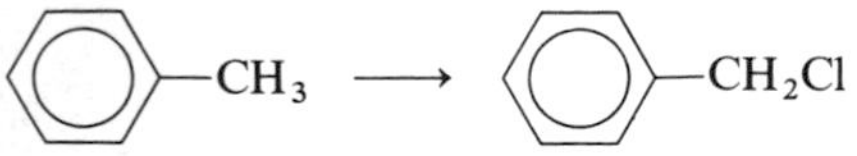 ⬡—CH_3 $\longrightarrow$ ⬡—CH_2Cl

(f)  ⬠ $\longrightarrow$ ⬠ (cyclopentene)

(g) $(CH_3)_2CHCH_2CH_2Br$ $\longrightarrow$ $(CH_3)_2CHCH_2CH_2SH$

(h)

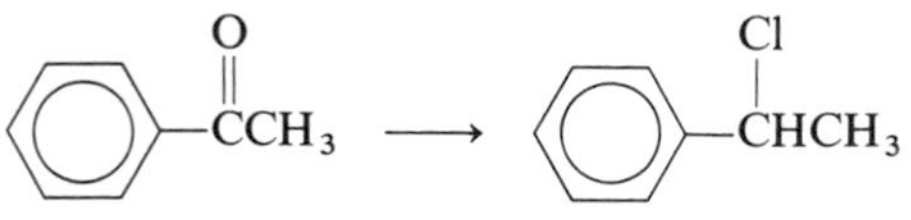

O
‖
⬡—CCH_3 $\longrightarrow$ ⬡—CHCH_3 with Cl

(i)

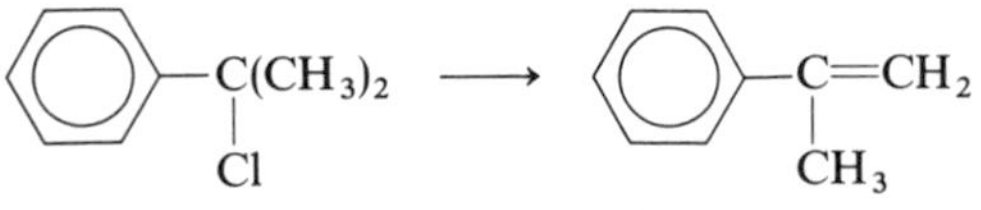

⬡—$C(CH_3)_2$ $\longrightarrow$ ⬡—C=CH_2
    | |
   Cl CH_3

(j) 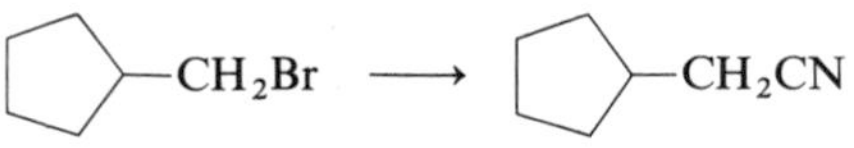 ⬠—CH_2Br $\longrightarrow$ ⬠—CH_2CN

(k) 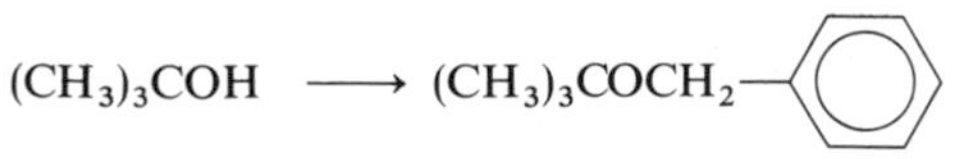 $(CH_3)_3COH$ $\longrightarrow$ $(CH_3)_3COCH_2$—⬡

(l) $CH_3CH_2CH_2Br \longrightarrow CH_3CH_2CH_2MgBr$

(m) $ClCH_2CH_2CH_2CH_2Cl \longrightarrow$ [cyclopentane-S ring structure]

(n)   [benzene]—$CH_2Br \longrightarrow$ [benzene]—CH_2D

8. Write equations for the preparation of ethyl phenyl ether from sodium phenoxide and ethyl bromide. Can bromobenzene and sodium ethoxide be used instead?

9. (a) Arrange the following compounds in order of increasing reactivity toward substitution by an S_N1 mechanism.

CH_3Br [cyclopentane]—Br [cyclopentane with CH_3 and Br]   [cyclopentane]—CH_2Br

   (b) Arrange the compounds in part (a) in order of increasing reactivity toward substitution by an S_N2 mechanism.
   (c) Which of the following compounds is most reactive in an S_N1 reaction?

$CH_2{=}CHCH_2Cl$ $CH_3CH_2CH_2Cl$

   (d) Arrange the following compounds in order of increasing reactivity toward substitution by an S_N1 mechanism.

[benzene]—CH_2Cl [cyclohexane]—CH_2Cl $($[benzene]$)_3CCl$

10. Explain the following observations.
   (a) Even though it is a primary halide, benzyl bromide often reacts via an S_N1 mechanism.
   (b) In some nucleophilic substitution reactions, p-methylbenzyl bromide is considerably more reactive than p-nitrobenzyl bromide.
   (c) In most nucleophilic substitution reactions, 1-bromo-2,2-dimethylpropane is much less reactive than 1-bromopropane.
   (d) The following tertiary chloride is very unreactive toward dehydrohalogenation or nucleophilic substitution reactions.

Constructing a model may be helpful.

11. Draw structures for the substitution products in the following reactions, indicating stereochemistry when necessary. (The structures of the organic reactants are projections.)

(a) 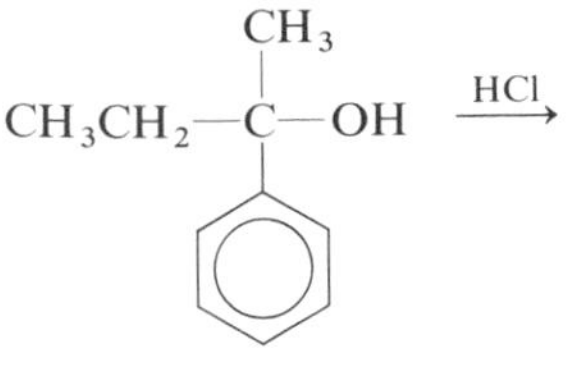 $CH_3CH_2-\underset{\underset{C_6H_5}{|}}{\overset{\overset{CH_3}{|}}{C}}-OH \xrightarrow{HCl}$

(b) 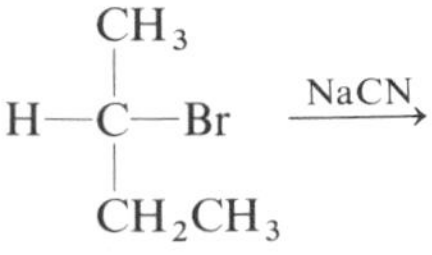 $H-\underset{\underset{CH_2CH_3}{|}}{\overset{\overset{CH_3}{|}}{C}}-Br \xrightarrow{NaCN}$

(c) $H-\underset{\underset{CH_2CH_3}{|}}{\overset{\overset{CH_3}{|}}{C}}-Br \xrightarrow{NaBr}$

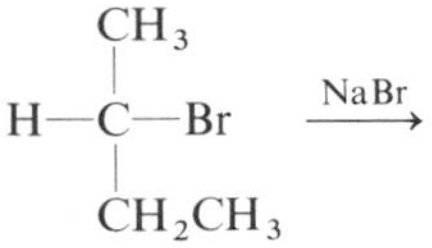

12. Why is it reasonable that carbocations are planar?

13. How could you demonstrate that nucleophilic substitution reactions of primary halides occur with inversion of configuration?

14. The following highly teratogenic compound is sometimes a contaminant of the herbicide 2,4,5-T.

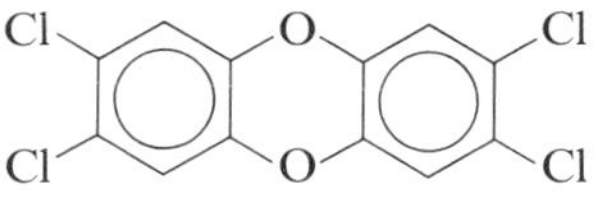

2,3,7,8-Tetrachlorodibenzo-*p*-dioxin

How could it arise in the following synthesis of 2,4,5-T (ethylene glycol is used as a solvent so a high temperature can be achieved)?

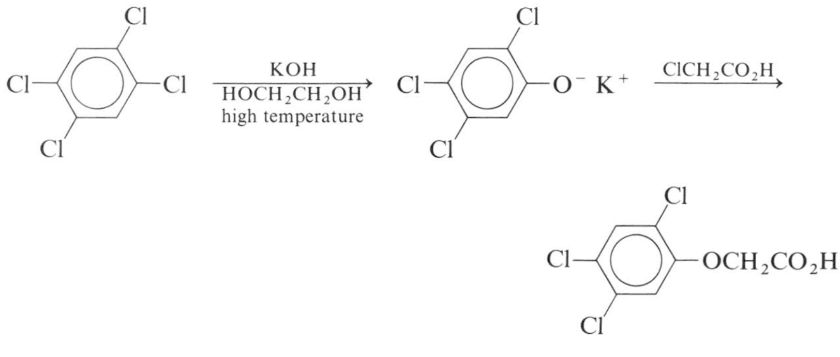

2,4,5-T

Amines

Amines are organic derivatives of ammonia (NH_3) in which one, two, or three of the hydrogens are replaced by alkyl or aryl groups. Just as some properties of alcohols and ethers resemble those of water, so some important properties of amines are related to those of ammonia. Amines are weak bases—as is ammonia—and are the principal organic bases.

Amino groups are among the most important functional groups in biological compounds. They play a pivotal role in the chemistry of proteins (Chapter 14) and nucleic acids (Chapter 15), which along with carbohydrates are the major classes of polymers in living systems. The characteristic odor of decaying fish is due to volatile amines and the bitter tastes of many plants are due to amines called alkaloids. Many metabolic intermediates and drugs are amines.

8.1
NOMENCLATURE OF AMINES

Amines with simple alkyl groups attached to the nitrogen are frequently named by combining the names of the alkyl groups with the word **amine**. Since "amine" in these names stands for ammonia, the compounds are named as ammonia derivatives; therefore the names are written as one word. The first name under each of the following amines is of this type.

$$CH_3CH_2CH_2CH_2—NH_2 \qquad CH_3CH_2\underset{\underset{NH_2}{|}}{CH}CH_3 \qquad CH_3—\underset{\underset{CH_3}{|}}{\overset{\overset{CH_3}{|}}{C}}—NH_2$$

Butylamine   *sec*-Butylamine   *tert*-Butylamine
1-Butanamine   2-Butanamine   2-Methyl-2-propanamine
(primary)   (primary)   (primary)

$$\text{⬠}—NH_2 \qquad CH_3CH_2—NH—CH_2CH_3 \qquad CH_3CH_2CH_2—\underset{\underset{CH_3}{|}}{N}—CH_3$$

Cyclopentylamine   Diethylamine   Dimethylpropylamine
Cyclopentanamine   *N*-Ethylethanamine   *N,N*-Dimethyl-1-propanamine
(primary)   (secondary)   (tertiary)

It is sometimes useful to classify amines according to the number of organic groups attached to nitrogen. As indicated in parentheses for the preceding amines, the designations primary (for one group), secondary (for two groups), and tertiary (for three groups) are used for this purpose. Note that butylamine, *sec*-butylamine, and *tert*-butylamine are all primary amines. When applied to amines, the terms primary, secondary, and tertiary are used to indicate different numbers of alkyl groups. Primary and secondary amines differ in the same way as do alcohols and ethers. When applied to other species (such as alcohols, alkyl halides, and carbocations), primary, secondary, and tertiary indicate the nature of a single alkyl group.

The second name appearing under each amine is a substitutive name. Such names of primary amines are formed by replacing the final -*e* of the parent hydrocarbon by -*amine* and, if necessary, adding a number to indicate the position of —NH$_2$. Secondary and tertiary amines are named as *N*-substituted derivatives of a parent primary amines. The largest or most complicated of the groups attached to nitrogen is chosen as the organic group of the primary amine. Other groups are named as substituents and, to make it clear that they are linked to the nitrogen, are labeled with an *N*-.

The —NH$_2$ group is named amino and sometimes that name is used as a prefix when other substituents are present.

$$H_2N—CH_2CH_2CH_2CH_2—OH \qquad \text{⬡}—\underset{}{\overset{\overset{OH}{|}}{C}H}—\underset{\underset{CH_3}{|}}{CH}—NH—CH_3$$

4-Amino-1-butanol   2-(*N*-Methylamino)-1-phenyl-1-propanol
(ephedrine)

Aromatic names are usually named as derivatives of the simplest aromatic amine, **aniline**.

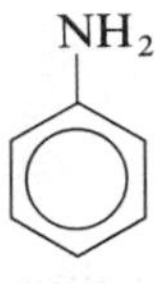

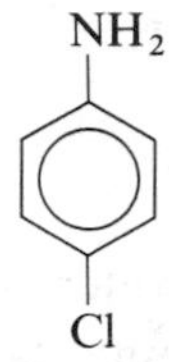

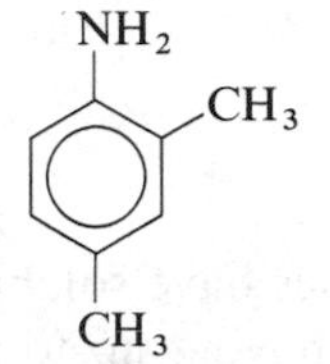

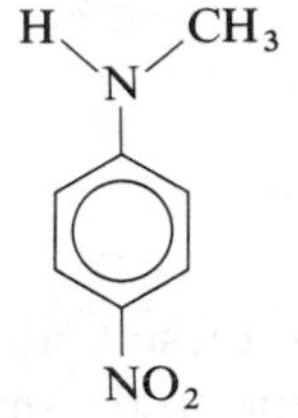

Aniline   *p*-Chloroaniline   2,4-Dimethylaniline   *p*-Nitro-*N*-
                                                        methylaniline

---

problem 8.1   Name the following amines and classify each as primary, secondary, or tertiary.

(a) 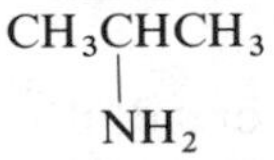 CH$_3$CHCH$_3$
  $\quad\quad\quad$ |
  $\quad\quad\quad$ NH$_2$

(b) (CH$_3$CH$_2$)$_3$N

(c) 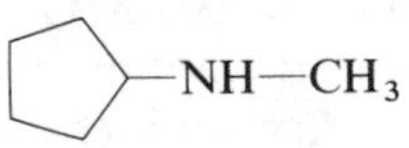 —NH—CH$_3$

(d) 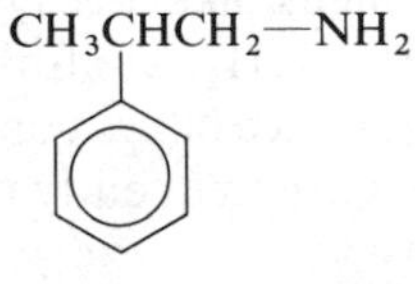 CH$_3$CHCH$_2$—NH$_2$

8.2
PHYSICAL PROPERTIES

Primary and secondary amines have both features needed to form a hydrogen bond: a relatively positive hydrogen (HN$\diagup$) and a nitrogen with an available electron pair (—N:). As a result, primary and secondary amines form intermolecular hydrogen bonds and exhibit boiling points higher than those of hydrocarbons of similar molecular weight (though not as high as comparable alcohols).

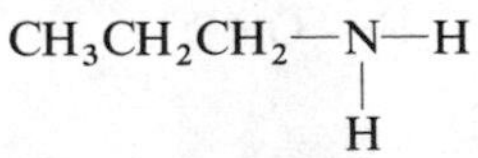

CH$_3$CH$_2$CH$_2$—N—H   CH$_3$CH$_2$—N—CH$_3$ CH$_3$—N—CH$_3$
  $\quad\quad\quad\quad$ | $\quad\quad\quad$ | $\quad\quad\quad$ |
  $\quad\quad\quad\quad$ H $\quad\quad\quad$ H $\quad\quad\quad$ CH$_3$

Propylamine   Ethylmethylamine   Trimethylamine
1-Propanamine   *N*-Methylethanamine   *N,N*-Dimethylmethanamine
bp 48°   bp 36°   bp 3°

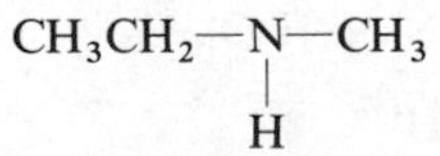

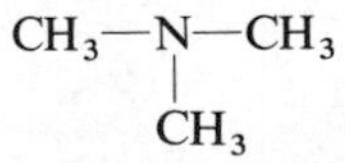

Tertiary amines lack a suitable hydrogen and have boiling points closer to those of hydrocarbons.

Even tertiary amines can form hydrogen bonds with the hydrogens of water.

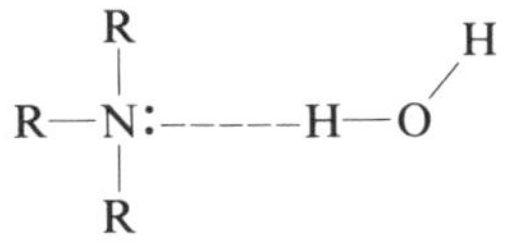

$$
\begin{array}{c}
R \\
| \\
R-N:\!-\!-\!-\!-\!-H-O \\
| \\
R
\end{array}
\quad
\begin{array}{c}
H \\
\diagup \\
\\
\end{array}
$$

As a result, all amines have solubilities in water similar to those of alcohols of the same size. Amines having up to three or four carbons are miscible in water, but solubility decreases rapidly with an increasing number of carbons.

---

problem 8.2 Arrange the following compounds in order of boiling point.
$CH_3CH_2N(CH_3)_2$, $CH_3CH_2CH_2CH_2NH_2$, $H_2NCH_2CH_2CH_2NH_2$

Many low molecular weight alkyl amines have dreadful odors. Trimethyl-amine smells like decaying fish. Putrescine, $H_2N(CH_2)_4NH_2$, and cadaverine, $H_2N(CH_2)_5NH_2$, are formed in decaying flesh from the amino acids ornithine and lysine and have odors consonant with their names. Aryl amines have more pleasant odors but are frequently toxic.

8.3
REACTIONS OF AMINES AS BASES; AMINE SALTS

Some of the most important behavior of amines is a consequence of their basicity. Although far less basic than hydroxide or alkoxide ions, amines are considerably more basic than other neutral organic compounds.

You know that ammonia dissolved in water reacts to form small amounts of ammonium and hydroxide ions.

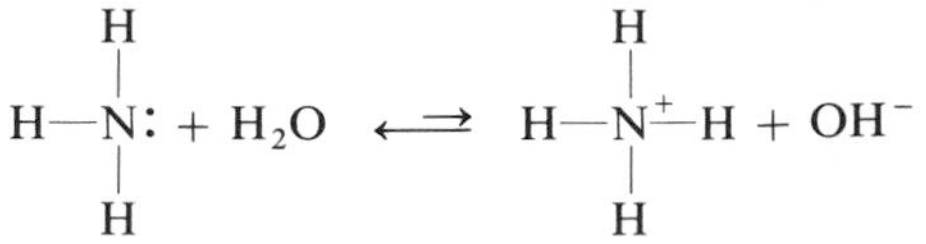

$$
\begin{array}{c}
H \\
| \\
H-N: + H_2O \\
| \\
H
\end{array}
\;\rightleftharpoons\;
\begin{array}{c}
H \\
| \\
H-\overset{+}{N}-H + OH^- \\
| \\
H
\end{array}
$$

The new N—H bond is formed by nitrogen sharing its unused electron pair with a proton donated by water. Because the reaction proceeds only to a small extent, ammonia is a weak base in water.

Basicity is indicated quantitatively by the numerical value of the equilibrium constant for the reaction with water. This equilibrium constant, called the **basicity**

constant or K_b, is related to the concentrations of the species involved in the reaction by this equation.*

$$K_b = \frac{[NH_4^+][OH^-]}{[NH_3]} = 2 \times 10^{-5}$$

The greater the basicity of a compound—and hence the greater its tendency to accept a proton from water—the greater its value of K_b.

An alkyl amine, such as methylamine, ionizes in a similar fashion in water.

$$CH_3-\overset{\overset{\displaystyle H}{|}}{\underset{\underset{\displaystyle H}{|}}{N}}{:} + H_2O \;\rightleftarrows\; CH_3-\overset{\overset{\displaystyle H}{|}}{\underset{\underset{\displaystyle H}{|}}{N}}{\overset{+}{}}{-}H + OH^-$$

The value of its basicity constant indicates that methylamine is somewhat more basic than ammonia.

$$K_b = \frac{[CH_3NH_3^+][OH^-]}{[CH_3NH_2]} = 4 \times 10^{-4}$$

Most alkyl amines, whether primary, secondary, or tertiary, have values of K_b similar to that of methylamine. Aryl amines are considerably less basic. Aniline has $K_b = 4 \times 10^{-10}$. Basicities of other aryl amines depend somewhat on the electron-attracting or electron-withdrawing nature of substituents, but all are far weaker bases than alkyl amines.

Alkyl and aryl amines are basic enough to react essentially quantitatively with strong acids, such as hydrogen halides, sulfuric acid, or nitric acid.

$$CH_3CH_2-\underset{\underset{\displaystyle CH_2CH_3}{|}}{N}-CH_2CH_3 + HCl \;\longrightarrow\; CH_3CH_2-\overset{\overset{\displaystyle H}{|}}{\underset{\underset{\displaystyle CH_2CH_3}{|}}{N}}{\overset{+}{}}{-}CH_2CH_3 \;\; Cl^-$$

Triethylamine   Triethylammonium chloride

$$Cl-\!\!\left\langle\bigcirc\right\rangle\!\!-NH_2 + HNO_3 \;\longrightarrow\; Cl-\!\!\left\langle\bigcirc\right\rangle\!\!-NH_3^+ \;\; NO_3^-$$

p-Chloroaniline   *p*-Chloroanilinium nitrate

---

* The concentration of the solvent water is so large that it does not change significantly when small amounts of a base are dissolved. By convention, the essentially constant value of the water concentration is incorporated into the value of K_b.

The resulting salts can be isolated by evaporating the solvent (usually water). They are often named by replacing -*amine* by -*ammonium** or -*aniline* by -*anilinium* and adding the name of the anion.

Salts of amines are ordinarily crystalline, nonvolatile solids that are very soluble in water but not too soluble in such nonpolar solvents as alkanes. Except for amines with only a few carbons, which are very water soluble, the salt of an amine generally is much more water soluble than the amine itself. We can take advantage of this solubility behavior to separate amines from other compounds. Consider a sample of cyclohexylamine that is contaminated with nonbasic organic compounds. If the sample is treated with aqueous hydrochloric acid, the relatively water-insoluble amine is converted into a salt that dissolves in the aqueous solution.

$$\text{\Large$\bigcirc$—NH}_2 \quad + \text{ HCl} \longrightarrow \quad \text{\Large$\bigcirc$—NH}_3^+ \quad \text{Cl}^-$$

Cyclohexylamine	Cyclohexylammonium chloride
(slightly water soluble)	(very water soluble)

The aqueous layer containing the salt can be separated from the undissolved organic material. Addition of sodium hydroxide to the aqueous solution regenerates cyclohexylamine.

$$\text{\Large$\bigcirc$—NH}_3^+ \quad \text{Cl}^- + \text{Na}^+ \quad \text{OH}^- \longrightarrow \quad \text{\Large$\bigcirc$—NH}_2 + \text{H}_2\text{O} + \text{Na}^+ \quad \text{Cl}^-$$

This reaction is the reverse of the reaction of an amine with water and proceeds essentially to completion.

8.4
ORGANIC ACIDS AND BASES

In the preceding section we considered the basicity of amines in terms of their ability to remove a proton from water molecules to produce hydroxide ions. Similarly, the acidity of phenols (Sec. 5.19) was considered in terms of their ability to donate protons to water to form hydronium ions. In practice, we are usually less concerned with the behavior of organic acids and bases in pure water than in more complex solutions, such as those in living systems, in which the hydronium ion and hydroxide ion concentrations are closely regulated.

To describe the acidity or basicity of an aqueous solution, we usually specify its hydronium ion concentration. This also specifies the hydroxide ion concentration, for they are related by the following equation.

---

* The ending -*aminium* is now frequently used for ions formed from protonating an amine.

$$K_w = [H_3O^+][OH^-] = 1.0 \times 10^{-14}$$

Expressing hydronium ion concentration in moles per liter often gives very small numbers. Therefore it is customary to use the **pH** scale. The pH of a solution is defined as the negative of the logarithm (to the base 10) of the hydronium ion concentration.

$$pH = -\log[H_3O^+]$$

In neutral water, $[H_3O^+] = 10^{-7}$ M and therefore pH = 7. In a 0.1 M aqueous solution of hydrochloric acid, $[H_3O^+] = 10^{-1}$ M and pH = 1. The relations between $[H_3O^+]$, pH, and $[OH^-]$ are shown at the bottom of Figure 8.1. pH decreases one unit for every tenfold increase in $[H_3O^+]$.

The pH of a biological fluid is usually closely controlled. Gastric juice sometimes has pH values considerably below 2 ($[H_3O^+]$ greater than 10^{-2} M) and some other specialized fluids have unusual values. Most biological fluids, however, have pH values near that of pure water (7.0). The pH of human blood plasma, for instance, is about 7.4.

In such solutions, in which the concentration of hydronium ion (and hence of hydroxide ion) is regulated, is methylamine present as such or as the methylammonium ion? The expression for K_b for methylamine

$$K_b = \frac{[CH_3NH_3^+][OH^-]}{[CH_3NH_2]}$$

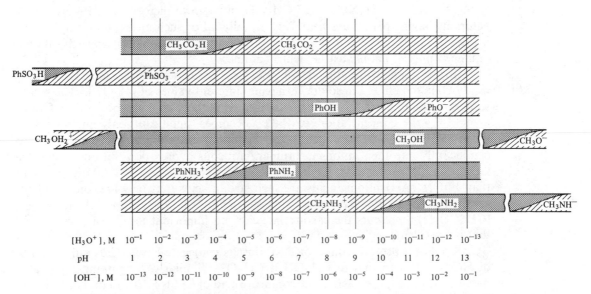

Figure 8.1   The species present in aqueous solutions of some representative organic acids and bases at different concentrations of H_3O^+.

Table 8.1 The ratio of $[CH_3NH_3{}^+]$ to $[CH_3NH_2]$ in aqueous solutions having different concentrations of OH^-

$\dfrac{[CH_3NH_3{}^+]}{[CH_3NH_2]}$	$[OH^-]$ M	$[H_3O^+]$ M	pH
10^2	4×10^{-6}	2.5×10^{-9}	8.6
10^1	4×10^{-5}	2.5×10^{-10}	9.6
1	4×10^{-4}	2.5×10^{-11}	10.6
10^{-1}	4×10^{-3}	2.5×10^{-12}	11.6
10^{-2}	4×10^{-2}	2.5×10^{-13}	12.6

can be rearranged to a form that gives the ratio of $[CH_3NH_3{}^+]$ to $[CH_3NH_2]$ as a function of $[OH^-]$.

$$\frac{[CH_3NH_3{}^+]}{[CH_3NH_2]} = \frac{K_b}{[OH^-]} = \frac{4 \times 10^{-4}}{[OH^-]}$$

The ratio of $[CH_3NH_3{}^+]$ to $[CH_3NH_2]$ is exactly 1 (these species are present in equal concentrations) when $[OH^-] = K_b = 4 \times 10^{-4}$.* Using the equation, we can calculate for other values of $[OH^-]$ the ratios of $[CH_3NH_3{}^+]$ to $[CH_3NH_2]$ shown in Table 8.1. Note that a tenfold increase in hydroxide concentration (a one-unit increase in pH) decreases tenfold the $[CH_3NH_3{}^+]$ to $[CH_3NH_2]$ ratio. The concentrations of $CH_3NH_3{}^+$ and CH_3NH_2 are shown pictorially in Figure 8.1. In solutions having pH values near 7, essentially all of such an equilibrating mixture is present as $CH_3NH_3{}^+$. A similar treatment, however, shows that in such solutions aniline is present mainly as the neutral amine.

From the K_a value for phenol (Sec. 5.19) we can similarly calculate the ratio of $[PhOH]$ to $[PhO^-]$ in solutions having different values of $[H_3O^+]$. The results, summarized in Figure 8.1, indicate that PhOH is the dominant species at pH values around 7—the concentration of PhO^- becomes significant only in quite basic solutions. Alcohols are so much less acidic than phenols that they are not converted significantly into their anions in ordinary aqueous hydroxide solutions. Note that $CH_3NH_3{}^+$ is considerably less acidic than $PhNH_3{}^+$, just as CH_3OH is less acidic than PhOH. Arylsulfonic acids are such strong acids that they are ionized almost completely even in very acidic aqueous solutions. In Figure 8.1 acetic acid (CH_3CO_2H) represents the carboxylic acids, an important family of organic acids that is considered in Chapter 11.

Loss of a proton from an acid forms its **conjugate base**. The conjugate base of HCl is Cl^- and that of CH_3OH is CH_3O^-. Similarly, by gaining a proton, a base forms its **conjugate acid**. The conjugate acid of Cl^- is HCl and that of CH_3NH_2

---

* Basicity is sometimes indicated by $pK_b = -\log K_b$. An amine with $K_b = 10^{-4}$ has $pK_b = 4$.

Table 8.2 Acids and their conjugate bases arranged in order of acidity or basicity of typical members of each class[a]

	Acid		Conjugate base	
	$ArSO_3H$	$\rightleftharpoons$	$ArSO_3^-$	
	$\begin{bmatrix} RCO_2H \\ ArNH_3^+ \end{bmatrix}$	$\rightleftharpoons$ $\rightleftharpoons$	$\begin{bmatrix} RCO_2^- \\ ArNH_2 \end{bmatrix}$	
Increasing acidity	$\begin{bmatrix} ArOH \\ RNH_3^+ \end{bmatrix}$	$\rightleftharpoons$ $\rightleftharpoons$	$\begin{bmatrix} ArO^- \\ RNH_2 \end{bmatrix}$	Increasing basicity
	H_2O	$\rightleftharpoons$	OH^-	
	ROH	$\rightleftharpoons$	RO^-	
	RNH_2	$\rightleftharpoons$	RNH^-	

[a] Species in brackets have similar acidities or basicities.

is $CH_3NH_3^+$.* As evident from Table 8.2, the weaker an acid, the stronger its conjugate base. An alcohol is weakly acidic because its tendency to donate a proton is low; it follows that the corresponding alkoxide ion has a strong tendency to accept a proton and is a strong base. Note that RNH_2 is the conjugate base of RNH_3^+ but the conjugate acid of RNH^-. The anions formed from reactions of amines with extremely strong bases (Sec. 7.4) are themselves strongly basic. Strongly acidic solutions (see Fig. 8.1) are required to form significant concentrations of $CH_3OH_2^+$, the conjugate acid of CH_3OH. Earlier we encountered such protonated alcohols as reaction intermediates (Secs. 3.9, 5.10, and 7.9).

8.5
ALKYLATION OF AMINES

Nucleophilic substitution reactions of halides with ammonia form alkylammonium ions (Sec. 7.5).

$$CH_3\overset{\frown}{CH_2}-I \xrightarrow{:NH_3} CH_3CH_2-NH_3^+ \ I^- \xrightarrow{OH^-} CH_3CH_2-NH_2$$

Ethylammonium iodide   Ethylamine

* The basic properties of an amine (RNH_2) are often described in terms of K_a for ionization of its conjugate acid (RNH_3^+) in water.

$$RNH_3^+ + H_2O \ \rightleftharpoons \ RNH_2 + H_3O^+ \qquad K_a = \frac{[RNH_2][H_3O^+]}{[RNH_3^+]}$$

K_a for RNH_3^+ and K_b for RNH_2 are simply related.

$$K_a \times K_b = K_w = 10^{-14}$$

Therefore if either K_a or K_b is known, the other can be simply calculated. Acidity is sometimes indicated by $pK_a = -\log K_a$.

When treated with base, the ammonium salts are converted to amines. These reactions provide a method for synthesizing primary alkyl amines.

Amines react with alkyl halides as does ammonia.

$$CH_3CH_2CH_2-NH_2 \xrightarrow{CH_3CH_2Br} CH_3CH_2CH_2-NH_2^+-CH_2CH_3 \quad Br^- \xrightarrow{OH^-}$$

Ethylpropylammonium bromide

$$CH_3CH_2CH_2-NH-CH_2CH_3$$

Ethylpropylamine

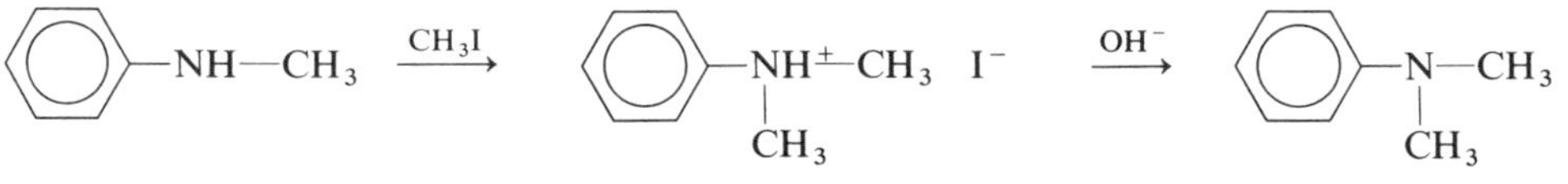

N,N-Dimethylanilinium iodide   N,N-Dimethylaniline

Both secondary and tertiary amines can be synthesized in this fashion. As is usual with nucleophilic substitution reactions, tertiary halides tend to give predominantly elimination rather than substitution products and most aryl halides do not react.

Tertiary amines also react with alkyl halides.

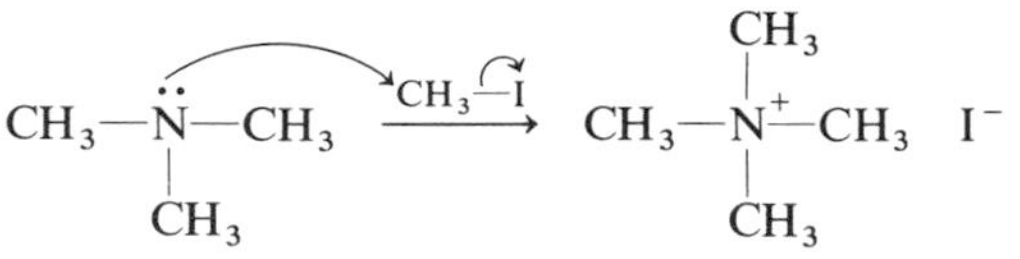

Tetramethylammonium iodide

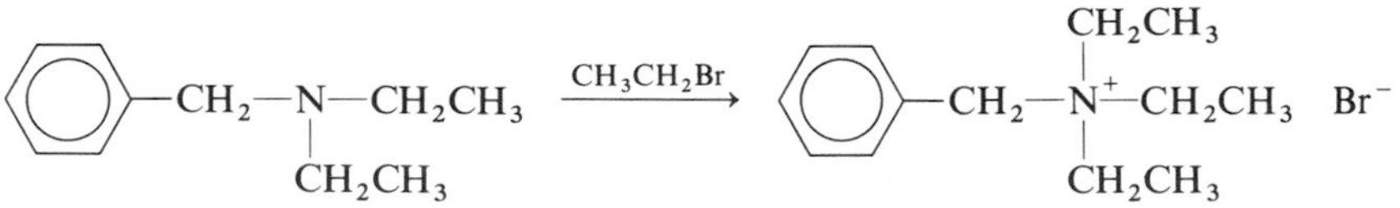

Benzyltriethylammonium bromide

The salts formed from alkylation of tertiary amines have four organic groups bonded to the nitrogen and are called **quaternary ammonium salts**. Having no hydrogen attached to nitrogen, quaternary ammonium salts do not react with bases to form neutral amines.

Some alkyl halides that are particularly reactive in nucleophilic substitution reactions are highly toxic. The following compounds are especially hazardous.

$ClCH_2-O-CH_2Cl$ $ClCH_2CH_2-S-CH_2CH_2Cl$

Di(chloromethyl) ether   Mustard gas

Their toxicity is attributed to an ability to react readily in nucleophilic displacement reactions with amine and hydroxyl groups of important biological compounds.

Many biological amines are methylated by enzyme-catalyzed nucleophilic substitution reactions with the reagent *S*-adenosylmethionine. Epinephrine (noradrenaline), for example, is converted to adrenaline.

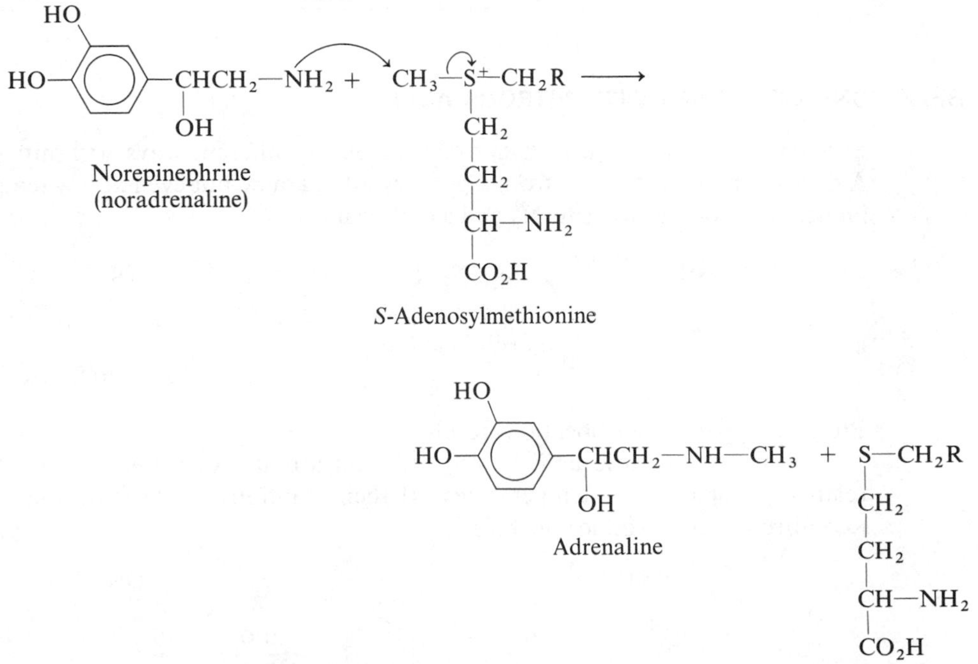

Norepinephrine
(noradrenaline)

S-Adenosylmethionine

Adrenaline

The leaving group is a sulfide. *S*-Adenosylmethionine is prepared by another nucleophilic substitution reaction in which the nucleophile is the sulfur of methionine and the leaving group is the triphosphate grouping of ATP.

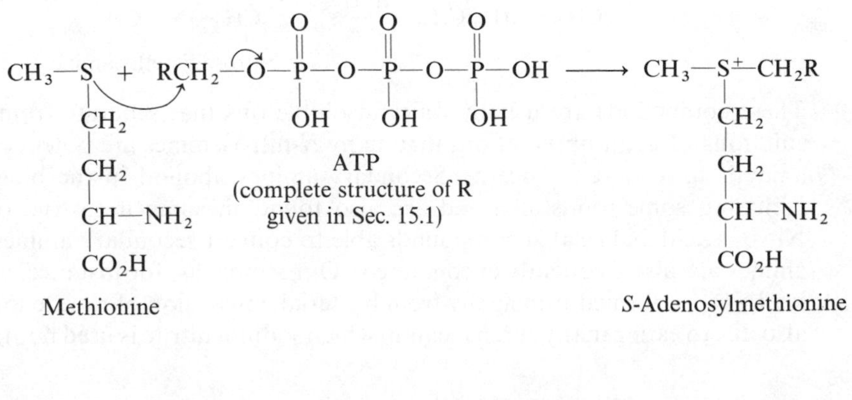

Methionine

ATP
(complete structure of R
given in Sec. 15.1)

S-Adenosylmethionine

problem 8.3 Draw structures for the products of the following reactions.

(a) $(CH_3CH_2)_2NH \xrightarrow{HCl}$

(b) $(CH_3CH_2)_2NH \xrightarrow{CH_3CH_2Br}$

(c) 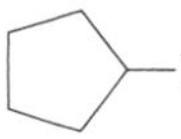 ⬠$-NH_3^+ \ Cl^- \xrightarrow{NaOH}$

(d) $CH_3CH_2CH_2N(CH_3)_2 \xrightarrow{CH_3I}$

8.6
REACTIONS OF AMINES WITH NITROUS ACID

Primary, secondary, and tertiary amines react in different ways with nitrous acid. A diazonium ion is formed from a primary alkyl amine but even at low temperature immediately decomposes to N_2 and a carbocation.

$$R-NH_2 \xrightarrow{HNO_2} [R-N_2^+] \longrightarrow N_2\uparrow + [R^+]$$

A diazonium ion
(rapidly loses N_2)

Alkenes, ROH, etc.

Products typical of carbocations result.

Primary aromatic amines also form diazonium ions, but usually these are relatively stable at low temperatures. If their solutions are heated, however, they lose nitrogen and produce phenols.

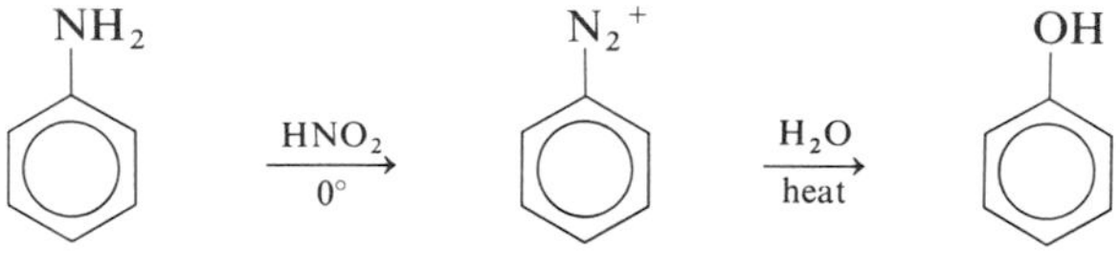

NH₂ $\xrightarrow[0°]{HNO_2}$ N_2^+ $\xrightarrow[heat]{H_2O}$ OH

Benzenediazonium ion

Secondary amines react with nitrous acid to form *N*-nitrosamines.

$$CH_3-NH-CH_3 \xrightarrow{HNO_2} CH_3-\overset{\displaystyle NO}{\underset{\displaystyle |}{N}}-CH_3$$

N-Nitrosodimethylamine

These compounds are usually water-insoluble oils that separate from the aqueous solutions. Recent observations that many *N*-nitrosamines are potent carcinogens in animals have raised concern. Secondary amines abound in the biological world, including some foodstuffs, and are also found in some industrial environments. Nitrous acid and related compounds able to convert secondary amines to *N*-nitrosamines are also frequently encountered. Our stomachs, for instance, contain nitrite ion (NO_2^-), formed principally from bacterial conversion of nitrate ion (NO_3^-) but also due to eating many cured meats in which sodium nitrite is used both as a preservative and to impart a red color. Under acidic conditions (e.g., in the stomach) nitrous acid (a weak acid) can be produced.

$$Na^+ \quad NO_2^- + H_3O^+ \; \rightleftharpoons \; HNO_2 + H_2O + Na^+$$

Tertiary amines dissolve in and react with nitrous acid solutions in a complex fashion but without visible signs of reaction, such as evolution of a gas or separation of an oil.

8.7
REACTIONS OF ARYLDIAZONIUM IONS

Because the $—N_2^+$ group can be replaced by a variety of other functions, aryldiazonium ions play an important role in the synthesis of aromatic compounds. As a result of their reactivity, aryldiazonium ions are usually prepared at low temperatures and reactions are carried out directly with the solutions in which they have been prepared.

As noted, heating a solution of a diazonium ion leads to evolution of nitrogen and formation of a phenol.

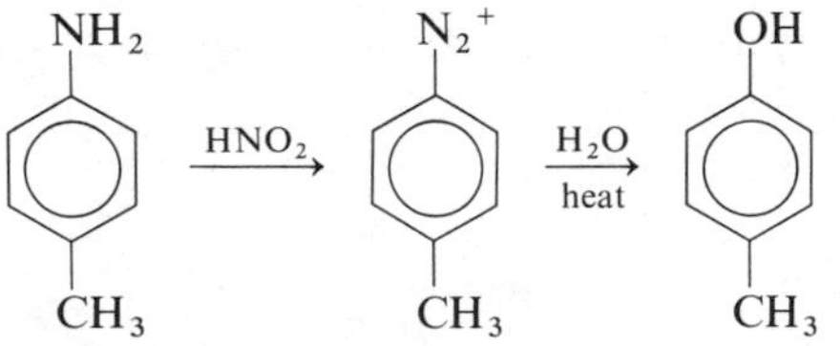

Chlorides and bromides are produced by reactions with cuprous chloride or bromide and iodides with potassium iodide.

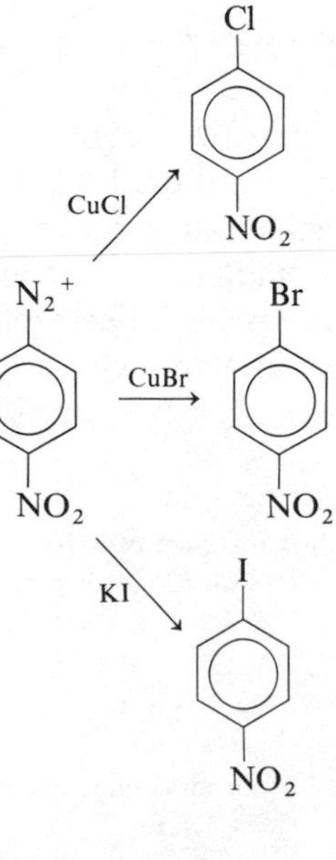

Reactions with cuprous cyanide produce nitriles.

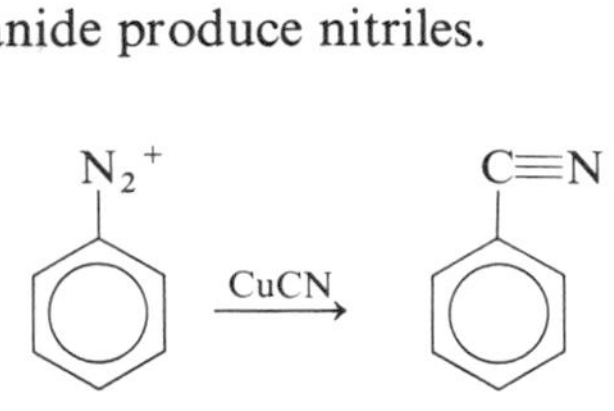

Benzonitrile

Some reducing agents, such as hypophosphorous acid (H_3PO_2), replace $-N_2^+$ by a hydrogen.

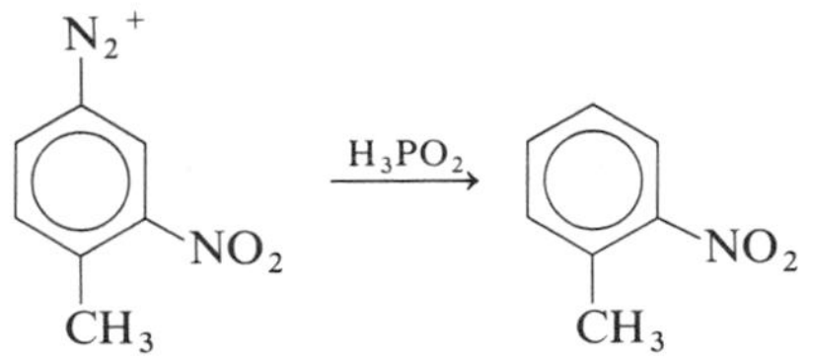

The preparation of 1,3,5-tribromobenzene is an example of the synthetic use of diazonium salts.

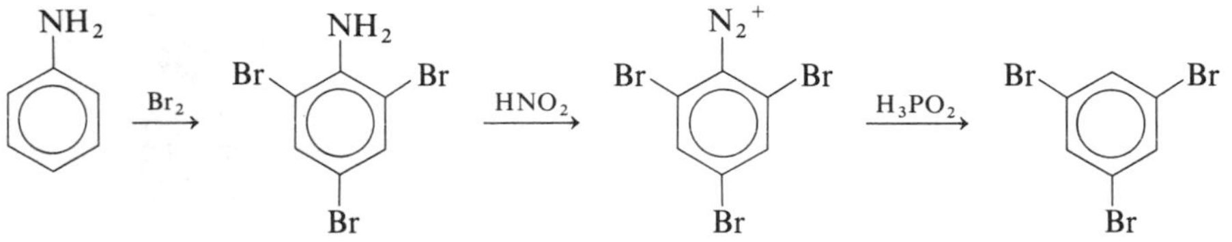

2,4,6-Tribromoaniline   1,3,5-Tribromobenzene

Bromine is ortho-para directing; consequently, bromination of benzene does not produce significant amounts of the 1,3,5-tribromo isomer. Because an amino group is ortho-para directing and strongly activating, however, the ortho and para positions of aniline are readily brominated.* The amine function then is removed.

Diazonium ions are weakly electrophilic and perform aromatic substitutions on particularily reactive aromatic compounds, such as phenols and amines. The products, containing the $-N=N-$ group, are called **azo compounds**.

---

* Bromination of aniline was part of a commercial process once used to obtain bromine from seawater. The water was treated with Cl_2 to oxidize Br^- (present in seawater to the extent of about 70 parts per million) to Br_2.

$$2 \, Br^- + Cl_2 \longrightarrow Br_2 + 2 \, Cl^-$$

Aniline was added and the 2,4,6-tribromoaniline that precipitated was subjected to reactions that removed the bromine from it.

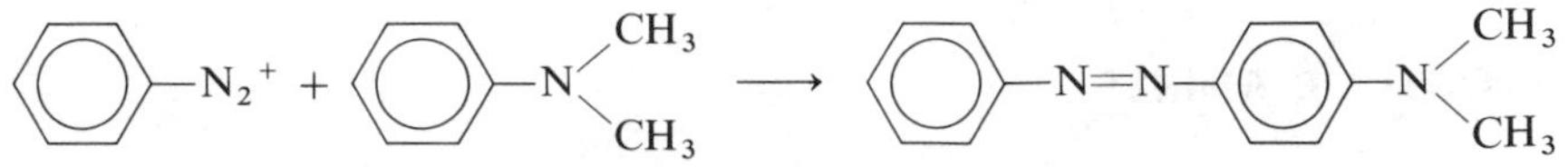

p-Dimethylaminoazobenzene
("butter yellow")

p-Dimethylaminoazobenzene was once used as a yellow food coloring. It no longer is, for it is now suspected to be a carcinogen. Many other azo compounds are used as dyes for fabrics.

8.8
PREPARATIONS OF AMINES

Nucleophilic substitution of alkyl halides by amines or ammonia, already discussed (Sec. 8.5) as an important reaction of amines, is also a preparation of amines.

Primary amines are also prepared by reduction of nitro compounds.

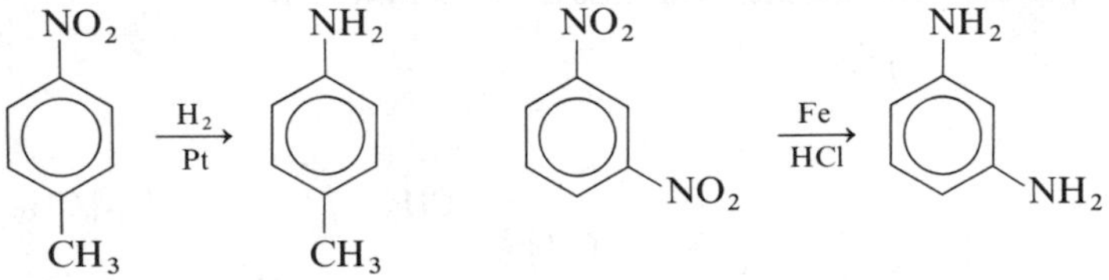

This is an important preparation of aryl amines, for aromatic nitro compounds are readily available by nitration. Common reducing agents are hydrogen in the presence of a platinum catalyst and iron or zinc metal in the presence of hydrochloric acid.*

Primary amines are also prepared by reduction of nitriles—for example, by use of hydrogen and a nickel catalyst.

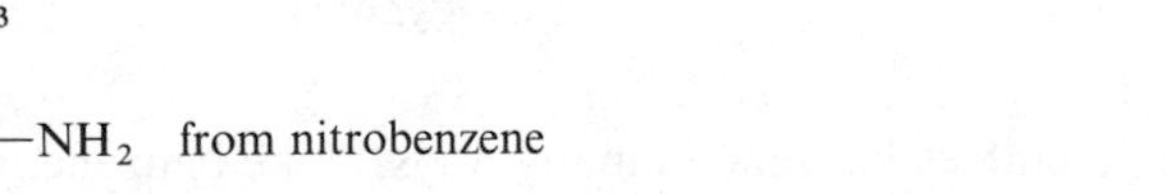

---

problem 8.4   How can the following amines be prepared?
(a) $CH_3CHCH_2CH_2NH_2$ from an alkyl halide
$\quad\quad$ |
$\quad\quad CH_3$

(b)

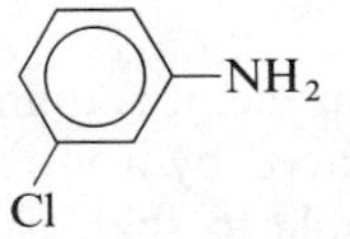

from nitrobenzene

---

* The amine product is present mainly as an ammonium ion until the acidic solution is neutralized by addition of base. Since the ratio of amine to ammonium ion depends on the particular acidity of the solution and is altered simply by addition of acid or base, we will usually write the amine even when the ammonium ion might actually be the dominant form.

8.9
HETEROCYCLIC AMINES

Heterocyclic amines have one or more nitrogen atoms incorporated into a ring. Most saturated heterocyclic amines behave chemically like acyclic amines.

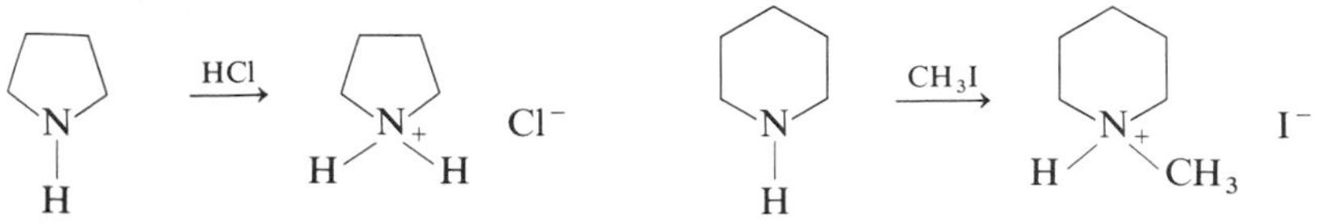

Pyrrolidine   Piperidine

Such ring systems are encountered in many naturally occurring compounds. Nicotine, a poison that occurs to the extent of 2 to 8% in tobacco plants and is used as an agricultural insecticide, contains a pyrrolidine ring.

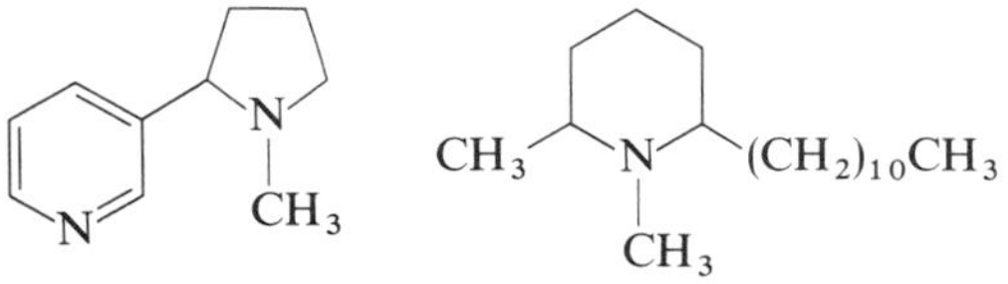

Nicotine   The active ingredient of the
                      venom of the red fire ant

A compound isolated from the red fire ant contains a piperidine ring. The six-membered ring of nicotine is a substituted pyridine ring.

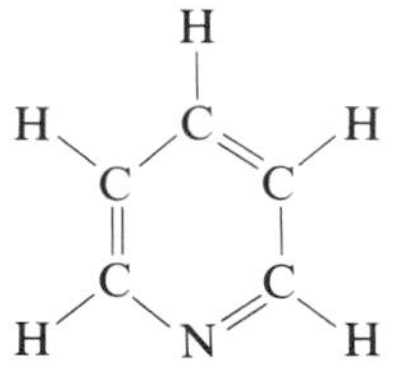

Pyridine

Pyridine resembles benzene in many ways. Like benzene, it is found in coal tar. Its structure is that of benzene with one CH unit replaced by a nitrogen. Pyridine is unusually stable—its resonance energy is comparable to that of benzene. It also resembles benzene in being unreactive toward many reagents that ordinarily attack double bonds and in undergoing substitution rather than addition reactions.

Just as for benzene, we can write two equivalent resonance structures for pyridine.

If the bonding is described instead in terms of overlapping of orbitals, pyridine has a π-electron cloud formed from overlap of six p-orbitals, one from each ring atom. Each ring atom contributes one electron to the π cloud (for bookkeeping purposes, the electron donated by each atom is indicated in its p orbital).

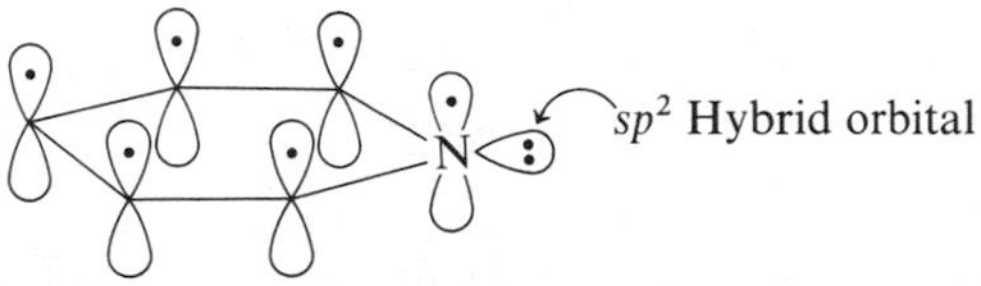

sp^2 Hybrid orbital

The unshared electron pair of nitrogen occupies an sp^2 hybrid orbital (just as does the electron that each carbon shares with a hydrogen).

Pyridine undergoes some characteristic amine reactions. Its basicity ($K_b = 2 \times 10^{-9}$) is somewhat greater than that of aniline

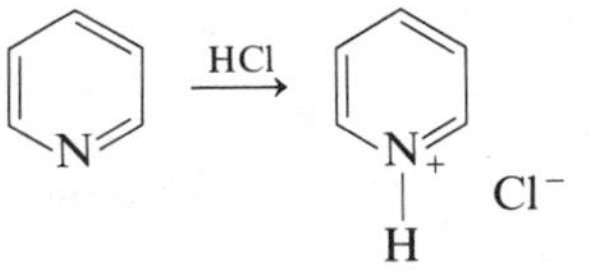

and it reacts with alkyl halides.

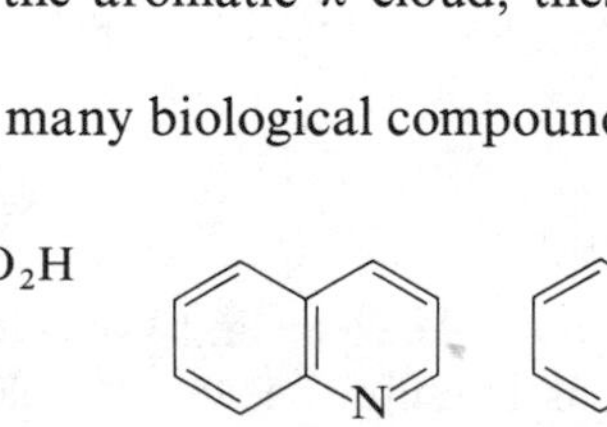

Because the electron pair of nitrogen that participates in reactions with acids or alkyl halides is not part of the aromatic π cloud, these reactions leave intact the stable aromatic system.

Pyridine rings occur in many biological compounds.

Nicotinic acid
(niacin)

Quinoline

Isoquinoline

A deficiency in human diets of the vitamin nicotinic acid leads to pellagra. Nicotinic acid becomes incorporated into the coenzyme NADH, which plays a major role in some biological oxidation-reduction processes (see Sec. 6.9 for an example). Quinoline and isoquinoline, compounds containing a fused pyridine ring, are found in coal tar. Many alkaloids (Sec. 8.11) contain their ring systems.

Pyrimidine also behaves as an aromatic system.

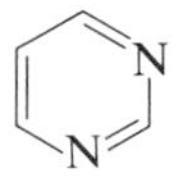

Pyrimidine

Substituted pyrimidines are important components of the nucleic acids (Chapter 15).

Although it has a five-membered ring, pyrrole also has aromatic properties.

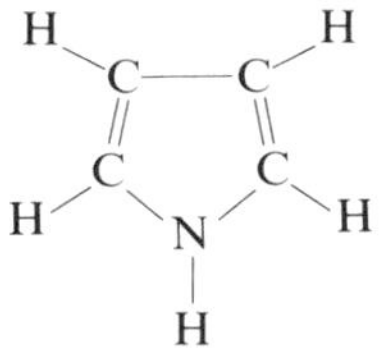

Pyrrole

Its resonance energy, however, is considerably smaller than that of benzene and its reactivity toward attack by electrophilic reagents, such as bromine, is greater. Pyrrole (K_b about 10^{-14}) is much less basic than pyridine and does not react with alkyl halides.

The aromatic character of pyrrole is due to its possession of a cyclic system containing six π electrons, the number found in benzene and pyridine. The π cloud is formed from overlap of five p orbitals, one contributed by each ring atom.

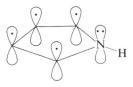

Each carbon contributes one electron to the cloud and the nitrogen contributes two. Pyrrole fails to react as an amine because the electron pair of nitrogen is incorporated into the π cloud. Reaction with this electron pair destroys the aromatic system.

The pyrrole ring system occurs in many naturally occurring compounds. Indole and skatole, compounds with a fused pyrrole ring, are found in coal tar and also in feces, to which they impart much of the characteristic fecal odor.

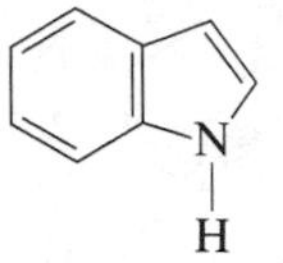

Indole 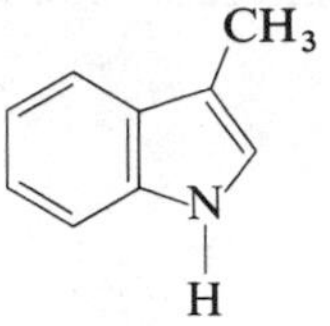   Skatole

Many alkaloids contain the indole ring system.
Imidazole also has some aromatic character.

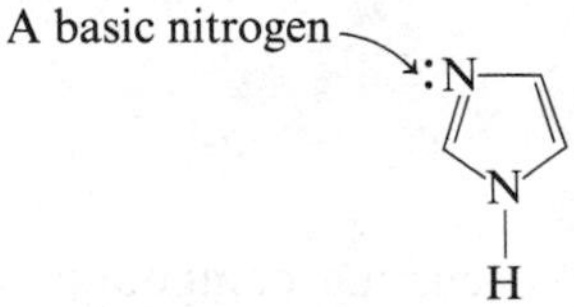

Imidazole

One of the nitrogens is basic, for its electron pair is not incorporated into the π cloud. This basicity is an important part of the behavior of the amino acid histidine when it is incorporated into proteins.

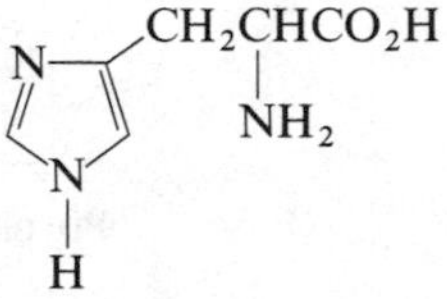

Histidine

Histamine is produced from histidine.

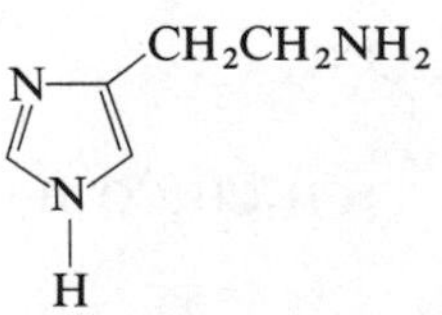

Histamine 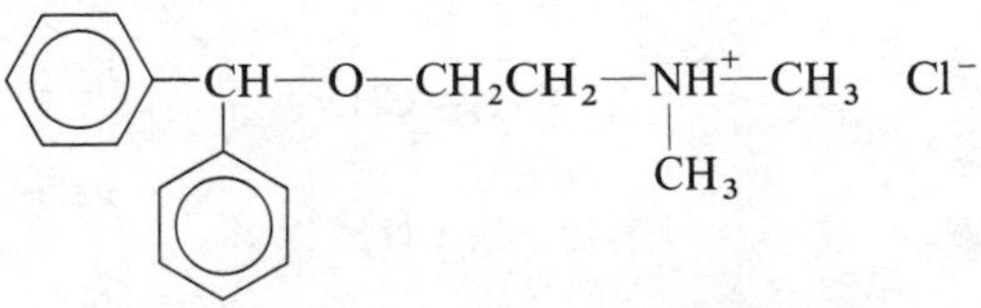   Diphenhydramine hydrochloride
(an antihistamine)

Excessive amounts probably form as a result of allergic reactions, leading to various symptoms associated with allergies and the common cold. Antihistamines, such as diphenhydramine hydrochloride, are drugs that reduce the effects of histamine.

Purine contains fused imidazole and pyrimidine rings.

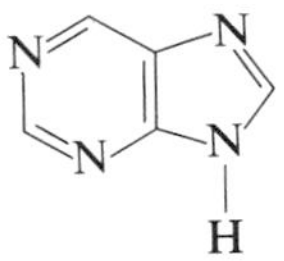

Purine

Its ring system is found in many biological compounds, including caffeine (Sec. 8.11) and nucleic acids (Chapter 15).

8.10 PORPHYRINS

Porphyrins, naturally occurring compounds related to pyrrole, are derivatives of the molecule porphine.

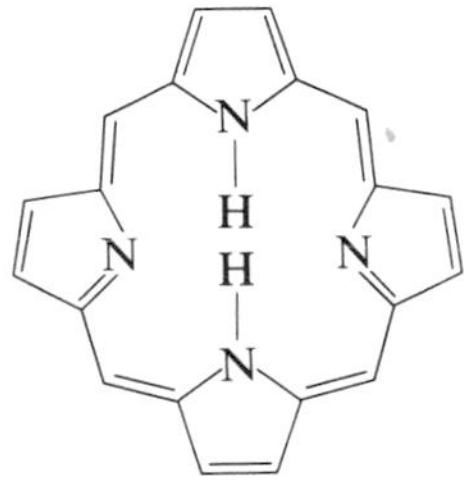

Porphine

Heme contains an Fe^{2+} bonded to the four N atoms (note that heme lacks two hydrogens of the parent ring structure).

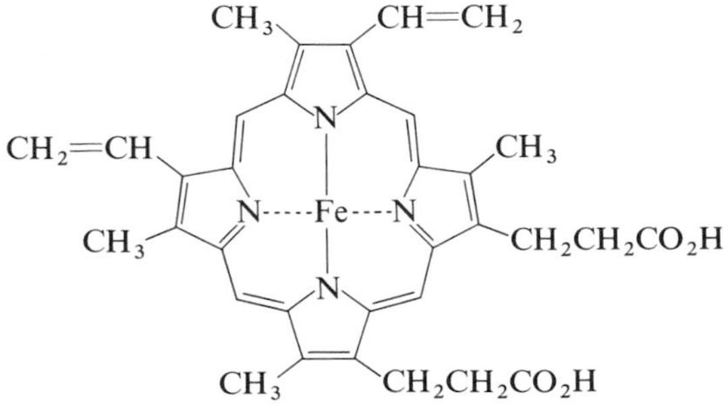

Heme

Hemoglobin, which transports O_2 in the blood of mammals, is a combination of heme and a protein (Chapter 14). An O_2 molecule bonds weakly to the iron. The equilibrium favors bonding in the lungs, where O_2 concentration is high, but dissociation in other tissues, where O_2 concentration is lower. Carbon monoxide (CO) is a poison because it combines more tightly than O_2 with the iron of hemoglobin.

Chlorophylls, pigments found in all photosynthetic cells, have structures very similar to that of heme. Chlorophylls *a* and *b* are found in green plants.

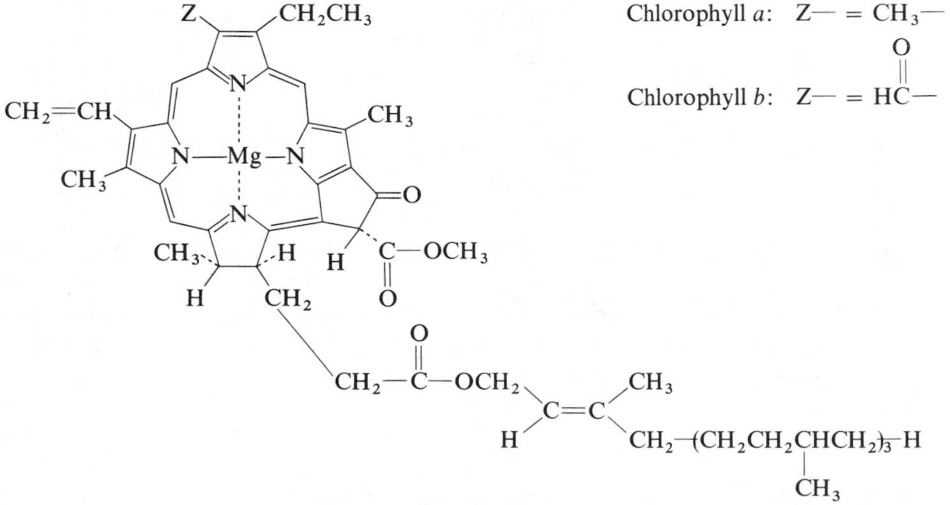

The energy of the sunlight absorbed by chlorophylls is converted into chemical energy used for the reduction of carbon dioxide (Sec. 2.16 and Chapter 10).

8.11
ALKALOIDS

Alkaloids are members of a family of amines, usually heterocyclic, found in plants. Their functions in plants are rarely known and many plants contain no alkaloids. Most alkaloids, however, have significant physiological effects on animals.

The simple alkaloid coniine, isolated from the poison hemlock,* was used to execute the Greek philosopher Socrates (as well as many lesser-known persons).

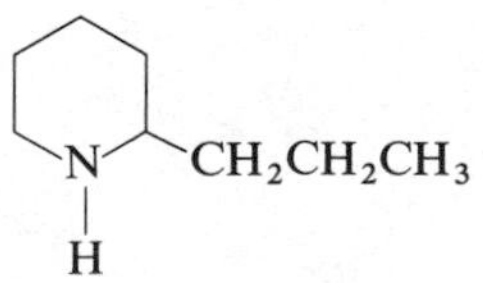

Coniine 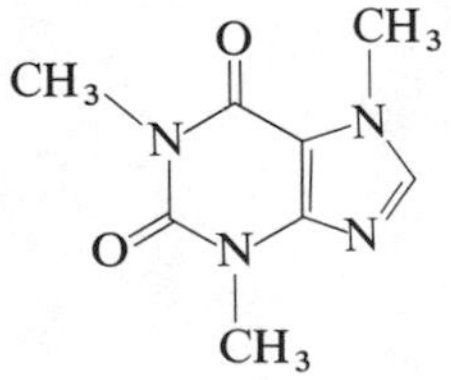   Caffeine

* A shrub, not the majestic state tree of Pennsylvania.

Caffeine, extracted from coffee beans or tea leaves simply by heating them in water, is the most widely used stimulant.

Cocaine, isolated from coca leaves, acts as a local anesthetic and central nervous system stimulant. It is habit forming.

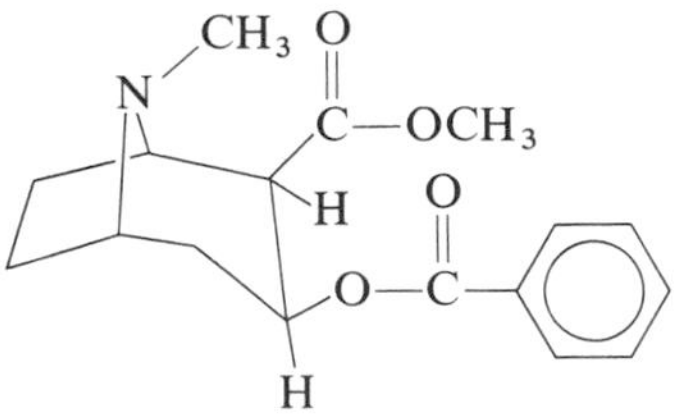

Cocaine 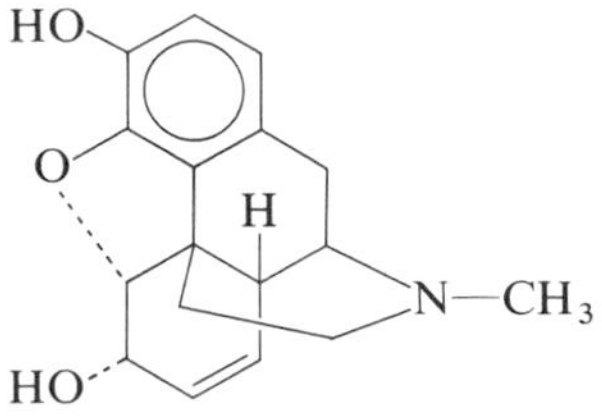   Morphine

Morphine is the compound principally responsible for the physiological effect of opium, a material exuded from cuts made in the seed capsule of the opium poppy. Extraordinarily effective in relieving severe pain, it is, unfortunately, habit forming. Papaverine is also found in opium. Synthetic papaverine is used as a smooth muscle relaxant, particularly for relieving arterial and bronchial spasms.

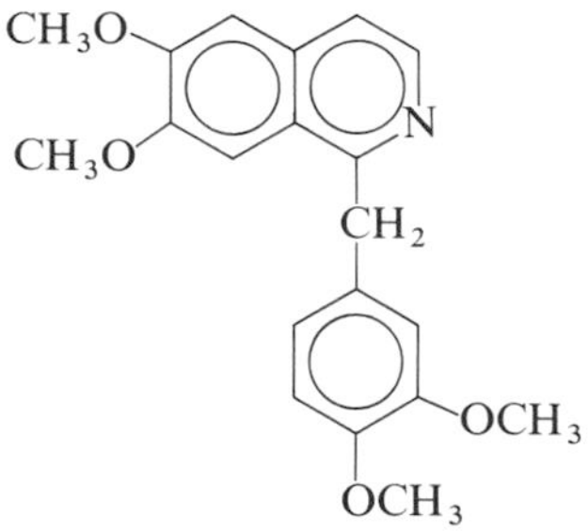

Papaverine

Quinine, isolated from the bark of the cinchona tree, in which it is present to the extent of about 8%, has been used for centuries to treat malaria.

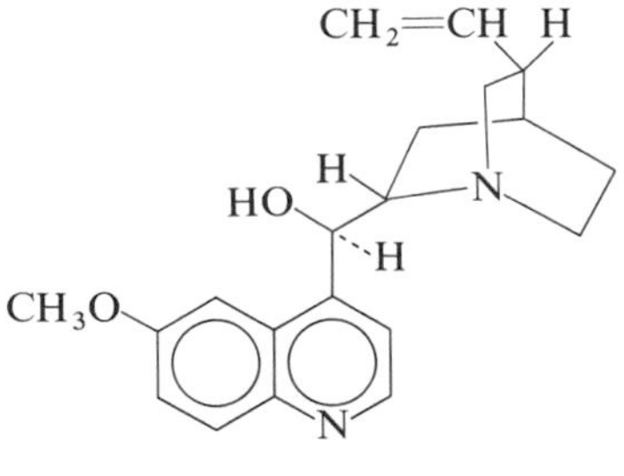

Quinine

Reserpine, isolated from the root of the Indian snakeroot plant, is a hypotensive (blood-pressure lowering) agent and a potent tranquilizer.

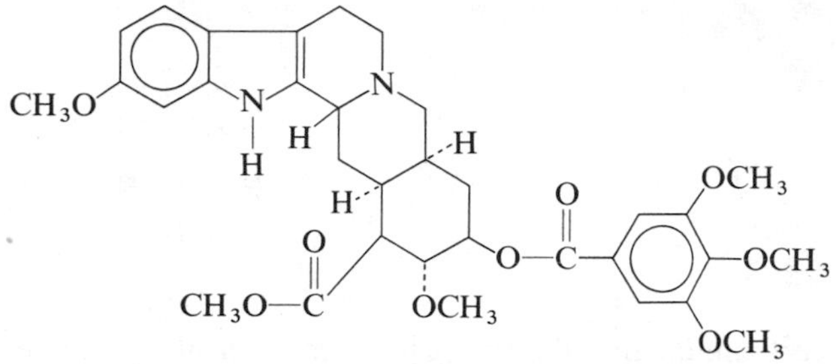

Reserpine

8.12
NEUROCHEMICALLY ACTIVE AMINES

An impulse is transmitted from one nerve cell to another by release of a compound called a neurotransmitter into the synapse, the space between the cells. This neuro-transmitter interacts with a specific receptor on the adjoining nerve cell, stimulating the generation of a new impulse. The action of the neurotransmitter is then rapidly terminated either by its recapture by the nerve cell and storage for later use or by its inactivation by chemical modification.

Neurotransmitters involved in different parts of the nervous system include the following amines.

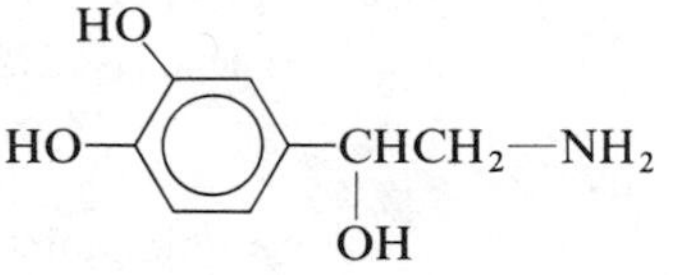

Noradrenaline
(norepinephrine)

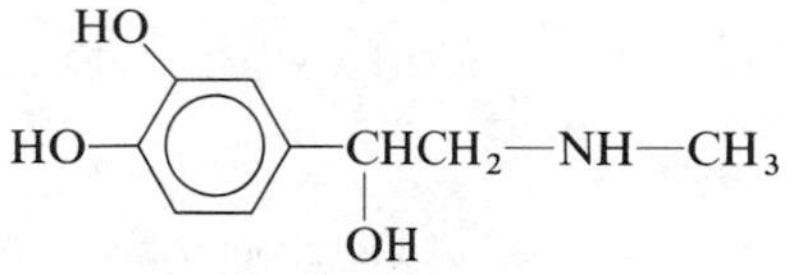

Adrenaline
(epinephrine)

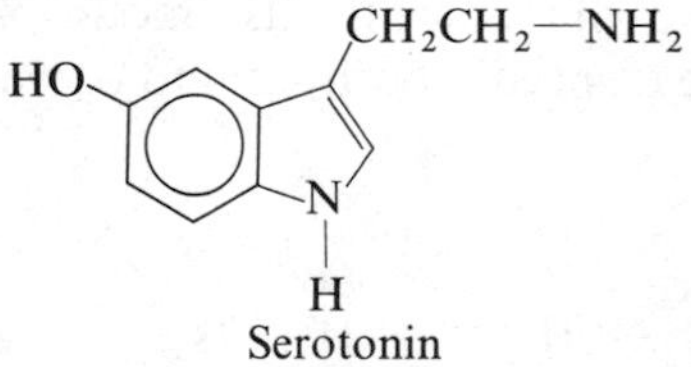

Serotonin

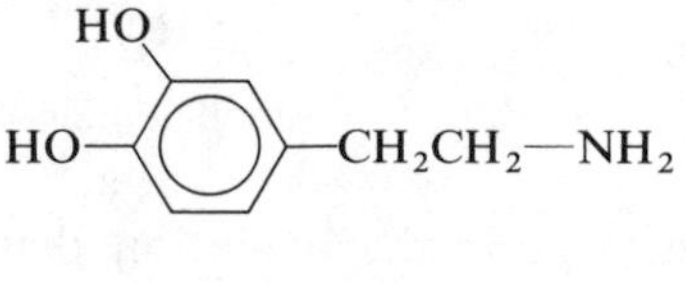

Dopamine

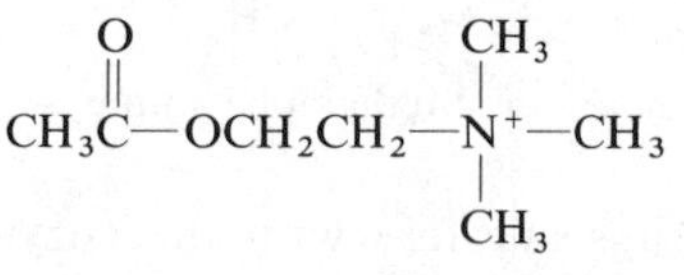

Acetylcholine

The symptoms of Parkinson's disease seem to be caused at least in part by a deficiency of dopamine in the central nervous system. Dopa has been used to relieve these symptoms.

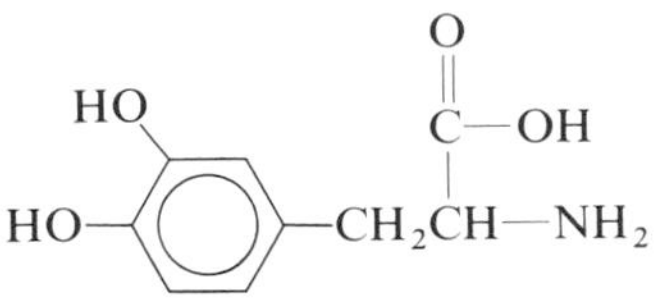

Dopa

It travels from the blood into the brain more easily than does dopamine; once there, it is converted into dopamine.

The effects of various drugs on the nervous system may often be related to their effects on the neurotransmitters. Norepinephrine seems to be involved particularly in mood elevation and depression. Tranquilizers reduce the effects due to norepinephrine. The alkaloid reserpine (Sec. 8.11), the first tranquilizer to be used clinically, reduces the amount of stored norepinephrine that is available for release. Chlorpromazine, a synthetic tranquilizer in wide use, may combine with receptors for norepinephrine and in that way reduce their capability to respond to released norepinephrine.

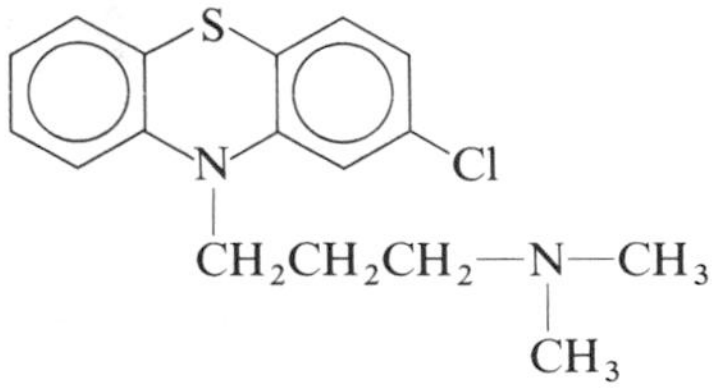

Chlorpromazine

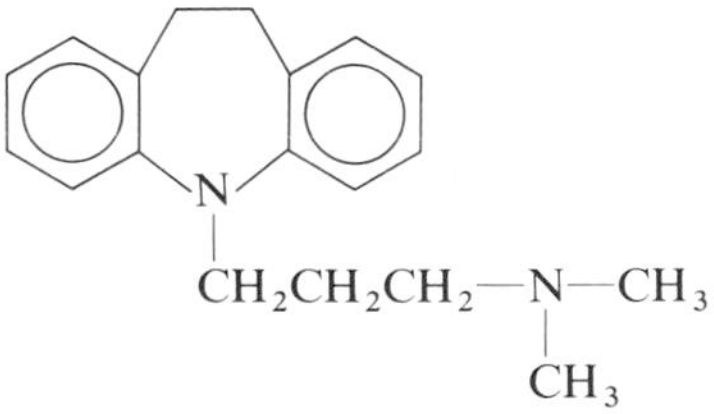

Imipramine

Some antidepressant drugs, such as imipramine, seem to block the recapture of norepinephrine and thus increase its amount at the receptor. Amphetamine and methamphetamine ("speed") both seem to prevent recapture and to promote release of norepinephrine.

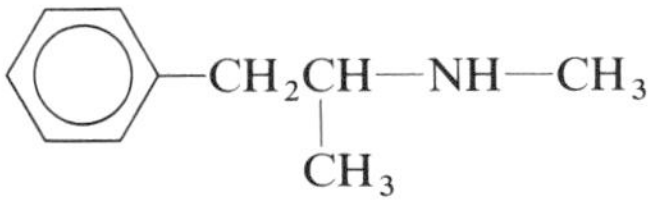

Methamphetamine

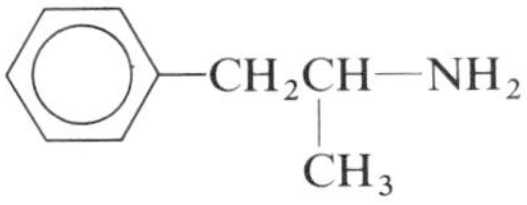

Amphetamine

Other drugs interfere with the enzyme catalyzing the oxidation that inactivates norepinephrine.

SUMMARY

Definitions and Ideas

Amines are derivatives of ammonia in which one, two, or three of the hydrogens are replaced by alkyl or aryl groups. Depending on the number of organic groups attached to the nitrogen, an amine is classified as **primary** (one group), **secondary** (two groups), or **tertiary** (three groups).

Amines with simple alkyl groups attached to the nitrogen are frequently named by combining the names of the alkyl groups with the word **amine**. Substitutive names of primary amines are formed by replacing the final -*e* of the name of the parent hydrocarbon by -*amine* and adding a number to indicate the position of attachment of the amino group. To name secondary and tertiary amines, the largest or most complex of the groups attached to nitrogen is chosen as the organic group of a parent primary amine; the other groups are named as substituents and labeled with an *N*-. Aromatic amines are usually named as derivatives of **aniline**, the simplest aromatic amine.

Amines hydrogen bond to water and have solubilities in water similar to those of alcohols. Primary and secondary amines have boiling points considerably higher than those of hydrocarbons of similar molecular weight. Molecules of tertiary amines cannot form hydrogen bonds to each other and tertiary amines do not exhibit unusually high boiling points.

Amines are weak bases in water. Alkyl amines are slightly more basic than ammonia, but aryl amines are considerably less basic. Both alkyl and aryl amines, however, are sufficiently basic to be converted almost quantitatively to salts by strong acids. In an aqueous solution that is maintained at neutrality (pH = 7), alkyl amines exist largely as ammonium ions but aryl amines largely as the neutral amines.

Salts of amines, usually very water soluble, can be converted to the neutral amines by treatment with bases. The salts are often named by replacing -*amine* by -*ammonium*, or -*aniline* by -*anilinium* and adding the name of the anion. Salts that have four organic groups bonded to nitrogen are called **quaternary ammonium salts**.

Most saturated heterocyclic amines behave like acyclic amines. Pyridine and pyrrole have aromatic properties.

Important Reactions Involving Amines

1. *Reactions as bases* $-\overset{|}{\underset{|}{N}} \xrightarrow{\text{HX}} -\overset{|}{\underset{|}{N}}{}^{\pm}\!-\!H\ X^- \xrightarrow{\text{OH}^-} -\overset{|}{\underset{|}{N}}$

   HX = HCl, HBr, HI, H_2SO_4, HNO_3, or other acids
   Addition of a strong base converts the salt back to the amine.

2. *Alkylation of ammonia or amines* $R_3N \xrightarrow{R'X} R_3NR'^+\ X^-$

   X = Cl, Br, or I
   Reactions with tertiary halides often give predominantly elimination. The R groups can be any combination of alkyl groups, aryl groups, or H's. If at least one H is present, then the salt can be converted to a neutral amine by treatment with base (see reaction 1).

3. *Preparation of primary amines by reduction of nitro compounds*

$$R-NO_2 \xrightarrow[\text{Fe (or Zn) and HCl}]{\text{H}_2 \text{ and Pt, or}} R-NH_2$$

R = alkyl or aryl

PROBLEMS

1. Draw structures for the following compounds.
   (a) diethylamine
   (b) butylamine
   (c) benzylamine
   (d) triethylamine
   (e) cyclohexylamine
   (f) ethylisopropylmethylamine
   (g) 1-hexanamine
   (h) 1,3-pentanediamine
   (i) 2-amino-3-hexanol
   (j) N-methyl-l-pentanamine
   (k) m-nitroaniline
   (l) p-bromo-N,N-dimethylaniline
   (m) isobutylammonium chloride
   (n) tetraethylammonium hydroxide
   (o) pyridine
   (p) pyrrole

2. Name the following compounds.
   (a) $CH_3CH_2CH_2NH_2$
   (b) CH_3NHCH_3

   (c) $CH_3NHCH_2CH_2CH_2CH_3$
   (d) 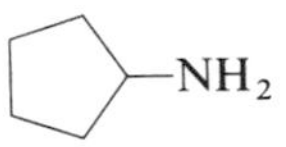 ⬠—NH$_2$

   (e) $CH_2{=}CHCH_2N(CH_3)_2$
   (f) $(CH_3)_3CNH_2$
   (g) $(CH_3)_3N$
   (h) 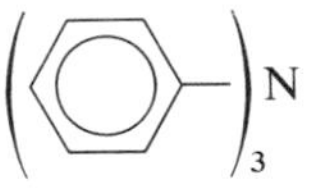 (⬡—)$_3$N

   (i) 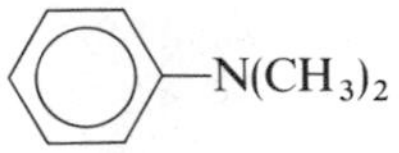 ⬡—N(CH$_3$)$_2$
   (j) ⬡—NH$_2$, Cl

   (k) $\overset{NH_2}{\underset{|}{CH_3CH_2CHCH_2CH_3}}$
   (l) $H_2NCH_2CH_2CH_2CH_2CH_2CH_2NH_2$

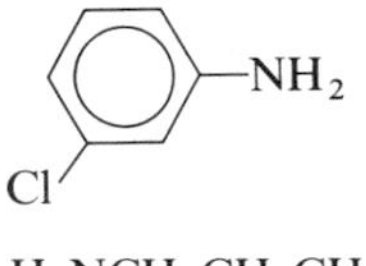

   (m) $CH_3CH_2N^+(CH_3)_3 \quad Br^-$
   (n) 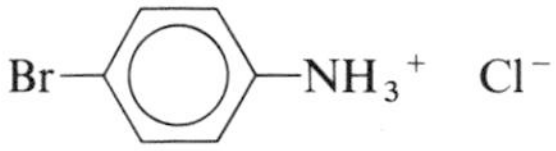 Br—⬡—NH$_3^+$ Cl$^-$

   (o) 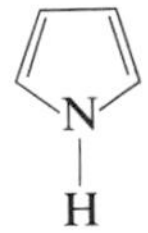 pyrrole structure
   (p) 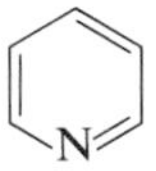 pyridine structure

3. Draw structures for all compounds that fit the following descriptions.
   (a) amines of molecular formula C_3H_9
   (b) amines of molecular formula C_4H_{11}
   (c) amines of molecular formula C_7H_9N that contain a benzene ring
   (d) salts of molecular formula $C_4H_{12}N^+ Cl^-$

4. Classify the following amines as primary, secondary, or tertiary.
  (a) compounds (a) to (k) in problem 2
  (b) compounds that are answers to problem 3(a)
  (c) compounds that are answers to problem 3(b)
  (d) compounds that are answers to problem 3(c)

5. Arrange the following compounds in order of increasing boiling point.

  propylamine,   propyl alcohol,   butane,   trimethylamine

6. Draw the structures of any organic products that result from the following reactions.

(a) $CH_3CH_2CH_2NH_2 \xrightarrow{CH_3I}$

(b) ⟨benzene ring⟩$-NH_2 \xrightarrow{HCl}$

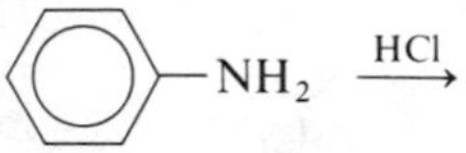

(c) ⟨cyclopentane⟩$-CH_2Br \xrightarrow{NH_3}$

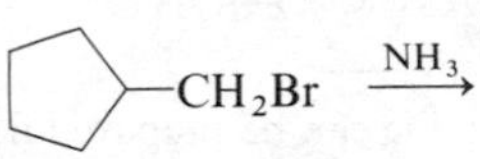

(d) $(CH_3)_3NH^+ \ Cl^- \xrightarrow{NaOH}$

(e) $(CH_3)_3N \xrightarrow{CH_3I}$

(f) CH_3-⟨benzene ring⟩$-NO_2 \xrightarrow[HCl]{Fe}$

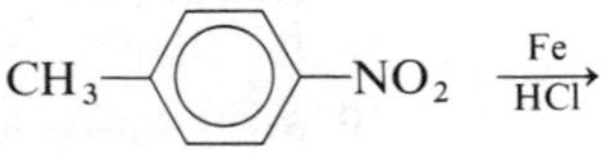

(g) ⟨pyrrolidine ring with NH⟩ $\xrightarrow{CH_3CH_2Br}$

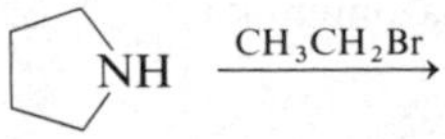

(h) ⟨benzene ring⟩$-NH_2 \xrightarrow{HNO_2}$

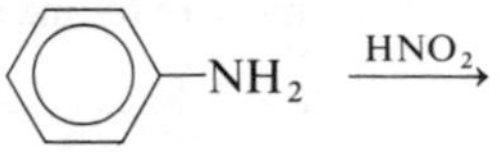

(i) ⟨benzene ring with NO_2 and NO_2⟩ $\xrightarrow[Pt]{H_2 \ (excess)}$

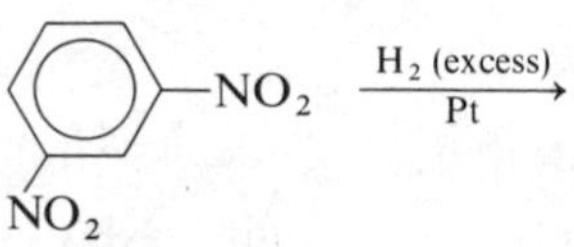

(j) ⟨benzene ring⟩$-CH_2C{\equiv}N \xrightarrow[Ni]{H_2}$

(k) CH_3O-⟨benzene ring⟩$-N_2^+ \xrightarrow{CuCN}$

(l) CH_3-⟨benzene ring with two Br⟩$-N_2^+ \xrightarrow{H_3PO_2}$

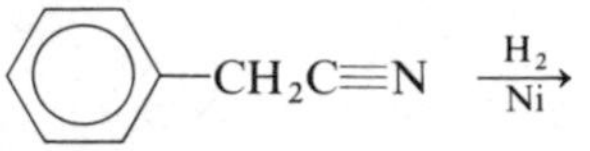

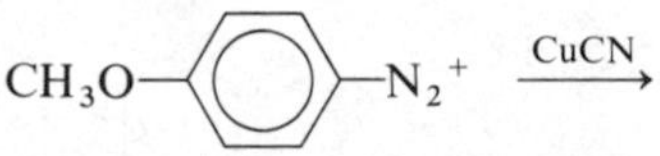

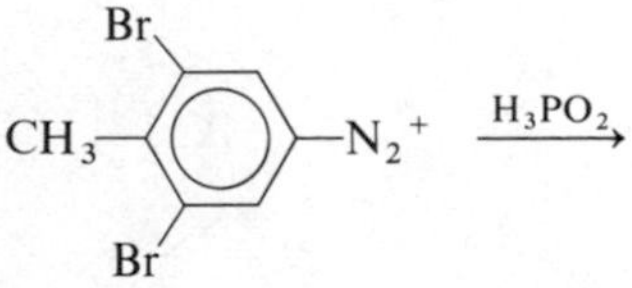

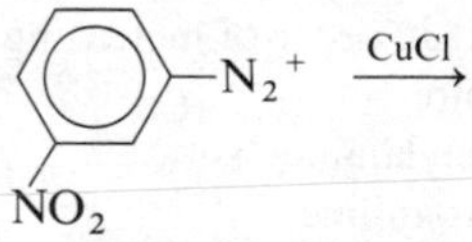

(m) ⟨benzene ring with NO_2⟩$-N_2^+ \xrightarrow{CuCl}$

(n) ⟨benzene ring with Cl⟩$-N_2^+ \xrightarrow[heat]{H_2O}$

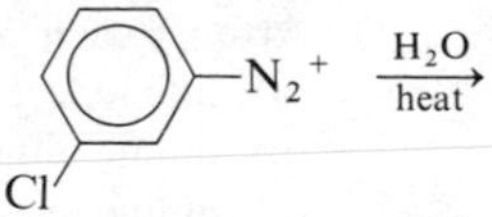

(o) $CH_3CH_2NHCH_2CH_3 \xrightarrow{HNO_2}$

(p) ⟨benzene ring with NO_2⟩$-N_2^+ \xrightarrow{KI}$

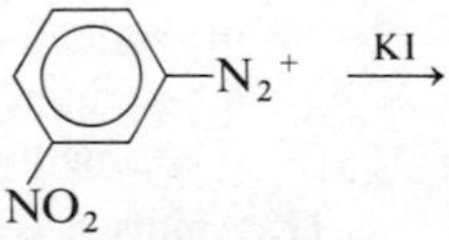

7. What is the predominant form of each of the following compounds when dissolved in aqueous solutions of pH 3, 7, or 12?
  (a) ethylamine   (b) aniline
  (c) phenol   (d) benzenesulfonic acid

8. Write equations to show how each conversion can be carried out. Many will require more than one step.

(a) ⟨benzene ring⟩$-NO_2 \Longrightarrow$ ⟨benzene ring⟩$-I$

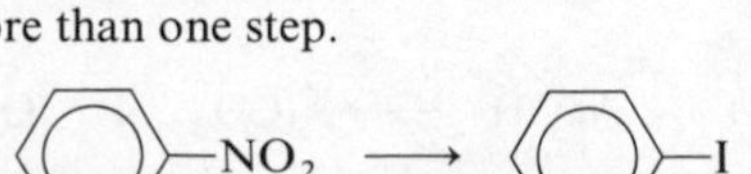

(b) $CH_3CH_2CH_2CH_2Br \longrightarrow CH_3CH_2CH_3CH_2NH_2$

(c)

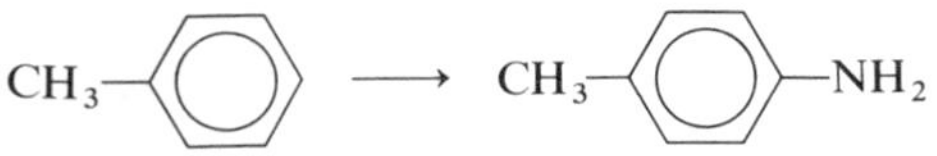

$CH_3-\text{(benzene ring)} \longrightarrow CH_3-\text{(benzene ring)}-NH_2$

(d)

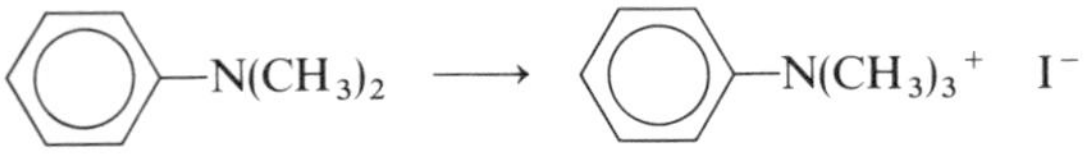

$\text{(benzene ring)}-N(CH_3)_2 \longrightarrow \text{(benzene ring)}-N(CH_3)_3{}^+ \quad I^-$

(e)

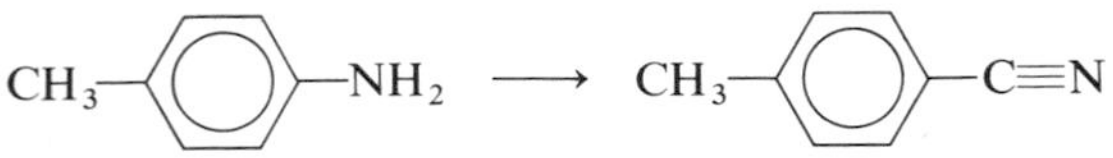

$CH_3-\text{(benzene ring)}-NH_2 \longrightarrow CH_3-\text{(benzene ring)}-C\equiv N$

(f)

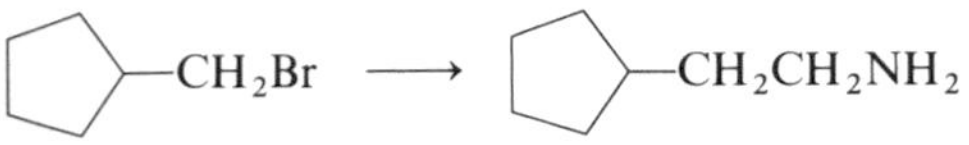

$\text{(cyclopentane ring)}-CH_2Br \longrightarrow \text{(cyclopentane ring)}-CH_2CH_2NH_2$

(g)

$CH_3O-\text{(benzene ring)} \longrightarrow CH_3O-\text{(benzene ring)}-OH$

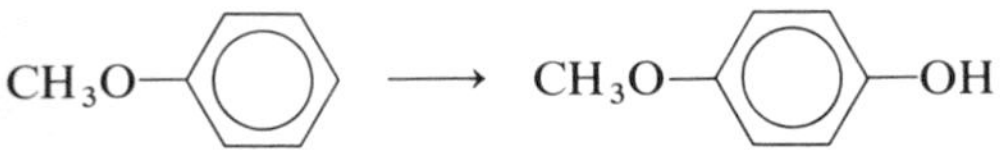

9. Write equations to show how butylamine can be prepared from each of the following compounds.
   (a) 1-bromopropane   (b) 1-bromobutane   (c) 1-nitrobutane

10. Write equations to show how each conversion can be carried out. Each will require more than one step and the use of a diazonium ion.

(a) NO_2 (benzene) $\longrightarrow$ I (benzene) Br

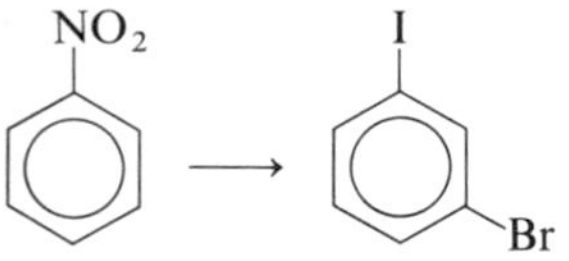

(b) (benzene) $\longrightarrow$ Br (benzene) OH

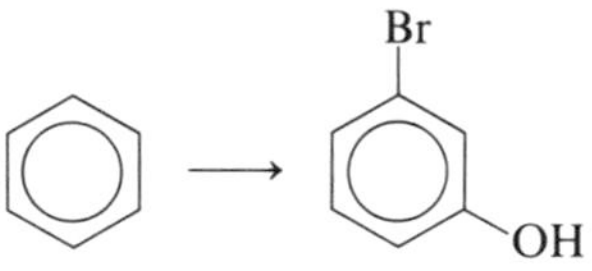

(c) NH_2 (benzene) NO_2 $\longrightarrow$ Br, Br (benzene) Br

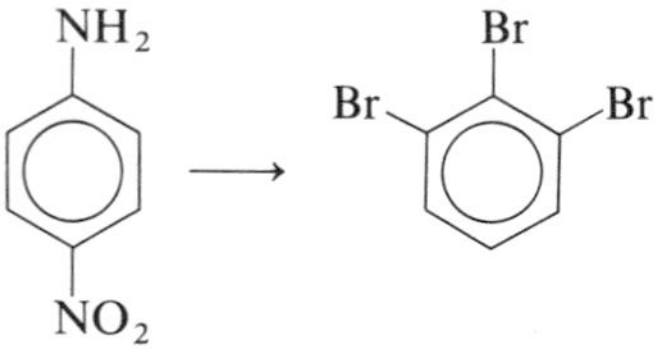

(d) NH_2 (benzene) $\longrightarrow$ CH_2NH_2 (benzene)

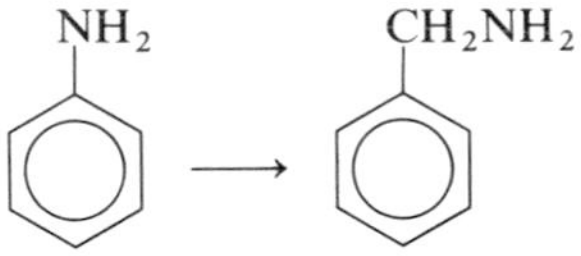

11. Arrange each group of compounds in order of increasing basicity.
    (a) aniline, cyclohexylamine, ammonia
    (b) aniline, diphenylamine, triphenylamine
    (c) aniline, *p*-methylaniline, *p*-nitroaniline
    (d) sodium methoxide, sodium phenoxide, sodium hydroxide
    (e) pyridine, pyrrole, diethylamine
    (f) trimethylamine, tetramethylammonium hydroxide

12. Aniline is considerably less basic than typical alkyl amines. The reduced basicity can be rationalized by considering small contributions to the structure of aniline by three resonance structures in which the unshared electron pair of nitrogen is used to form a double bond to the adjoining carbon. Draw these structures and explain why there are not equivalent structures stabilizing the anilinium ion.

13. Quaternary ammonium halides can be converted to hydroxide salts by treatment with silver hydroxide.

$$(CH_3)_4N^+ \quad Cl^- + AgOH \longrightarrow (CH_3)_4N^+ \quad OH^- + AgCl\downarrow$$

The precipitation of silver halides drives the reactions to completion. Quaternary ammonium hydroxides are strong bases, comparable to sodium hydroxide. Why don't similar reactions with silver hydroxide convert primary, secondary, or tertiary ammonium halides to the corresponding hydroxide salts?

14. Describe a procedure for separating a mixture of phenol, aniline, and toluene without distilling or crystallizing.

15. The effectiveness of the synthesis of amines by reaction of alkyl halides with ammonia or amines is limited by the following problem. Even if the ammonia or amine is present in excess, products form in which more than one alkyl group has become attached to the nitrogen. For example, reaction of ethyl bromide and ammonia produces not only ethylamine but also diethylamine and even some triethylamine and tetraethylammonium bromide. Explain how this problem arises.

16. Indicate a simple chemical test that will distinguish between the members of each of the following pairs of compounds. Describe the observations that would be made with each compound.
(a) aniline and cyclohexylamine
(b) 1-hexanamine and dipropylamine
(c) benzylamine and triethylamine

17. Bromination of aniline (to form 2,4,6-tribromoaniline) occurs readily. Why, then, is sulfonation of aniline exceedingly difficult?

18. As shown in the following example, quaternary ammonium hydroxides undergo an elimination on heating to form an alkene and a tertiary amine.

$$CH_3CH_2N(CH_3)_3{}^+ \ OH^- \ \xrightarrow{\text{heat}} \ CH_2{=}CH_2 + N(CH_3)_3$$

Tertiary amines do not react when heated with hydroxide ion.

$$CH_3CH_2N(CH_3)_2 + OH^- \ \xrightarrow{\text{heat}} \ \text{no reaction}$$

Explain this difference in behavior.

19. Explain why the unsaturated heterocyclic ether furan is less reactive than typical alkenes and tends to undergo substitution rather than addition reactions.

Furan

20. Why is synthetic nicotine, a racemic mixture, only about 50% as active an insecticide as nicotine isolated from tobacco plants?

Aldehydes and Ketones

Aldehydes and ketones both have a **carbonyl** group, $-\overset{\overset{\displaystyle O}{\|}}{C}-$, as their characteristic function. Aldehydes have at least one hydrogen attached to the carbonyl group; the other group may be alkyl, aryl, or hydrogen (R$-\overset{\overset{\displaystyle O}{\|}}{C}-H, Ar-\overset{\overset{\displaystyle O}{\|}}{C}-H, H-\overset{\overset{\displaystyle O}{\|}}{C}-$H).

Ketones have only alkyl or aryl groups attached to the carbonyl group (R$-\overset{\overset{\displaystyle O}{\|}}{C}-R, Ar-\overset{\overset{\displaystyle O}{\|}}{C}-R, Ar-\overset{\overset{\displaystyle O}{\|}}{C}-$Ar).

Aldehyde and ketone functions occur frequently in biological compounds, including carbohydrates (next chapter) and many important metabolic intermediates. Because their reactions provide useful ways of forming carbon-carbon bonds, aldehydes and ketones also play an important role in organic synthesis.

9.1
NOMENCLATURE

The simplest aldehyde, and the only one with just hydrogens attached to the carbonyl group, is usually called formaldehyde. This name arose because oxidation of formaldehyde produces the carboxylic acid called formic acid.

$$\underset{\text{Formaldehyde}}{\overset{\displaystyle\overset{O}{\|}}{HCH}} \quad \xrightarrow{\text{oxidation}} \quad \underset{\text{Formic acid}}{\overset{\displaystyle\overset{O}{\|}}{HCOH}}$$

Some of the trivial names of this sort that are accepted by the IUPAC for simple aldehydes are listed in Table 9.1.

Substitutive names for aldehydes are based on the longest chain that contains the carbonyl group. The ending of the corresponding alkane is changed from -*e* to -*al*. Because the carbonyl carbon of an aldehyde must come at the end of a chain, it is assumed to be carbon number 1 and is not numbered in the name.

$$\underset{\text{Butanal}}{\overset{\displaystyle\overset{O}{\|}}{CH_3CH_2CH_2CH}} \qquad \underset{\underset{\displaystyle Cl}{|}}{\overset{\displaystyle\overset{O}{\|}}{CH_3CH_2CH_2CHCH}} \qquad \underset{\underset{\displaystyle OH}{|}}{\overset{\displaystyle\overset{O}{\|}}{HOCH_2CHCH}}$$

Butanal   2-Chloropentanal   2,3-Dihydroxypropanal

Table 9.1   Trivial names of some aldehydes and carboxylic acids

| | Aldehyde | | Carboxylic acid | |
Name	Structural formula	Name	Structural formula
Formaldehyde	$\overset{\overset{O}{\|}}{HCH}$	Formic acid	$\overset{\overset{O}{\|}}{HCOH}$
Acetaldehyde	$\overset{\overset{O}{\|}}{CH_3CH}$	Acetic acid	$\overset{\overset{O}{\|}}{CH_3COH}$
Propionaldehyde	$\overset{\overset{O}{\|}}{CH_3CH_2CH}$	Propionic acid	$\overset{\overset{O}{\|}}{CH_3CH_2COH}$
Butyraldehyde	$\overset{\overset{O}{\|}}{CH_3CH_2CH_2CH}$	Butyric acid	$\overset{\overset{O}{\|}}{CH_3CH_2CH_2COH}$
Valeraldehyde	$\overset{\overset{O}{\|}}{CH_3CH_2CH_2CH_2CH}$	Valeric acid	$\overset{\overset{O}{\|}}{CH_3CH_2CH_2CH_2COH}$
Benzaldehyde	$C_6H_5\overset{\overset{O}{\|}}{CH}$	Benzoic acid	$C_6H_5\overset{\overset{O}{\|}}{COH}$

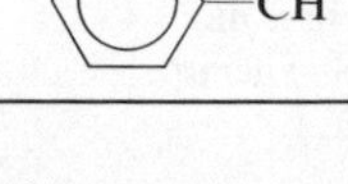

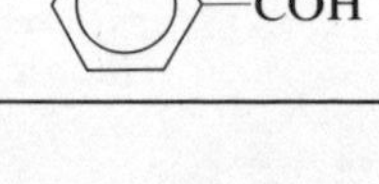

Aromatic aldehydes are usually named as derivatives of benzaldehyde.

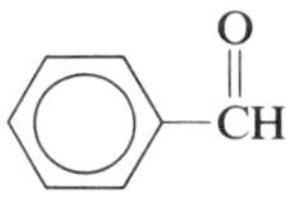

Benzaldehyde 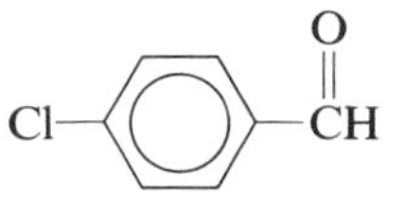   p-Chlorobenzaldehyde

The simplest ketone is almost always called by the trivial name acetone. Radicofunctional names are assigned to other simple ketones in a manner similar to that used for ethers: the organic groups attached to the carbonyl groups are named and the word **ketone** added.

$$CH_3CCH_3 \qquad CH_3CH_2CCH_3 \qquad \text{—}CH_2CCH_3$$

Acetone   Ethyl methyl ketone   Benzyl methyl ketone
Dimethyl ketone

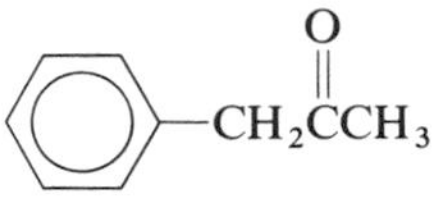

Substitutive names of ketones are similar to those of alcohols except that the ending is -one.

$$CH_3CH_2CH_2CH_2CCH_3 \qquad CH_3CHCCH_3$$
$$\qquad\qquad\qquad\qquad\qquad\qquad\quad | $$
$$\qquad\qquad\qquad\qquad\qquad\qquad\quad Cl$$

2-Hexanone   3-Chloro-2-butanone   2-Chlorocyclohexanone

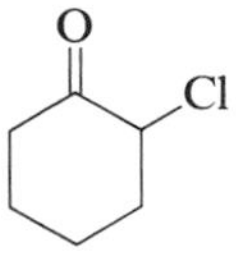

The carbonyl carbon of a cyclic ketone is assumed to be carbon number 1. In numbering chains or rings, a carbonyl group takes precedence over alkene functions.

$$CH_2{=}CHCH_2CCH_3$$

4-Penten-2-one (not 1-penten-4-one)
Allyl methyl ketone

Aromatic ketones are often assigned names ending in -phenone (from phenyl and ketone). The prefix is derived from the trivial name of the carboxylic acid that contains the fragment attached to the aromatic ring.

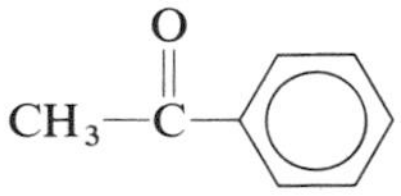

$$CH_3{-}C{-} \qquad CH_3CH_2{-}C{-} \qquad {-}C{-}$$

Acetophenone   Propiophenone   Benzophenone
(acetic + phenone)   (propionic + phenone) (benzoic + phenone)

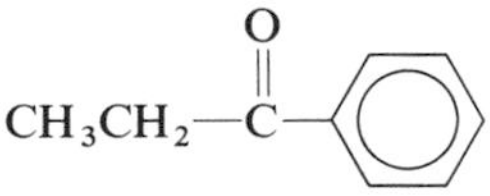

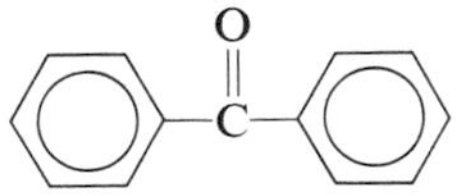

As a residue of an older nomenclature, Greek letters are sometimes used to refer to the positions of carbons (or groups attached to them) relative to a carbonyl group.

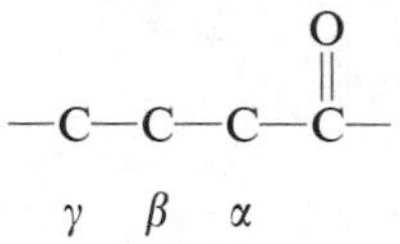

$$-C-C-C-\overset{\overset{\displaystyle O}{\|}}{C}-$$
$$\gamma \quad \beta \quad \alpha$$

Note that the carbon labeled α is *adjacent* to the carbonyl group.

---

problem 9.1 Name the following compounds.

(a)

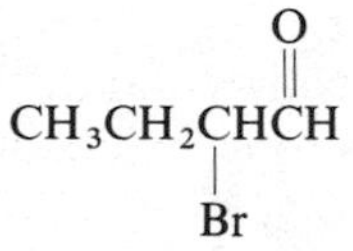

$$\underset{\underset{Br}{|}}{CH_3CH_2CHCH}\overset{\overset{\displaystyle O}{\|}}{}$$

(b)

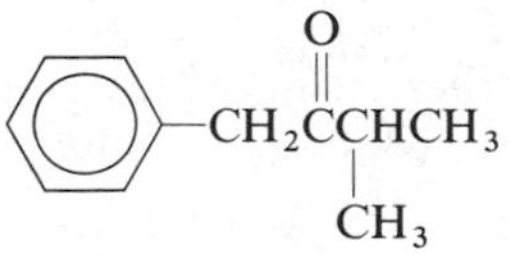

$$\bigcirc\!\!-CH_2\overset{\overset{\displaystyle O}{\|}}{C}\underset{\underset{CH_3}{|}}{CHCH_3}$$

9.2
PHYSICAL PROPERTIES

The hydrogens of water can hydrogen bond to the carbonyl group of an aldehyde or ketone.

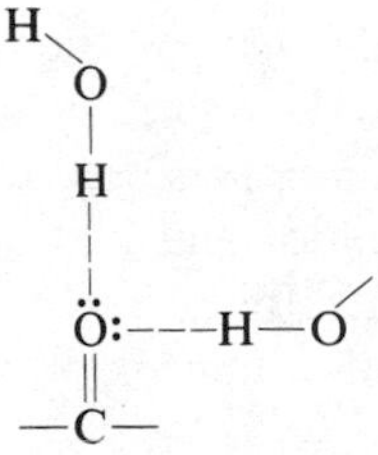

$$
\begin{array}{c}
H\backslash \\
\quad O \\
| \\
H \\
\vdots \\
\ddot{O}:\text{---}H-O \overset{H}{\diagup} \\
\| \\
-C-
\end{array}
$$

As a result, solubilities of aldehydes and ketones are similar to those of alcohols and ethers having the same number of carbons. Acetone is miscible with water and propionaldehyde is soluble to the extent of 16 g in 100 g of water. Water solubility decreases with increasing number of carbon atoms.

Lacking a hydrogen bonded to oxygen, an aldehyde or ketone molecule does not hydrogen bond to another molecule of the same kind. Therefore aldehydes and ketones boil considerably below alcohols with the same number of carbons. Nevertheless, aldehydes and ketones boil somewhat higher than hydrocarbons of similar molecular weight. Formaldehyde, boiling point $-21°$, and acetaldehyde, boiling point $20°$, are the only aldehydes that are gases at normal temperatures. Acetone, the simplest ketone, boils at $56°$.

9.3
REACTIONS OF ALDEHYDES AND KETONES

The reactions of aldehydes and ketones discussed in the following sections fall into three groups:

1. oxidations

2. additions to the carbonyl group (including additions that result in reduction)

3. reactions at an α carbon (a carbon adjacent to the carbonyl group)

9.4
OXIDATION OF ALDEHYDES

Earlier we noted (Sec. 5.8) an important difference between aldehydes and ketones: aldehydes are readily oxidized; ketones are not. Strong oxidizing agents, such as $K_2Cr_2O_7$ or $KMnO_4$, oxidize aldehydes to carboxylic acids but do not react rapidly with ketones. Even very mild oxidizing agents oxidize aldehydes. An example is Tollens' reagent, a basic aqueous solution of silver ion.*

$$\underset{\substack{\| \\ RCH}}{O} + 2\,Ag^+ + 2\,OH^- \longrightarrow \underset{\substack{\| \\ RCOH}}{O} + 2\,Ag + H_2O$$

The silver ion is reduced to silver metal. If the reaction is done in a clean glass vessel, the metallic silver deposits on the walls as a "silver mirror." Such reactions were once used to produce the reflecting surfaces of mirrors.

Tollens' reagent is sometimes used to detect the presence of an aldehyde or to distinguish between aldehydes and ketones. It is more useful for these purposes than strong oxidizing agents, for it does not react with most other organic functions. A stronger oxidizing agent would oxidize the alcohol and alkene functions of the following compound.

$$HO-CH_2-CH=CH-CH_2-\underset{\substack{\| \\ CH}}{O} \xrightarrow{Ag^+} HO-CH_2-CH=CH-CH_2-\underset{\substack{\| \\ COH}}{O}$$

Fehling's and Benedict's reagents are also used to selectively oxidize aldehyde functions, particularly in carbohydrates (Chapter 10). These reagents contain a basic solution of cupric ion (Cu^{2+}) complexed with the anion of tartaric acid (Fehling's reagent) or citric acid (Benedict's reagent). In oxidizing an aldehyde to a

---

* Because AgOH is extremely insoluble, the silver ion is actually kept in solution by complexing it with ammonia to form $Ag(NH_3)_2^+$.

carboxylic acid, a blue Cu^{2+} complex is reduced to cuprous ion (Cu^+), which in the basic solution forms a precipitate of red-brown Cu_2O.

$$
\underset{\text{Blue}}{\overset{\displaystyle O}{\overset{\|}{R C H}}} + 2\,Cu^{2+} + 4\,OH^- \longrightarrow \underset{\text{Red-brown}}{\overset{\displaystyle O}{\overset{\|}{R C O H}}} + Cu_2O\downarrow + 2\,H_2O
$$

9.5
ADDITIONS TO THE CARBONYL GROUPS
OF ALDEHYDES AND KETONES

It is useful to compare and contrast the carbonyl group to the carbon-carbon double bond. The three groups attached to the carbon of a carbonyl group lie in a plane at angles of about 120°. The carbon can be considered to be sp^2 hybridized and to participate in three σ bonds and one π bond, just as do the carbons of a carbon-carbon double bond.

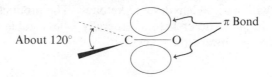

In contrast to a carbon-carbon double bond, however, the carbon-oxygen double bond is polar. Oxygen is far more electronegative than carbon and the electrons shared by oxygen and carbon are, on the average, somewhat nearer oxygen than carbon. The resulting charge distribution can be represented by the symbols δ^+ and δ^-

$$
\overset{\delta^-}{\overset{\displaystyle O}{\overset{\|}{\underset{\delta^+}{-C-}}}}
$$

or by a contribution from a charged resonance structure.

$$
\overset{\displaystyle O}{\overset{\|}{-C-}} \longleftrightarrow \overset{\displaystyle O^-}{\overset{|}{-C^+-}}
$$

Weak attractions between polar carbonyl groups result in the boiling points of aldehydes and ketones being somewhat higher than those of hydrocarbons.

Additions, the characteristic reactions of alkenes, are also one of the two major classes of reactions of aldehydes and ketones (reactions at the α carbon are the other class). The additions that we will discuss involve attack by a nucleophilic group at the relatively positive carbon of the carbonyl group.

$$Nu:^- \curvearrowright -\overset{\overset{\displaystyle O{\scriptstyle\delta^-}}{\|}}{\underset{\delta^+}{C}}- \longrightarrow -\overset{\overset{\displaystyle O^-}{|}}{\underset{Nu}{C}}- \xrightarrow{H^+} -\overset{\overset{\displaystyle OH}{|}}{\underset{Nu}{C}}-$$

An electrophilic group, often a proton, becomes attached to oxygen. The attack by a nucleophile is sometimes assisted by prior protonation of the carbonyl group.

$$-\overset{\overset{\displaystyle :O:}{\|}}{C}- \underset{}{\overset{H^+}{\rightleftarrows}} -\overset{\overset{\displaystyle :\overset{+}{O}:H}{\|}}{C}- \longleftrightarrow -\overset{\overset{\displaystyle :\overset{..}{O}:H}{|}}{\underset{}{C^+}}- \xrightarrow{Nu^-} -\overset{\overset{\displaystyle OH}{|}}{\underset{Nu}{C}}-$$

Protonation makes the carbon more positive and even more receptive to attack by a nucleophile.

The susceptibility of a carbonyl group to attack by nucleophiles permits additions by some reagents that do not add readily to alkenes. On the other hand, some reagents whose additions to alkenes are important do not form stable addition products with most aldehydes and ketones. The product of addition of a hydrogen halide to an aldehyde or ketone generally is less stable than the reactants and the equilibrium lies toward the reactants.

$$-\overset{\overset{\displaystyle O}{\|}}{C}- + H-X \rightleftharpoons -\overset{\overset{\displaystyle OH}{|}}{\underset{X}{C}}-$$

Water adds to an aldehyde or ketone, but again the product is relatively unstable.

$$-\overset{\overset{\displaystyle O}{\|}}{C}- + HOH \rightleftharpoons -\overset{\overset{\displaystyle OH}{|}}{\underset{OH}{C}}-$$

Although the same reagents add both to aldehydes and to ketones, additions to aldehydes are generally more rapid. During the addition a fourth group becomes attached to the carbonyl carbon, reducing bond angles at that carbon from 120 to 109°. The resultant crowding together of groups is less for an aldehyde, since one group is the uniquely small hydrogen.

9.6
ADDITION OF ALCOHOLS

In reactions that can be catalyzed by traces of acid or base, alcohols add to aldehydes and ketones. The products are **hemiacetals**, compounds with a hydroxyl and alkoxyl group attached to the same carbon.

$$\underset{\substack{\| \\ \text{O}}}{\text{CH}_3\text{—C—H}} + \text{CH}_3\text{CH}_2\text{OH} \rightleftharpoons \underset{\substack{| \\ \text{OCH}_2\text{CH}_3}}{\overset{\overset{\displaystyle \text{OH}}{|}}{\text{CH}_3\text{—C—H}}}$$

1-Ethoxyethanol
(a hemiacetal)

$$\underset{\substack{\| \\ \text{O}}}{\text{CH}_3\text{—C—CH}_3} + \text{CH}_3\text{CH}_2\text{OH} \rightleftharpoons \underset{\substack{| \\ \text{OCH}_2\text{CH}_3}}{\overset{\overset{\displaystyle \text{OH}}{|}}{\text{CH}_3\text{—C—CH}_3}}$$

2-Ethoxy-2-propanol
(a hemiacetal)

Like the addition products of water to aldehydes and ketones, most hemiacetals revert readily to the reactants and are rarely isolated.

Hydroxyaldehydes and hydroxyketones exist in equilibrium with cyclic hemiacetals.

$$\underset{\substack{| \\ \text{OH}}}{\text{CH}_2\text{—CH}_2\text{—CH}_2\text{—}\underset{\substack{\| \\ \text{O}}}{\text{CH}}} \rightleftharpoons \text{(ring structure)}$$

A cyclic hemiacetal

Particularly when their rings are five- or six-membered, cyclic hemiacetals are more stable than their acyclic analogs. Carbohydrates (Chapter 10) contain both hydroxyl and aldehyde or ketone functions and often adopt cyclic hemiacetal structures.

Treating aldehydes or ketones with an excess of an alcohol in the presence of an acid leads to formation of **acetals**, compounds with two alkoxy groups attached to the same carbon.

$$\underset{\substack{\| \\ \text{O}}}{\text{CH}_3\text{—C—H}} + 2\ \text{CH}_3\text{CH}_2\text{OH} \underset{}{\overset{\text{H}^+}{\rightleftharpoons}} \underset{\substack{| \\ \text{OCH}_2\text{CH}_3}}{\overset{\overset{\displaystyle \text{OCH}_2\text{CH}_3}{|}}{\text{CH}_3\text{—C—H}}} + \text{H}_2\text{O}$$

1,1-Diethoxyethane
(an acetal)

$$\underset{\substack{\| \\ \text{O}}}{\text{CH}_3\text{—C—CH}_3} + 2\ \text{CH}_3\text{CH}_2\text{OH} \underset{}{\overset{\text{H}^+}{\rightleftharpoons}} \underset{\substack{| \\ \text{OCH}_2\text{CH}_3}}{\overset{\overset{\displaystyle \text{OCH}_2\text{CH}_3}{|}}{\text{CH}_3\text{—C—CH}_3}} + \text{H}_2\text{O}$$

2,2-Diethoxypropane
(an acetal)

An acetal arises from further reaction of a hemiacetal and an alcohol. Cyclic hemiacetals react to form cyclic acetals.

$$
\underset{\substack{\text{CH}_2\text{--CH}_2}}{\overset{\substack{\text{OH}\\|\\\text{CH}\\ \diagup\qquad\diagdown}}{\text{CH}_2\qquad\text{O}}} + \text{CH}_3\text{OH} \overset{\text{H}^+}{\rightleftharpoons} \underset{\substack{\text{CH}_2\text{--CH}_2}}{\overset{\substack{\text{OCH}_3\\|\\\text{CH}\\ \diagup\qquad\diagdown}}{\text{CH}_2\qquad\text{O}}} + \text{H}_2\text{O}
$$

The mechanism of conversion of a hemiacetal to an acetal closely resembles that for the reaction of a tertiary alcohol with a hydrogen halide (Sec. 5.10).

$$
\underset{\substack{\text{OCH}_2\text{CH}_3}}{\overset{\substack{\text{OH}\\|}}{\text{CH}_3\text{--C--H}}} \underset{-\text{H}^+}{\overset{\text{H}^+}{\rightleftharpoons}} \underset{\substack{\text{OCH}_2\text{CH}_3}}{\overset{\substack{^+\text{OH}_2\\|}}{\text{CH}_3\text{--C--H}}} \underset{\text{H}_2\text{O}}{\overset{-\text{H}_2\text{O}}{\rightleftharpoons}}
$$

$$
\begin{array}{c}
\underset{\substack{\text{:OCH}_2\text{CH}_3}}{\overset{\substack{\text{CH}_3\text{--C--H}\\|}}{}}\\
\updownarrow\\
\underset{\substack{_+\overset{..}{\text{O}}\text{CH}_2\text{CH}_3}}{\overset{\substack{\text{CH}_3\text{--C--H}\\\|}}{}}
\end{array}
\overset{\text{CH}_3\text{CH}_2\text{OH}}{\underset{-\text{CH}_3\text{CH}_2\text{OH}}{\longrightarrow}}
$$

(resonance structures of the intermediate carbocation)

$$
\underset{\substack{\text{OCH}_2\text{CH}_3}}{\overset{\substack{\overset{+}{\text{H}}\text{OCH}_2\text{CH}_3\\|}}{\text{CH}_3\text{--C--H}}} \underset{\text{H}^+}{\overset{-\text{H}^+}{\rightleftharpoons}} \underset{\substack{\text{OCH}_2\text{CH}_3}}{\overset{\substack{\text{OCH}_2\text{CH}_3\\|}}{\text{CH}_3\text{--C--H}}}
$$

Protonation of the hydroxyl group leads to loss of water and formation of a particularly stable carbocation, stabilized by a resonance interaction involving the unshared electron pair of oxygen.* Combination of the carbocation with a molecule of alcohol followed by loss of a proton furnishes the acetal.

By treatment with aqueous acid, an acetal is readily converted back to the alcohol and the aldehyde or ketone. The mechanism, of course, is exactly the reverse of that for acetal formation. As with other readily reversible reactions, conditions can be chosen to shift the equilibrium in either direction.

$$
\underset{}{\overset{\substack{\text{O}\\\|}}{\text{R--C--H}}} + 2\,\text{R}'\text{OH} \rightleftharpoons \underset{\substack{\text{OR}'}}{\overset{\substack{\text{OR}'\\|}}{\text{R--C--H}}} + \text{H}_2\text{O}
$$

Acetal formation is favored by using an excess of alcohol or removing water as it forms. A large excess of water favors acetal hydrolysis.

---

* This is analogous to the stabilization by —OH or —OR (Sec. 4.8) of the carbocations formed in electrophilic aromatic substitution reactions.

Acetals resemble ethers structurally and share the inertness of ethers to many reagents, including bases and most oxidizing and reducing agents.

A polymeric acetal made from polyvinyl alcohol and butyraldehyde is used as the adhesive layer between sheets of plate glass to make safety glass, such as that used in automobile windshields.

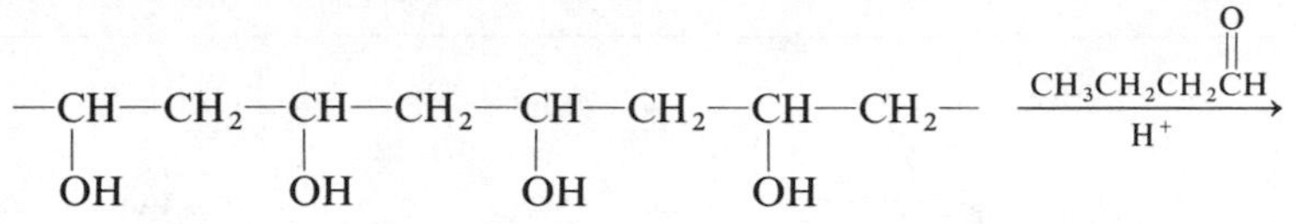

$$-CH-CH_2-CH-CH_2-CH-CH_2-CH-CH_2- \quad \xrightarrow[\text{H}^+]{\text{CH}_3\text{CH}_2\text{CH}_2\overset{\overset{\displaystyle O}{\|}}{\text{CH}}}$$

OH   OH   OH   OH

Polyvinyl alcohol

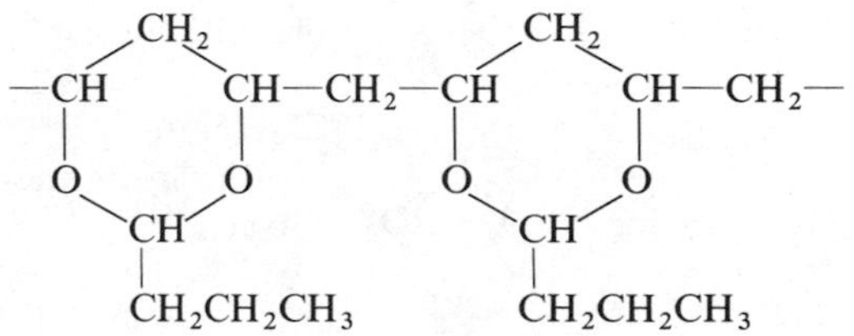

Polyvinyl butyral

9.7
REACTIONS WITH AMMONIA DERIVATIVES

In the presence of acid catalysts, aldehydes and ketones react with compounds containing primary amino groups to form addition products similar to those formed by alcohols.

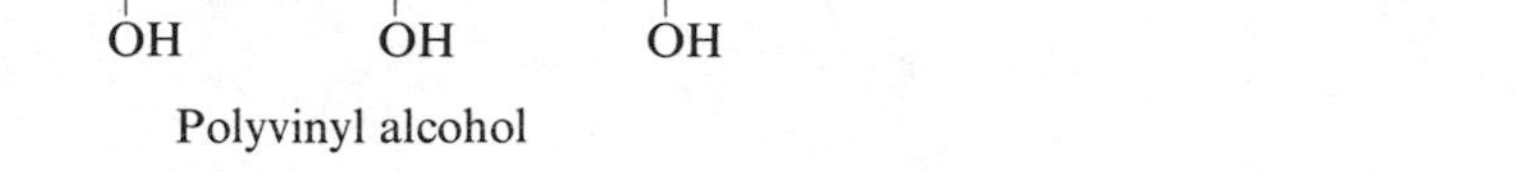

$$\text{C}{=}\text{O} + \text{H}_2\text{N}{-}\text{Y} \longrightarrow \text{C} \overset{\text{OH}}{\underset{\text{NH}{-}\text{Y}}{}} \longrightarrow \text{C}{=}\text{N}{-}\text{Y} + \text{H}_2\text{O}$$

The addition products dehydrate readily, so the compounds ordinarily isolated contain a carbon-nitrogen double bond.

$$\text{CH}_3{-}\text{CH}{=}\text{O} + \text{H}_2\text{N}{-}\text{OH} \xrightarrow{\text{H}^+} \text{CH}_3{-}\text{CH}{=}\text{N}{-}\text{OH}$$

Hydroxylamine   Acetaldehyde oxime

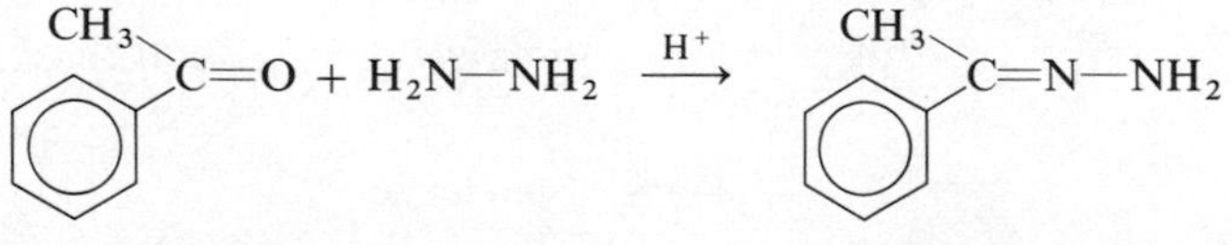

Acetophenone   Hydrazine   Acetophenone hydrazone

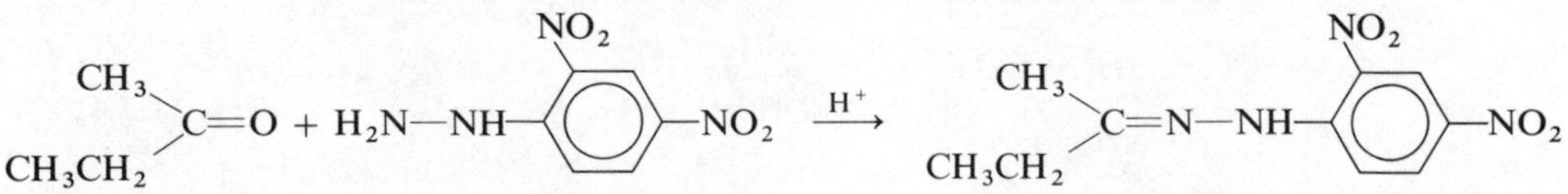

2-Butanone   2,4-Dinitrophenylhydrazine   2-Butanone 2,4-dinitrophenylhydrazone

Table 9.2   Reactions of aldehydes and ketones with ammonia derivatives

$$\text{>C=O} + \text{H}_2\text{N—Y} \longrightarrow \text{>C=N—Y} + \text{H}_2\text{O}$$

| Ammonia derivative | | Product | |
Structure	Name	Structure	Name
H₂N—OH	hydroxylamine	C=N—OH	oxime
H₂N—NH₂	hydrazine	C=N—NH₂	hydrazone
H₂N—NH—⬡	phenylhydrazine	C=N—NH—⬡	phenylhydrazone
H₂N—NH—⬡(NO₂)—NO₂	2,4-dinitro-phenylhydrazine (yellow)	C=N—NH—⬡(NO₂)—NO₂	2,4-dinitro-phenylhydrazone (orange or red)
H₂N—NH—C(=O)—NH₂	semicarbazide	C=N—NH—C(=O)—NH₂	semicarbazone
H₂N—R, H₂N—Ar	primary amine	C=N—R, C=N—Ar	imine

Table 9.2 lists some of the most commonly used ammonia derivatives and the products that they form. Oximes, the various hydrazones, and semicarbazones are usually easily isolated solids. Acetone (melting point $-95°$), for example, is readily converted to a solid oxime ($59°$), 2,4-dinitrophenylhydrazone ($125°$), and semicarbazone ($190°$). It is easier to purify solids than liquids and to determine accurately melting points than boiling points. Therefore liquid aldehydes and ketones are sometimes identified by conversion to such solid "derivatives" whose melting points can be compared with already known values.

Imines, the compounds formed from primary amines, are readily hydrolyzed back to amines and aldehydes or ketones. Not ordinarily used as derivatives, they are of value as chemical intermediates. They can, for instance, be hydrogenated to secondary amines.

$$\text{⬠—CH=O} \underset{\text{H}_2\text{O}}{\overset{\text{H}_2\text{N—CH}_3}{\rightleftharpoons}} \text{⬠—CH=N—CH}_3 \xrightarrow[\text{Pd}]{\text{H}_2} \text{⬠—CH}_2\text{—NH—CH}_3$$

(an imine)

Formation and reduction of an imine is a common biological reaction sequence.

$$\underset{\text{Glutamic acid semialdehyde}}{\overset{\text{O}}{\overset{\|}{\text{CH}}}\text{—CH}_2\text{—CH}_2\text{—CH—CO}_2\text{H} \atop \qquad\qquad \text{NH}_2} \longrightarrow \underset{}{\overset{\text{CH}_2\text{—CH}_2}{\text{CH} \quad \text{CH}} \atop \text{N} \quad \text{CO}_2\text{H}} \xrightarrow{\text{NADPH}} \underset{\text{Proline}}{\overset{\text{CH}_2\text{—CH}_2}{\text{CH}_2 \quad \text{CH}} \atop \overset{\text{N} \quad \text{CO}_2\text{H}}{\text{H}}}$$

Reaction together of the aldehyde and primary amine functions of glutamic acid semialdehyde leads to a cyclic imine. This imine is reduced to the amino acid proline by the biological reducing agent **NADPH** (reduced nicotinamide adenine dinucleotide phosphate).

Formation and hydrolysis of imines are steps in biological "transamination" reactions. In an enzyme-catalyzed synthesis of the amino acid valine, pyridoxamine phosphate and α-ketoisovaleric acid condense to form an imine.

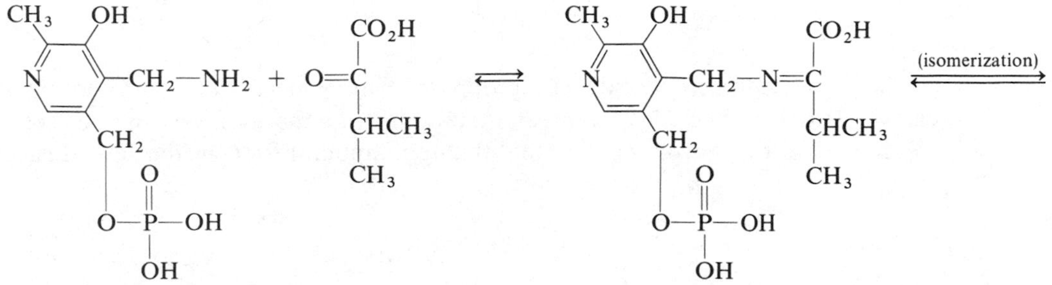

Pyridoxamine phosphate   α-Ketoisovaleric acid

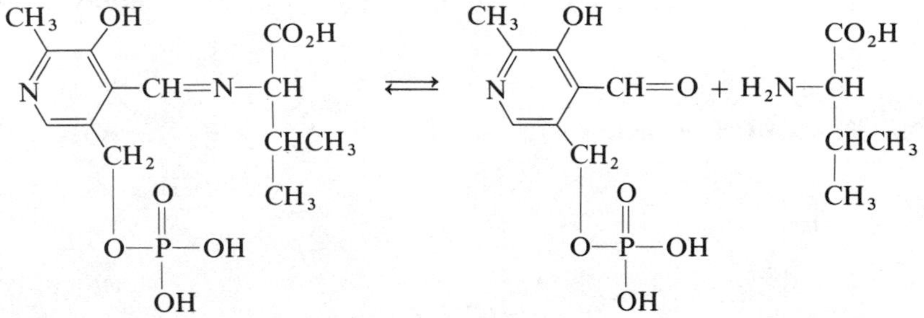

Pyridoxal phosphate   Valine

Isomerization of the carbon-nitrogen double bond gives a new imine that is then hydrolyzed to valine and pyridoxal phosphate. Transamination reactions effect the interchange of carbonyl and primary amino functions.

problem 9.2 Draw structures of any organic products of the following reactions.

(a) 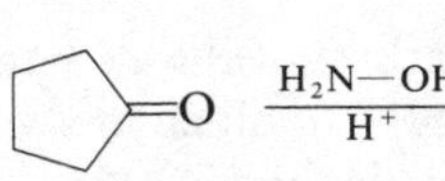 $\xrightarrow[\text{H}^+]{\text{CH}_3\text{OH}}$

(b) $(CH_3)_2CHCH \overset{\text{O}}{\overset{\|}{}} \xrightarrow{K_2Cr_2O_7}$

(c) $\xrightarrow[\text{H}^+]{\text{H}_2\text{N}-\text{OH}}$

(d) $CH_3CH_2 \overset{\text{O}}{\overset{\|}{C}} CH_2CH_3 \xrightarrow{KMnO_4}$

9.8
ADDITION OF ORGANOMETALLIC COMPOUNDS

Grignard reagents (Sec. 7.4) add readily to aldehydes and ketones.

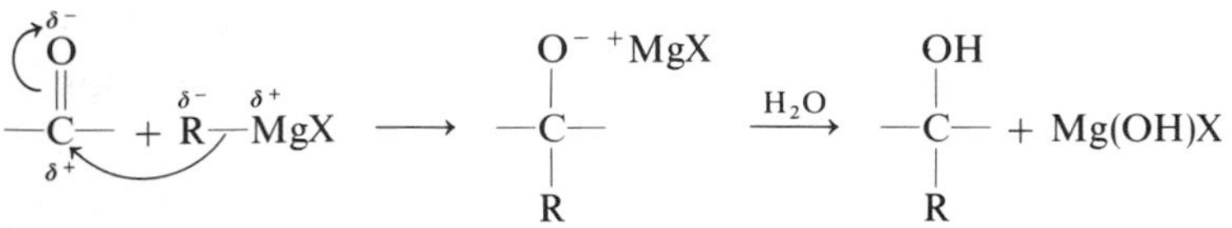

$$
\begin{array}{ccc}
\overset{\delta-}{\underset{\delta+}{\overset{\displaystyle O}{\underset{|}{\overset{\|}{\underset{R}{\overset{\delta-}{-C-}}}}}}} + \overset{\delta-}{R}\overset{\delta+}{-MgX} & \longrightarrow & -\overset{O^-\ ^+MgX}{\underset{R}{\underset{|}{\overset{|}{C}}}}- \xrightarrow{H_2O} -\overset{OH}{\underset{R}{\underset{|}{\overset{|}{C}}}}- + Mg(OH)X
\end{array}
$$

As in other additions to carbonyl groups, the more negative group adds to the carbonyl carbon. The magnesium salt that results from the addition can be converted to an alcohol by hydrolysis. Organolithium compounds add in the same fashion.

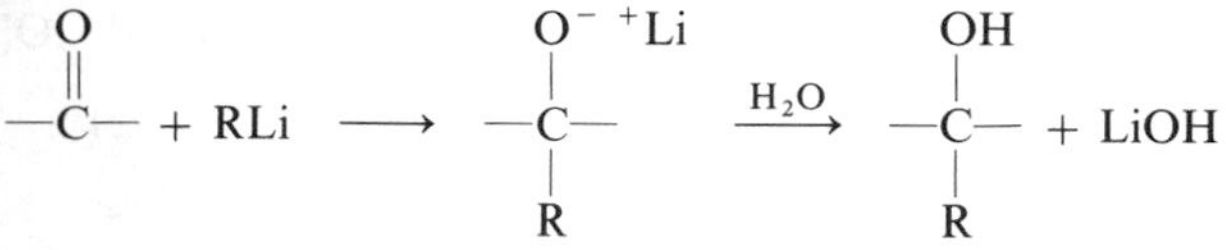

$$
-\overset{O}{\overset{\|}{C}}- + RLi \longrightarrow -\overset{O^-\ ^+Li}{\underset{R}{\underset{|}{\overset{|}{C}}}}- \xrightarrow{H_2O} -\overset{OH}{\underset{R}{\underset{|}{\overset{|}{C}}}}- + LiOH
$$

Such reactions can be used to synthesize a variety of structures.

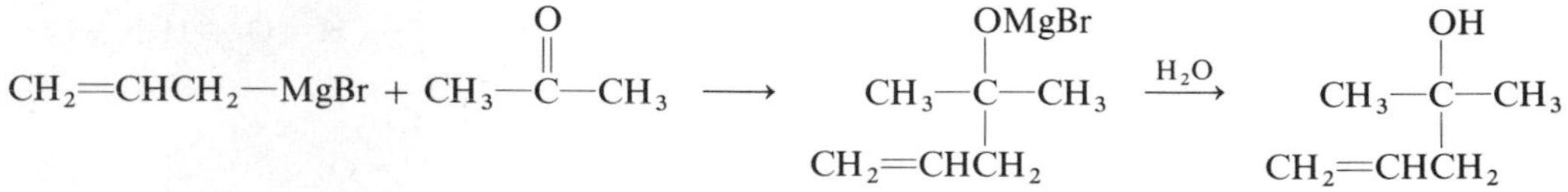

$$
CH_2{=}CHCH_2{-}MgBr + CH_3{-}\overset{O}{\overset{\|}{C}}{-}CH_3 \longrightarrow CH_3{-}\underset{CH_2{=}CHCH_2}{\underset{|}{\overset{OMgBr}{\overset{|}{C}}}}{-}CH_3 \xrightarrow{H_2O} CH_3{-}\underset{CH_2{=}CHCH_2}{\underset{|}{\overset{OH}{\overset{|}{C}}}}{-}CH_3
$$

A tertiary alcohol

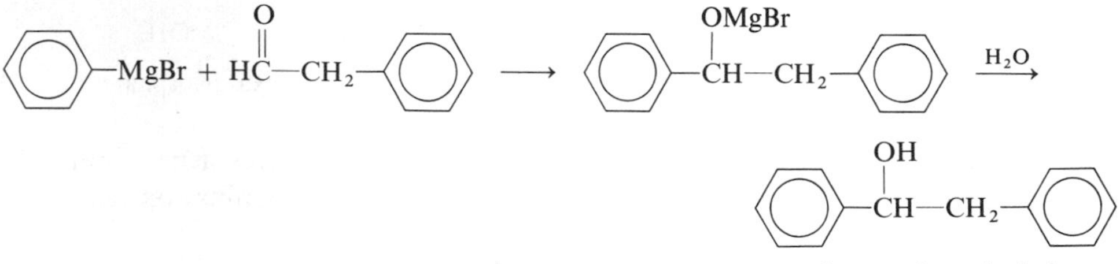

$$
\langle\bigcirc\rangle{-}MgBr + H\overset{O}{\overset{\|}{C}}{-}CH_2{-}\langle\bigcirc\rangle \longrightarrow \langle\bigcirc\rangle{-}\overset{OMgBr}{\underset{|}{\overset{|}{CH}}}{-}CH_2{-}\langle\bigcirc\rangle \xrightarrow{H_2O}
$$

$$
\langle\bigcirc\rangle{-}\overset{OH}{\underset{|}{\overset{|}{CH}}}{-}CH_2{-}\langle\bigcirc\rangle
$$

A secondary alcohol

$$
\langle\bigcirc\rangle{-}MgCl + CH_2{=}O \longrightarrow \langle\bigcirc\rangle{-}CH_2{-}OMgCl \xrightarrow{H_2O} \langle\bigcirc\rangle{-}CH_2OH
$$

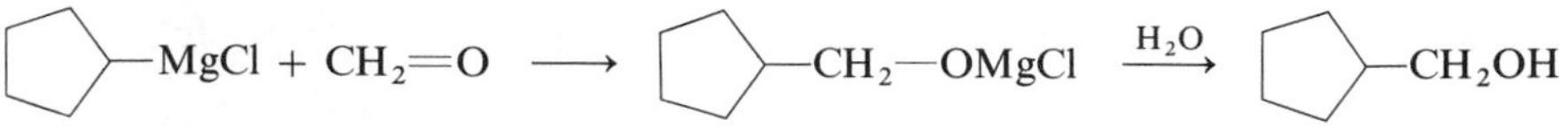

A primary alcohol

These reactions add one organic group to the carbon of the carbonyl group. Therefore additions to ketones produce tertiary alcohols, to aldehydes produce secondary alcohols, and to formaldehyde produce primary alcohols.

Grignard reagent additions to aldehydes and ketones provide an important way of linking organic molecules to form larger ones. The alcohols that result from these reactions can be transformed to a variety of other compounds, even to the components for a new Grignard synthesis. The secondary alcohol synthesized above, for instance, can be oxidized to a ketone. The primary alcohol can be converted to a halide and then to a Grignard reagent. Combination of the new ketone and Grignard reagent leads to a larger molecule.

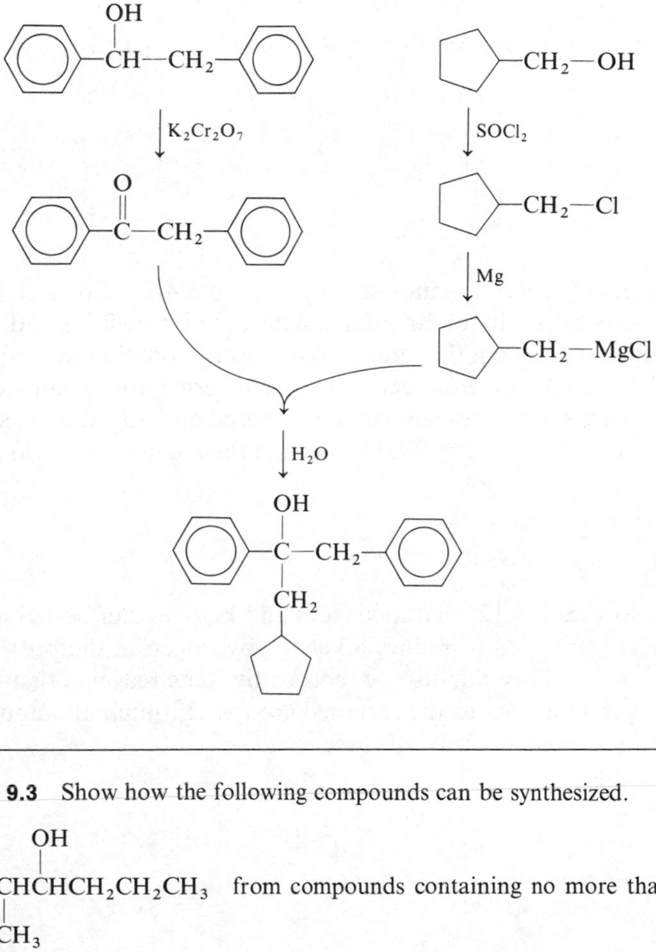

---

problem 9.3 Show how the following compounds can be synthesized.

(a)

$$\underset{\underset{CH_3}{|}}{\overset{\overset{OH}{|}}{CH_3CHCHCH_2CH_2CH_3}}$$ from compounds containing no more than four carbon atoms.

(b)

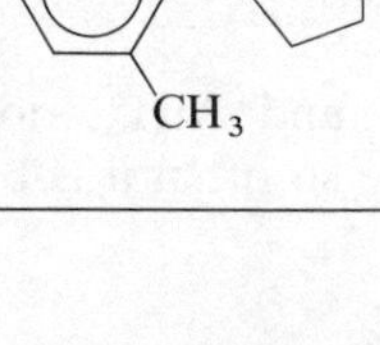

from cyclopentanone and any other organic compounds.

A synthesis of 5-hydroxyhexanal illustrates the use of a "protecting group."

$$BrCH_2CH_2CH_2\overset{\overset{\displaystyle O}{\|}}{C}H \xrightarrow[H^+]{CH_3CH_2OH} BrCH_2CH_2CH_2\overset{\overset{\displaystyle OCH_2CH_3}{|}}{\underset{\underset{\displaystyle OCH_2CH_3}{|}}{C}}H \xrightarrow{Mg}$$

4-Bromobutanal

$$BrMgCH_2CH_2CH_2\overset{\overset{\displaystyle OCH_2CH_3}{|}}{\underset{\underset{\displaystyle OCH_2CH_3}{|}}{C}}H \xrightarrow[\;]{CH_3\overset{\overset{\displaystyle O}{\|}}{C}H} \xrightarrow{H_2O} CH_3\overset{\overset{\displaystyle OH}{|}}{C}HCH_2CH_2CH_2\overset{\overset{\displaystyle OCH_2CH_3}{|}}{\underset{\underset{\displaystyle OCH_2CH_3}{|}}{C}}H \xrightarrow[H^+]{H_2O}$$

$$CH_3\overset{\overset{\displaystyle OH}{|}}{C}HCH_2CH_2CH_2\overset{\overset{\displaystyle O}{\|}}{C}H$$

5-Hydroxyhexanal

A Grignard reagent cannot be prepared from 4-bromobutanal and added to acetaldehyde because the organomagnesium function will instead react rapidly with the aldehyde function in the same molecule or a bromobutanal molecule. The offending aldehyde function, however, can be converted to an unreactive acetal (Sec. 9.6). Then the Grignard reagent can be prepared and added to acetaldehyde. Hydrolysis with dilute aqueous acid will regenerate the aldehyde function.

9.9
REDUCTION

We know (Sec. 5.12) that aldehydes and ketones can be reduced to alcohols by the same reagent used to reduce alkenes—hydrogen in the presence of a catalyst such as Ni or Pt. They can also be reduced by some reagents that provide a nucleophilic hydrogen that adds to the carbonyl groups. Lithium aluminum hydride ($LiAlH_4$) is the most commonly used of these reagents.

$$CH_3\overset{\overset{\displaystyle O}{\|}}{C}HCH_2\overset{\overset{\displaystyle O}{\|}}{C}CH_2CH_3 \xrightarrow[\;]{LiAlH_4} \xrightarrow{H_2O} CH_3\overset{\overset{\displaystyle OH}{|}}{C}HCH_2\overset{\overset{\displaystyle OH}{|}}{C}HCH_2CH_3$$
$$\underset{\displaystyle CH_3}{|} \qquad\qquad\qquad\qquad\qquad\qquad \underset{\displaystyle CH_3}{|}$$

Lithium aluminum hydride consists of Li^+ and AlH_4^- ions. We tend to associate positive character with hydrogen because so often it is bonded to a more electronegative atom, such as oxygen, nitrogen, or a halogen. When bonded to a metal like aluminum (electronegativity = 1.6), however, hydrogen (2.2) is relatively negative and behaves as though it were a hydride ion, $:H^-$. A hydrogen from AlH_4^- adds as a nucleophile to a carbonyl carbon, and aluminum becomes attached to the carbonyl oxygen.

$$H-\overset{\overset{\displaystyle H}{|}}{\underset{\underset{\displaystyle H}{|}}{Al}} \overset{H}{=} H + O = \overset{H}{\underset{}{C}} - CH_3 \longrightarrow H - \overset{\overset{\displaystyle H}{|}}{\underset{\underset{\displaystyle H}{|}}{Al}} = O - \overset{\overset{\displaystyle H}{|}}{\underset{\underset{\displaystyle H}{|}}{C}} - CH_3$$

Note that both electrons of the new C—H bond are provided by the hydrogen. Just as in additions of Grignard reagents, a nucleophilic group bonded to a metal becomes attached to the carbonyl carbon. The remaining Al—H bonds can add to additional molecules of the carbonyl compound.

$$H-\overset{\overset{\displaystyle H}{|}}{\underset{\underset{\displaystyle H}{|}}{Al}} = O - CH_2CH_3 \xrightarrow{CH_3CH=O} CH_3CH_2O - \overset{\overset{\displaystyle H}{|}}{\underset{\underset{\displaystyle H}{|}}{Al}} = OCH_2CH_3 \xrightarrow{CH_3CH=O}$$

$$CH_3CH_2O - \overset{\overset{\displaystyle H}{|}}{\underset{\underset{\displaystyle OCH_2CH_3}{|}}{Al}} = OCH_2CH_3 \xrightarrow{CH_3CH=O} CH_3CH_2O - \overset{\overset{\displaystyle OCH_2CH_3}{|}}{\underset{\underset{\displaystyle OCH_2CH_3}{|}}{Al}} = OCH_2CH_3 \xrightarrow{H_2O}$$

$$4\ CH_3CH_2OH + Al(OH)_4^-$$

Hydrolysis of the aluminum salt releases the alcohol.

Lithium aluminum hydride does not ordinarily react with alkene functions. As a result, it can be used for the selective reduction of an unsaturated aldehyde or ketone.

$$CH_2=CHCH_2\overset{\overset{\displaystyle O}{\|}}{C}H \begin{array}{c} \xrightarrow{LiAlH_4} \xrightarrow{H_2O} CH_2=CHCH_2CH_2OH \\ \\ \xrightarrow[Pt]{H_2} CH_3CH_2CH_2CH_2OH \end{array}$$

Sodium borohydride ($NaBH_4$), a similar but less reactive reducing agent, can be used in aqueous solutions.

Although biological reagents that reduce carbonyl groups are very different than lithium aluminum hydride, they share with it the property of hydride addition to the carbonyl carbon. Many biological reductions of aldehydes and ketones use NADH (or its phosphate ester NADPH) as the reducing agent. Acetaldehyde is the simplest carbonyl compound to be reduced in this fashion (see Sec. 6.9 for another example).

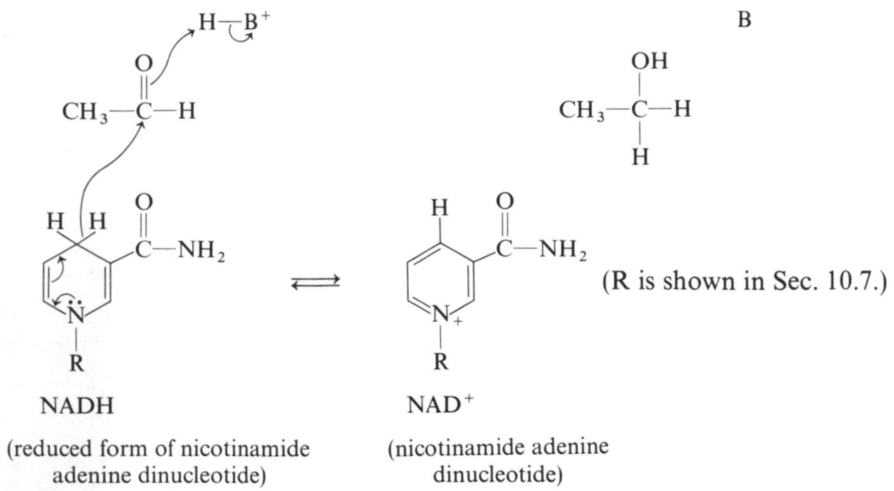
H—B+
B
O
CH3—C—H
OH
CH3—C—H
H
H H
O
C—NH2
N
R
H
O
C—NH2
N+
R
(R is shown in Sec. 10.7.)
NADH
(reduced form of nicotinamide adenine dinucleotide)
NAD+
(nicotinamide adenine dinucleotide)

A hydrogen is transferred as a nucleophilic hydride ion (H^-) from the ring of NADH to the carbonyl carbon. At the same time, a proton is added to the carbonyl oxygen by a group (H—B^+) in the enzyme (alcohol dehydrogenase) that catalyzes this interconversion of acetaldehyde and ethanol. The dietary requirement for the vitamin nicotinic acid (Sec. 8.9) occurs because it is needed to construct NADH.

9.10 ENOLATE ANIONS AND ENOLS

Hydrogens on a carbon adjacent to a carbonyl group (an α carbon) are considerably more acidic than hydrogens of alkanes. The enhanced acidity is due to the formation of a particularly stable anion, often called an **enolate anion**.

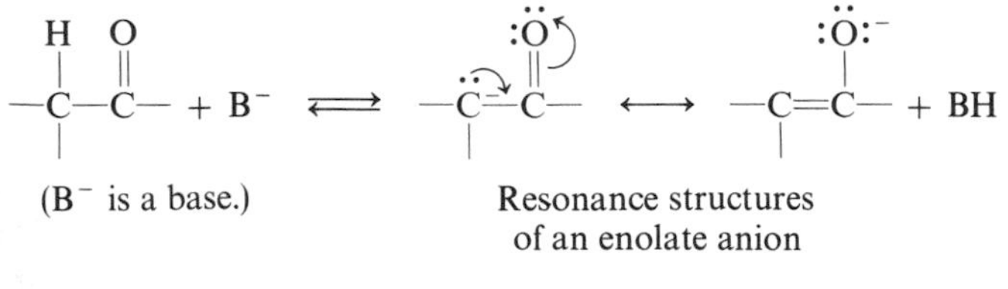
H O
—C—C— + B⁻
:O:
—C—C—
:O:⁻
—C=C— + BH
(B⁻ is a base.)
Resonance structures of an enolate anion

Simply by moving electrons, we can draw a resonance structure in which the negative charge is favorably located on an electronegative oxygen rather than a carbon. An enolate anion resembles this second structure more than it does the first.

The hydroxide ion is able to form low concentrations of the enolate anions of simple aldehydes and ketones.

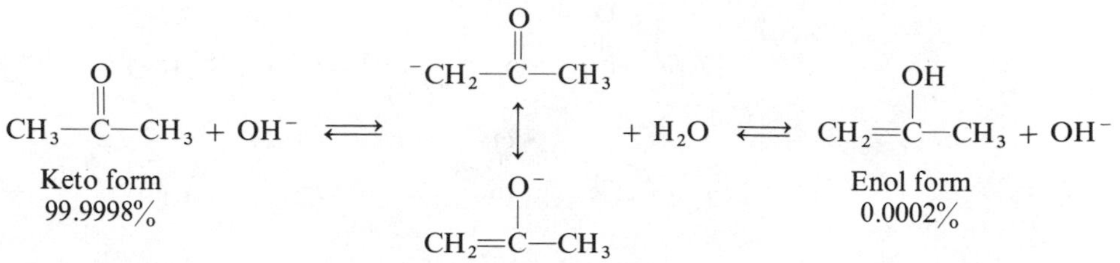

$$CH_3-\overset{\overset{\displaystyle O}{\|}}{C}-CH_3 + OH^- \rightleftharpoons$$

Keto form
99.9998%

$$\begin{array}{c} ^-CH_2-\overset{\overset{\displaystyle O}{\|}}{C}-CH_3 \\ \updownarrow \\ CH_2=\overset{\overset{\displaystyle O^-}{|}}{C}-CH_3 \end{array} + H_2O \rightleftharpoons CH_2=\overset{\overset{\displaystyle OH}{|}}{C}-CH_3 + OH^-$$

Enol form
0.0002%

Protonation of the enolate anion can occur either on carbon, to regenerate the ketone, or on oxygen. Protonation on oxygen forms an isomer of the carbonyl compound called an **enol**, for it contains alkene (-*ene*) and alcohol (-*ol*) functions. Carbonyl compounds with α hydrogens are in equilibrium with the corresponding enols. One way that keto and enol forms are interconverted is by formation and reprotonation of the enolate anion (the conjugate base of both). Enols of simple aldehydes and ketones are much less stable than the keto forms and at equilibrium are present in only small amounts. In special instances, however, the enol form is the more stable. Because it has the stabilization associated with an aromatic ring, phenol is a particularly stable enol.

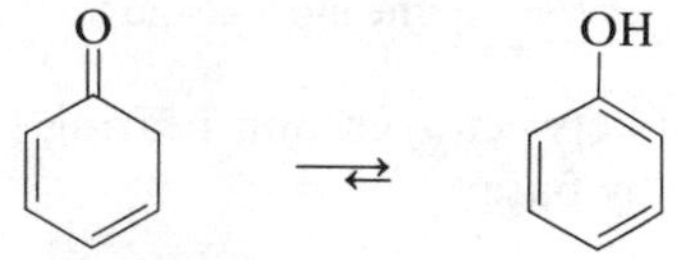

Negligible concentration
at equilibrium

The next two sections present reactions of aldehydes and ketones that are due to the formation of enolate anions (or enols).

9.11
α HALOGENATION

Ketones and aldehydes react with basic solutions of chlorine, bromine, or iodine to form α-halo compounds.

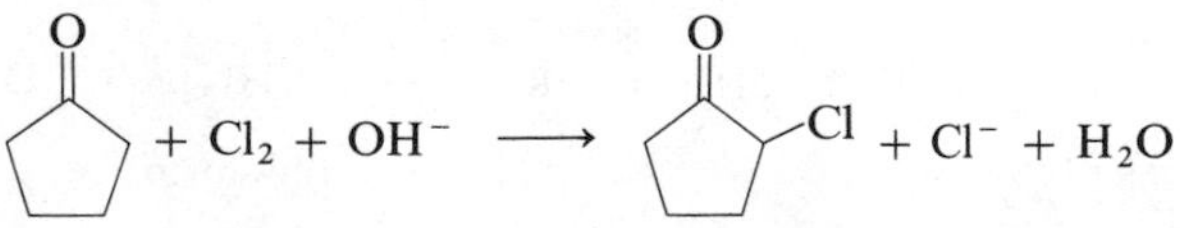

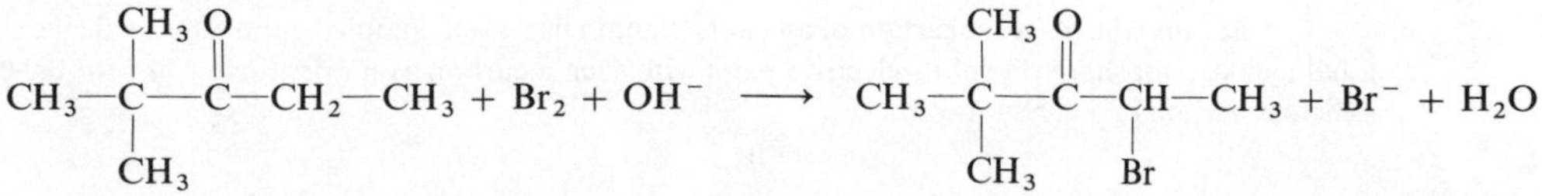

$$+ Cl_2 + OH^- \longrightarrow \qquad Cl + Cl^- + H_2O$$

$$CH_3-\overset{\overset{\displaystyle CH_3}{|}}{\underset{\underset{\displaystyle CH_3}{|}}{C}}-\overset{\overset{\displaystyle O}{\|}}{C}-CH_2-CH_3 + Br_2 + OH^- \longrightarrow CH_3-\overset{\overset{\displaystyle CH_3}{|}}{\underset{\underset{\displaystyle CH_3}{|}}{C}}-\overset{\overset{\displaystyle O}{\|}}{C}-\overset{}{\underset{\underset{\displaystyle Br}{|}}{CH}}-CH_3 + Br^- + H_2O$$

These reactions proceed by formation of an enolate anion, which then reacts with the halogen.

$$
\underset{\overset{\mid}{\text{H}}}{\overset{\overset{\displaystyle\text{O}}{\parallel}}{-\text{C}-\text{C}-}} + \text{OH}^- \;\rightleftharpoons\; \underset{\overset{\mid}{}}{\overset{\overset{\displaystyle\text{O}}{\parallel}}{-\text{C}=\text{C}-}} + \text{H}_2\text{O}
$$

$$
\text{Cl}-\text{Cl} + -\overset{..}{\text{C}}=\overset{\overset{\displaystyle\text{O}}{\parallel}}{\text{C}}- \;\longrightarrow\; \underset{\overset{\mid}{\text{Cl}}}{\overset{}{-\text{C}}}-\overset{\overset{\displaystyle\text{O}}{\parallel}}{\text{C}}- + \text{Cl}^-
$$

Only α hydrogens are easily removed by base; consequently, halogen is introduced specifically at α carbons.

Being more electronegative than carbon, a halogen enhances the acidity of hydrogens attached to the same carbon.* Therefore additional substitution by halogen tends to take place at the same carbon as the first.

$$
\text{CH}_3-\overset{\overset{\displaystyle\text{O}}{\parallel}}{\text{C}}-\text{CH}_3 \;\xrightarrow[\text{OH}^-]{\text{Br}_2}\; \text{BrCH}_2-\overset{\overset{\displaystyle\text{O}}{\parallel}}{\text{C}}-\text{CH}_3 \;\xrightarrow[\text{OH}^-]{\text{Br}_2}\; \text{Br}_2\text{CH}-\overset{\overset{\displaystyle\text{O}}{\parallel}}{\text{C}}-\text{CH}_3 \;\xrightarrow[\text{OH}^-]{\text{Br}_2}\;
$$

$$
\text{Br}_3\text{C}-\overset{\overset{\displaystyle\text{O}}{\parallel}}{\text{C}}-\text{CH}_3
$$

These hydrogens are
the most acidic

The bond between a carbon bearing three halogens and a carbonyl carbon is readily cleaved by base.

$$
\text{Br}_3\text{C}-\overset{\overset{\displaystyle\text{O}}{\parallel}}{\text{C}}-\text{CH}_3 + \text{OH}^- \;\longrightarrow\; \text{Br}_3\text{CH} + {}^-\text{O}-\overset{\overset{\displaystyle\text{O}}{\parallel}}{\text{C}}-\text{CH}_3
$$

Bromoform   Acetate ion
(the anion of
acetic acid)

When iodine is used, halogenation and cleavage constitute the "iodoform test" for methyl ketones. A compound is treated with an excess of iodine and base. Iodoform, a bright yellow, water-insoluble solid, precipitates from the aqueous solution if the compound is a methyl ketone.†

$$
\text{CH}_3-\overset{\overset{\displaystyle\text{O}}{\parallel}}{\text{C}}-\text{R} \;\xrightarrow[\text{OH}^-]{\text{I}_2}\; \text{CHI}_3\downarrow + {}^-\text{O}-\overset{\overset{\displaystyle\text{O}}{\parallel}}{\text{C}}-\text{R}
$$

Iodoform

* Because the negative carbon of an enolate anion has a full complement of eight valence electrons, a halogen cannot share one of its electron pairs with such a carbon as it does (Sec. 4.8) with the electron-deficient carbon of a carbocation.

† Alcohols with the structure $\overset{\overset{\displaystyle\text{OH}}{\mid}}{\text{CH}_3\text{CHR}}$ are oxidized to methyl ketones by alkaline solutions of halogens and also give a positive iodoform test. Acetaldehyde, the only aldehyde with a methyl group next to the carbonyl group, also gives a positive test.

9.12
ALDOL CONDENSATION

In the presence of a basic catalyst two aldehyde or ketone molecules can be linked together. This reaction involves addition of an α carbon-hydrogen bond of one molecule to the carbonyl group of another.

$$
\underset{\text{3-Hydroxybutanal}}{CH_3-\overset{\displaystyle O}{\overset{\|}{CH}} + CH_3-\overset{\displaystyle O}{\overset{\|}{CH}} \xrightarrow{OH^-} CH_3-\overset{\displaystyle OH}{\overset{|}{CH}}-CH_2-\overset{\displaystyle O}{\overset{\|}{CH}} + H_2O}
$$

$$
CH_3-CH_2-\overset{\displaystyle O}{\overset{\|}{CH}} + CH_3-CH_2-\overset{\displaystyle O}{\overset{\|}{CH}} \longrightarrow CH_3-CH_2-\overset{\displaystyle OH}{\overset{|}{CH}}-\underset{\underset{\displaystyle CH_3}{|}}{CH}-\overset{\displaystyle O}{\overset{\|}{CH}} + H_2O
$$

3-Hydroxy-2-methylpentanal

This reaction, called the **aldol condensation** because the products from aldehydes have both *ald*ehyde and alcoh*ol* functions, provides another important way of linking organic molecules to form larger ones.

The mechanism of the aldol condensation combines the two general types of reactions of aldehydes and ketones: reaction at an α carbon and addition to a carbonyl group.

Step 1
$$
\underset{\underset{\displaystyle R}{|}}{CH_2}-\overset{\displaystyle O}{\overset{\|}{CH}} + OH^- \longleftrightarrow {}^-\underset{\underset{\displaystyle R}{|}}{CH}-\overset{\displaystyle O}{\overset{\|}{CH}} + HOH
$$

Step 2
$$
\underset{\underset{\displaystyle R}{|}}{CH_2}-\overset{\displaystyle \ddot{O}:}{\overset{\|}{CH}} + {}^-\ddot{C}H-\underset{\underset{\displaystyle R}{|}}{\overset{\displaystyle O}{\overset{\|}{CH}}} \rightleftharpoons \underset{\underset{\displaystyle R}{|}}{CH_2}-\underset{}{\overset{\displaystyle :\ddot{O}:^-}{\overset{|}{CH}}}-\underset{\underset{\displaystyle R}{|}}{CH}-\overset{\displaystyle O}{\overset{\|}{CH}}
$$

Step 3
$$
\underset{\underset{\displaystyle R}{|}}{CH_2}-\overset{\displaystyle O^-}{\overset{|}{CH}}-\underset{\underset{\displaystyle R}{|}}{CH}-\overset{\displaystyle O}{\overset{\|}{CH}} + HOH \rightleftharpoons \underset{\underset{\displaystyle R}{|}}{CH_2}-\overset{\displaystyle OH}{\overset{|}{CH}}-\underset{\underset{\displaystyle R}{|}}{CH}-\overset{\displaystyle O}{\overset{\|}{CH}} + OH^-
$$

Removal of an α hydrogen by the base (step 1) generates a low concentration of enolate anion. The enolate anion is nucleophilic and, just as other nucleophiles, adds (step 2) to the carbonyl group of another molecule. The resulting alkoxide ion abstracts a proton from water (step 3), regenerating the hydroxide catalyst.

problem 9.4 Draw a structure for the aldol condensation product of acetone.

problem 9.5 How can 2-ethylhexane-1,3-diol, once a commercial insect repellent ("6-12"), be synthesized from a four-carbon compound?

The products of aldol condensations dehydrate when treated with dilute acid or sometimes merely on heating.

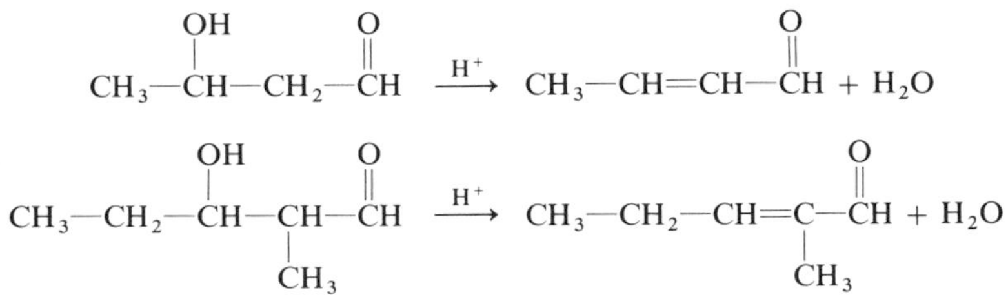

$$
\underset{\substack{\;\\ \;}}{CH_3\!-\!\overset{\displaystyle OH}{\overset{|}{CH}}\!-\!CH_2\!-\!\overset{\displaystyle O}{\overset{\|}{CH}}} \;\xrightarrow{H^+}\; CH_3\!-\!CH\!=\!CH\!-\!\overset{\displaystyle O}{\overset{\|}{CH}} + H_2O
$$

$$
CH_3\!-\!CH_2\!-\!\underset{\underset{\textstyle CH_3}{|}}{\overset{\displaystyle OH}{\overset{|}{CH}}}\!-\!CH\!-\!\overset{\displaystyle O}{\overset{\|}{CH}} \;\xrightarrow{H^+}\; CH_3\!-\!CH_2\!-\!CH\!=\!\underset{\underset{\textstyle CH_3}{|}}{C}\!-\!\overset{\displaystyle O}{\overset{\|}{CH}} + H_2O
$$

Dehydration occurs in the direction that gives conjugated carbon-carbon and carbon-oxygen double bonds.

An aldehyde or ketone, such as benzaldehyde, that lacks an α hydrogen cannot produce an enolate anion. It can, however, react with an enolate anion from another aldehyde or ketone.

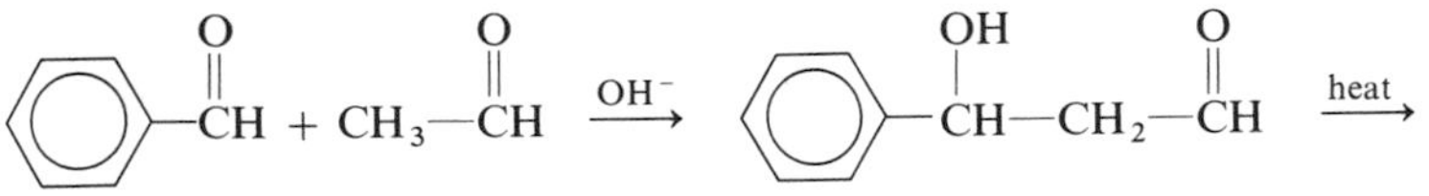

$$
\underset{\text{Benzaldehyde}}{\text{C}_6\text{H}_5\!-\!\overset{\displaystyle O}{\overset{\|}{CH}}} + CH_3\!-\!\overset{\displaystyle O}{\overset{\|}{CH}} \;\xrightarrow{OH^-}\; \underset{\text{3-Hydroxy-3-phenylpropanal}}{\text{C}_6\text{H}_5\!-\!\overset{\displaystyle OH}{\overset{|}{CH}}\!-\!CH_2\!-\!\overset{\displaystyle O}{\overset{\|}{CH}}} \;\xrightarrow{heat}
$$

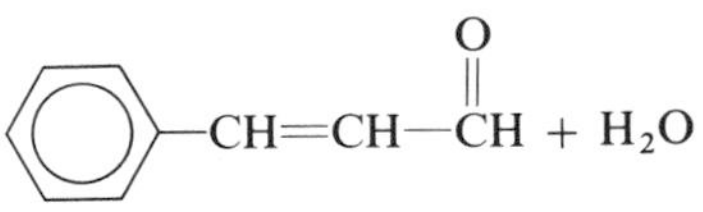

$$
\underset{\text{Cinnamaldehyde}}{\text{C}_6\text{H}_5\!-\!CH\!=\!CH\!-\!\overset{\displaystyle O}{\overset{\|}{CH}}} + H_2O
$$

In this reaction acetaldehyde is added slowly to a basic solution of benzaldehyde. With such addition, the enolate of acetaldehyde is more likely to encounter benzaldehyde than acetaldehyde molecules, minimizing formation of the aldol condensation product of acetaldehyde.

The aldol condensation and related condensations are important in biological systems. The following reaction, catalyzed by the enzyme fructose diphosphate aldolase, is important in carbohydrate metabolism (Chapter 10).

Dihydroxyacetone phosphate

$$CH_2-O-\overset{\displaystyle O}{\overset{\|}{P}}-OH$$

$$C=O \quad OH$$

$$H-C-OH$$

$$H$$

D-Glyceraldehyde 3-phosphate

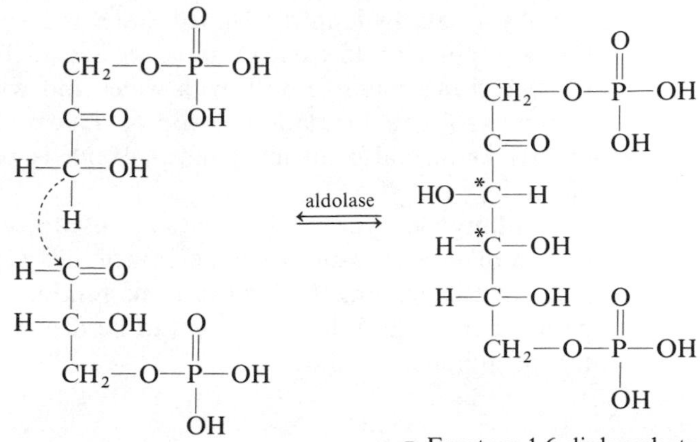

$$H-C=O$$

$$H-C-OH$$

$$CH_2-O-\overset{\displaystyle O}{\overset{\|}{P}}-OH$$

$$OH$$

aldolase

$$CH_2-O-\overset{\displaystyle O}{\overset{\|}{P}}-OH$$

$$C=O \quad OH$$

$$HO-\overset{*}{C}-H$$

$$H-\overset{*}{C}-OH$$

$$H-C-OH \quad O$$

$$CH_2-O-\overset{\displaystyle O}{\overset{\|}{P}}-OH$$

$$OH$$

D-Fructose 1,6-diphosphate

Two new asymmetric carbons (starred) are created. As is generally the case in enzyme-catalyzed reactions, only one of the possible stereoisomers is formed.

9.13
SYNTHESIS OF ALDEHYDES AND KETONES

Several methods for synthesizing aldehydes and ketones have already been given. Oxidation of secondary alcohols is a particularly important ketone synthesis (Sec. 5.8).

$$(CH_3)_3CCH_2\overset{\displaystyle OH}{\overset{|}{C}}HCH_3 \xrightarrow{K_2Cr_2O_7} (CH_3)_3CCH_2\overset{\displaystyle O}{\overset{\|}{C}}CH_3$$

Oxidation of primary alcohols can be used to prepare aldehydes, although many oxidizing agents produce carboxylic acids instead (Sec. 5.8).

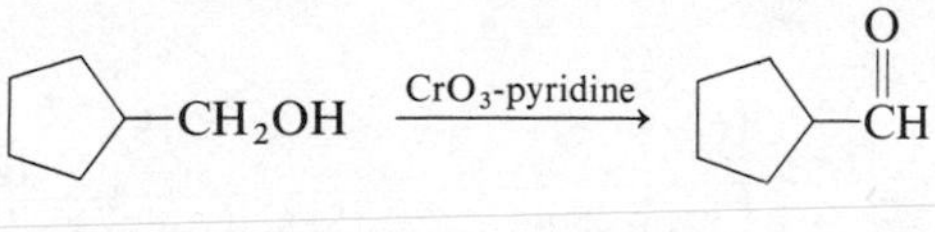

$$\text{—CH}_2\text{OH} \xrightarrow{\text{CrO}_3\text{-pyridine}} \text{—}\overset{\displaystyle O}{\overset{\|}{C}}\text{H}$$

Aldehydes and ketones also result from the ozonolysis of alkenes (Sec. 3.10). Complex aldehydes and ketones can be constructed from simpler ones by the aldol condensation (preceding section).

9.14
SOME IMPORTANT ALDEHYDES AND KETONES

Formaldehyde has been detected in many areas of interstellar space. On earth it has long been used as a preservative of biological specimens and as an embalming fluid. These uses depend on the reactions of this reactive aldehyde with hydroxyl and amino groups, which are abundant in biological compounds. Current annual use of formaldehyde in the United States is 5 kg per person, mostly for incorporation into polymers (for one example see Sec. 12.10).

Acetone, miscible both with water and with many organic compounds, is a common solvent. Oxidation of isopropyl alcohol is one synthesis of acetone. Ethyl methyl ketone, also an important solvent, is prepared by oxidation of *sec*-butyl alcohol.

Many aldehydes and ketones are used because of their odors or tastes. The simple diketone 2,3-butanedione contributes significantly to the aroma of butter and, for this reason, is added to some food products that have never had contact with a cow. Benzaldehyde occurs in kernels of bitter almonds and has a characteristic almondlike odor.

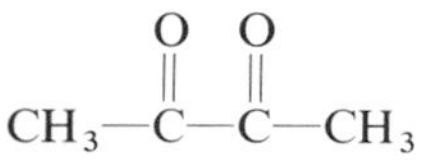

2,3-Butanedione 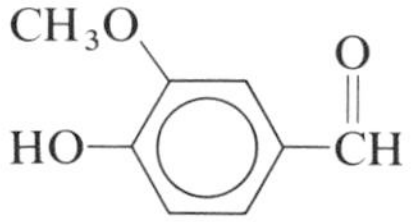   Vanillin 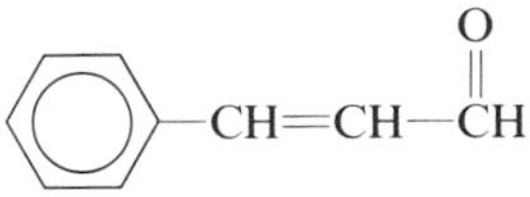   Cinnamaldehyde

Vanillin, present in vanilla beans, has the characteristic vanilla odor and taste and is one of the flavoring agents used most extensively in the food and beverage industries. Most commercial vanillin is now obtained from the wastes discarded in the manufacture of paper pulp. Cinnamaldehyde, a constituent of oil of cinnamon, has a strong cinnamon odor and is also a common flavoring agent.

The terpene citronellal, the major constituent of citronella oil, is used as a perfume in soaps and as an insect repellent. The *S* enantiomer of the terpene carvone is present in caraway seed and dill seed oils and contributes significantly to the odor of dill pickles.

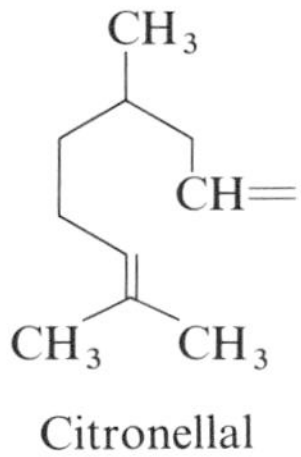

Citronellal 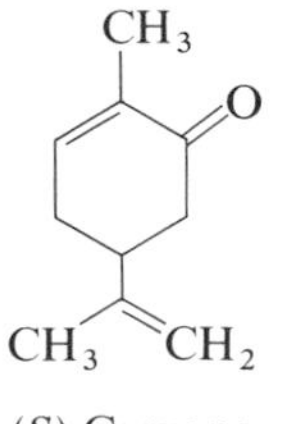   (*S*)-Carvone 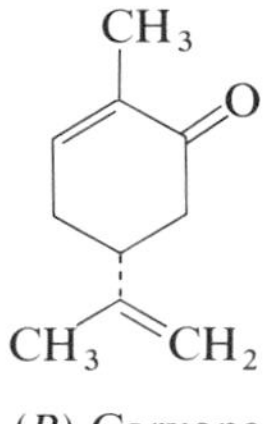   (*R*)-Carvone

The *R* enantiomer is found in spearmint oil and contributes to its characteristic aroma.

Civetone is present in the glandular secretion of male and female civet cats, and muscone is a sex attractant secreted by the musk glands of the male musk deer.

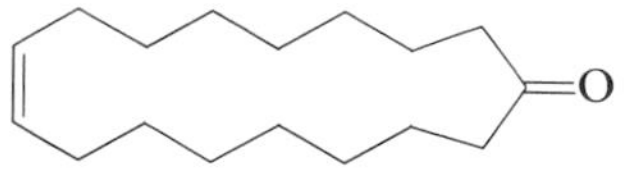

Civetone   Muscone
9-Cycloheptadecen-1-one   3-Methylcyclopentadecanone

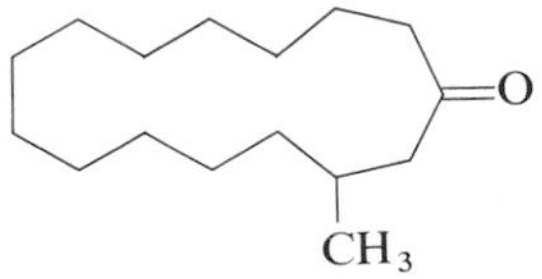

These cyclic ketones have musky odors and are used as constituents of expensive perfumes.

9.15
STEROIDS

The steroids, a large family of biological compounds distributed throughout the plant and animal kingdoms, contain a framework of four fused rings.

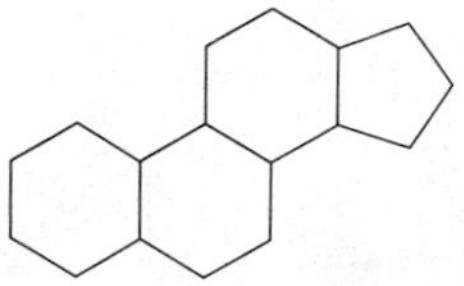

This ring system and alkyl groups attached to it are ornamented with functional groups. Often they are carbonyl groups, but alkene, hydroxyl, and carboxyl groups are also common. The steroids include the sex hormones, the adrenal cortex hormones, the bile acids (Sec. 13.6), and some vitamins.

Cholesterol is found in most living organisms. In animals it is not only the most abundant steroid but also one of the most abundant compounds of any type.

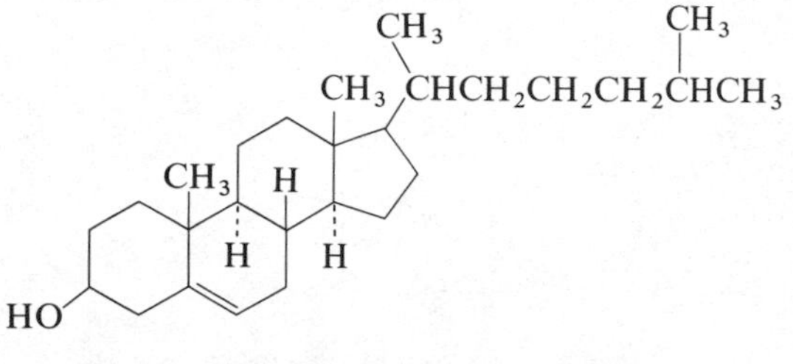

Cholesterol

Organisms synthesize their other steroids from cholesterol. Cholesterol is particularly concentrated in the brain and spinal cord. Human blood serum contains approximately 0.2 g/100 ml of cholesterol, either free or combined with carboxylic acids in esters (Chapter 12). Deposition of cholesterol and some other compounds in the inner walls of blood vessels leads to the condition known as atherosclerosis.

Testosterone, produced in the testes, is the most important of the male sexual hormones, known as androgens.

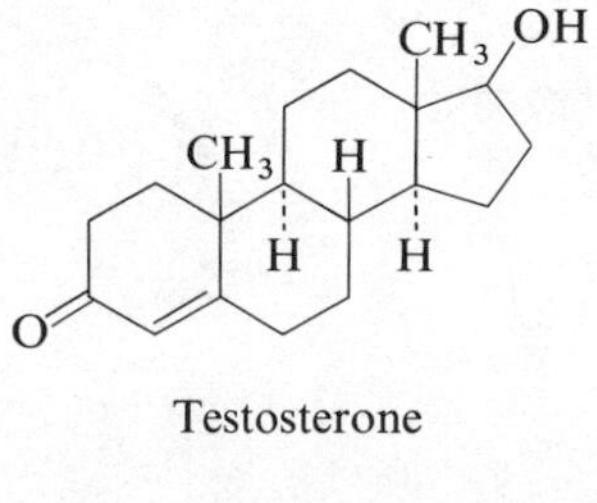

Testosterone

The androgens are responsible for male secondary sexual characteristics and for the functioning of male sexual organs.

Two categories of female sex hormones are produced in the ovaries. Estradiol is the most important of the hormones known as estrogens.

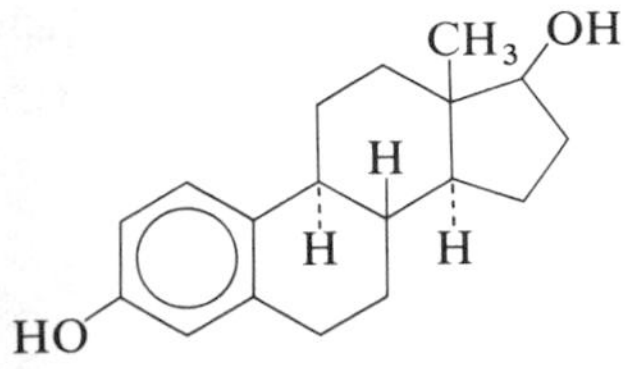

Estradiol 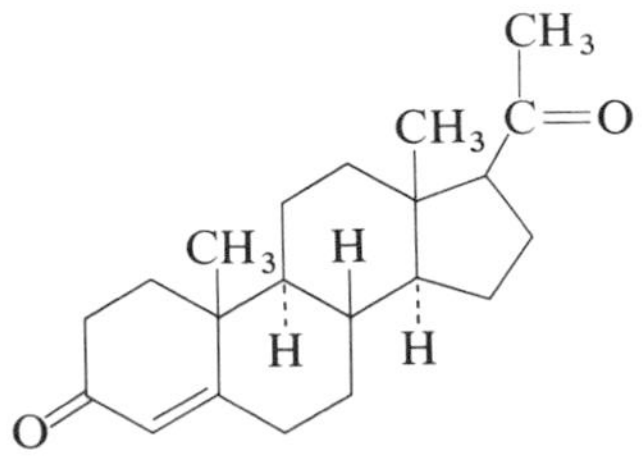   Progesterone

The estrogens are responsible for the development of female sexual characteristics and play an important role in stimulating ovulation. Release of progesterone, another type of hormone, begins just prior to ovulation to prepare the uterus for implantation and maintenance of a fertilized egg and to prevent other ova from ripening. Progesterone is responsible for preventing ovulation during pregnancy. Contraceptive drugs contain as their principal component a synthetic steroid with progesteronelike activity. Norethindrone is one of the most widely used.

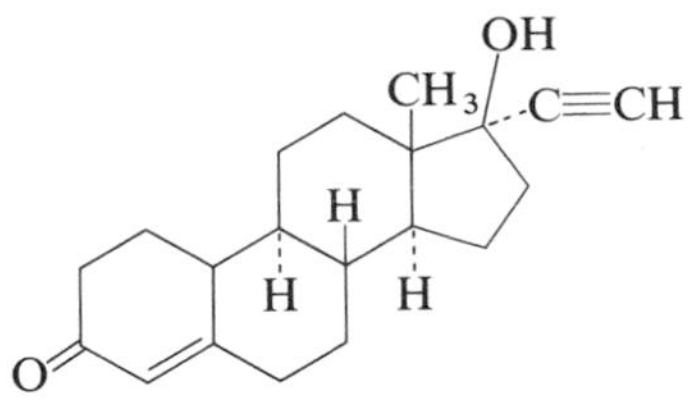

Norethindrone

The adrenal gland produces a group of hormones that influence various physiological functions, including metabolism of carbohydrates, proteins, and lipids and maintenance of sodium ion concentration and blood volume. Among these hormones is hydrocortisone (cortisol), used medically as an anti-inflammatory agent for treating a variety of conditions, including rheumatoid arthritis.

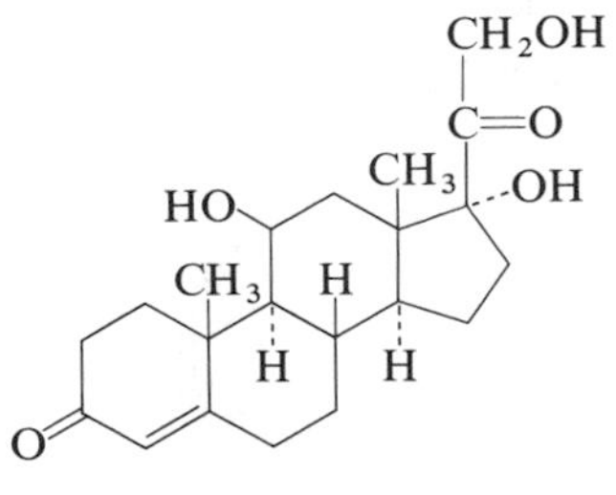

Hydrocortisone (cortisol)

Some synthetic steroids are even more effective and have fewer undesirable side effects.

Rickets is due to a deficiency of the D vitamins, which play a role in regulating calcium and phosphorus metabolism. The most important D vitamin in mammals is ordinarily vitamin D_3.

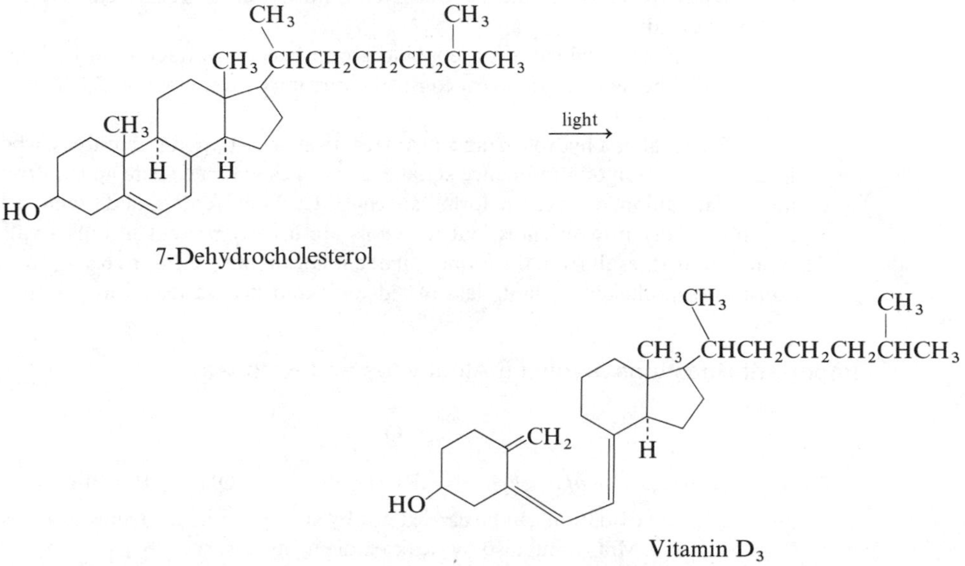

7-Dehydrocholesterol

Vitamin D_3

This vitamin arises from the action of sunlight on 7-dehydrocholesterol in the skin. In the absence of sufficient sunlight, a D vitamin must be supplied in the diet.

SUMMARY

Definitions and Ideas

Aldehydes are compounds having one organic group and one hydrogen (or two hydrogens) bonded to a carbonyl group. **Ketones** are compounds having two organic groups bonded to a carbonyl group.

Some simple aldehydes are given trivial names (formaldehyde, acetaldehyde, propionaldehyde, butyraldehyde, valeraldehyde, benzaldehyde) related to trivial names of the corresponding carboxylic acids. Substitutive names of aldehydes are based on the longest chain that contains the carbonyl group. The ending of the name of the corresponding hydrocarbon is changed from -e to -al. The carbonyl carbon is assumed to be number 1 and is not numbered in the name. Aromatic aldehydes are usually named as derivatives of benzaldehyde.

Radicofunctional names of ketones are formed by naming the groups attached to the carbonyl group and adding the word **ketone**. Substitutive names of ketones are similar to those of alcohols except that the ending is *-one*. Aromatic ketones are often assigned names (such as acetophenone) ending in *-phenone*. The prefix is derived from the trivial name of the carboxylic acid that contains the fragment attached to the aromatic ring.

Aldehydes and ketones form hydrogen bonds with water and have solubilities in water similar to those of alcohols with the same number of carbons.

An important difference between the chemical behavior of aldehydes and ketones is the ready oxidation of aldehydes. Even mild oxidizing agents convert aldehydes to carboxylic acids.

Additions in which a nucleophilic group becomes attached to carbon (reactions 2 to 5 and 7 below are examples) constitute one important class of aldehyde and ketone reactions.

Removal of a hydrogen (an **α hydrogen**) from a carbon adjacent to a carbonyl group leads to formation of a resonance-stabilized anion called an **enolate anion**. Protonation of an enolate anion on oxygen forms an **enol**. Carbonyl compounds with α hydrogens are in equilibrium with enols, but the enols are usually present in only small amounts. Reactions that result from the formation of enolate anions (reactions 6 and 7 are examples) constitute another important class of aldehyde and ketone reactions.

Important Reactions Involving Aldehydes and Ketones

1. *Oxidation of aldehydes*
$$\underset{\text{RCH}}{\overset{O}{\|}} \longrightarrow \underset{\text{RCOH}}{\overset{O}{\|}} \qquad R = \text{alkyl or aryl}$$

The oxidation can be carried out by strong oxidizing agents, such as $K_2Cr_2O_7$ or $KMnO_4$, but also by weak oxidizing agents, such as Ag^+.

2. *Reactions with alcohols*
$$-\underset{}{\overset{O}{\underset{\|}{C}}}- \underset{\overset{ROH}{\rightleftharpoons}}{} -\underset{\overset{|}{OR}}{\overset{OH}{\underset{|}{C}}}-$$
(a hemiacetal)

Hemiacetals revert readily to the reactants.

$$-\underset{}{\overset{O}{\underset{\|}{C}}}- \xrightleftharpoons{ROH \text{ (excess), } H^+} -\underset{\overset{|}{OR}}{\overset{OR}{\underset{|}{C}}}- + H_2O$$
(an acetal)

Acetals are easily hydrolyzed back to aldehydes or ketones by using an excess of water.

3. *Reactions with ammonia derivatives*
$$-\underset{}{\overset{O}{\underset{\|}{C}}}- \xrightarrow[H^+]{Y-NH_2} -\underset{}{\overset{N-Y}{\underset{\|}{C}}}-$$

Table 9.2 summarizes these reactions.

4. *Addition of Grignard reagents*

$$\overset{\overset{\displaystyle O}{\|}}{-C-} \xrightarrow{\text{RMgX}} \xrightarrow{\text{H}_2\text{O}} -\overset{\overset{\displaystyle OH}{|}}{\underset{\underset{\displaystyle R}{|}}{C}}-$$

X = Cl, Br, or I   R = alkyl or aryl

5. *Reduction*

$$\overset{\overset{\displaystyle O}{\|}}{-C-} \xrightarrow{\text{LiAlH}_4} \xrightarrow{\text{H}_2\text{O}} -\overset{\overset{\displaystyle OH}{|}}{\underset{\underset{\displaystyle H}{|}}{C}}-$$

An alternative reduction using H_2 and a Ni or Pt catalyst was described in Chapter 5.

6. *α Halogenation*

$$-\overset{\overset{\displaystyle }{|}}{\underset{\underset{\displaystyle H}{|}}{C}}-\overset{\overset{\displaystyle O}{\|}}{C}- \xrightarrow[\text{OH}^-]{\text{X}_2} -\overset{\overset{\displaystyle }{|}}{\underset{\underset{\displaystyle X}{|}}{C}}-\overset{\overset{\displaystyle O}{\|}}{C}-$$

X = Cl, Br, or I

7. *Aldol condensation*

$$\text{H}-\overset{\overset{\displaystyle }{|}}{\underset{\underset{\displaystyle }{|}}{C}}-\overset{\overset{\displaystyle O}{\|}}{C} + \text{H}-\overset{\overset{\displaystyle }{|}}{\underset{\underset{\displaystyle }{|}}{C}}-\overset{\overset{\displaystyle O}{\|}}{C}- \xrightarrow{\text{OH}^-} \text{H}-\overset{\overset{\displaystyle }{|}}{\underset{\underset{\displaystyle }{|}}{C}}-\overset{\overset{\displaystyle OH}{|}}{\underset{\underset{\displaystyle }{|}}{C}}-\overset{\overset{\displaystyle }{|}}{\underset{\underset{\displaystyle }{|}}{C}}-\overset{\overset{\displaystyle O}{\|}}{C}-$$

8. *Synthesis by oxidation of alcohols*

Secondary alcohol $\text{R}\overset{\overset{\displaystyle OH}{|}}{C}\text{HR}$ $\xrightarrow{\text{K}_2\text{Cr}_2\text{O}_7 \text{ or KMnO}_4}$ $\text{R}\overset{\overset{\displaystyle O}{\|}}{C}\text{R}$ Ketone

Primary alcohol RCH_2OH $\xrightarrow{\text{CrO}_3\text{-pyridine}}$ $\text{R}\overset{\overset{\displaystyle O}{\|}}{C}\text{H}$ Aldehyde

(discussed in Chapter 5)   R = alkyl or aryl

PROBLEMS

1. Draw structures for the following compounds.
   (a) 2-pentanone
   (b) hexanal
   (c) diethyl ketone
   (d) valeraldehyde
   (e) 2-methylbutanal
   (f) 2-methylcyclopentanone
   (g) benzyl phenyl ketone
   (h) *trans*-4-phenyl-3-buten-2-one
   (i) phenylethanal
   (j) dicyclopropyl ketone
   (k) *cis*-2-butenal
   (l) diallyl ketone

(m) 3-cyclopentenone   (n) 2,4-pentanedione
(o) acetophenone   (p) 1,1-dimethoxypropane
(q) 3,3-diethoxypentane   (r) pentanal oxime
(s) cyclopentanone semicarbazone   (t) acetone 2,4-dinitrophenylhydrazone

2. Name the following compounds.

(a)
$$\overset{\overset{\displaystyle O}{\|}}{CH_3CH_2CH}$$

(b)
$$\overset{\overset{\displaystyle O}{\|}}{CH_3CCH_2CH_3}$$

(c)

(d)
$$\overset{\overset{\displaystyle O}{\|}}{CH_3CHCH}$$
$$\underset{CH_3}{\,}$$

(e)

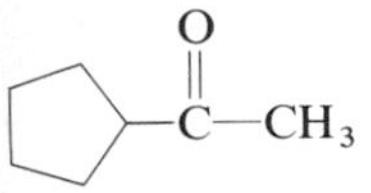

(f)
$$\overset{\overset{\displaystyle O}{\|}}{CH_3CH_2CCH_2CH_3}$$

(g)
$$\overset{\overset{\displaystyle O}{\|}}{CH_3CH{=}CHCH}$$

(h)

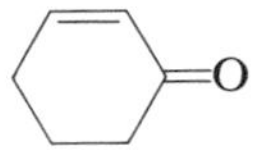

(i)
$$\overset{\overset{\displaystyle O}{\|}}{CH_3CH_2CHCH}$$
$$\underset{Cl}{\,}$$

(j)

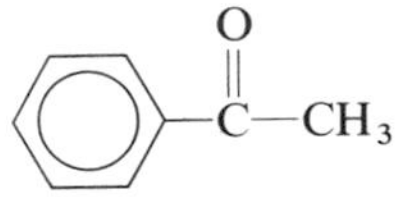

(k)

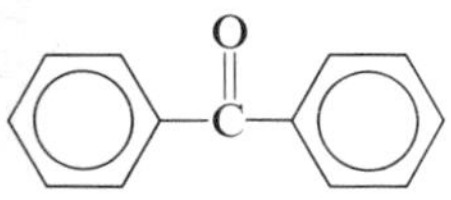

(l)

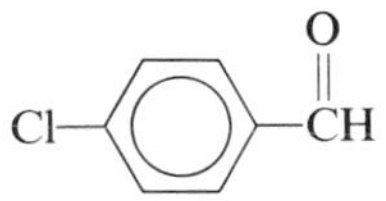

(m)
$$\underset{\overset{|}{OCH_2CH_3}}{\overset{\overset{OCH_2CH_3}{|}}{CH_3CH_2CH}}$$

(n)

(o)
$$\overset{\overset{\displaystyle NOH}{\|}}{CH_3CH_2CCH_3}$$

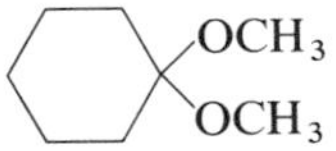

(p)

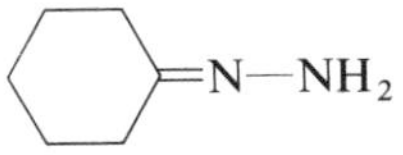

(q)

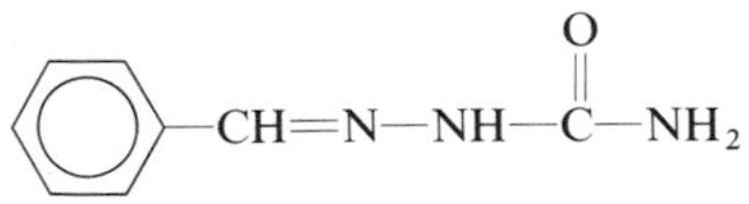

(r) 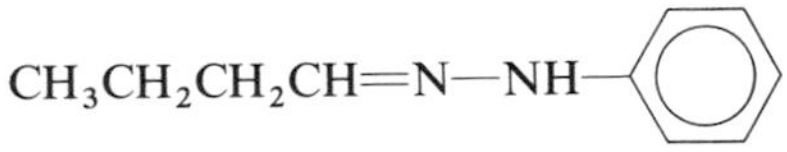
$$CH_3CH_2CH_2CH{=}N{-}NH{-}$$

3. Give another name for each compound.
(a) dipropyl ketone   (b) pentanal
(c) 3-pentanone   (d) butyraldehyde

4. Show by a structural drawing what is meant by each of the following terms.
(a) acetal   (b) hemiacetal
(c) oxime   (d) hydrazone
(e) 2,4-dinitrophenylhydrazone   (f) semicarbazone
(g) enol   (h) enolate anion

5. Draw structures of all carbonyl compounds that have these molecular formulas.
   (a) C_4H_8O (b) $C_5H_{10}O$

6. Draw the structural formulas of any organic products that result from treating acetone with the following reagents. Do the same for propionaldehyde.
   (a) Ag^+ (Tollens' reagent)   (b) CH_3OH
   (c) CH_3CH_2OH (excess), H^+ (d) $KMnO_4$
   (e) H_2N—OH, H^+ (f) H_2N—NH_2, H^+
   (g)

   $$H_2N\text{—}NH\text{—}\underset{}{\bigcirc}\text{—}NO_2, H^+$$
   with NO_2 substituent

   (h) $LiAlH_4$, followed by H_2O

   (i) CH_3MgI, followed by H_2O (j) H_2, Ni
   (k) OH^- (l) Br_2, OH^-
   (m) I_2 (excess), OH^-

7. Draw structures of the organic products that result from each reaction.
   (a)

   $$CH_3\overset{O}{\overset{\|}{C}}CH_3 \xrightarrow[OH^-]{Cl_2}$$

   (b)

   $$\bigcirc\text{—}\overset{O}{\overset{\|}{C}}H \xrightarrow[H^+]{H_2N\text{—}NH_2}$$

   (c)

   $$CH_3\overset{}{\underset{CH_3}{CH}}\overset{O}{\overset{\|}{C}}H \xrightarrow{CH_3MgI} \xrightarrow{H_2O}$$

   (d)

   $$CH_3\overset{O}{\overset{\|}{C}}H \xrightarrow{CH_3OH}$$

   (e)

   $$CH_2{=}CH\overset{O}{\overset{\|}{C}}H \xrightarrow{Ag^+}$$

   (f)

   $$\bigcirc{=}O \xrightarrow[H^+]{H_2N\text{—}OH}$$

   (g)

   $$CH_3CH_2CH_2\overset{O}{\overset{\|}{C}}H \xrightarrow{OH^-}$$

   (h)

   $$\square{=}O \xrightarrow[H^+]{CH_3CH_2OH \text{ (excess)}}$$

   (i)

   $$CH_3CH{=}CH\overset{O}{\overset{\|}{C}}CH_3 \xrightarrow[Pt]{H_2}$$

   (j)

   $$\bigcirc\text{—}\overset{O}{\overset{\|}{C}}H \xrightarrow{LiAlH_4} \xrightarrow{H_2O}$$

   (k)

   $$\bigcirc\text{—}\overset{O}{\overset{\|}{C}}\text{—}\bigcirc \xrightarrow{CH_3CH_2MgBr} \xrightarrow{H_2O}$$

   (l)

   $$\bigcirc\text{—}\overset{O}{\overset{\|}{C}}CH_3 \xrightarrow[OH^-]{I_2 \text{ (excess)}}$$

   (m)

   $$CH_3\overset{O}{\overset{\|}{C}}CH_2CH_3 \xrightarrow[H^+]{H_2N\text{—}NH\text{—}\bigcirc}$$

   (n)

   $$(CH_3)_3C\overset{O}{\overset{\|}{C}}CH_3 \xrightarrow[OD^-]{D_2O \text{ (excess)}}$$

8. Draw structures of all products that result from each reaction.
   (a) treating a mixture of acetaldehyde and propionaldehyde with aqueous base
   (b) treating cyclopentanone with aqueous base

9. Draw structures for the products of each reaction.

(a)
$$\underset{\text{HCCH}_2\text{CH}_2\text{CH}_2\text{CH}_2\text{CH}}{\overset{\text{O}\qquad\qquad\qquad\text{O}}{||\qquad\qquad\qquad||}} \xrightarrow{\text{OH}^-} \xrightarrow{\text{warm}} \text{C}_6\text{H}_8\text{O}$$

(b)
$$\underset{\text{CH}_3\text{CH}}{\overset{\text{O}}{||}} \xrightarrow[\text{H}^+]{\text{HOCH}_2\text{CH}_2\text{OH}} \text{C}_4\text{H}_8\text{O}_2$$

(c)
$$\underset{\text{CH}_3\text{CCH}_3}{\overset{\text{O}}{||}} \xrightarrow{\text{OH}^-} \text{C}_6\text{H}_{10}\text{O} + \text{C}_9\text{H}_{14}\text{O}$$

10. Write equations to show how acetaldehyde can be converted to each compound.
(a) acetic acid   (b) acetaldehyde phenylhydrazone
(c) 1-phenylethanol   (d) 3-hydroxybutanal
(e) 1,1-dimethoxyethane

11. Write equations to show how acetone can be converted to each compound.
(a) iodoform   (b) *tert*-butyl alcohol
(c) acetone oxime   (d) 2-propanol
(e) 1-chloro-2-propanone

12. Write equations to show how the following conversions can be carried out. Many will require more than one step.

(a)
$$\text{C}_6\text{H}_5\text{—Br} \longrightarrow \text{C}_6\text{H}_5\text{—CH}_2\text{OH}$$

(b)
$$\underset{\text{C}_6\text{H}_5\text{—CH}}{\overset{\text{O}}{||}} \longrightarrow \text{C}_6\text{H}_5\text{—CO}_2\text{H}$$

(c)
$$\text{cyclopentene} \longrightarrow \text{cyclopentanone}$$

(d)
$$\underset{\text{CH}_2{=}\text{CHCH}_2\text{CH}_2\text{CH}}{\overset{\text{O}}{||}} \longrightarrow \text{CH}_2{=}\text{CHCH}_2\text{CH}_2\text{CH}_2\text{OH}$$

(e)
$$\underset{\text{CH}_3\text{CH}_2\text{CCH}_3}{\overset{\text{O}}{||}} \longrightarrow \underset{\underset{\text{OCH}_3}{|}}{\overset{\overset{\text{OCH}_3}{|}}{\text{CH}_3\text{CH}_2\text{CCH}_3}}$$

(f)
$$\text{cyclopentanone} \longrightarrow \text{cyclopentene}$$

(g)
$$\underset{\text{CH}_3\text{CH}_2\text{CHCH}_2\text{CH}_3}{\overset{\overset{\text{OH}}{|}}{}} \longrightarrow \underset{\underset{\text{CH}_2\text{CH}_3}{|}}{\overset{\overset{\text{OH}}{|}}{\text{CH}_3\text{CH}_2\text{CCH}_2\text{CH}_3}}$$

(h)

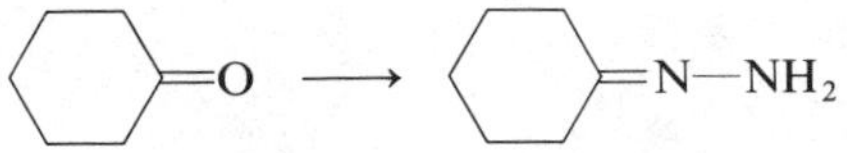

(i)

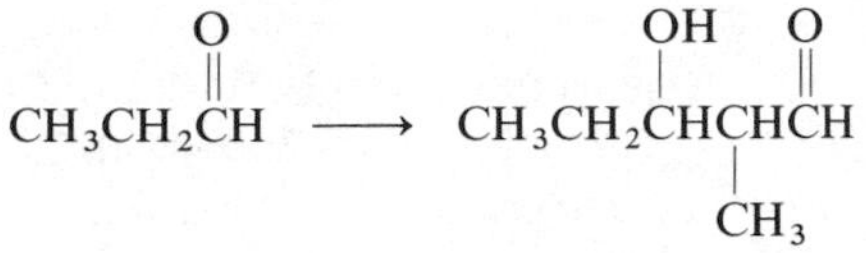

$$CH_3CH_2\overset{O}{\overset{\|}{C}}H \longrightarrow CH_3CH_2\overset{OH}{\overset{|}{C}}H\overset{O}{\overset{\|}{C}}HCH_3$$

$$\underset{CH_3}{}$$

(j)

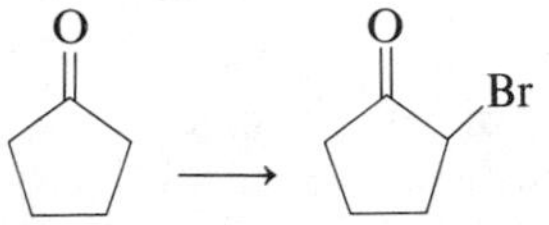

(k)

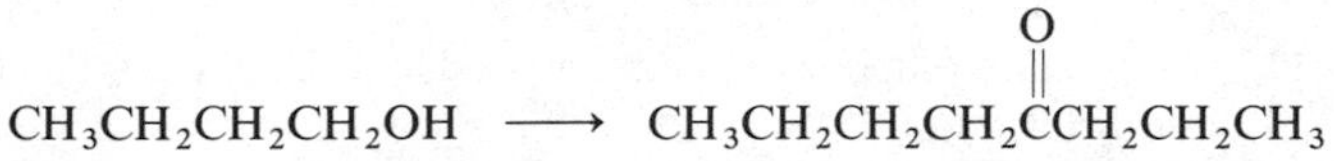

$$CH_3CH_2CH_2CH_2OH \longrightarrow CH_3CH_2CH_2CH_2\overset{O}{\overset{\|}{C}}CH_2CH_2CH_3$$

(l)

$$\longrightarrow HCCH_2CH_2CH_2CH_2CH$$

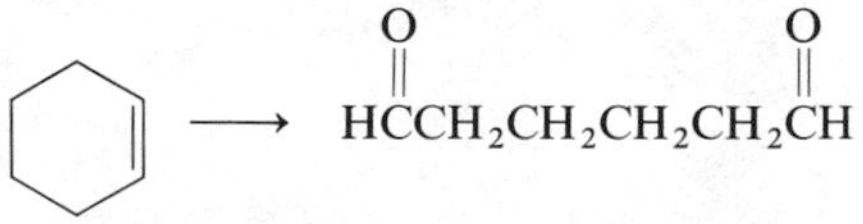

13. Write equations to show how each compound can be synthesized by using an aldol condensation. Other steps will sometimes be needed.

(a)

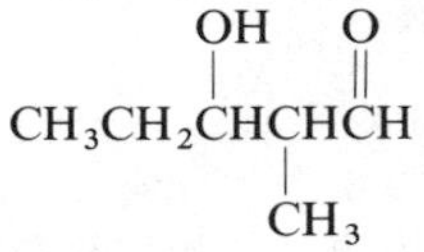

$$CH_3CH_2\overset{OH}{\overset{|}{C}}H\overset{O}{\overset{\|}{C}}H$$

$$\underset{CH_3}{}$$

(b)

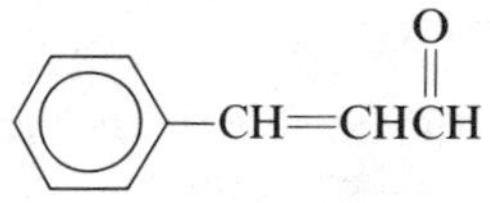

—CH=CHCH

(c)

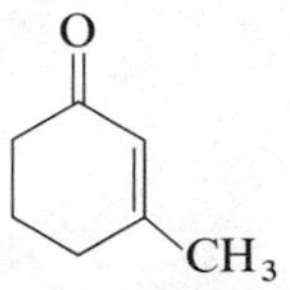

(d)

$$CH_3CHCH_2\overset{OH}{\overset{|}{C}}HCH_3$$

$$\underset{CH_3}{}$$

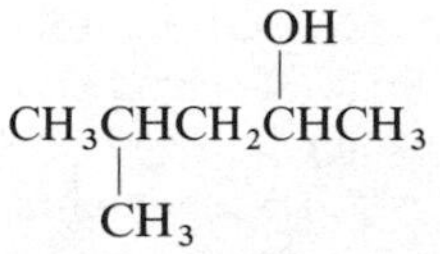

14. Write equations to show how each compound can be synthesized by using the reaction of a Grignard reagent with an aldehyde or ketone. Other steps will sometimes be needed.

(a)

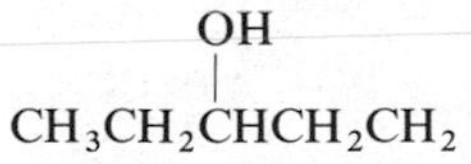

$$CH_3CH_2\overset{OH}{\overset{|}{C}}HCH_2CH_2$$

(b)

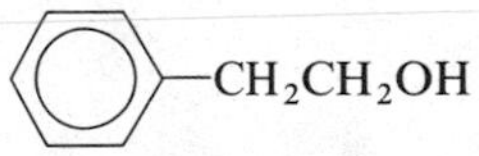

—CH$_2$CH$_2$OH

(c)

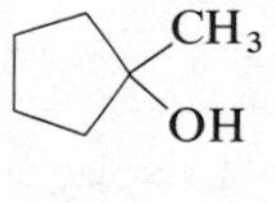

(d)

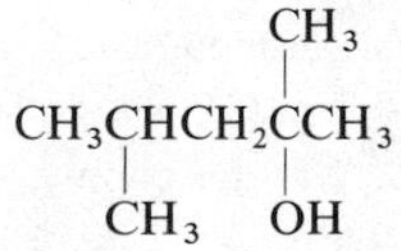

$$CH_3\overset{CH_3}{\overset{|}{C}}HCH_2\overset{CH_3}{\overset{|}{C}}CH_3 \quad \text{(two ways)}$$

$$\underset{CH_3 \quad OH}{}$$

(e)

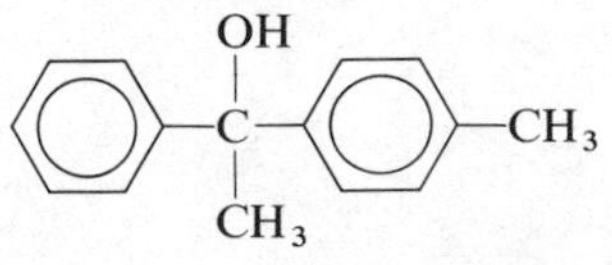

(three ways)

(f)

$$CH_3CH_2CH_2\overset{\overset{\textstyle O}{\|}}{C}CH_2CH_2CH_3$$

(g)

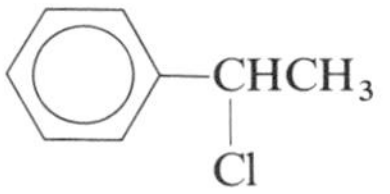

15. Compound A, molecular formula $C_5H_{12}O$, reacts with $K_2Cr_2O_7$ to form compound B, molecular formula $C_5H_{10}O$. Compound B reacts with 2,4-dinitrophenylhydrazine to form a yellow solid but does not give a silver mirror when treated with Ag^+ (Tollens' reagent) or a precipitate of iodoform when treated with a basic solution of I_2. Draw the structures of A and B.

16. Draw the structure of a compound of molecular formula C_8H_8O that forms a solid precipitate when treated with phenylhydrazine and a precipitate of iodoform when treated with a basic solution of I_2.

17. What stable compound or compounds does each of the following unstable compounds readily form?

(a)

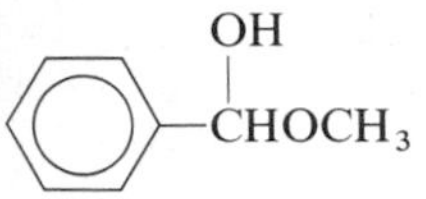

(b)

(c) $CH_2{=}CH{-}OH$

(d)

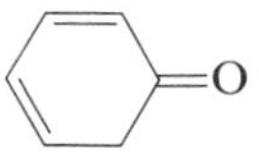

18. Indicate a chemical test, simple to carry out and with easily observable results, that can be used to distinguish between the members of each pair of compounds. Describe the observations that would be made with each compound.
    (a) acetaldehyde and acetone
    (b) 2-heptanone and benzaldehyde
    (c) benzyl alcohol and benzaldehyde
    (d) 3-pentanone and 2-pentanone
    (e) diethyl ketone and ethyl vinyl ketone

19. Provide an explanation for the slow incorporation of ^{18}O into acetone containing the normal isotope of oxygen (^{16}O) when it is dissolved in $H_2{}^{18}O$.

$$CH_3\overset{\overset{\textstyle {}^{16}O}{\|}}{C}CH_3 \xrightarrow{H_2{}^{18}O} CH_3\overset{\overset{\textstyle {}^{18}O}{\|}}{C}CH_3$$

20. Explain why acetals are much more readily cleaved by acids than are ethers.

21. Reaction of propyne with a strongly acidic aqueous solution forms acetone. Propose a mechanism for this reaction. Why isn't propionaldehyde formed?

22. Why are conjugated double bonds more stable than isolated double bonds?

10

Carbohydrates

The name carbohydrate arose because many simple members of this class of compounds have the formula $(CH_2O)_n$ and hence appear to be "hydrates" of carbon. Carbohydrates are more appropriately defined as polyhydroxy aldehydes and ketones and some related derivatives and polymers. Their chemistry is typical of hydroxyl, aldehyde, and ketone functions. Carbohydrates are among the most abundant of organic compounds in living systems, particularly in plants where they sometimes constitute as much as 75% of the dry weight.

Some carbohydrates are small molecules, usually having three to six carbons, although occasionally a few more. These simple carbohydrates, of which glucose is the most common, have important functions in all living cells as sources of energy, as components of more complex carbohydrates or carbohydrate-containing molecules, and as intermediates for the synthesis of essential noncarbohydrate molecules.

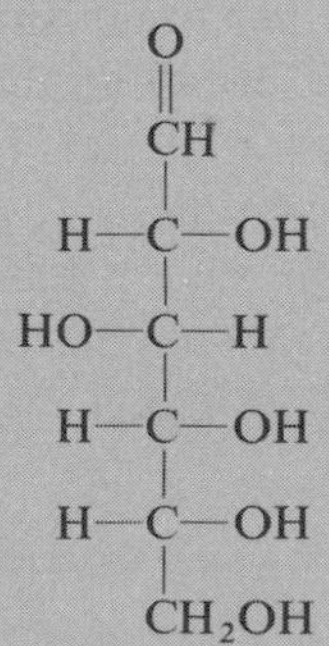

$$
\begin{array}{c}
\overset{\displaystyle O}{\underset{\displaystyle \parallel}{}} \\
CH \\
H-C-OH \\
HO-C-H \\
H-C-OH \\
H-C-OH \\
CH_2OH
\end{array}
$$

Glucose

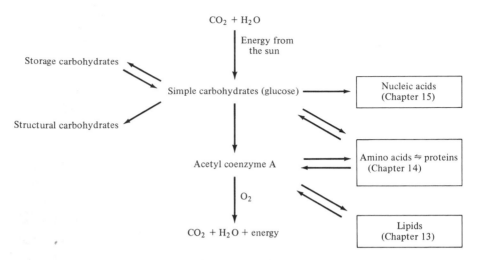

Figure 10.1 Some relationships between carbohydrates, energy, and other important groups of biological molecules.

Some carbohydrates are large molecules—molecular weights can be as high as several million. These large carbohydrates, however, are polymers constructed of simple carbohydrates to which they can be converted by acid-catalyzed or enzyme-catalyzed hydrolysis. The high molecular weight polymers include important energy-storage molecules of plants (starch) and animals (glycogen) and important structural materials of cells (e.g., much of the material in rigid cell walls).

Figure 10.1 shows the biological importance of carbohydrates. As noted (Sec. 2.16), the synthesis of carbohydrates from carbon dioxide and water is the ultimate source of organic compounds for all organisms on earth. This process, in which plants, blue green algae, or some bacteria obtain the necessary energy from sunlight,* is called **photosynthesis**. This complex process results ultimately in the production of simple carbohydrates. Glucose, the most common simple carbohydrate, is converted in plants to the polymer starch, used for energy storage and transfer, and to structural carbohydrates, particularly cellulose.

Humans and many other organisms cannot perform a comparable reduction of carbon dioxide and must obtain in their diets the organic compounds needed as sources of energy and as raw materials for constructing other compounds. We eat organisms capable of photosynthesis or animals that have fed on them. Glycogen, a polymer of glucose, is the major energy source stored in our bodies. In both plants and animals simple carbohydrates are converted by a series of reactions to the key biological intermediate acetyl coenzyme A (a derivative of acetic acid). This

---

* A few groups of bacteria obtain the energy from oxidation-reduction reactions of inorganic materials, such as sulfur or ammonia.

process evolves some energy; considerably more is released in a further chain of reactions that oxidizes acetyl coenzyme A to carbon dioxide and water.

10.1
MONOSACCHARIDES

Because low molecular weight carbohydrates generally have sweet tastes, carbohydrates are often called **saccharides** (from the Latin *saccharum*, sugar). Simple carbohydrates that cannot by hydrolyzed to smaller carbohydrates are called **monosaccharides**. Those carbohydrates that contain several monosaccharide units linked together are called **oligosaccharides** (from the Greek *oligos*, a few). Carbohydrates containing many monosaccharide units (hundreds, thousands, or even tens of thousands) are called **polysaccharides**. Monosaccharides and small oligosaccharides are also frequently called "sugars." The word sugar is also used in everyday speech for the specific disaccharide sucrose (table sugar).

The most common monosaccharides have six carbons and are called **hexoses**. Other monosaccharides include **trioses** (three carbons), **tetroses** (four carbons), **pentoses** (five carbons), and **heptoses** (seven carbons). Monosaccharides that contain an aldehyde function are sometimes called **aldoses** and those with a ketone function **ketoses**.

D-Glucose, an **aldohexose**, is the most important of the monosaccharides.

$$
\begin{array}{cc}
1 & \overset{\displaystyle O}{\overset{\displaystyle \|}{CH}} \\
2 & H-C-OH \\
3 & HO-C-H \\
4 & H-C-OH \\
5 & H-C-OH \\
6 & CH_2OH
\end{array}
$$

Projection formula of D-glucose

Recall that when we indicate that a drawing is a projection, we assign specific directional properties to its bonds (Sec. 6.2). If its important polymers—starch, glycogen, and cellulose—are included, glucose is the most abundant organic compound in the biological world. An optically active molecule, it is dextrorotatory and, for this reason, is sometimes called dextrose. It is also known as blood sugar or grape sugar because it is found in those sources. Glucose constitutes about 0.1% of the blood of normal humans and is a major source of energy for cells.

exercise with models 10.1 Construct a model of D-glucose. Note how the carbons in the chain must be stretched and the substituents at each carbon flattened to arrive at the drawing shown.

Glucose has four asymmetric carbons and is one of $2^4 = 16$ aldohexoses. All the aldohexoses are shown in Table 10.1. The absolute configuration of any aldohexose can be specified by using R or S to describe the configuration at each of its four asymmetric carbons. In carbohydrate chemistry, however, it is customary to define configuration by using an earlier system. This system is based on the enantiomeric glyceraldehydes, the simplest aldoses.

$$
\begin{array}{cc}
\overset{\displaystyle O}{\underset{\displaystyle |}{\overset{\displaystyle \|}{C}}} H & \overset{\displaystyle O}{\underset{\displaystyle |}{\overset{\displaystyle \|}{C}}} H \\
\end{array}
$$

O   O
‖   ‖
CH   CH
| |
H—C—OH   HO—C—H
| |
CH_2OH CH_2OH

D-Glyceraldehyde   L-Glyceraldehyde
(R)-Glyceraldehyde (S)-Glyceraldehyde

One of the glyceraldehydes is named D and the other L. The naturally occurring isomer of glucose is designated D because C-5, its highest-numbered asymmetric carbon (the asymmetric carbon farthest from the carbonyl function), has a configuration resembling that of D-glyceraldehyde. All other aldohexoses having this configuration at C-5 are also labeled D. The enantiomer of D-glucose has the opposite configuration at each of the four chiral carbons. It is called L-glucose, for its configuration at C-5 resembles that of L-glyceraldehyde. Note that D and L specify configuration at only one carbon. Aldohexoses that are not enantiomers of one another have different names. The compound that differs from glucose in configuration at C-3, for instance, is named allose; allose is a diastereomer of glucose.

In addition to D-glucose, two other aldohexoses occur commonly in foods. D-Mannose is a constituent of the polysaccharides of many vegetables, and D-galactose is part of the disaccharide called lactose or milk sugar. Mannose and galactose are converted in the body to glucose, the carbohydrate actually metabolized to provide energy. Infants suffering from galactosemia lack one of the enzymes necessary for the isomerization of galactose to glucose. Accumulation of galactose in their bodies leads to mental disorders and eye damage. This condition can be treated by eliminating milk and other sources of galactose from their diets.

D-Ribose and 2-deoxy-D-ribose (deoxy means lacking an oxygen) are important aldopentoses.

Table 10.1 Projection formulas of the 16 isomeric aldohexoses

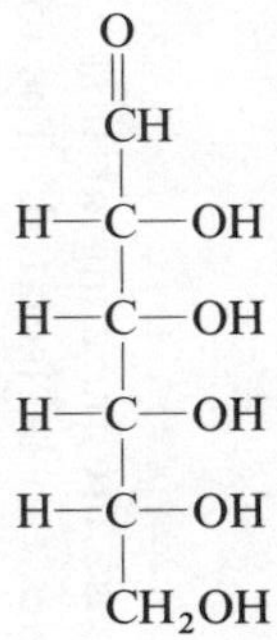

D-allose
```
    CHO
H—C—OH
H—C—OH
H—C—OH
H—C—OH
    CH₂OH
```

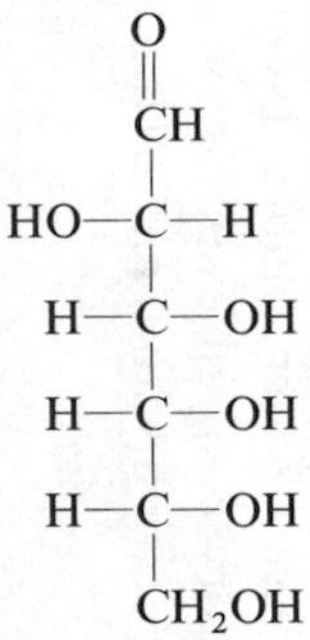

D-altrose
```
    CHO
HO—C—H
H—C—OH
H—C—OH
H—C—OH
    CH₂OH
```

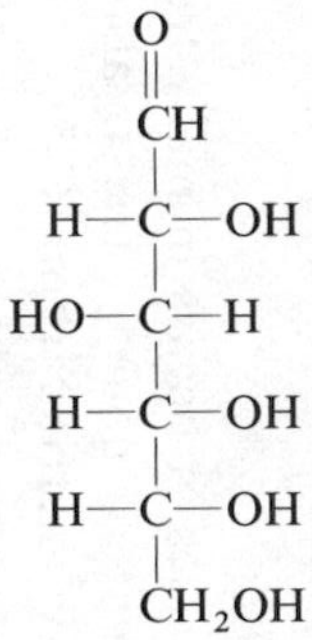

D-glucose
```
    CHO
H—C—OH
HO—C—H
H—C—OH
H—C—OH
    CH₂OH
```

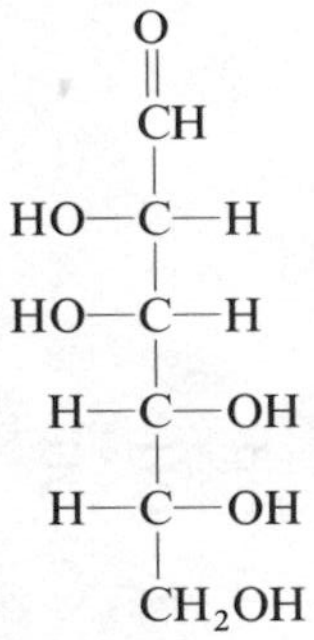

D-mannose
```
    CHO
HO—C—H
HO—C—H
H—C—OH
H—C—OH
    CH₂OH
```

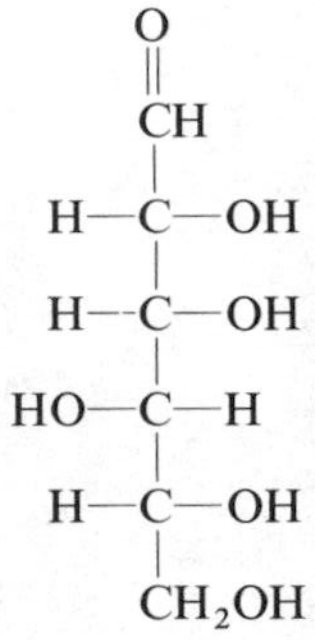

D-gulose
```
    CHO
H—C—OH
H—C—OH
HO—C—H
H—C—OH
    CH₂OH
```

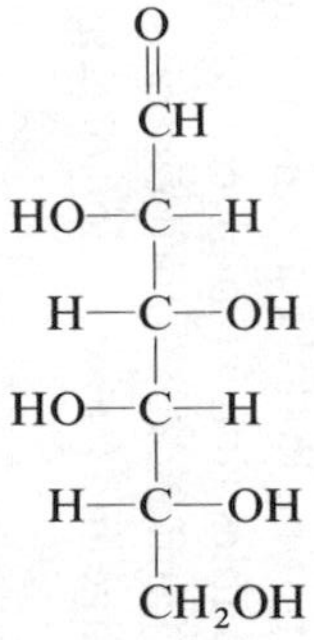

D-idose
```
    CHO
HO—C—H
H—C—OH
HO—C—H
H—C—OH
    CH₂OH
```

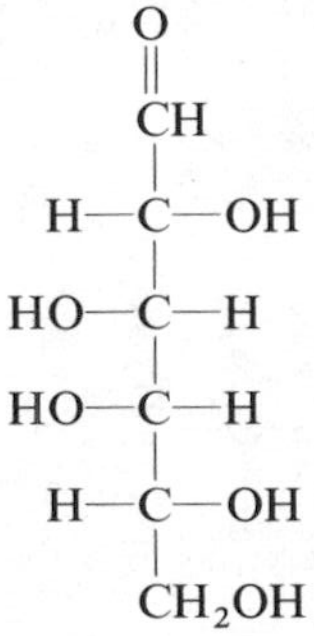

D-galactose
```
    CHO
H—C—OH
HO—C—H
HO—C—H
H—C—OH
    CH₂OH
```

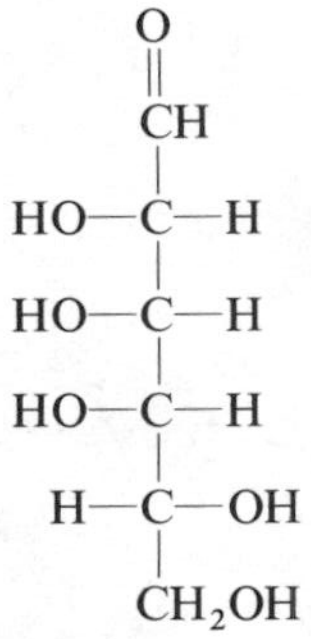

D-talose
```
    CHO
HO—C—H
HO—C—H
HO—C—H
H—C—OH
    CH₂OH
```

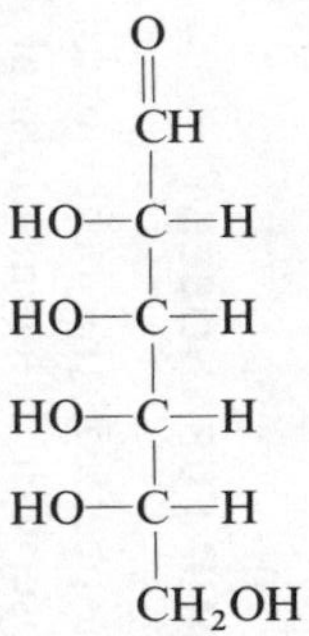

L-allose
```
    CHO
HO—C—H
HO—C—H
HO—C—H
HO—C—H
    CH₂OH
```

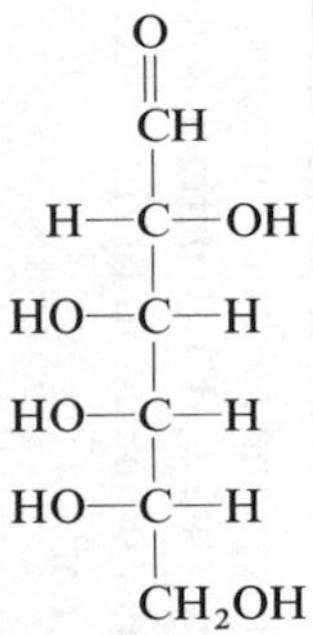

L-altrose
```
    CHO
H—C—OH
HO—C—H
HO—C—H
HO—C—H
    CH₂OH
```

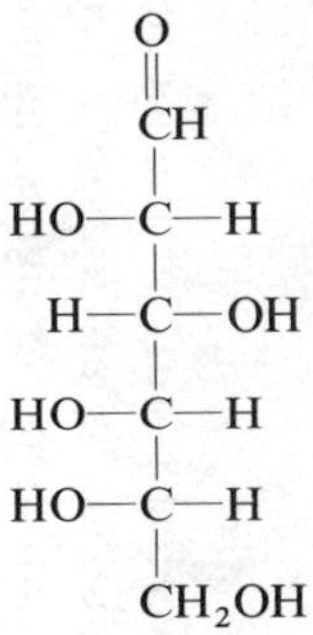

L-glucose
```
    CHO
HO—C—H
H—C—OH
HO—C—H
HO—C—H
    CH₂OH
```

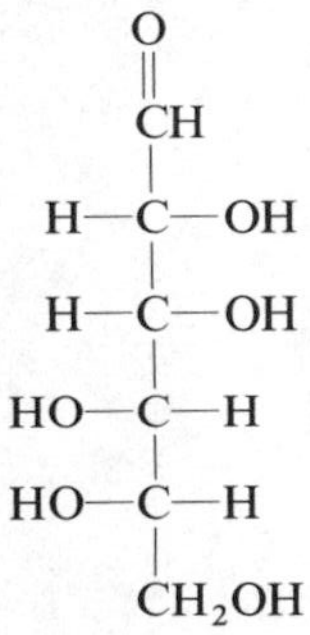

L-mannose
```
    CHO
H—C—OH
H—C—OH
HO—C—H
HO—C—H
    CH₂OH
```

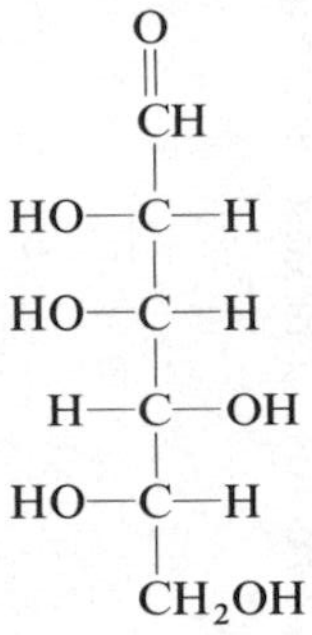

L-gulose
```
    CHO
HO—C—H
HO—C—H
H—C—OH
HO—C—H
    CH₂OH
```

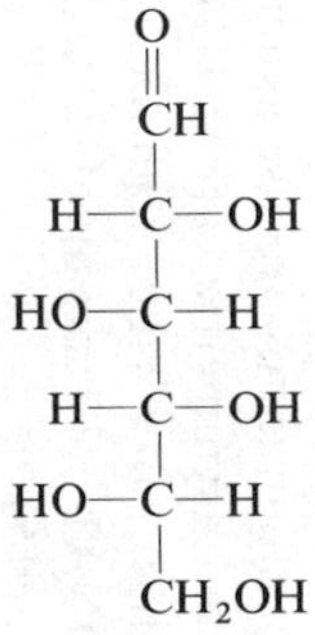

L-idose
```
    CHO
H—C—OH
HO—C—H
H—C—OH
HO—C—H
    CH₂OH
```

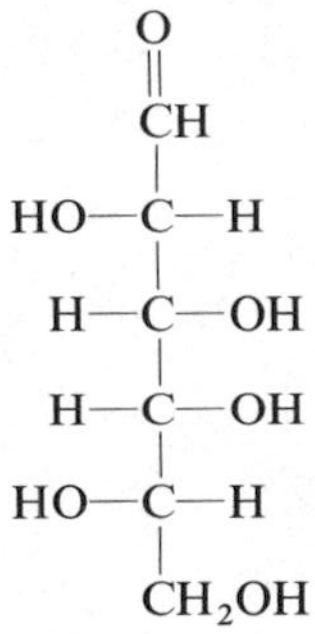

L-galactose
```
    CHO
HO—C—H
H—C—OH
H—C—OH
HO—C—H
    CH₂OH
```

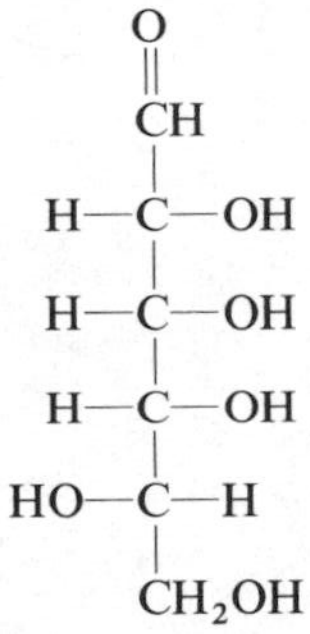

L-talose
```
    CHO
H—C—OH
H—C—OH
H—C—OH
HO—C—H
    CH₂OH
```

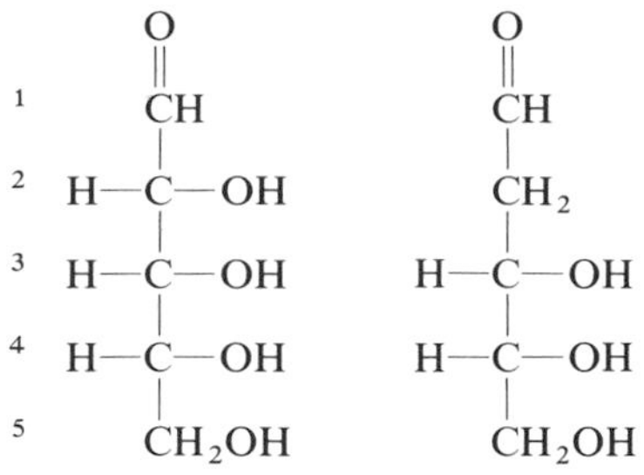

$$
\begin{array}{cc}
& \underset{\|}{O} \qquad\qquad \underset{\|}{O} \\
1 \quad & CH \qquad\qquad CH \\
2 \quad & H-C-OH \qquad CH_2 \\
3 \quad & H-C-OH \qquad H-C-OH \\
4 \quad & H-C-OH \qquad H-C-OH \\
5 \quad & CH_2OH \qquad CH_2OH
\end{array}
$$

D-Ribose   2-Deoxy-D-ribose

Their configurations are also assigned by noting that the arrangement at the highest-numbered asymmetric carbon (C-4) resembles that of D-glyceraldehyde. D-Ribose is a component of ribonucleic acid (RNA) and 2-deoxy-D-ribose of deoxyribonucleic acid (DNA), polymers involved in the storage and transmission of genetic information (Chapter 15).

D-Fructose, a hexose, is the most abundant of the ketoses.

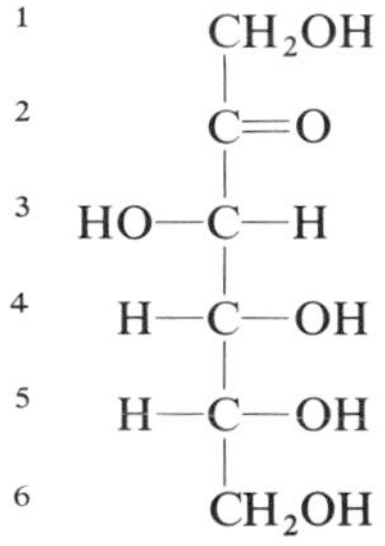

$$
\begin{array}{cl}
1 \quad & CH_2OH \\
2 \quad & C{=}O \\
3 \quad & HO-C-H \\
4 \quad & H-C-OH \\
5 \quad & H-C-OH \\
6 \quad & CH_2OH
\end{array}
$$

D-Fructose

The sweetest tasting of the monosaccharides, it occurs in many fruits and, along with glucose, is an important constituent of honey. The D also refers to the configuration at the highest-numbered asymmetric carbon (C-5).

Most monosaccharides are solids. As you would guess from their abundance of hydroxyl and carbonyl groups, they are very water soluble. D-Glucose is dextro-rotatory (+) and D-fructose is levorotatory (−), an example of the lack of any particular connection of the direction of rotation with D and L (just as there is none with R and S).

10.2
PYRANOSES AND FURANOSES

Some aspects of the behavior of D-glucose are puzzling at first. Two forms of D-glucose with different physical properties can be isolated. The form called α-D-glucopyranose has $[\alpha]_{589\,nm}^{20°} = +112.2°$ in water and the form called β-D-glucopyranose has $[\alpha]^{20°}_{589\,\text{nm}} = +18.7°$. When either is dissolved in water, the rotation slowly changes and finally reaches a value of $[\alpha]^{20°}_{589\,\text{nm}} = +52.5°$. This change, due to the formation of an equilibrium mixture of about one-third α-D-glucose and two-thirds β-D-glucose, is called **mutarotation**.

D-Glucose undergoes reactions typical of aldehyde and hydroxyl groups. It can be reduced, for example, with sodium borohydride (Sec. 9.9) to D-glucitol (also known as sorbitol).

D-Glucitol (sorbitol) D-Glucose D-Glucose oxime

D-Gluconic acid

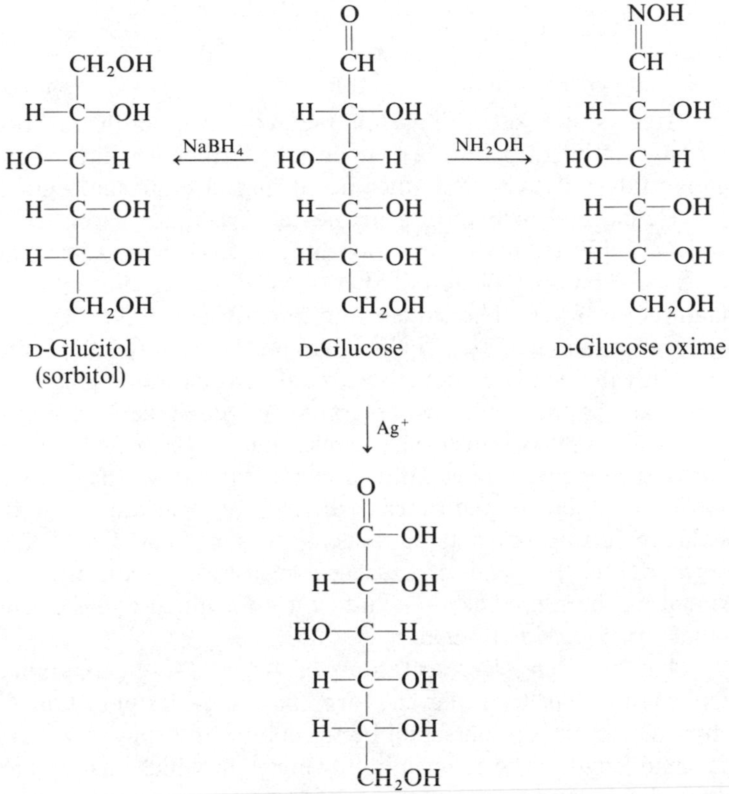

This compound, found naturally in many fruits and other natural sources, is almost as sweet as sugar. It is used as the sweetening agent in some candies made for diabetics and, because it produces very viscous solutions, in the formulation of food products, cosmetics, and pharmaceuticals. D-Glucose can also be converted to an oxime or oxidized by Tollens' reagent to a carboxylic acid. These and other reactions involving the aldehyde function, however, are unexpectedly slow.

The existence of two forms of D-glucose and its lethargic behavior in typical aldehyde reactions have a common origin. D-Glucose exists in large part as two cyclic hemiacetals.

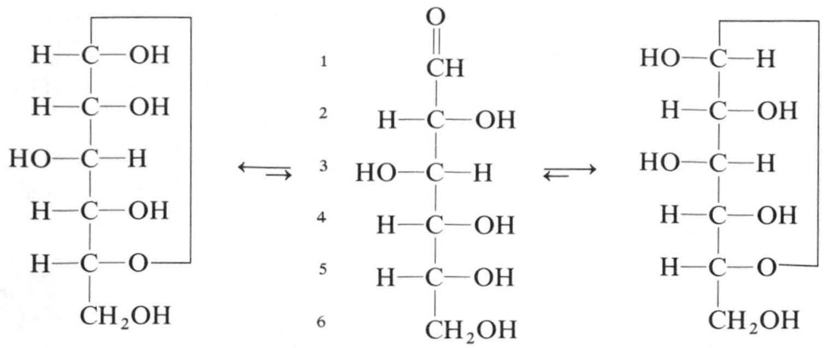

$$
\begin{array}{ccc}
\text{H—C—OH} & & \text{HO—C—H} \\
\text{H—C—OH} & \quad 1 \quad \overset{\text{O}}{\underset{\text{CH}}{\parallel}} & \text{H—C—OH} \\
\text{HO—C—H} & 2 \quad \text{H—C—OH} & \text{HO—C—H} \\
\text{H—C—OH} & 3 \quad \text{HO—C—H} & \text{H—C—OH} \\
\text{H—C—O} & 4 \quad \text{H—C—OH} & \text{H—C—O} \\
\text{CH}_2\text{OH} & 5 \quad \text{H—C—OH} & \text{CH}_2\text{OH} \\
& 6 \quad \text{CH}_2\text{OH} &
\end{array}
$$

α-D-Glucopyranose* β-D-Glucopyranose

This fact is not surprising, for D-glucose has both functions—hydroxyl and carbonyl—needed to form a hemiacetal. Earlier we learned (Sec. 9.6) that cyclic hemiacetals with five- or six-membered rings are unusually stable.

Hemiacetal formation generates a new chiral carbon (C-1) and hence two isomers. Differing in configuration only at C-1, these isomers are not enantiomers but diastereomers. Therefore they have different properties, including different magnitudes of $[\alpha]$. The equilibrium mixture (that rotates at $+52°$) contains 37% α-D-glucopyranose, 63% β-D-glucopyranose, and only 0.02% free aldehyde.

Only the small amount of the free aldehyde form of glucose undergoes carbonyl reactions. As that small concentration of aldehyde is depleted in the course of some reaction, it is replenished by cleavage of the α and β forms. Reactions that involve the aldehyde, however, can be no faster than this relatively slow hemiacetal cleavage. D-Glucose and other carbohydrates that are oxidized (Sec. 9.4) by such weakly oxidizing reagents as Tollens' reagent (Ag^+) or Fehling's or Benedict's reagents (Cu^{2+}) are called **reducing** (because they reduce the reagents). Fehling's reagent has been used as a qualitative test for glucose in the urine, an indication of diabetes or kidney malfunction.

The preceding representations of α- and β-D-glucopyranose, drawn to show their relationships to the acyclic form, have a grotesquely long C—O bond. In the actual molecules, of course, all bonds connecting ring atoms are of approximately the same length. The following drawings,[†] in which shaded lines indicate the side of the ring toward the viewer, are more realistic.

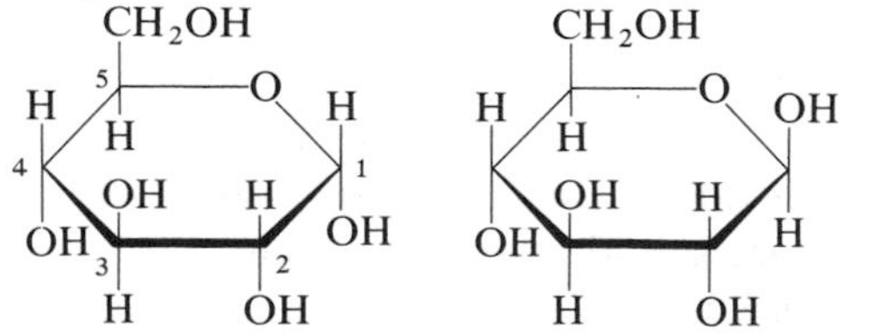

α-D-Glucopyranose β-D-Glucopyranose

* The isomer that in its projection formula has the same arrangement at the new asymmetric carbon (former carbonyl carbon) as at the highest-numbered asymmetric carbon is designated α.

[†] You will often see carbohydrates represented by such drawings, known as Haworth representations. In these representations it is the convention to place the ring oxygen in back and C-1 to its right.

Such six-membered ring structures resemble pyran and are called **pyranoses**.

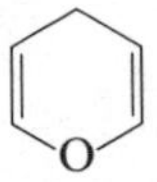

Pyran

The six-membered rings, of course, are not planar and are more accurately represented as chair conformations.

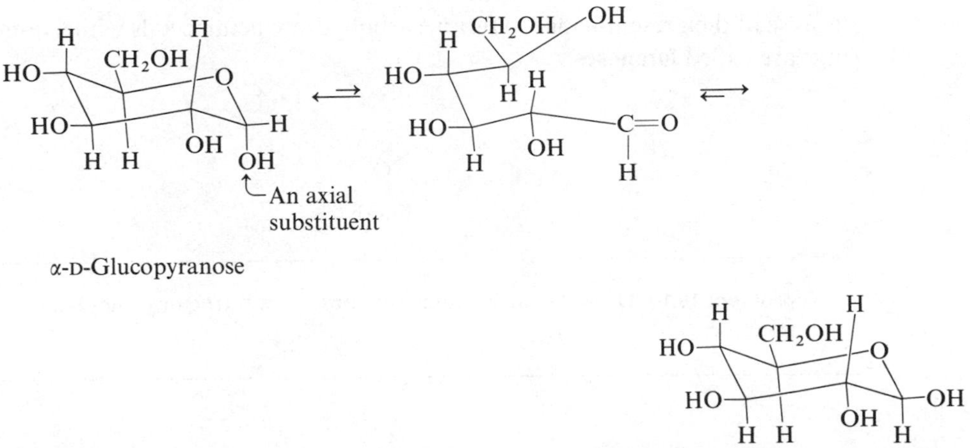

α-D-Glucopyranose

An axial substituent

β-D-Glucopyranose

Each chair conformation shown above is in equilibrium with another in which the equatorial and axial groups at each carbon have been interchanged. The particular conformation shown for the β isomer has no axial substituent, however, and that for the α isomer has only one. Therefore most α and β molecules have the particular conformations illustrated.

---

exercise with models 10.2   Construct a model of acylic D-glucose and arrange it in a conformation that brings the hydroxyl group at C-5 near the aldehyde group. Note that whether hemiacetal formation will produce α-D-glucopyranose or β-D-glucopyranose depends on the side of the carbonyl function that faces the C-5 hydroxyl group.

Other aldoses and ketoses that can form cyclic hemiacetals containing five- or six-membered rings also exist in part in such cyclic forms. An aqueous solution of D-fructose, for example, consists mainly of a pair of hemiacetals with six-membered rings and another pair with five-membered rings. One member of each pair is shown.

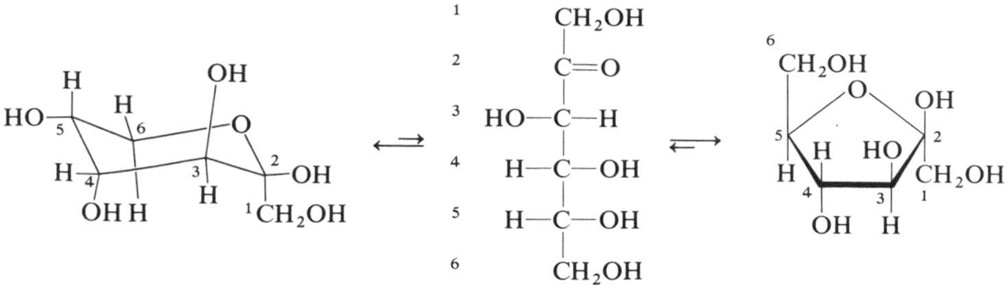

β-D-Fructopyranose   D-Fructose   β-D-Fructofuranose

Because of their resemblance to furan, carbohydrate hemiacetals with five-membered rings are called **furanoses**.

Furan

---

problem 10.1   Draw suitable representations of α-D-fructopyranose and α-D-fructofuranose.

10.3
GLYCOSIDES

D-Glucose reacts with methanol in the presence of an acidic catalyst to form two stereoisomeric acetals.

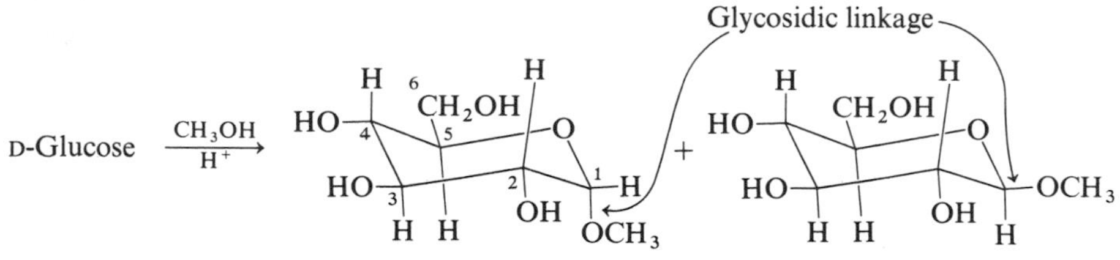

Methyl α-D-glucopyranoside   Methyl β-D-glucopyranoside

Note that one oxygen of the acetal function comes internally from the C-5 hydroxyl group. Formation of a cyclic acetal is to be expected, for we know that glucose exists largely as a cyclic hemiacetal that incorporates the C-5 hydroxyl group (see Sec. 9.6 for a simpler example of cyclic acetal formation). Except in the presence of acid, an acetal function is stable and not in equilibrium with the free aldehyde. Therefore glucopyranosides do not mutarotate or undergo aldehyde reactions. Not being oxidized by Tollens', Benedict's, or Fehling's reagents, glucopyranosides are nonreducing.

Other monosaccharides that exist largely as hemiacetals also react with alcohols in the presence of an acidic catalyst to form cyclic acetals. As a group they are called **glycosides** and the bond between the carbohydrate and the alcohol is called a **glycosidic linkage**. Specific compounds are named for the monosaccharide from which they are derived: glucosides from glucose, fructosides from fructose, galactosides from galactose, and so on. Many naturally occurring glycosides are known. In the vanilla bean, vanillin exists both in the free form and as a β-D-glucoside, which can be hydrolyzed to D-glucose and vanillin.

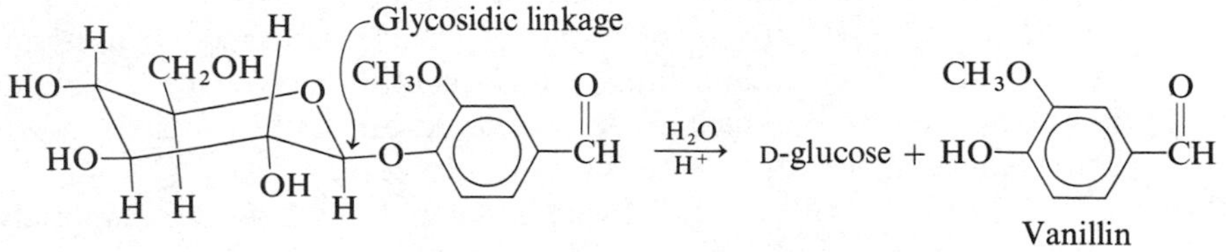

As shown in the following sections, it is glycosidic linkages that attach monosaccharides together to form oligosaccharides and polysaccharides.

10.4
DISACCHARIDES

Oligosaccharides that consist of just two monosaccharides linked together are called **disaccharides**. The most abundant disaccharide is sucrose, familiar to us as table sugar.

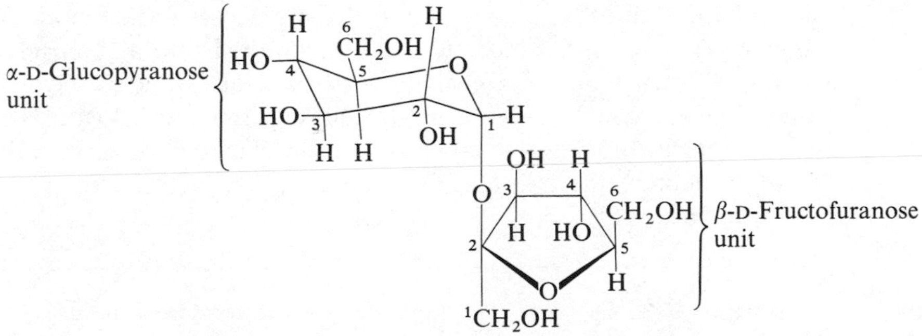

Sucrose
β-D-Fructofuranosyl-$(2\rightarrow1)$-α-D-glucopyranoside

Sucrose, found in every photosynthetic plant, is particularly abundant in sugar cane, sugar beets, sorghum, and sugar maple sap. It consists of an α-D-glucopyranose unit and a β-D-fructofuranose unit joined by a glycosidic link.

Acid-catalyzed hydrolysis cleaves sucrose to its monosaccharide components.

$$\text{Sucrose} \xrightarrow[\text{H}^+ \text{ or sucrase}]{\text{H}_2\text{O}} \text{D-glucose} + \text{D-fructose}$$

$$[\alpha]_{589\ nm}^{20°} = +66.5° \qquad\qquad [\alpha]_{589\ nm}^{20°} = +52.5° \qquad [\alpha]_{589\ nm}^{20°} = -92°$$

Alternatively, the hydrolysis can be catalyzed by the enzyme sucrase (invertase). The mixture of D-glucose and D-fructose that results is known as **invert sugar** because its rotation $[(52 - 92)/2 = -20°]$ is opposite in sign to that of sucrose. Somewhat sweeter than sucrose and with less tendency to crystallize, invert sugar is produced on a large scale for use in the manufacture of candy, ice cream, and other food products. In its fermentation to ethanol by yeast or in its digestion in our bodies, sucrose is first hydrolyzed to fructose and glucose.

The carbonyl carbons of both the glucose and fructose units in sucrose are involved in the glycosidic link. Consequently, sucrose does not give carbonyl reactions and is nonreducing. As we would expect from its structure, and as we know from experience, sucrose is very water soluble.

Lactose or "milk sugar," found in the milk of almost all mammals, constitutes about 5% of cows' milk and 7% of human milk.

β-D-Galactopyranose unit   D-Glucopyranose unit

The β configuration is shown, but some molecules have the α configuration.

Lactose
β-D-Galactopyranosyl-(1→4)-D-glucopyranose

It is synthesized in the mammary glands from glucose in the blood. The aldehyde carbon (C-1) of the D-galactose unit is incorporated into a glycosidic link (acetal function) and hence does not give aldehyde reactions. The aldehyde carbon (C-1) of the D-glucose unit is involved in a hemiacetal function. Just as with D-glucose itself, this ring is found as a mixture of α- and β-glucopyranoses in equilibrium with a small amount of the free aldehyde. As a result, lactose exhibits mutarotation and is reducing.

---

problem 10.2 Draw the acyclic and the α-D-glucopyranose forms of lactose.

In the presence of an acid catalyst, lactose can be hydrolyzed to its monosaccharide constituents. In the body this reaction is catalyzed by the enzyme lactase. Some adults lack this enzyme and must avoid milk in their diets.

Maltose or "malt sugar" is obtained by hydrolysis of starch catalyzed by enzymes present in malt.

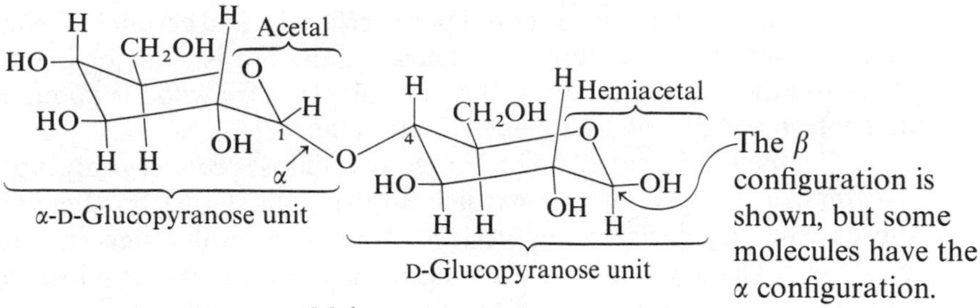

Maltose
α-D-Glucopyranosyl-(1→4)-D-glucopyranose

In the presence of an acid catalyst, maltose is hydrolyzed to D-glucose as the only product. The hydrolysis can also be catalyzed by maltase, an enzyme that catalyzes hydrolysis of α-D-glucopyranosides. The aldehyde carbon of one of the D-glucose units is usually incorporated into a hemiacetal function. This ring is found as α and β isomers in equilibrium with a small amount of the free aldehyde. Therefore maltose is a reducing sugar.

Cellobiose, one of the fragments obtained by partial acid-catalyzed hydrolysis of the polysaccharide cellulose, is identical to maltose in all but one detail. Both contain two D-glucose units linked from C-1 of one unit to C-4 of the other. Both are reducing sugars and can be hydrolyzed in the presence of an acid catalyst to form only D-glucose. Hydrolysis of cellobiose is catalyzed not by maltase, however, but by emulsin, an enzyme (isolated from almonds) known to be specific for hydrolysis of β-D-glucopyranosides. Cellobiose differs from maltose only in having a β-glycosidic linkage.

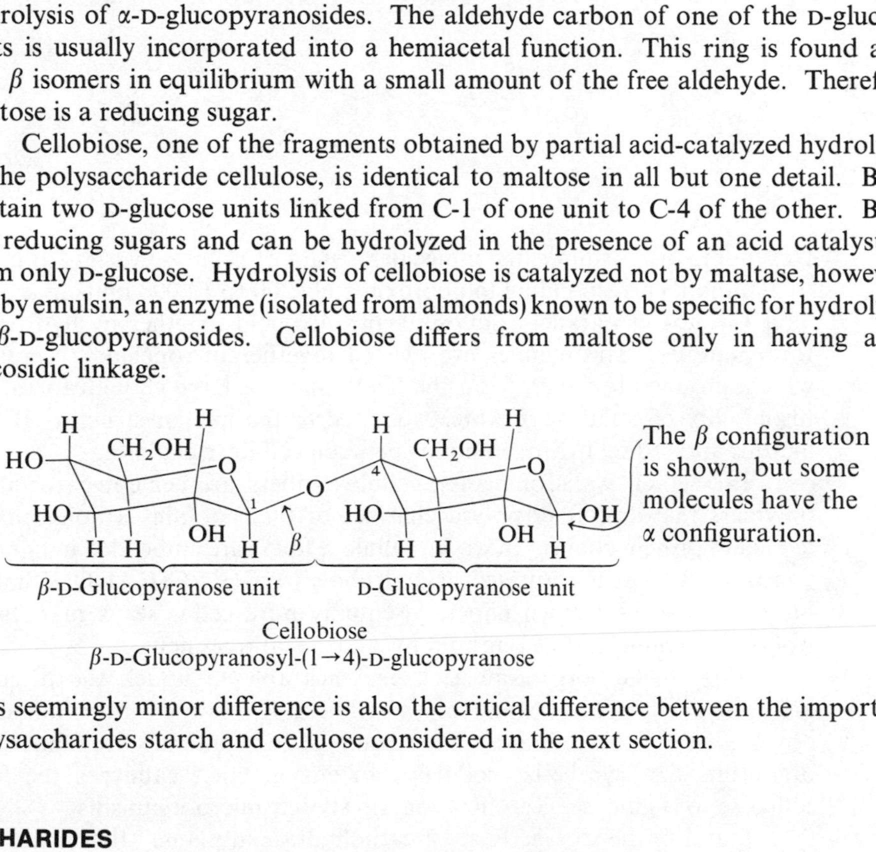

Cellobiose
β-D-Glucopyranosyl-(1→4)-D-glucopyranose

This seemingly minor difference is also the critical difference between the important polysaccharides starch and celluose considered in the next section.

10.5
POLYSACCHARIDES

The carbohydrate materials found in living organisms consist mainly of high molecular weight polysaccharides. These polysaccharides serve either as structural elements or as storage reservoirs from which carbohydrates are released as needed to provide energy.

The most abundant structural polysaccharide, and certainly the most abundant organic substance in nature, is cellulose. Cellulose is a major component of cell walls and the chief structural material of plants. Dry wood is about 50% cellulose, and cotton and flax fibers are nearly pure cellulose.

Complete acid-catalyzed hydrolysis of cellulose produces only D-glucose. When hydrolysis is carried out only partially so that some larger fragments remain, cellobiose is the only disaccharide produced. As these results suggest, cellulose is composed of D-glucopyranose units connected together, as in cellobiose, by a β linkage from C-1 of each unit to C-4 of the next.

Cellulose

Depending on its source, the molecular weight of cellulose varies from about 50,000 to 2 million, corresponding to approximately 300 to 12,000 units in a chain.

Groups of parallel cellulose chains are held together by hydrogen bonds to form bundles. The bundles are twisted together in ropelike structures, many of which, grouped together, form the fibers that are large enough for us to see. The insolubility of cellulose in water is related to the regular structure that favors numerous and strong hydrogen bonds between cellulose chains.

In the cell walls of plants, cellulose fibers are cemented together by other polymers that are either polysaccharides or that contain carbohydrate groups attached to protein chains. In wood, cellulose fibers are embedded in lignin, a polymer constructed from unsaturated aryl alcohols ($ArCH=CHCH_2OH$) linked together in complex ways. When paper, essentially pure cellulose, is manufactured from wood, the lignin must be removed by a chemical treatment.

Most higher organisms lack enzymes able to attack the β-glucopyranosyl linkages of cellulose and cannot digest cellulose. Cows and other ruminants, however, obviously thrive on cellulose. The rumens of these animals are inhabited by microorganisms that synthesize cellulase, an enzyme that catalyzes the hydrolysis of cellulose to D-glucose. Termites contain similar microorganisms.

Starch is the storage form of carbohydrate in plants. It constitutes the major part of cereal grains and plant tubers and is the largest food source for humans. Starch can be divided into two components: 10 to 25% is a water-soluble* portion

---

* Starch is not soluble in a strict sense, but it does go into water in the form of very small (colloidal) particles that are extensively hydrogen bonded to water.

called amylose, and the remainder a less water-soluble portion called amylopectin.

α-Amylase, an enzyme present in saliva and pancreatic juice, catalyzes the hydrolysis of amylose to D-glucose and to the disaccharide maltose. As the results of this hydrolysis suggest, amylose is a linear polymer containing D-glucopyranose units linked, as in maltose, by an α linkage from C-1 of each unit to C-4 of the next.

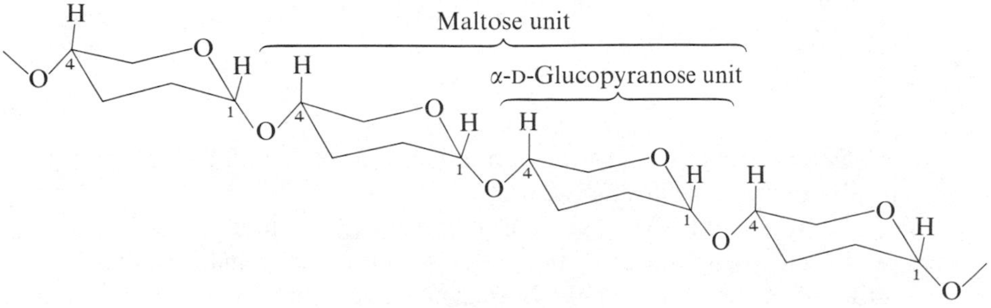

Amylose

Amylose has a molecular weight range of 10,000 to 100,000, corresponding to about 60 to 600 glucose units. Amylose and cellulose differ only in the stereochemistry of the glycosidic linkage. Yet this seemingly small difference is responsible for their different solubilities in water and for the stability of cellulose toward enzymes that catalyze hydrolysis of amylose.

Amylopectin is also constructed completely of D-glucopyranose units, joined mostly by α linkages from C-1 to C-4 as in amylose. About every 20 to 30 units, however, a unit is linked to the end of another chain by a bond from C-6 to C-1.

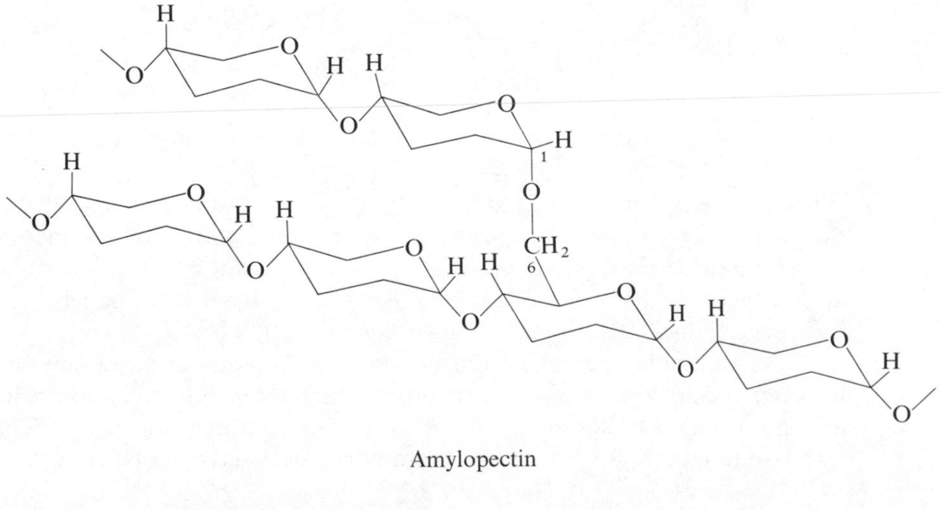

Amylopectin

The branched structure of amylopectin can be represented diagrammatically.

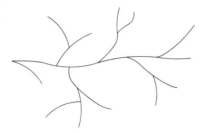

It has a molecular weight range of a hundred thousand to 1 million, corresponding to about 600 to 6000 glucose units.

Glycogen, the storage form of carbohydrate in animals, is found particularly in muscle tissue and the liver. Its structure is similar to that of amylopectin, although even more highly branched. Glycogen molecules can have molecular weights of many million. Glucose obtained by an animal in its diet is converted in the liver to glycogen. Controlled hydrolysis of glycogen maintains the normal glucose concentration of the blood.

10.6
COMMERCIAL CELLULOSE AND STARCH DERIVATIVES

Cellulose is converted to a number of commercial polymers. Reaction with nitric acid produces nitrate esters.

Cellulose	$\xrightarrow{HNO_3}$	Cellulose nitrate

When an average of about 2.5 of the 3 hydroxyl groups per glucose unit are nitrated, the product has explosive properties and is known as guncotton. Guncotton is used in the manufacture of smokeless gunpowder. If about 2.0 to 2.5 hydroxyl groups are nitrated, the material is known as pyroxylin. Mixed with the terpene camphor, it forms celluloid, the first widely used commercial polymer.

Natural cellulose fibers from such sources as wood or straw can be converted to better-quality fibers known as rayon, in which the cellulose chains are less tangled and more aligned. Rayon manufacture requires converting the cellulose into a water-soluble derivative. The most common process involves treatment with a basic solution of carbon disulfide (CS_2) to convert some of the hydroxyl groups to xanthate groups.

$$R{-}OH \xrightarrow[NaOH]{CS_2} R{-}O{-}\overset{\overset{\displaystyle S}{\|}}{C}{-}S^-\,Na^+ \xrightarrow[H^+]{H_2O} R{-}OH + CS_2$$

Cellulose   Cellulose xanthate   Rayon
(regenerated cellulose)

The resulting solution of cellulose xanthate in water is extruded through small holes into a bath of aqueous acid to regenerate the cellulose as fine threads. Extruding the solution through a narrow slot gives sheets known as cellophane.

Reaction of cellulose with acetic anhydride (a reaction of alcohols discussed in Sec. 12.3) to convert an average of 2.0 to 2.8 hydroxyl groups per glucose unit into acetate functions gives the material known as cellulose acetate.

Cellulose   Cellulose acetate

Cellulose acetate is used for motion picture film and for manufacturing a thin film used for wrapping food products. Acetate rayon is prepared by passing a solution of cellulose acetate in acetone and ethanol through a perforated plate into a hot air stream that evaporates the volatile solvent molecules, leaving long fibers.

Partial hydrolysis of starch forms shorter chains called dextrins. Because they form sticky solutions when water is added, dextrins are used in the manufacture of adhesives (e.g., the glue on postage stamps). More complete hydrolysis of corn starch produces corn syrup, a mixture of glucose, maltose, and oligosaccharides of glucose used in enormous quantities in the food-processing industry.

As the demand for fossil fuels increases and their supply diminishes, plants are receiving increasing attention as a renewable source of organic materials. Obviously combustion of wood or other plant matter can produce energy and cellulose can be used to make a variety of polymeric materials. Production of simple organic raw materials from plant sources is also being considered. Cellulose, for instance, can be hydrolyzed to glucose, which can be fermented (Sec. 5.14) to produce ethanol, an easily handled liquid fuel and important chemical intermediate.

10.7
GLYCOSYLAMINES

D-Glucose reacts with secondary amines and an appropriate catalyst to form derivatives similar to acetals (glycosides) but containing —NR$_2$ instead of —OR.

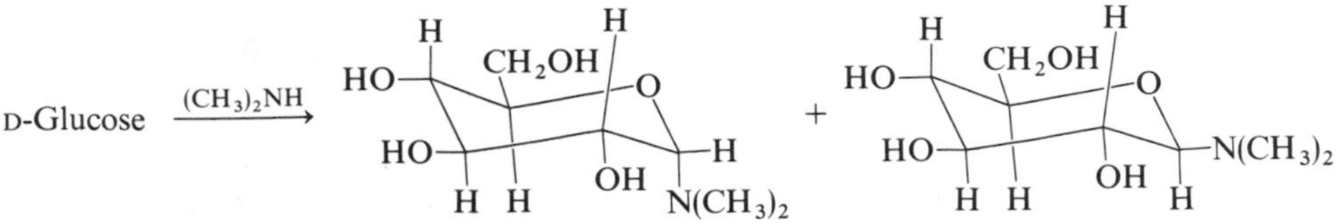

Similar derivatives, as a class called **glycosylamines** (or *N*-glycosides), are formed from other aldoses and ketoses. Glycosylamines formed from D-ribose or 2-deoxy-D-ribose and some heterocyclic amines are found in several important biological compounds. These include ATP (structure in Sec. 15.1), the polymers DNA and RNA (Chapter 15), and the biological reducing agent NADH (Sec. 9.9).

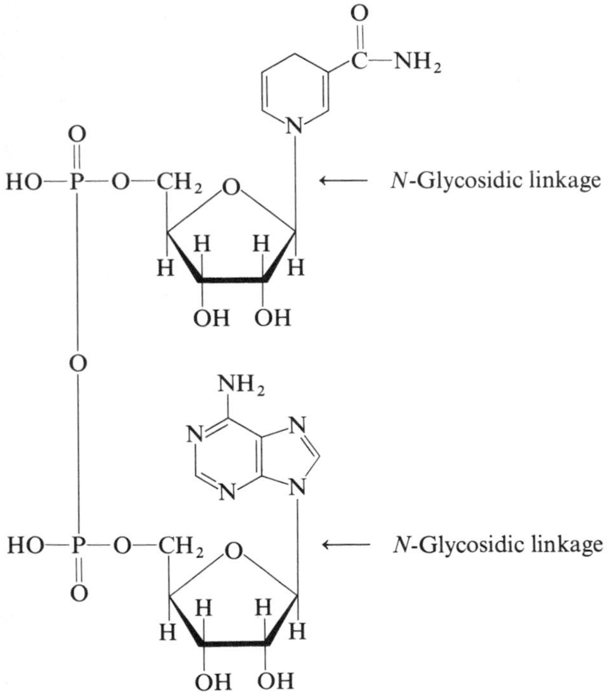

NADH
reduced form of nicotinamide adenine dinucleotide

SUMMARY

Definitions and Ideas

Carbohydrates or **saccharides**, an important family of biological compounds, are polyhydroxy aldehydes and ketones and some related derivatives and polymers.

Simple carbohydrates that cannot be hydrolyzed to smaller carbohydrates are called **monosaccharides**. Carbohydrates that contain several monosaccharide units linked together are called **oligosaccharides**, and carbohydrates containing many monosaccharide units are called **polysaccharides**.

Monosaccharides include **trioses** (three carbons), **tetroses** (four carbons), **pentoses** (five carbons), **hexoses** (six carbons), and **heptoses** (seven carbons). Monosaccharides containing an aldehyde function are called **aldoses** and those with a ketone function are called **ketoses**.

The enantiomeric glyceraldehydes (aldotrioses), assigned the names D and L, are used as standards for specifying configurations of other carbohydrates. A carbohydrate is named D or L, depending on whether the configuration of its highest-numbered asymmetric carbon resembles that of D- or L-glyceraldehyde.

The chemistry of carbohydrates is characteristic of hydroxyl, aldehyde, and ketone functions. Carbohydrates that can form internal hemiacetals with five-membered rings (**furanoses**) or six-membered rings (**pyranoses**) often exist largely in such forms. Hemiacetal formation creates an additional asymmetric carbon, giving rise to two diastereomers (α and β) that differ in configuration at the former carbonyl carbon. Interconversion of such isomeric hemiacetals is called **mutarotation**.

Carbohydrates that contain aldehyde functions (often in equilibrium with cyclic hemiacetals) are called **reducing**, for they react with mild oxidizing agents.

Carbohydrates that exist significantly as hemiacetals react with alcohols in the presence of an acid catalyst to form acetals known as **glycosides**. One oxygen of the acetal function of a glycoside comes from an internal hydroxyl group. The bond between the carbohydrate and the external alcohol molecule is called a **glycosidic linkage**. Two configurations (α and β) are possible for a glycosidic linkage.

In disaccharides and larger saccharides, monosaccharide units are joined by glycosidic linkages.

Glycosylamines, formed from carbohydrates and secondary amines, are similar to glycosides but contain $-NR_2$ instead of $-OR$.

PROBLEMS

1. Define each of the following terms. Where helpful, use structural drawings as illustrations.

   (a) aldose

   (b) ketose

   (c) hexose

   (d) pentose

   (e) D-hexose

   (f) L-tetrose

   (g) ketopentose

   (h) aldohexose

   (i) furanose

   (j) pyranose

   (k) glycosidic linkage

   (l) mutarotation

    (m) monosaccharide   (n) reducing sugar   (o) oligosaccharide

    (p) disaccharide   (q) *N*-glycosidic linkage   (r) polysaccharide

2. Write substitutive names for the open-chain forms of these compounds.

    (a) glucose   (b) fructose

3. Explain the meaning of each part of the name methyl α-D-glucopyranoside.

4. Draw suitable representations of all compounds that fit the following descriptions.

    (a) aldotetroses

    (b) D-aldopentoses

    (c) D-hexoses with the carbonyl group at C-2

5. Draw structures of the organic products of the following reactions.

    (a) D-glucose $\xrightarrow[\text{H}^+]{\text{CH}_3\text{OH}}$ (b) D-fructose $\xrightarrow{\text{NH}_2\text{OH}}$

    (c) D-fructose $\xrightarrow{\text{NaBH}_4}$ (d) D-glucose $\xrightarrow{\text{Cu}^{2+}}$

    (e) D-fructose $\xrightarrow[\text{H}^+]{\text{CH}_3\text{OH}}$ (f) D-galactose $\xrightarrow{\text{NH}_2\text{OH}}$

    (g) sucrose $\xrightarrow[\text{H}_2\text{O}]{\text{invertase}}$ (h) lactose $\xrightarrow{\text{NH}_2\text{OH}}$

6. Draw suitable representations of the acyclic form of cellobiose and of the form that has one α-D-glucopyranose unit.

7. The trisaccharide raffinose occurs abundantly in sugar beets.

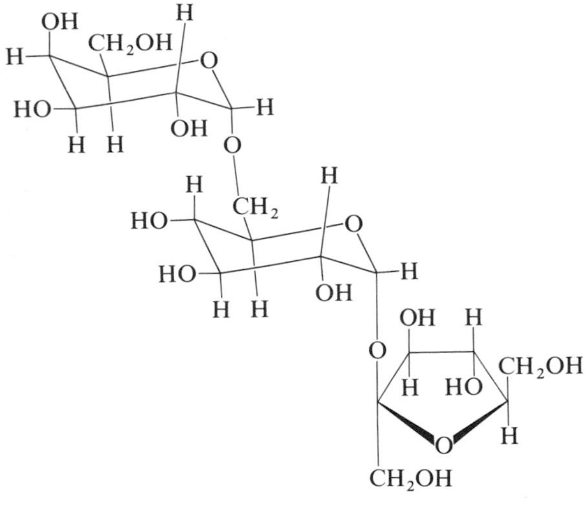

    (a) What monosaccharides are formed by its hydrolysis?

    (b) Is it reducing?

    (c) Does it exhibit mutarotation?

8. Identify any of the following structural elements that occur in raffinose (problem 7).

    (a) a hemiacetal function   (b) an acetal function

    (c) a furanose   (d) a pyranose

    (e) an α-glycosidic linkage   (f) a β-glycosidic linkage

9. D-Ribose exists in water as an equilibrium mixture of two furanoses and two pyranoses. Draw representations of these species.

10. Trehalose is a disaccharide found in some fungi and the cocoons of a parasitic beetle. It does not react with Cu^{2+} (Fehling's or Benedict's reagents) or form a derivative with hydroxylamine. D-Glucose is the only product formed from its hydrolysis. What are possible structures for trehalose?

11. Salicin, obtained by extracting poplar or willow bark with hot water, was once used as a pain-reliever. Salicin gives negative tests for a phenol function but can be hydrolyzed to give D-glucose and *o*-hydroxybenzyl alcohol. The hydrolysis is catalyzed by an enzyme known as β-glucosidase. Draw a suitable representation of salicin.

12. Draw a suitable representation of any aldotetrose that is reduced with sodium borohydride to form the meso isomer of 1,2,3,4-butanetetraol (HOCH$_2$—CHOH—CHOH—CH$_2$OH).

13. Why is β-D-glucopyranose more stable than α-D-glucopyranose?

14. Inositol, a sugar alcohol widely distributed in plants and animals, is one of nine stereoisomeric 1,2,3,4,5,6-hexahydroxycyclohexanes. Draw suitable representations of the nine stereoisomers.

Carboxylic Acids

As their name suggests, carboxylic acids are acidic compounds. They are charac-
terized by the presence of the **carboxyl group**, $-\overset{\overset{\displaystyle O}{\|}}{C}-OH$. Although the carboxyl
group contains both a carbonyl and a hydroxyl portion (carboxyl = *carbo*nyl +
hydro*xyl*), the chemistry of carboxylic acids differs significantly from that of aldehydes, ketones, or alcohols.

Carboxylic acids are the major organic acids. Many important biological
compounds contain carboxyl groups. Moreover, carboxylic acids are closely related
to other important families of compounds. The next chapter is devoted to these
families—esters, amides, acyl halides, and acid anhydrides—that have other nucleophilic groups in place of the —OH of the carboxyl group.

11.1
NOMENCLATURE

Because they have been known for a long time, many carboxylic acids have well-
entrenched trivial names that are accepted by the IUPAC. We are familiar with
some of the trivial names of carboxylic acids listed in Table 11.1, for names of
aldehydes are derived from them. A trivial name often reveals a source from which
a carboxylic acid can be obtained. Formic acid (from the Latin *formica*, ant), the
irritating agent in ant bites, was obtained by distillation of ants. Some ants contain

Table 11.1 Names of some carboxylic acids

Structural formula	Trivial name	Substitutive name
HCOOH	formic acid	methanoic acid
CH_3COOH	acetic acid	ethanoic acid
CH_3CH_2COOH	propionic acid	propanoic acid
$CH_3CH_2CH_2COOH$	butyric acid	butanoic acid
$CH_3CH_2CH_2CH_2COOH$	valeric acid	pentanoic acid
$CH_3CH_2CH_2CH_2CH_2COOH$	caproic acid	hexanoic acid
$CH_3(CH_2)_{10}COOH$	lauric acid	dodecanoic acid
$CH_3(CH_2)_{14}COOH$	palmitic acid	hexadecanoic acid
$CH_3(CH_2)_{16}COOH$	stearic acid	octadecanoic acid

surprising amounts of this acid, which blisters human skin and attacks most other tissues. Acetic acid (from the Latin *acetum*, vinegar) is the principal ingredient of vinegar. Many carboxylic acids with unbranched chains can be isolated from fats and, for this reason, are sometimes called fatty acids. Propionic acid (from the Greek *protos*, first and *pion*, fat) received its name because this simple acid was obtained from some fats. Butyric acid (from the Latin *butyrum*, butter) contributes to the odor of rancid butter and caproic acid (from the Latin *caper*, goat) to the odor of goats. Valeric acid is obtained from valerian roots, lauric acid from laurel, palmitic acid from palm oil, and stearic acid (from the Greek *stear*, tallow) from beef fat.

Substitutive names of carboxylic acids are based on the longest chain that contains the carboxyl group. The ending *-oic acid* replaces the *-e* of the name of the corresponding alkane. Because the carboxyl carbon must come at the end of a chain, it is assumed to be number 1.

$$CH_3CH_2CH_2CH_2CH_2-\overset{\overset{\textstyle O}{\|}}{C}-OH \qquad CH_3\underset{\underset{\textstyle Br}{|}}{CH}-\overset{\overset{\textstyle O}{\|}}{C}-OH$$

Hexanoic acid   2-Bromopropanoic acid
Caproic acid   2-Bromopropionic acid

$$CH_3\underset{\underset{\textstyle Cl}{|}}{CH}CH_2-COOH \qquad \underset{H}{\overset{CH_3}{>}}C=C\underset{CO_2H}{\overset{H}{<}}$$

3-Chlorobutanoic acid   *trans*-2-Butenoic acid
3-Chlorobutyric acid

Note the condensed ways (—COOH and —CO$_2$H) in which the structure of a carboxyl group is ordinarily written.

Particularly when the carboxyl group is attached to a ring, substitutive names are also derived by adding *-carboxylic acid* to the name of the compound that would result from replacing the carboxyl group by a hydrogen.

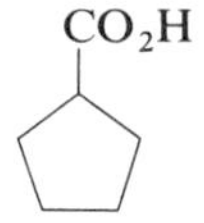
CO2H

Cyclopentanecarboxylic acid

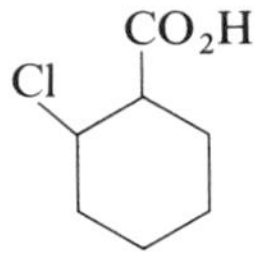
CO2H
Cl

2-Chlorocyclohexanecarboxylic acid

The substitutive name of the smallest dicarboxylic acid is ethanedioic acid. Usually, however, it is called oxalic acid, a trivial name. Trivial names are often used for other common dicarboxylic acids.

$HO_2C—CO_2H$	$HO_2C—CH_2—CO_2H$	$HO_2C—CH_2CH_2—CO_2H$
Oxalic acid Ethanedioic acid	Malonic acid Propanedioic acid	Succinic acid Butanedioic acid

$$HO_2C—CH_2CH_2CH_2CH_2—CO_2H$$

Adipic acid
Hexanedioic acid

Aromatic acids are often named as derivatives of benzoic acid, the simplest aromatic acid, although trivial names are frequently used for some commonly encountered aromatic acids.

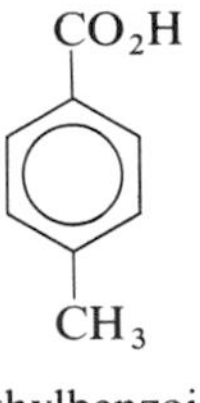
CO2H
CH3

p-Methylbenzoic acid
p-Toluic acid

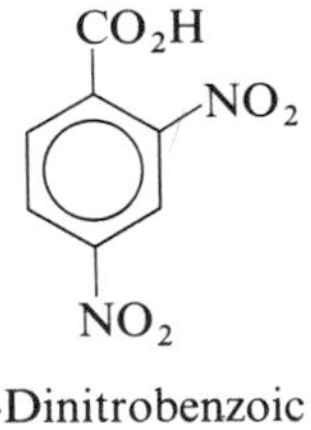
CO2H
NO2
NO2

2,4-Dinitrobenzoic acid

---

problem 11.1 Name the following compounds.

(a) 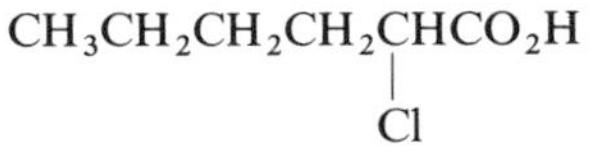
CH3CH2CH2CH2CHCO2H
Cl

(b) 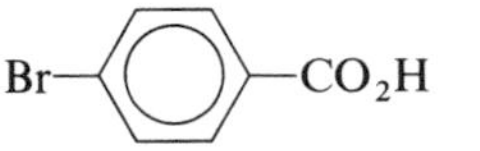
Br
CO2H

problem 11.2 Draw structures for the following compounds.

(a) dichloroacetic acid
(b) 4-*tert*-butylcyclohexanecarboxylic acid
(c) *o*-chlorobenzoic acid
(d) 2-bromo-4-pentenoic acid

11.2
PHYSICAL PROPERTIES

Formic and acetic acids have sharp acrid odors. Carboxylic acids with 3 to 10 carbons often are extremely disagreeable. The stench of the secretion of the "stinkpot turtle" is due principally to two malodorous carboxylic acids:

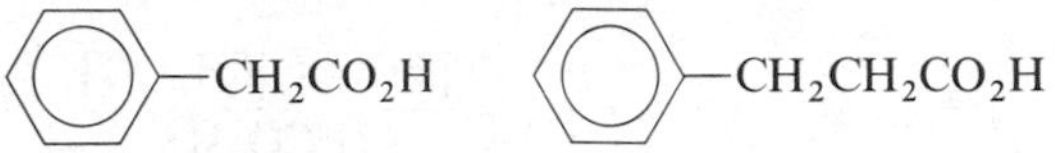

$$\text{\large{\bigcirc}}\!-\!CH_2CO_2H \qquad \text{\large{\bigcirc}}\!-\!CH_2CH_2CO_2H$$

Phenylacetic acid	3-Phenylpropanoic acid
	3-Phenylpropionic acid

Carboxylic acids with more than about 10 carbons have low volatilities and hence usually exhibit little odor.

The carboxyl group participates in hydrogen bonds as both a hydrogen donor and an acceptor.

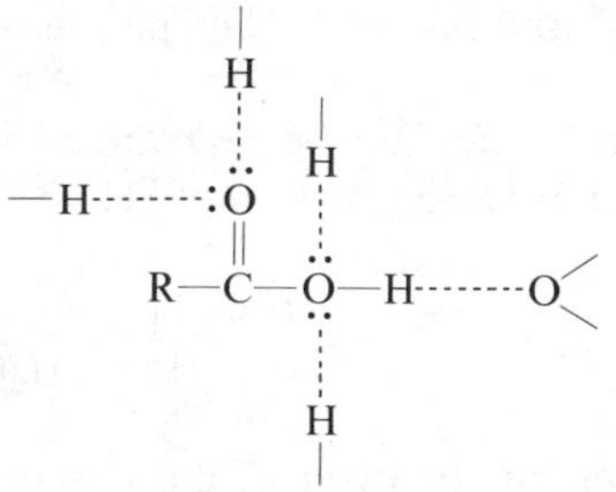

Carboxylic acids are even somewhat more soluble in water than are alcohols with the same number of carbon atoms. Formic, acetic, propanoic, and butanoic acids are completely miscible with water, and the solubility of pentanoic acid (5 g in 100 g of water) is similar to that of 1-butanol. The boiling points of carboxylic acids are somewhat higher than those of alcohols of comparable molecular weight. Acetic acid (molecular weight 60), for example, boils at 118° but propyl alcohol (also molecular weight 60) at 98°.

11.3
REACTIONS AS ACIDS

Carboxylic acids exhibit two particularly important types of chemical behavior. One is transformation to derivatives that have some other nucleophilic group in place of the —OH of the carboxyl group. This important group of reactions is considered in the next chapter. The other is acidity, described here.

Carboxylic acids ionize only partially in water and thus are weak acids.

$$\underset{\displaystyle R-\overset{\displaystyle O}{\overset{\|}{C}}-O-H}{} + H_2O \; \rightleftharpoons \; R-\overset{\displaystyle O}{\overset{\|}{C}}-O^- + H_3O^+$$

In a 0.1 M aqueous solution, only about 1% of acetic acid molecules are ionized. Nearly 99% of the molecules are present in the undissociated form (RCO_2H). Nevertheless, carboxylic acids are considerably more acidic than alcohols or phenols, which also have O—H bonds that ionize.

Acidity is indicated quantitatively by the numerical value of the equilibrium constant, called K_a or the acidity constant, for the reaction with water.

$$K_a = \frac{[RCO_2^-][H_3O^+]}{[RCO_2H]}$$

The larger the value of K_a, the more extensively ionized and hence the more acidic the acid. K_a is 1.8×10^{-5} for acetic acid and 6.3×10^{-5} for benzoic acid. These values are typical of most carboxylic acids.

Acidity is also frequently expressed as pK_a, which is defined as $-\log K_a$. For an acid having $K_a = 10^{-5}$, $pK_a = -\log(10^{-5}) = -(-5) = 5$. A change in pK of one unit corresponds to a change in acidity of a factor of 10. A tenfold stronger acid has $K_a = 10^{-4}$ and $pK_a = 4$. The pK_a of acetic acid is 4.7 and of benzoic acid 4.3.

The equation for K_a can be rearranged into a form that gives the ratio of $[RCO_2^-]$ to $[RCO_2H]$ as a function of $[H_3O^+]$.

$$\frac{[RCO_2^-]}{[RCO_2H]} = \frac{K_a}{[H_3O^+]}$$

(A similar equation for the ratio of an amine and the corresponding ammonium ion was considered in Sec. 8.4.) For acetic acid, this expression becomes

$$\frac{[CH_3CO_2^-]}{[CH_3CO_2H]} = \frac{1.8 \times 10^{-5}}{[H_3O^+]}$$

We can see that $CH_3CO_2^-$ and CH_3CO_2H are present in equal concentrations in a solution in which $[H_3O^+]$ is maintained at 1.8×10^{-5} M (pH = 4.7). As was shown in Figure 8.1, CH_3CO_2H is the dominant species in more acidic solutions and $CH_3CO_2^-$ in more basic solutions. In a neutral solution ($[H_3O^+] = 10^{-7}$ M, pH = 7), $CH_3CO_2^-$ greatly exceeds CH_3CO_2H.

$$\frac{[CH_3CO_2^-]}{[CH_3CO_2H]} = \frac{1.8 \times 10^{-5}}{10^{-7}} = 180$$

In most physiological environments anions of carboxylic acids rather than the free acids are the dominant species.

Typical carboxylic acids react only partially with water, a very weak base. They react, however, essentially completely with hydroxide ion, a strong base, and even with weaker bases, such as carbonate ion (CO_3^{2-}), bicarbonate ion (HCO_3^-), and amines.

$$2 \text{ CH}_3\overset{\overset{\displaystyle O}{\|}}{\text{C}}\text{OH} + 2 \text{ Na}^+ + \text{CO}_3{}^{2-} \longrightarrow 2 \text{ CH}_3\overset{\overset{\displaystyle O}{\|}}{\text{C}}\text{O}^- \text{ Na}^+ + \text{H}_2\text{O} + \text{CO}_2$$

Sodium acetate

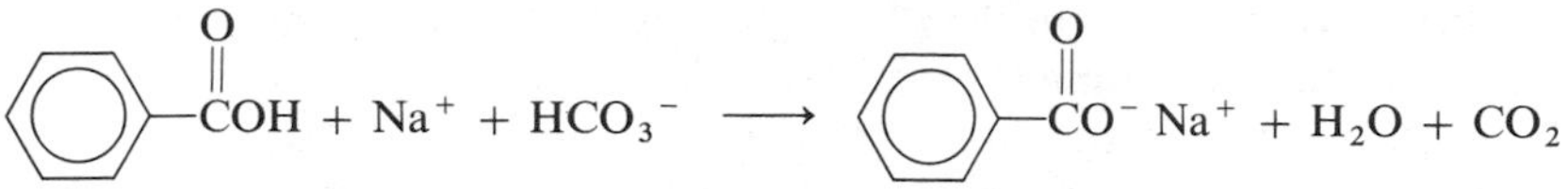

Sodium benzoate

$$\text{CH}_3\text{CH}_2\overset{\overset{\displaystyle O}{\|}}{\text{C}}\text{OH} + \text{CH}_3\text{NH}_2 \longrightarrow \text{CH}_3\text{CH}_2\overset{\overset{\displaystyle O}{\|}}{\text{C}}\text{O}^- \text{ CH}_3\text{NH}_3{}^+$$

Methylammonium propanoate
Methylammonium propionate

Alka-Seltzer bubbles are CO_2 gas produced by reaction of a solid carboxylic acid (citric acid) and sodium bicarbonate when they dissolve in water. Baking powders produce CO_2 by a similar reaction. Phenols, although more acidic than alcohols, are generally not sufficiently acidic to react with bicarbonate, although they do react with carbonate and hydroxide. In considering the conversion of a racemic mixture of a carboxylic acid to diastereomeric salts, we introduced (Sec. 6.8) the reaction of a carboxylic acid and an amine to form a salt. The anion of a carboxylic acid is converted back to the carboxylic acid on treatment with a strong acid.

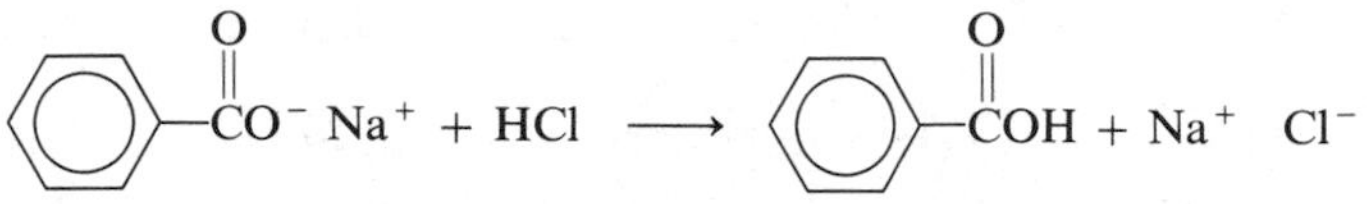

Although not strong acids, carboxylic acids are approximately 10 trillion times more acidic than alcohols.

$$\text{R}-\text{O}-\text{H} + \text{H}_2\text{O} \rightleftharpoons \text{R}-\text{O}^- + \text{H}_3\text{O}^+ \qquad K_a \simeq 10^{-18}$$

$$\text{R}-\overset{\overset{\displaystyle O}{\|}}{\text{C}}-\text{O}-\text{H} + \text{H}_2\text{O} \rightleftharpoons \text{R}-\overset{\overset{\displaystyle O}{\|}}{\text{C}}-\text{O}^- + \text{H}_3\text{O}^+ \qquad K_a \simeq 10^{-5}$$

With the acidic behavior of both carboxylic acids and alcohols due to ionization of O—H bonds, what is the origin of this enormous difference in acidity? It is due principally to the stability of the anion of a carboxylic acid. Two equivalent resonance structures can be written for this ion.

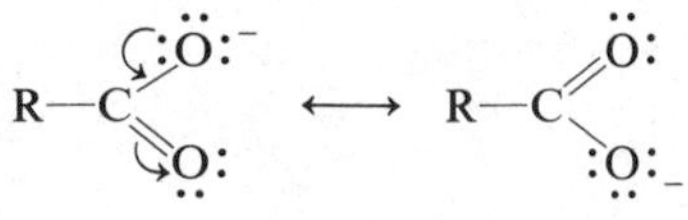

Instead of being concentrated on a single oxygen as in an alkoxide ion, the negative charge of a carboxylate anion is divided equally between two oxygens. Because of dispersion of the charge over a larger volume, the equilibrium for formation of RCO_2^- from RCO_2H is much more favorable than for formation of RO^- from ROH. Recall that the acidity of phenol is also enhanced by resonance stabilization of the resulting anion (Sec. 5.19).

Reaction with a solution of a base converts a carboxylic acid into a salt. The pure salt, ordinarily a nonvolatile solid, can be obtained by evaporating the solvent. Salts of carboxylic acids are named in the same manner as inorganic salts; the name of the cation is followed by the name of the anion. The carboxylate anion is named by replacing *-ic acid* of the acid name by *-ate*.

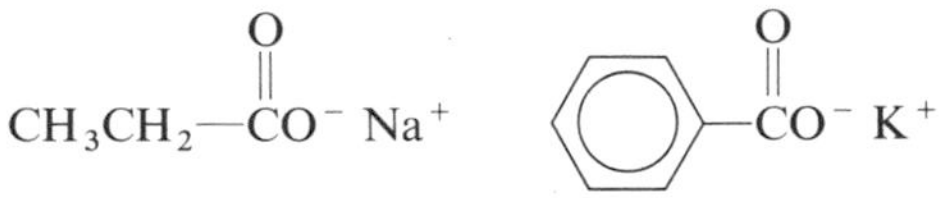

$$CH_3CH_2-\overset{\overset{O}{\|}}{C}O^-\ Na^+$$

Sodium propanoate
Sodium propionate

Potassium benzoate

---

problem 11.3   Draw structures and give names for the products of the following reactions.
(a) benzoic acid + sodium hydroxide →
(b) sodium acetate + hydrochloric acid →
(c) propanoic acid + trimethylamine →

Salts of carboxylic acids, particularly those of Na^+, K^+, and NH_4^+ ions, ordinarily are water soluble but not too soluble in nonpolar solvents, such as alkanes. Except for the water-miscible acids containing only a few carbons, the salt of an acid is generally more soluble in water than the acid itself. This difference can be used to separate carboxylic acids from other compounds. Consider a mixture of benzyl bromide and phenylacetic acid, a relatively water-insoluble carboxylic acid. This mixture can arise in the preparation (Sec. 11.6) of phenylacetic acid from benzyl bromide. Treating the mixture with aqueous base converts the phenylacetic acid into a salt that dissolves in the aqueous solution.

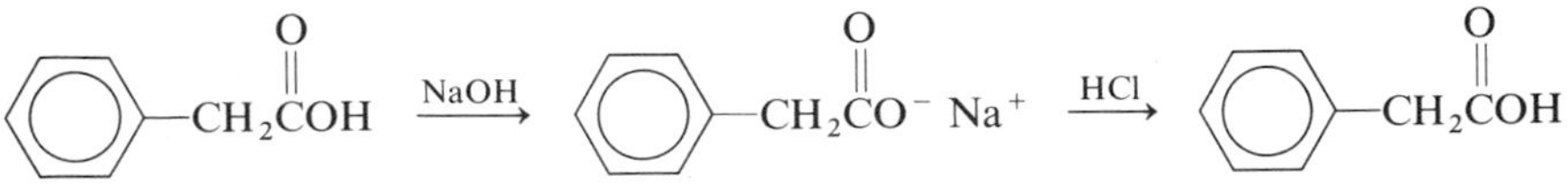

Phenylacetic acid   →(NaOH)→   Sodium phenylacetate   →(HCl)→

This solution can be separated from unreacted benzyl bromide. Adding a strong acid to the aqueous solution regenerates the phenylacetic acid, which precipitates from the solution.

EFFECT OF SUBSTITUENTS ON ACIDITY

As shown by the data in Table 11.2, chloro substituents significantly increase the acidity of carboxylic acids. We know (Sec. 1.3) that chlorine is considerably more electronegative than carbon and tends to attract more than its share of the electrons in a C—Cl bond. The chloro substituent makes the adjoining carbon in the chloro-acetate ion more positive and more able to attract the electron pair that it shares with the carboxylate carbon.

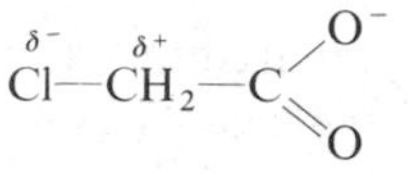

$$\overset{\delta^-}{Cl}-\overset{\delta^+}{CH_2}-C\overset{\displaystyle O^-}{\underset{\displaystyle O}{}}$$

The drift of electrons toward chlorine can be shown by arrows.

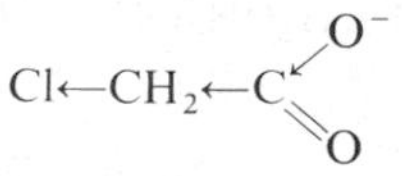

$$Cl\leftarrow CH_2\leftarrow C\overset{\displaystyle O^-}{\underset{\displaystyle O}{}}$$

By helping to disperse the negative charge, chlorine stabilizes the anion and thereby favors ionization of the carboxylic acid.

Other substituent groups more electronegative than carbon also are acid strengthening. Substituent effects that can be ascribed to a drift of electrons along a chain of single bonds are called **inductive effects**. Inductive effects in carboxylic acids are largest when substituents are closest to the carboxyl group. 2-Chloropro-panoic acid is a hundred times more acidic than propanoic acid, but 3-chloro-propanoic acid only seven times more acidic. Dichloroacetic and trichloroacetic

Table 11.2 Acidities of some carboxylic acids

Structure	Name	K_a	pK_a
CH_3CO_2H	acetic acid	1.8×10^{-5}	4.7
$ClCH_2CO_2H$	chloroacetic acid	1.4×10^{-3}	2.9
Cl_2CHCO_2H	dichloroacetic acid	5.7×10^{-2}	1.2
Cl_3CCO_2H	trichloroacetic acid	2.2×10^{-1}	0.7
$CH_3CH_2CO_2H$	propanoic acid	1.3×10^{-5}	4.9
$CH_3\underset{\displaystyle Cl}{CHCO_2H}$	2-chloropropanoic acid	1.3×10^{-3}	2.9
$CH_2\underset{\displaystyle Cl}{CH_2CO_2H}$	3-chloropropanoic acid	9×10^{-5}	4.0
HCO_2H	formic acid	1.8×10^{-4}	3.7

acids are even more acidic than chloroacetic acid because the inductive effect increases with the number of chloro substituents. Since formic acid is 10 times more acidic than acetic acid, the methyl group of acetic acid decreases acidity. This indication that methyl is more electron donating than hydrogen is consistent with earlier observations that methyl is more able than hydrogen to stabilize carbocations.

11.5
PREPARATIONS OF CARBOXYLIC ACIDS
BY OXIDATION REACTIONS

Carboxylic acids can be prepared by oxidation of several classes of compounds. We know that strong oxidizing agents convert primary alcohols to carboxylic acids (Sec. 5.8).

$$(CH_3)_2CHCH_2CH_2CH_2OH \xrightarrow{K_2Cr_2O_7} (CH_3)_2CHCH_2CH_2\overset{\overset{\displaystyle O}{\parallel}}{C}OH$$

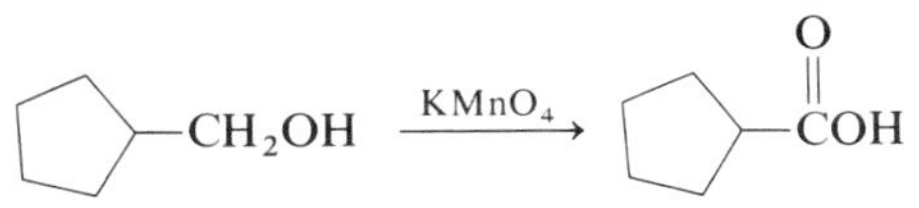

Aldehydes are readily oxidized to carboxylic acids (Sec. 9.4). Even very mild reagents, such as Tollens' reagent, are effective, but stronger oxidizing agents can also be used.

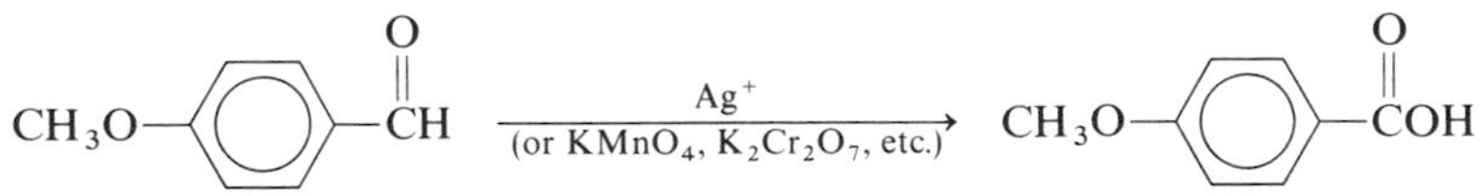

Alkanes are not easily oxidized. Alkyl groups adjacent to aromatic rings, however, are oxidized to carboxyl groups by such strong oxidizing agents as $KMnO_4$ or $K_2Cr_2O_7$.

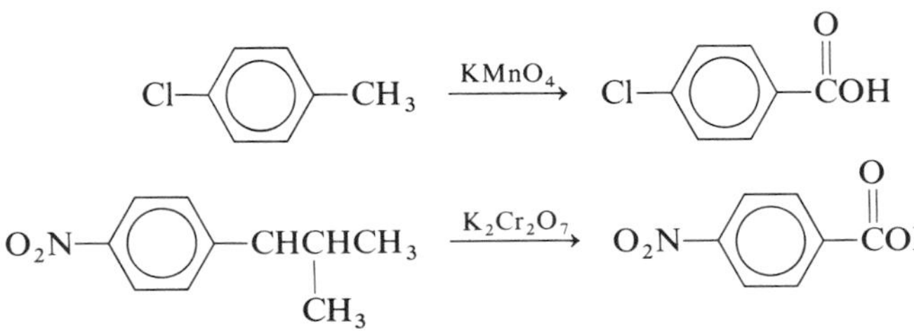

Note that oxidation of an alkyl group containing several carbons leaves only one carbon attached to the aromatic ring. Terephthalic acid, a component of the polymer

Dacron (Sec. 12.16), is prepared commercially by high-temperature oxidation of *p*-xylene by oxygen in the presence of a cobalt catalyst.

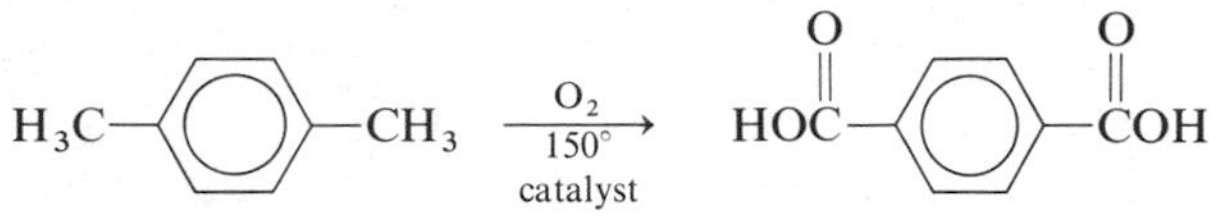

$$H_3C-\text{⟨⟩}-CH_3 \xrightarrow[\text{catalyst}]{O_2 \atop 150°} HO\overset{O}{\overset{\|}{C}}-\text{⟨⟩}-\overset{O}{\overset{\|}{C}}OH$$

p-Xylene   Terephthalic acid
p-Dimethylbenzene   *p*-Benzenedicarboxylic acid

11.6
PREPARATION OF CARBOXYLIC ACIDS
BY CARBONATION OF GRIGNARD REAGENTS

Grignard reagents react with one $C{=}O$ bond of carbon dioxide in exactly the same fashion as with the $C{=}O$ bond of an aldehyde or ketone (Sec. 9.8).

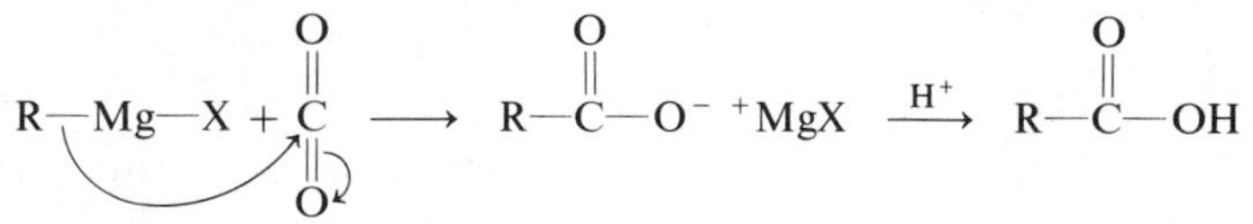

$$R-Mg-X + \overset{O}{\overset{\|}{C}}\underset{O}{\|} \longrightarrow R-\overset{O}{\overset{\|}{C}}-O^- {}^+MgX \xrightarrow{H^+} R-\overset{O}{\overset{\|}{C}}-OH$$

Addition of a strong acid to the resulting magnesium salt converts it to the carboxylic acid. Remember that a Grignard reagent is prepared from a reaction of an organic halide with magnesium (Sec. 7.4).

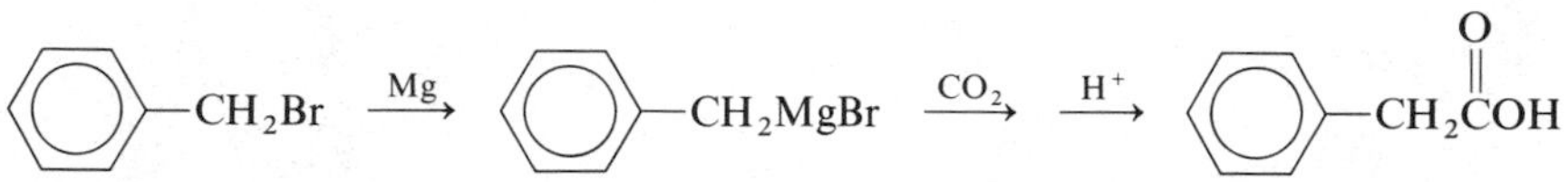

$$\text{⟨⟩}-CH_2Br \xrightarrow{Mg} \text{⟨⟩}-CH_2MgBr \xrightarrow{CO_2} \xrightarrow{H^+} \text{⟨⟩}-CH_2\overset{O}{\overset{\|}{C}}OH$$

Benzyl bromide   Phenylacetic acid

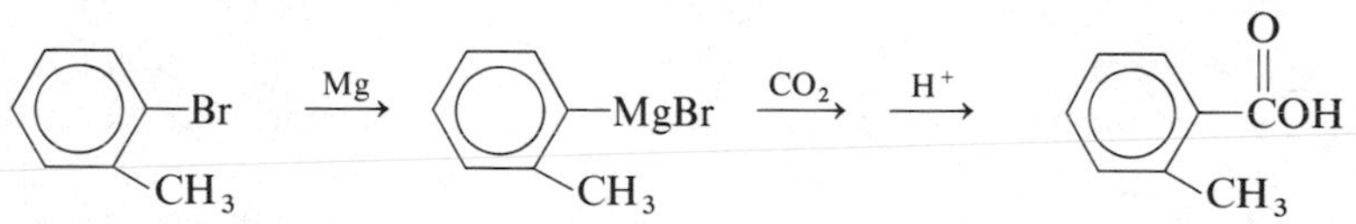

$$\text{⟨⟩}\underset{CH_3}{-Br} \xrightarrow{Mg} \text{⟨⟩}\underset{CH_3}{-MgBr} \xrightarrow{CO_2} \xrightarrow{H^+} \text{⟨⟩}\underset{CH_3}{-\overset{O}{\overset{\|}{C}}OH}$$

o-Bromotoluene   *o*-Methylbenzoic acid
                                                                       o-Toluic acid

Therefore Grignard reagent formation and carbonation converts an organic halide to a carboxylic acid containing one additional carbon.

---

problem 11.4   Write equations to show how cyclopentanecarboxylic acid can be prepared from an alkyl halide or an alcohol.

11.7
SOME IMPORTANT CARBOXYLIC ACIDS

Acetic acid, in the form of its derivative acetyl coenzyme A, is a key biochemical building block (Fig. 10.1). Acetic acid is also a large-scale industrial chemical. Vinegar is mainly acetic acid (4–5%) and water.

Small amounts of heptanoic acid are in the material excreted by the potato tube worm as it chews its way through a potato. The heptanoic acid attracts a parasite of the tube worm, which then seeks its prey.

Oxalic acid, found in rhubarb and spinach leaves, is one of the strongest naturally occurring acids (K_a for ionization of one carboxyl group $= 5.4 \times 10^{-2}$, $pK_a = 1.3$).

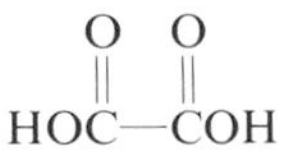

$$\underset{\text{Oxalic acid}}{\overset{\displaystyle O \quad\; O}{\underset{\displaystyle \| \quad\; \|}{HOC-COH}}} \qquad \underset{\text{Calcium oxalate}}{\overset{\displaystyle O \quad\; O}{\underset{\displaystyle \| \quad\; \|}{^-OC-CO^-} \quad Ca^{2+}}}$$

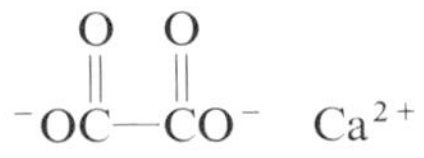

Kidney stones often consist chiefly of calcium oxalate. Potassium oxalate has been used to prevent clotting of blood samples drawn for clinical analysis. It functions by precipitating Ca^{2+}, which is necessary for the coagulation process.

Citric acid is widely distributed in plants, particularly in citrus fruits and berries.

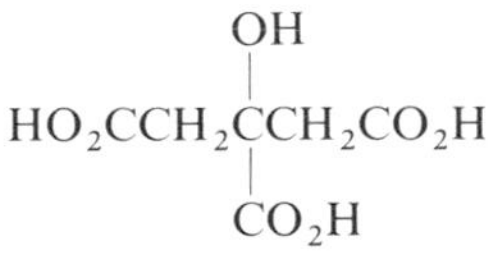

$$\overset{\displaystyle OH}{\underset{\displaystyle CO_2H}{HO_2CCH_2\overset{\displaystyle |}{\underset{\displaystyle |}{C}}CH_2CO_2H}}$$

Citric acid

It is used in large amounts as a food additive, especially in carbonated beverages, to which it imparts a tangy taste. Citric acid is produced commercially (about 200×10^6 kg a year) by fermentation of molasses by a mold. Various other organic compounds, including many antibiotics, are also produced commercially by fermentation.

Salts of acids have practical uses. Sodium benzoate is added to many foods as a preservative. Calcium propionate is added to bread to prevent molding. Sorbic acid and its potassium salt are used for the same purpose, particularly in cheese.

$$CH_3CH{=}CHCH{=}CHCO_2H$$

Sorbic acid
2,4-Hexadienoic acid

Such additives are most effective when the food is sufficiently acidic that the carboxylic acid rather than its anion is the major species present. 10-Undecenoic acid and its zinc salt are used as fungicides in foot powders.

$$CH_2=CH(CH_2)_8CO_2H \qquad (CH_2=CH(CH_2)_8CO_2^-)_2\,Zn^{2+}$$

10-Undecenoic acid   Zinc 10-undecenoate

Lithium stearate and other salts of long-chain carboxylic acids are blended with oils to form lubricating greases.

$$CH_3(CH_2)_{16}CO_2^-\,Li^+$$

Lithium stearate

Indole-3-acetic acid, an important hormone in higher plants, promotes elongation of growing plant cells.

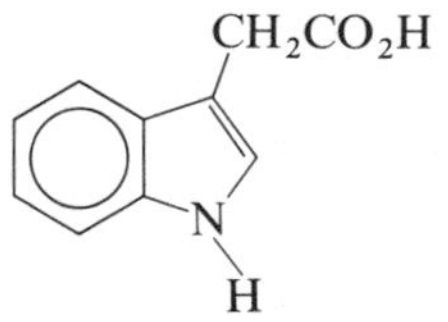

Indole-3-acetic acid

Some herbicides (compounds that kill plants) used extensively by farmers to control weeds have somewhat related structures.

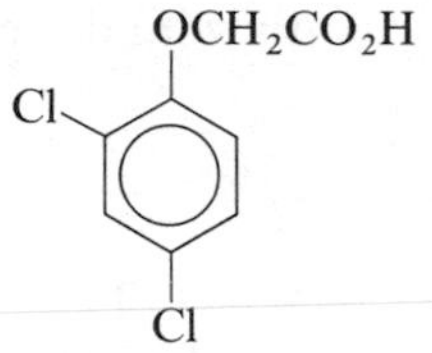

2,4-D   Dicamba
2,4-Dichlorophenoxyacetic acid   3,6-Dichloro-2-methoxybenzoic acid

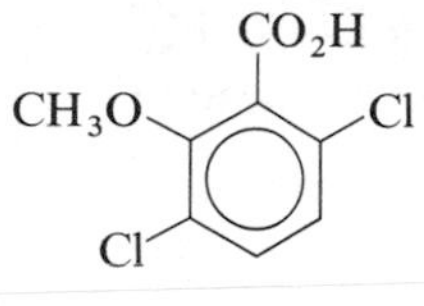

These herbicides have effects similar to the natural hormone but are not as rapidly inactivated. When present in sufficient quantities, they induce unrestrained growth that leads to death of a plant.

Abcissic acid, found in a wide range of plant species, inhibits the activity of growth hormones, such as indole-3-acetic acid, and operates to defend plants against the effects of physiological stress.

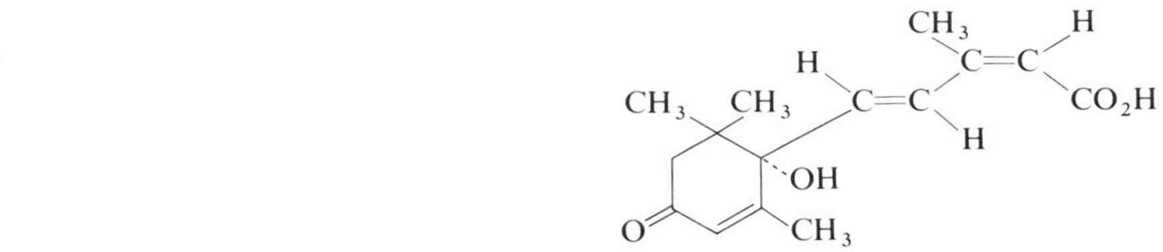

Abcissic acid

When a tomato plant receives insufficient water, its leaves produce greatly increased amounts of abcissic acid. The abcissic acid inhibits growth and, by promoting closure of the leaf stomata, reduces loss of water.

Ethylenediaminetetraacetic acid (EDTA) is a complexing agent for many metal ions, although not for the group 1 ions (Na^+, K^+) present in large amounts in living organisms.

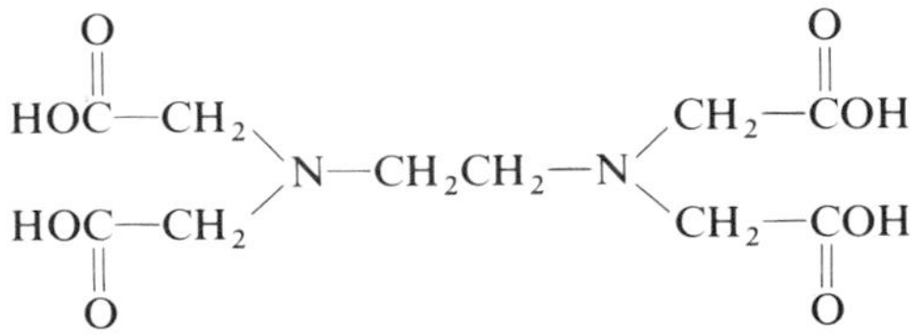

EDTA
Ethylenediaminetetraacetic acid

When its carboxyl groups are ionized (as is the case in neutral aqueous solutions), EDTA has six groups that can complex effectively with positive ions: four negative carboxylate groups and two amine nitrogens. A medical use of EDTA is to remove harmful ions like Pb^{2+}. EDTA complexes tightly with these ions and the complexes are excreted in the urine.

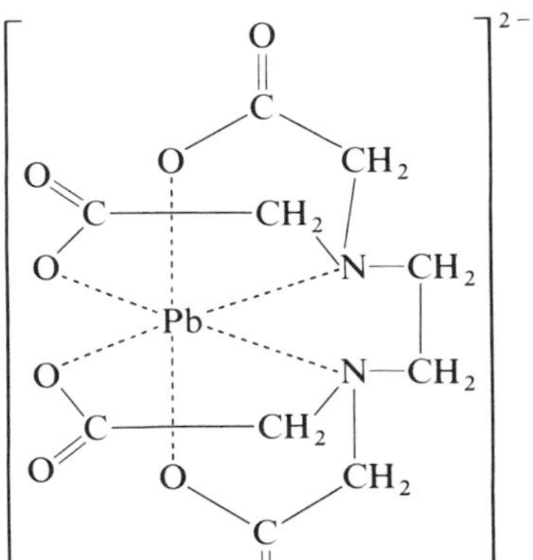

Another use is to stabilize some drug preparations that rapidly deteriorate in the presence of trace quantities of certain metal ions.

The prostaglandins are a family of C-20 carboxylic acids found in most mammalian tissues. Prostaglandins have a prostanoic acid skeleton but differ in substitution by various hydroxyl, carbonyl, and alkene functions.

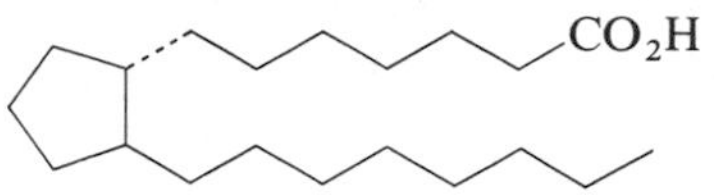

Prostanoic acid

Examples are PGE_2, the most common of the prostaglandins, and $PGF_{2\alpha}$.

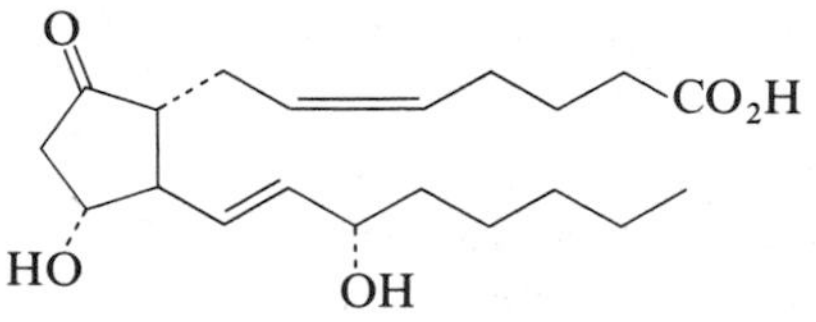

PGE$_2$ 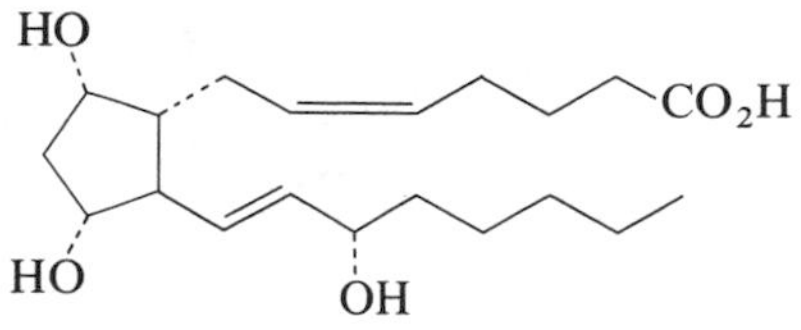   PGF$_{2\alpha}$

Although present in only minute amounts, prostaglandins seem to have regulatory effects on a variety of functions, including metabolism, blood pressure, rate of gastric secretion, contraction and relaxation of smooth muscles, and reproductive processes. PGE_2 and $PGF_{2\alpha}$ have been used clinically for inducing labor and terminating pregnancy. The effects of aspirin may be due at least in part to an inhibitory action on prostaglandin synthesis.

SUMMARY

Definitions and Ideas

Carboxylic acids are compounds containing a **carboxyl group**, —CO_2H.

Trivial names (e.g., formic acid, acetic acid, propionic acid, butyric acid, valeric acid) are accepted by the IUPAC for many common carboxylic acids. Substitutive names are based on the longest chain that contains the carboxyl group. The ending -*e* of the name of the corresponding hydrocarbon is replaced by -*oic acid*. The carboxyl carbon is assumed to be number 1 and is not numbered in the name. When the carboxyl group is attached to a ring, substitutive names are frequently formed by adding -*carboxylic acid* to the name of the compound that would result from replacing the carboxyl group by a hydrogen. Aromatic carboxylic acids are often named as derivatives of benzoic acid, the simplest aromatic acid.

Carboxylic acids are even somewhat more water soluble and higher boiling than comparable alcohols.

Although considerably more acidic than alcohols or phenols, carboxylic acids are weak acids. They are sufficiently acidic, however, to be converted almost quantitatively to salts, not only by hydroxide ion but even by considerably weaker bases, such as bicarbonate ion and amines. In an aqueous solution maintained at neutrality (pH = 7), carboxylic acids are almost completely ionized.

The salts of carboxylic acids usually are nonvolatile solids and more water soluble than the acids. They can be converted to the undissociated carboxylic acids by treatment with strong acids. The salts are named by giving the name of the cation, followed by a name for the anion derived by replacing *-ic acid* of the acid name by *-ate*.

Important Reactions Involving Carboxylic Acids

1. *Reactions as acids*

$$\underset{\text{RCOH}}{\overset{\overset{\displaystyle O}{\parallel}}{}} \xrightarrow{\text{NaOH}} \underset{\text{RCO}^-\text{Na}^+}{\overset{\overset{\displaystyle O}{\parallel}}{}} \xrightarrow{\text{H}^+} \underset{\text{RCOH}}{\overset{\overset{\displaystyle O}{\parallel}}{}}$$

R = alkyl or aryl

Even bases like bicarbonate ion or amines, which are considerably weaker than hydroxide ion, can convert a carboxylic acid essentially quantitatively to its anion. Addition of a strong acid converts the anion back to the carboxylic acid.

2. *Preparation by oxidation of primary alcohols*

$$\text{RCH}_2\text{OH} \xrightarrow{\text{K}_2\text{Cr}_2\text{O}_7 \text{ or KMnO}_4} \underset{\text{RCOH}}{\overset{\overset{\displaystyle O}{\parallel}}{}}$$

(discussed in Chapter 5)   R = alkyl or aryl

3. *Preparation by oxidation of aldehydes*

$$\underset{\text{RCH}}{\overset{\overset{\displaystyle O}{\parallel}}{}} \xrightarrow{\text{Ag}^+,\ \text{K}_2\text{Cr}_2\text{O}_7,\ \text{KMnO}_4,\ \text{etc.}} \underset{\text{RCOH}}{\overset{\overset{\displaystyle O}{\parallel}}{}}$$

(discussed in Chapter 9)   R = alkyl or aryl

4. *Preparation by oxidation of alkyl-substituted aromatic compounds*

$$\text{ArR} \xrightarrow{\text{K}_2\text{Cr}_2\text{O}_7 \text{ or KMnO}_4} \underset{\text{ArCOH}}{\overset{\overset{\displaystyle O}{\parallel}}{}}$$

5. *Preparation by carbonation of Grignard reagents*

$$\text{RMgX} \xrightarrow{\text{CO}_2} \xrightarrow{\text{H}^+} \underset{\text{RCOH}}{\overset{\overset{\displaystyle O}{\parallel}}{}}$$

R = alkyl or aryl   X = Cl, Br, or I

PROBLEMS

1. Draw structures for the following compounds.
   (a) propanoic acid
   (b) butyric acid
   (c) 2-methylpentanoic acid
   (d) 2-chloropropionic acid
   (e) valeric acid
   (f) phenylacetic acid
   (g) 2,3-dichlorocaproic acid
   (h) trifluoroacetic acid
   (i) 4-methylpentanoic acid
   (j) 3,4-dimethyl-2-ethylhexanoic acid
   (k) *trans*-3-pentenoic acid
   (l) cyclobutanecarboxylic acid
   (m) *p*-aminobenzoic acid
   (n) 3,5-dimethylbenzoic acid
   (o) oxalic acid
   (p) succinic acid
   (q) sodium acetate
   (r) diethylammonium propanoate

2. Name the following compounds.
   (a) $CH_3CH_2CH_2CO_2H$
   (b) $ClCH_2CH_2CO_2H$
   (c) $CH_3CH_2CH_2CH_2CH_2CO_2H$
   (d) $CH_3CH_2CH_2\overset{\underset{\displaystyle |}{Br}}{C}HCO_2H$

   (e)

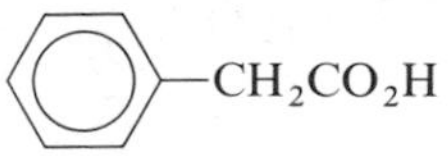

   ⬡—CH_2CO_2H

   (f) $CH_3\overset{\underset{\displaystyle |}{CHCH_3}}{C}HCH_2CH_2CO_2H$
   $\qquad\qquad\;\; \underset{\displaystyle CH_3}{|}$

   (g)

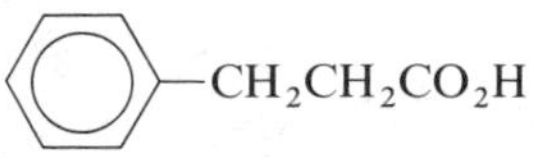

   ⬡—$CH_2CH_2CO_2H$

   (h)

   $\underset{CH_3CH_2}{H}\!\!\diagdown\!\!C\!\!=\!\!C\!\!\diagup\!\!\overset{CO_2H}{_H}$

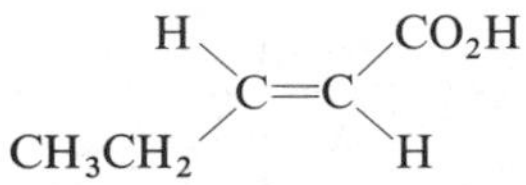

   (i) $CH_3\overset{\underset{\displaystyle |}{CH_3}}{C}HCH_2CH_2CO_2H$

   (j) $CH_3(CH_2)_{16}CO_2H$

   (k)

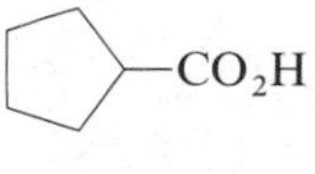

   ⬠—CO_2H

   (l)

   ⬡—CO_2H
   $\underset{Cl}{}$

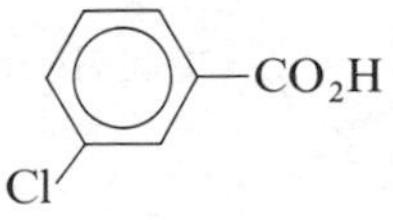

   (m) $HO_2CCH_2CO_2H$

   (n) $CH_3CH_2CO_2^-\ Na^+$

   (o)

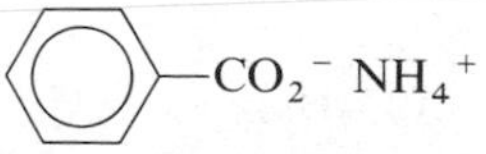

   ⬡—$CO_2^-\ NH_4^+$

   (p) $CH_3CO_2^-\ CH_3CH_2NH_3^+$

3. Give another name for each of these compounds.
   (a) propionic acid
   (b) 2-chloropentanoic acid
   (c) hexanedioic acid
   (d) malonic acid

4. Arrange each group of compounds in order of increasing solubility in water.

   (a) 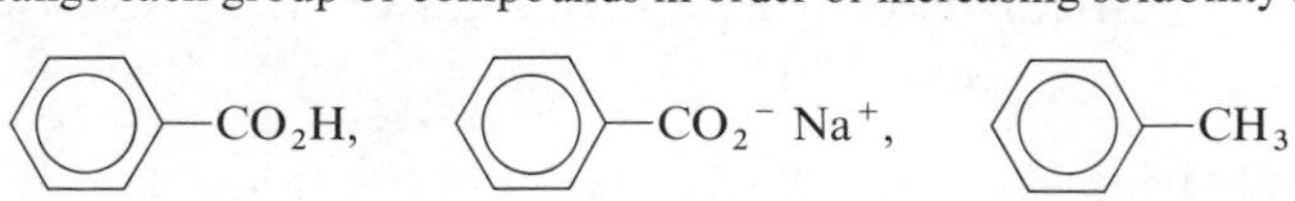 ⬡—CO_2H, ⬡—$CO_2^-\ Na^+$, ⬡—CH_3

   (b) $CH_3CH_2CH_2CH_2CH_3$, $CH_3CH_2CH_2CH_2CO_2H$, $CH_3CH_2CH_2CH_2CH_2OH$

5. Draw the structures of any organic products that result from the following reactions or reaction sequences.

(a) $CH_3CH_2CO_2H$ $\xrightarrow{KOH}$

(b) $(CH_3)_2CHCH_2OH$ $\xrightarrow{K_2Cr_2O_7}$

(c)

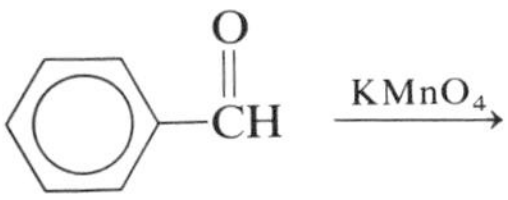

$\xrightarrow{KMnO_4}$

(d)

$-CO_2^-\ Na^+$ $\xrightarrow{HCl}$

(e) $CH_2{=}CHCH_2Br$ $\xrightarrow{Mg}$ $\xrightarrow{CO_2}$ $\xrightarrow{H^+}$

(f) CH_3CO_2H $\xrightarrow{NaHCO_3}$

(g) O_2N- $-CH_3$ $\xrightarrow{K_2Cr_2O_7}$

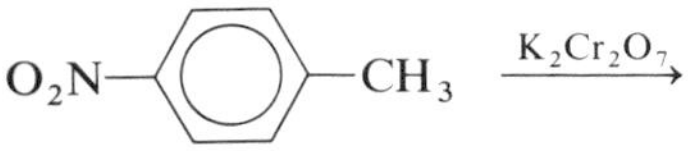

(h) $-CO_2H$ $\xrightarrow{CH_3NH_2}$

(i) $-CO_2H$ $\xrightarrow[FeBr_3]{Br_2}$

(j)

$\xrightarrow[OH^-]{I_2}$

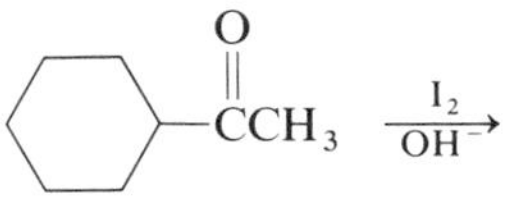

6. Write equations to show how each conversion can be carried out. More than one step will sometimes be necessary.

(a) $(CH_3)_2CHCH_2Br$ $\longrightarrow$ $(CH_3)_2CHCH_2CO_2H$

(b) $CH_3CH_2CO_2H$ $\longrightarrow$ $CH_3CH_2CO_2^-\ K^+$

(c)

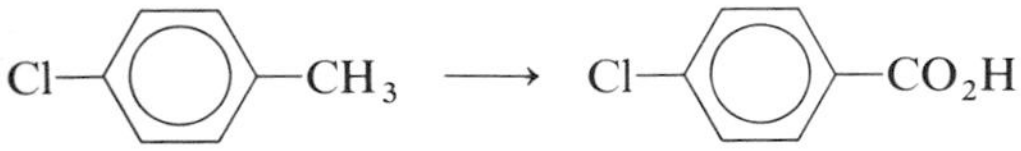

$Cl-$ $-CH_3$ $\longrightarrow$ $Cl-$ $-CO_2H$

(d)

CH_3CO_2H $\longrightarrow$ $CH_3CO_2^-$ $H_3\overset{+}{N}-$

(e)

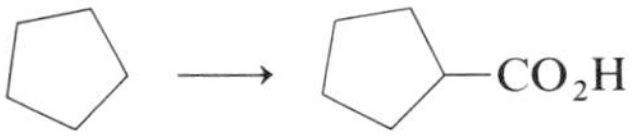

$\longrightarrow$ $-CO_2H$

(f)

$CH_3CH{=}C(CH_3)_2$ $\longrightarrow$ $CH_3CH_2\overset{\underset{\displaystyle CH_3}{|}}{\underset{\underset{\displaystyle CH_3}{|}}{C}}CO_2H$

(g) $-CH_2OH$ $\longrightarrow$ $-CH_2CO_2H$

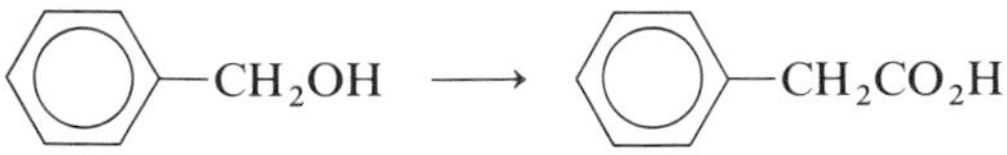

(h) $\longrightarrow$ O_2N- $-CO_2H$

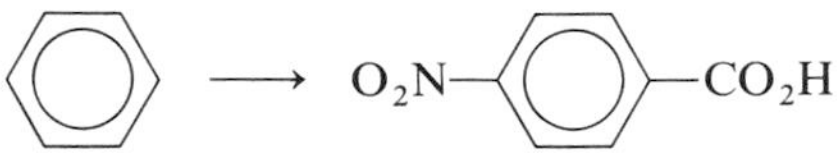

7. Arrange each group of compounds in order of increasing acidity.

(a) CH_3CO_2H, $\quad$ $BrCH_2CO_2H$, $\quad$ Br_2CHCO_2H

(b) HCO_2H, $\quad$ CH_3CO_2H

(c) $CH_3CH_2\overset{\underset{\displaystyle Cl}{|}}{C}HCO_2H$, $\quad$ $CH_3\overset{\underset{\displaystyle Cl}{|}}{C}HCH_2CO_2H$, $\quad$ $ClCH_2CH_2CH_2CO_2H$

(d) $-CO_2H$, $\quad$ O_2N- $-CO_2H$, $\quad$ CH_3- $-CO_2H$

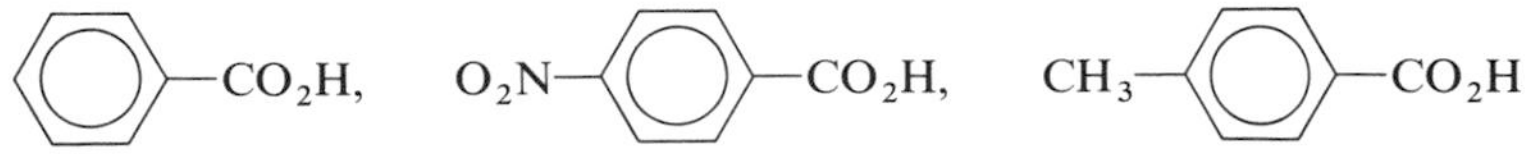

(e) 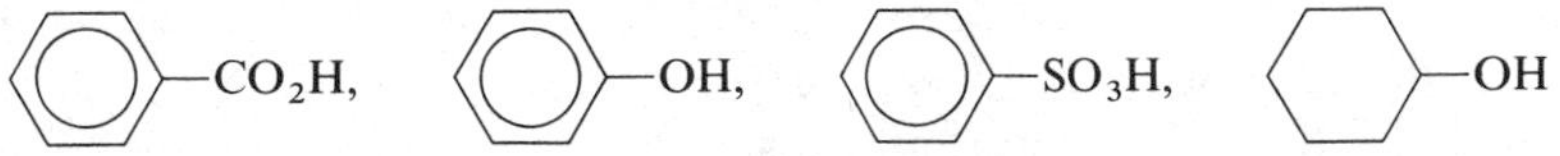
—CO$_2$H,   —OH,   —SO$_3$H,   —OH

(f) 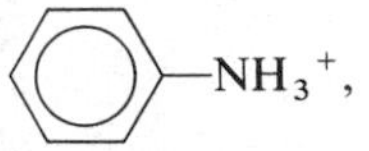
—NH$_3^+$,   CH$_3$NH$_3^+$,   CH$_3$NH$_2$

8. What is the predominant form of each compound when dissolved in aqueous solutions of pH 2, 7, or 12?
   (a) acetic acid   (b) phenol
   (c) methanol   (d) benzoic acid

9. Calculate the ratio of $[PhCO_2^-]$ to $[PhCO_2H]$ in aqueous solutions of $PhCO_2H$ (benzoic acid) having pH values of 3, 5, and 7. The K_a of benzoic acid is 6.3×10^{-5}.

10. The anion of a carboxylic acid is stabilized by two resonance structures but the anion of a phenol by four resonance structures. Why, then, is resonance stabilization less important for the anion of a phenol?

11. Describe a procedure for separating each of the following mixtures into its components without distilling or crystallizing.
    (a) hexanoic acid, 1-hexanol, and 1-hexanamine
    (b) *p*-chlorobenzoic acid, *p*-chlorophenol, and *p*-dichlorobenzene

12. Draw the structure of an optically active, hydroxyl-containing carboxylic acid of molecular formula $C_6H_{12}O_3$ that, on oxidation with potassium permanganate, gives an optically inactive dicarboxylic acid of molecular formula $C_6H_{10}O_4$.

13. Explain why the two carbon-oxygen bonds have different lengths in formic acid (1.36 Å and 1.23 Å) but the same length in sodium formate (1.27 Å).

14. A student needed to prepare *p*-hydroxybenzoic acid and decided to do so by carbonation of an appropriate Grignard reagent. He took *p*-bromophenol, treated it with Mg in ether, then added CO_2, and finally added aqueous acid. Instead of isolating *p*-hydroxybenzoic acid, however, he obtained a mixture of phenol and *p*-bromophenol. Why did his synthesis fail?

Derivatives of Carboxylic Acids: Esters and Amides

12

Several families of compounds are related to carboxylic acids.

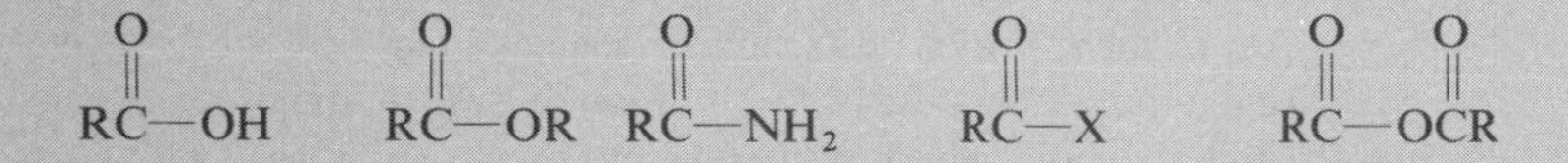

$$\underset{\text{Carboxylic acids}}{\overset{\displaystyle O}{\overset{\|}{RC}}-OH} \qquad \underset{\text{Esters}}{\overset{\displaystyle O}{\overset{\|}{RC}}-OR} \qquad \underset{\text{Amides}}{\overset{\displaystyle O}{\overset{\|}{RC}}-NH_2} \qquad \underset{\substack{\text{Acyl halides}\\ \text{(X is a halogen)}}}{\overset{\displaystyle O}{\overset{\|}{RC}}-X} \qquad \underset{\text{Acid anhydrides}}{\overset{\displaystyle O}{\overset{\|}{RC}}-O\overset{\displaystyle O}{\overset{\|}{C}}R}$$

All contain an **acyl group**, $\overset{\displaystyle O}{\overset{\|}{RC}}-$ or $\overset{\displaystyle O}{\overset{\|}{ArC}}-$, attached to a nucleophilic group ($-OH$,

$-OR$, $-NH_2$, $-X$, or $-O\overset{\displaystyle O}{\overset{\|}{C}}R$).

Many of the reactions of these compounds involve their interconversion by substitution of one nucleophilic group by another. In the chemistry of alcohols, ethers, amines, and alkyl halides, we encountered similar interchanges of nucleophilic groups.

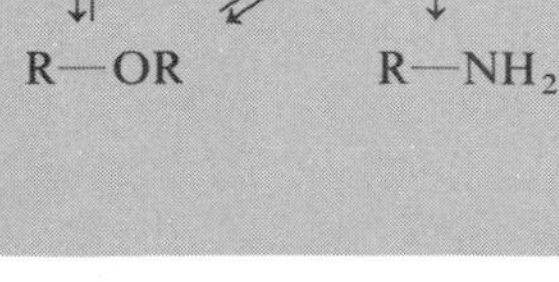

$$
\begin{array}{ccc}
R{-}OH & \rightleftharpoons & R{-}X \\
\updownarrow & \diagup & \downarrow \\
R{-}OR & & R{-}NH_2
\end{array}
$$

The nucleophilic substitution reactions of the acyl compounds, however, occur by different mechanisms and often more easily. Acyl compounds usually show the following order of reactivity.

$$
\underset{\substack{\| \\ RC-X}}{O} > \underset{\substack{\| \\ RC-OCR}}{O\ \ O} > \underset{\substack{\| \\ RC-OR}}{O} > \underset{\substack{\| \\ RC-NH_2}}{O}
$$

Substitution reactions ordinarily convert a more reactive to a less reactive compound. An acyl halide is converted readily to an ester; the reverse reaction is not easily achieved.

Esters and amides are particularly important families of compounds. Earlier we defined esters as compounds formed from an alcohol and an acid with loss of water. The esters discussed (Sec. 5.7) were of inorganic acids. The esters considered here are of carboxylic acids. Ester functions are important in lipids (Chapter 13) and many other biological compounds and are found in commercially important molecules, including such important polymers as Dacron. The amino group of amides may be —NH_2 but may also have organic groups in place of the hydrogens.

$$
\underset{\substack{\| \\ RC-NH_2}}{O} \qquad \underset{\substack{\| \\ RC-NHR}}{O} \qquad \underset{\substack{\| \\ RC-NR_2}}{O}
$$

It is amide functions that link amino acids together into proteins (Chapter 14). Amide functions are also found in many biological and commercial compounds, including the polymers known as nylons.

Acyl halides are not found in biological systems and acid anhydrides are rare. These highly reactive compounds are important mainly for their use in preparing esters and amides. The word **anhydride** refers to a compound that on reaction with water produces an acid. SO_3, for instance, is the anhydride of sulfuric acid (H_2SO_4). The acid anhydrides that we consider are formed in principle, although not usually in practice, from two molecules of a carboxylic acid.

$$
\underset{\substack{\| \\ RC-OH}}{O} + \underset{\substack{\| \\ HO-CR}}{O} \longrightarrow \underset{\substack{\| \\ RC-O-CR}}{O\ \ \ \ O} + H_2O
$$

12.1
NOMENCLATURE OF ACYL HALIDES AND ACID ANHYDRIDES

Acyl halides are given radicofunctional names formed by naming the acyl group, followed by the name of the halide (usually chloride). The names of acyl groups are constructed by replacing the *-ic acid* ending of either the trivial or substitutive name of the corresponding carboxylic acid by *-yl.*

$$
\underset{\substack{\text{Acetyl chloride}\\ \text{Ethanoyl chloride}}}{CH_3\overset{\overset{\displaystyle O}{\|}}{C}-Cl}
\qquad
\underset{\substack{\text{2-Bromobutyryl chloride}\\ \text{2-Bromobutanoyl chloride}}}{CH_3CH_2\underset{\underset{\displaystyle Br}{|}}{C}H\overset{\overset{\displaystyle O}{\|}}{C}-Cl}
\qquad
\underset{\text{Benzoyl chloride}}{\bigcirc\!\!-\overset{\overset{\displaystyle O}{\|}}{C}-Cl}
$$

Acid anhydrides are named by adding the word anhydride to the name (or names) of the corresponding carboxylic acid (or acids).

$$
\underset{\text{Acetic anhydride}}{CH_3\overset{\overset{\displaystyle O}{\|}}{C}-O-\overset{\overset{\displaystyle O}{\|}}{C}CH_3}
\qquad
\underset{\text{Benzoic anhydride}}{\bigcirc\!\!-\overset{\overset{\displaystyle O}{\|}}{C}-O-\overset{\overset{\displaystyle O}{\|}}{C}\!\!-\bigcirc}
\qquad
\underset{\text{Acetic propionic anhydride}}{CH_3\overset{\overset{\displaystyle O}{\|}}{C}-O-\overset{\overset{\displaystyle O}{\|}}{C}CH_2CH_3}
$$

12.2
PREPARATIONS OF ACYL HALIDES AND ACID ANHYDRIDES

Acyl halides can be prepared from carboxylic acids by use of phosphorus trihalides (PCl_3, PBr_3, or PI_3) or thionyl chloride ($SOCl_2$).

$$
3\ \underset{\substack{\text{Butyric acid}\\ \text{Butanoic acid}}}{CH_3CH_2CH_2\overset{\overset{\displaystyle O}{\|}}{C}-OH} + PCl_3 \longrightarrow 3\ \underset{\substack{\text{Butyryl chloride}\\ \text{Butanoyl chloride}}}{CH_3CH_2CH_2\overset{\overset{\displaystyle O}{\|}}{C}-Cl} + H_3PO_3
$$

$$
\underset{\text{Phenylacetic acid}}{\bigcirc\!\!-CH_2\overset{\overset{\displaystyle O}{\|}}{C}-OH} + SOCl_2 \longrightarrow \underset{\text{Phenylacetyl chloride}}{\bigcirc\!\!-CH_2\overset{\overset{\displaystyle O}{\|}}{C}-Cl} + SO_2\uparrow + HCl\uparrow
$$

The same reagents are also used to convert alcohols to alkyl halides (Sec. 5.9).

If the ring that forms is five or six membered, a dicarboxylic acid can usually be converted to a cyclic anhydride by heating.

$$
\underset{\text{Phthalic acid}}{\bigcirc\!\!\begin{array}{l}\overset{\overset{\displaystyle O}{\|}}{C}OH\\ \overset{\underset{\displaystyle O}{\|}}{C}OH\end{array}}
\xrightarrow{\text{heat}}
\underset{\text{Phthalic anhydride}}{\bigcirc\!\!\begin{array}{l}\overset{\overset{\displaystyle O}{\|}}{C}\\ \overset{\underset{\displaystyle O}{\|}}{C}\end{array}\!\!O}
+ H_2O
$$

$$\underset{\text{Maleic acid}}{\begin{array}{c} \text{H} \quad \quad \text{COH} \\ \diagdown \text{C} \diagup \quad \| \\ \| \quad \quad \text{O} \\ \text{C} \\ \diagup \quad \diagdown \\ \text{H} \quad \quad \text{COH} \\ \| \\ \text{O} \end{array}} \quad \xrightarrow{\text{heat}} \quad \underset{\text{Maleic anhydride}}{\begin{array}{c} \text{H} \quad \quad \text{O} \\ \diagdown \text{C} \diagup \| \\ \| \quad \text{C} \quad \text{O} \\ \text{C} \diagdown \quad \diagup \\ \diagup \quad \text{O} \\ \text{H} \quad \| \\ \text{O} \end{array}} + \text{H}_2\text{O}$$

Acyclic anhydrides must be prepared in other, less direct ways. The most general preparation, although not often used, is the reaction of an acyl halide and a salt of a carboxylic acid.

$$\underset{\substack{\text{Propionyl chloride} \\ \text{Propanoyl chloride}}}{\text{CH}_3\text{CH}_2\overset{\text{O}}{\overset{\|}{\text{C}}}\text{—Cl}} + \underset{\substack{\text{Sodium propionate} \\ \text{Sodium propanoate}}}{\text{CH}_3\text{CH}_2\overset{\text{O}}{\overset{\|}{\text{C}}}\text{O}^-\,\text{Na}^+} \longrightarrow \underset{\substack{\text{Propionic anhydride} \\ \text{Propanoic anhydride}}}{\text{CH}_3\text{CH}_2\overset{\text{O}}{\overset{\|}{\text{C}}}\text{—O—}\overset{\text{O}}{\overset{\|}{\text{C}}}\text{CH}_2\text{CH}_3} + \text{Na}^+\,\text{Cl}^-$$

12.3
REACTIONS OF ACYL HALIDES
AND ACID ANHYDRIDES

Acyl halides and acid anhydrides are readily converted to esters and amides and their most common use is for this purpose.

$$\underset{\text{Acetyl chloride}}{\text{CH}_3\overset{\text{O}}{\overset{\|}{\text{C}}}\text{—Cl}} + \text{HOCH}_2\text{CH}_2\text{CH}_3 \longrightarrow \underset{\substack{\text{Propyl acetate} \\ \text{Propyl ethanoate}}}{\text{CH}_3\overset{\text{O}}{\overset{\|}{\text{C}}}\text{—OCH}_2\text{CH}_2\text{CH}_3} + \text{HCl}$$

$$\underset{\text{Benzoyl chloride}}{\bigcirc\!\!-\overset{\text{O}}{\overset{\|}{\text{C}}}\text{—Cl}} + \text{NH}_3 \longrightarrow \underset{\text{Benzamide}}{\bigcirc\!\!-\overset{\text{O}}{\overset{\|}{\text{C}}}\text{—NH}_2} + \text{HCl}$$

$$\underset{\text{Acetic anhydride}}{\text{CH}_3\overset{\text{O}}{\overset{\|}{\text{C}}}\text{—O—}\overset{\text{O}}{\overset{\|}{\text{C}}}\text{CH}_3} + \text{HO—}\bigcirc \longrightarrow \underset{\text{Cyclopentyl acetate}}{\text{CH}_3\overset{\text{O}}{\overset{\|}{\text{C}}}\text{—O—}\bigcirc} + \text{CH}_3\overset{\text{O}}{\overset{\|}{\text{C}}}\text{—OH}$$

$$\underset{\text{Acetic anhydride}}{\text{CH}_3\overset{\text{O}}{\overset{\|}{\text{C}}}\text{—O—}\overset{\text{O}}{\overset{\|}{\text{C}}}\text{CH}_3} + \text{H}_2\text{N—}\bigcirc \longrightarrow \underset{\textit{N}\text{-Cyclopentylacetamide}}{\text{CH}_3\overset{\text{O}}{\overset{\|}{\text{C}}}\text{—NH—}\bigcirc} + \text{CH}_3\overset{\text{O}}{\overset{\|}{\text{C}}}\text{—OH}$$

The reagents that effect these conversions are those that you would probably guess might do the job. Alcohols are used to prepare esters; ammonia, primary amines, and secondary amines are used to prepare amides. There is no equivalent reaction of tertiary amines, for they lack the proton that must be lost from the nucleophilic reagent in these substitution reactions. A reaction in which an acyl compound is converted to an ester by reaction with an alcohol is called an **alcoholysis**. Similarly, a reaction in which an acyl compound is converted to an amide by reaction with ammonia or an amine is called an **ammonolysis**.

Acyl halides and acid anhydrides react readily with water to form carboxylic acids.

$$CH_3\overset{O}{\overset{\|}{C}}-O-\overset{O}{\overset{\|}{C}}CH_3 + H_2O \longrightarrow 2\ CH_3\overset{O}{\overset{\|}{C}}-OH$$

$$\text{Ph}-\overset{O}{\overset{\|}{C}}-Cl + H_2O \longrightarrow \text{Ph}-\overset{O}{\overset{\|}{C}}-OH + HCl$$

These **hydrolysis** reactions are rarely useful, for acyl halides and acid anhydrides are ordinarily prepared from carboxylic acids. To prevent hydrolysis, acyl halides and acid anhydrides must be kept from contacting water or even moist air. Many low molecular weight acyl halides are lachrymators, substances that irritate the eyes and produce tears. This action is due in part to their ready hydrolysis to produce the strongly acidic hydrogen halides.

In the presence of an appropriate catalyst an acyl halide reacts with an aromatic compound to give a ketone.

$$CH_3\overset{O}{\overset{\|}{C}}-Cl + \text{C}_6\text{H}_6 \xrightarrow{\text{AlCl}_3} CH_3\overset{O}{\overset{\|}{C}}-\text{C}_6\text{H}_5 + HCl$$

Acetophenone
Methyl phenyl ketone

This **acylation** reaction is analogous to the **alkylation** reaction of an alkyl halide with an aromatic compound (Sec. 4.4). Cyclic ketones, particularly with five- or six-membered rings, can be prepared from acyl halides containing an aryl group.

$$\text{Ph}-CH_2CH_2\overset{O}{\overset{\|}{C}}-OH \xrightarrow{\text{SOCl}_2} \text{Ph}-CH_2CH_2\overset{O}{\overset{\|}{C}}-Cl \xrightarrow{\text{AlCl}_3} \text{(indanone)}$$

12.4
NOMENCLATURE AND PHYSICAL PROPERTIES
OF ESTERS

Names of esters are formed by naming the organic group attached directly to oxygen, followed by the name of the corresponding acid with the *-ic acid* ending changed to *-ate*.

$$\underset{\substack{\text{Methyl acetate} \\ \text{Methyl ethanoate}}}{CH_3\overset{\displaystyle O}{\overset{\|}{C}}-OCH_3} \qquad \underset{\text{Ethyl benzoate}}{\bigcirc-\overset{\displaystyle O}{\overset{\|}{C}}-OCH_2CH_3} \qquad \underset{\substack{\text{Phenyl butyrate} \\ \text{Phenyl butanoate}}}{CH_3CH_2CH_2\overset{\displaystyle O}{\overset{\|}{C}}-O-\bigcirc}$$

This nomenclature is similar to that used for salts of carboxylic acids.

Most low molecular weight esters are pleasant smelling liquids. The fragrances of fruits and flowers are often due to volatile esters, although contributions from aldehydes, ketones, alcohols, and hydrocarbons are sometimes important. Some common esters with characteristic odors are listed here.

$$\underset{\substack{\text{Ethyl formate} \\ \text{(rum)}}}{HC\overset{\displaystyle O}{\overset{\|}{}}-OCH_2CH_3} \qquad \underset{\substack{\text{Ethyl butyrate} \\ \text{(pineapple)}}}{CH_3CH_2CH_2\overset{\displaystyle O}{\overset{\|}{C}}-OCH_2CH_3} \qquad \underset{\substack{\text{Isobutyl formate} \\ \text{(raspberry)}}}{HC\overset{\displaystyle O}{\overset{\|}{}}-OCH_2\underset{\underset{\displaystyle CH_3}{|}}{CH}CH_3}$$

$$\underset{\substack{\text{Methyl salicylate} \\ \text{(wintergreen)}}}{\underset{\underset{\displaystyle OH}{|}}{\bigcirc}-\overset{\displaystyle O}{\overset{\|}{C}}-OCH_3} \qquad \underset{\substack{\text{Pentyl butyrate} \\ \text{(apricot)}}}{CH_3CH_2CH_2\overset{\displaystyle O}{\overset{\|}{C}}-OCH_2CH_2CH_2CH_2CH_3} \qquad \underset{\substack{\text{Octyl acetate} \\ \text{(orange)}}}{CH_3\overset{\displaystyle O}{\overset{\|}{C}}-O(CH_2)_7CH_3}$$

Esters cannot hydrogen bond to one another and hence have boiling points not much higher than those of hydrocarbons of equivalent molecular weight. Their oxygens, however, can hydrogen bond to the hydrogens of water. As a result, esters have solubilities comparable to those of alcohols with the same numbers of carbon atoms.

12.5
PREPARATIONS OF ESTERS

Esters can be prepared by heating carboxylic acids and alcohols in the presence of an acid catalyst. Acetic acid and ethanol heated with a small amount of sulfuric acid produce ethyl acetate and water.

$$\underset{\substack{\|\\ O}}{CH_3C}-OH + HOCH_2CH_3 \underset{}{\overset{H^+}{\rightleftharpoons}} \underset{\substack{\|\\ O}}{CH_3C}-OCH_2CH_3 + H_2O$$

Such reactions, called **esterifications**, are readily reversible. Appreciable quantities of both the reactants and the products usually are present at equilibrium. Ester formation is favored by using a large excess of one of the reactants (whichever is most available) or by removing one of the products as it forms.

The reactions (described in Sec. 12.3) of acyl chlorides and acid anhydrides with alcohols offer the advantage of being essentially irreversible and, for this reason, are often used to prepare esters.

---

problem 12.1   Give the structure and name of the ester formed from
(a) butyl alcohol and benzoic acid.
(b) benzyl alcohol and butyric acid.
(c) allyl alcohol and acetic anhydride.
(d) isopropyl alcohol and propionyl chloride.

12.6
REACTIONS OF ESTERS

The most important reaction of esters is hydrolysis—cleavage with water to produce a carboxylic acid and an alcohol. An ester can be hydrolyzed by heating with water in the presence of an acid catalyst.

$$\underset{\substack{\|\\ O}}{CH_3C}-OCH_2CH_3 + H_2O \underset{}{\overset{H^+}{\rightleftharpoons}} \underset{\substack{\|\\ O}}{CH_3C}-OH + HOCH_2CH_3$$

This reaction is simply the reverse of the acid-catalyzed esterification described in the preceding section. Formation of hydrolysis products is favored by using a large excess of water.

Alternatively, an ester can be hydrolyzed by heating in an aqueous hydroxide solution.

$$\underset{\substack{\|\\ O}}{CH_3C}-OCH_2CH_3 + Na^+ OH^- \longrightarrow \underset{\substack{\|\\ O}}{CH_3C}-O^- Na^+ + CH_3CH_2OH$$

In the basic reaction solution the carboxylic acid is converted to its anion, rendering the hydrolysis essentially irreversible. Therefore use of base is more effective for completely hydrolyzing esters, although at the cost of using a mole of base for each mole of ester.

Esters can be converted to amides by heating with ammonia or amines.

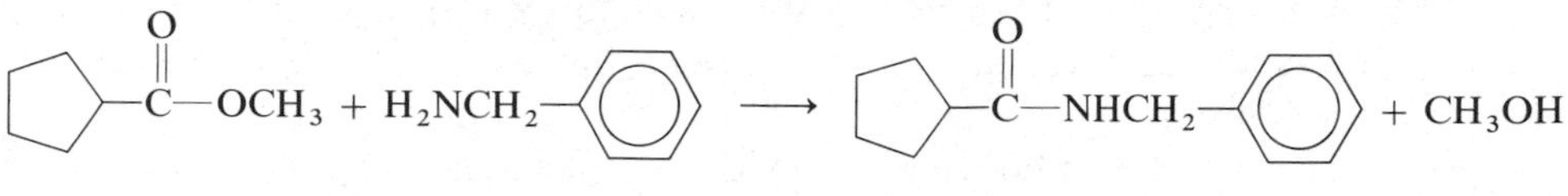

A related reaction is the synthesis of barbituric acid by heating the diethyl ester of malonic acid with urea.

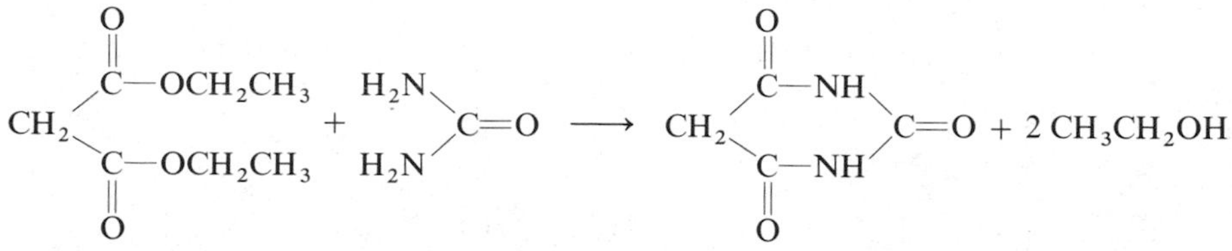

| Diethyl malonate | Urea | Barbituric acid |

Similar reactions of urea with substituted malonic esters yield the substituted barbituric acids called barbituates. This family of compounds, some of which are shown below, includes some of the most extensively used drugs.

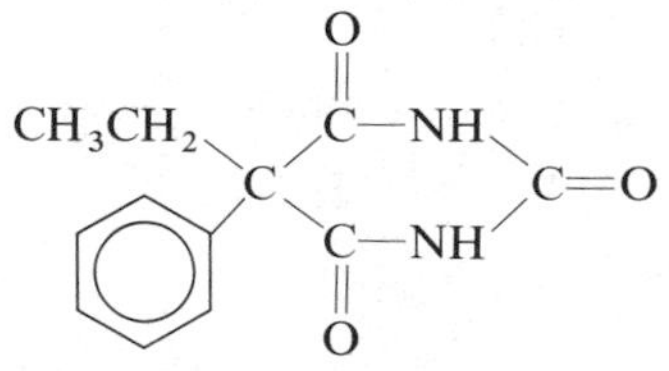

Phenobarbital
(luminal)

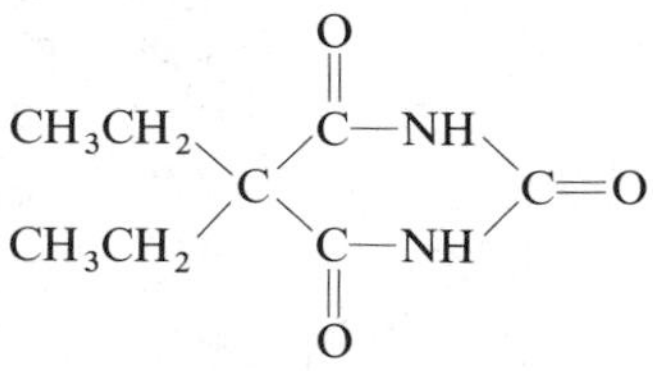

Barbital
(veronal)

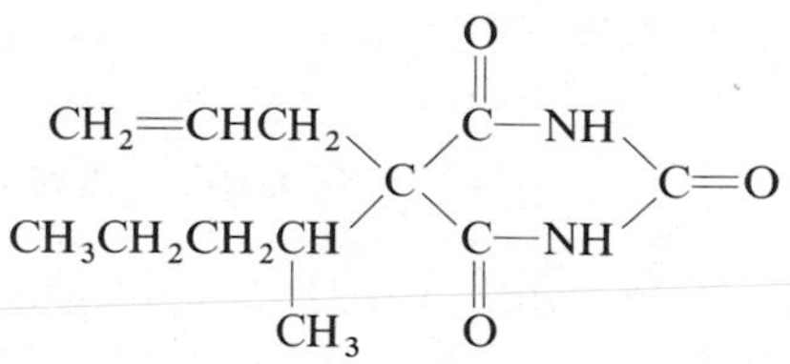

Secobarbital
(seconal)

Barbituates have a depressing effect on the central nervous system that ranges from mild sedation to deep anesthesia.

By heating with an alcohol (alcoholysis) in the presence of an acidic or basic catalyst (usually a salt of R'O⁻), an ester can be converted to an ester of the new alcohol.

$$\underset{O}{\overset{O}{\underset{\parallel}{RC}}}-OR + R'OH \rightleftharpoons \underset{O}{\overset{O}{\underset{\parallel}{RC}}}-OR' + ROH$$

Because such **ester interchange reactions** are readily reversible, high yields of the new ester can be obtained only if the alcohol ($R'OH$) is used in excess or one of the products is removed as it forms. Use of this reaction to form Dacron is illustrated in Section 12.16.

An ester can be reduced to yield two alcohols. One is the alcohol from which the ester is prepared, the other a primary alcohol corresponding to the acyl group.

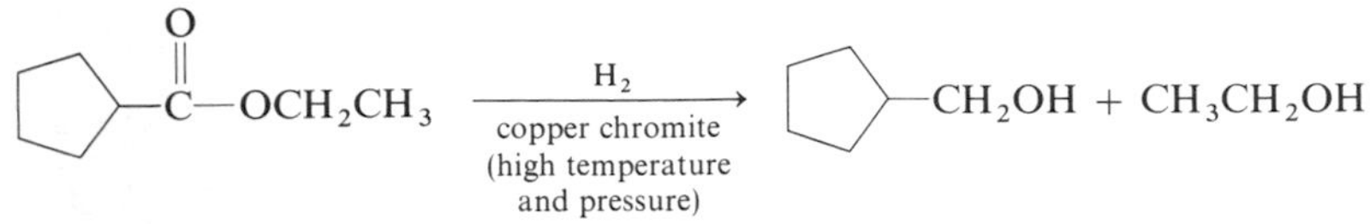

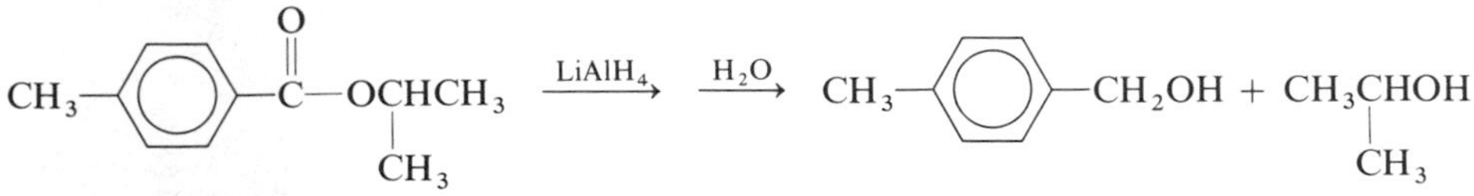

Reduction of an ester with H_2 and a catalyst is called **hydrogenolysis**. The catalyst is often a mixture of oxides of copper and chromium called copper chromite. Considerably more severe conditions are needed than for reductions of alkenes, aldehydes, or ketones. Reduction can also be effected by $LiAlH_4$.

Cleavage of an ester to form two alcohols is easily understood in terms of already familiar reaction steps. Consider the reaction of an ester with H_2.

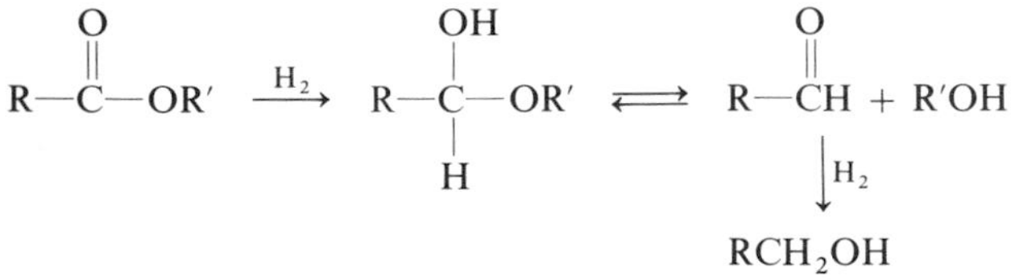

Just as in the reduction of aldehydes and ketones, the initial step is addition of H_2 to $C=O$. This addition produces a hemiacetal, which we know to be in equilibrium with an aldehyde and an alcohol. The aldehyde is easily reduced to a primary alcohol.

12.7
MECHANISMS OF ESTER HYDROLYSIS

The mechanism of ester hydrolysis in either acid or base usually involves addition of water to the carbonyl group.

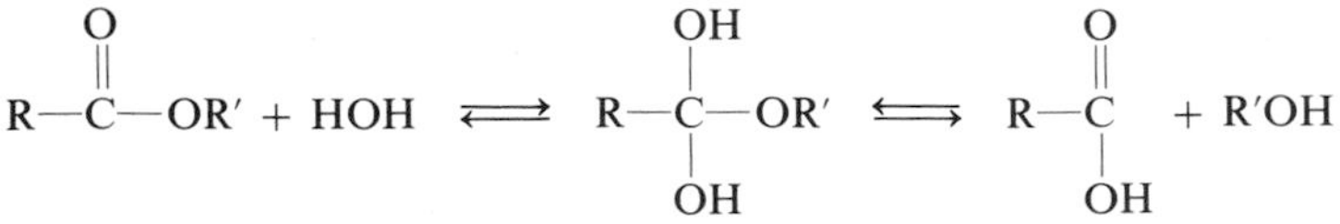

This addition is analogous to the formation of hydrates from addition of water to aldehydes and ketones. Just as hydrates readily revert to aldehydes or ketones, so the intermediate from addition to an ester readily loses water to regenerate the ester. This intermediate, however, can also lose R′OH to form the carboxylic acid.

Either acid or base catalyzes the addition of water to an ester. A water molecule is too weakly nucleophilic to rapidly attack the carbonyl group of an ester. The hydroxide ion, however, is more nucleophilic and attacks more readily.

$$
\begin{array}{c}
\underset{\displaystyle \text{R--C--OR'}}{\overset{\displaystyle O}{\parallel}}
\;\underset{-\text{OH}^-}{\overset{\text{OH}^-}{\rightleftharpoons}}\;
\underset{\displaystyle \underset{\text{OH}}{\overset{O^-}{|}}}{\text{R--C--OR'}}
\;\underset{-\text{H}^+}{\overset{\text{H}^+}{\rightleftharpoons}}\;
\underset{\displaystyle \underset{\text{OH}}{\overset{\text{OH}}{|}}}{\text{R--C--OR'}}
\end{array}
$$

A ⇅   B

$$
\underset{\displaystyle \text{R--C--OH}}{\overset{\displaystyle O}{\parallel}} + \text{R'O}^- \longrightarrow
\underset{\displaystyle \text{R--C--O}^-}{\overset{\displaystyle O}{\parallel}} + \text{R'OH}
$$

Intermediate A, resulting from hydroxide addition, is in equilibrium with B, the unstable product of addition of water. A can lose OH⁻ to regenerate the ester or lose R′O⁻ to form the carboxylic acid. Proton transfer rapidly converts the very basic R′O⁻ to R′OH and the acidic RCO₂H to RCO₂⁻. Because the anions of carboxylic acids do not react readily with nucleophiles, the essentially quantitative conversion of the carboxylic acid to its anion makes the reaction irreversible.

The function of an acid catalyst is to protonate the carbonyl group (step 1).

Step 1   Step 2   Step 3

$$
\underset{\displaystyle \text{R--C--OR'}}{\overset{\displaystyle O}{\parallel}} + \text{H}^+ \rightleftharpoons
\underset{\displaystyle C}{\underset{\displaystyle \text{R--}\overset{+}{\text{C}}\text{--OR'}}{\overset{\text{OH}}{|}}}
\;\underset{-\text{H}_2\text{O}}{\overset{\text{H}_2\text{O}}{\rightleftharpoons}}\;
\underset{\displaystyle \text{R--C--OR'}}{\overset{\text{OH}}{|}}
\;\rightleftharpoons\;
\underset{\displaystyle B}{\underset{\displaystyle \text{R--C--OR'}}{\overset{\text{OH}}{|}}} + \text{H}^+
$$

Step 6   Step 5   Step 4

$$
\underset{\displaystyle \text{R--C--OH}}{\overset{\displaystyle O}{\parallel}} + \text{R'OH} + \text{H}^+ \rightleftharpoons
\underset{\displaystyle \text{OH}}{\underset{\displaystyle \text{R--}\overset{+}{\text{C}}}{}} + \text{R'OH} \rightleftharpoons
\underset{\displaystyle \text{OH H}}{\underset{\displaystyle \text{R--C--}\overset{+}{\text{O}}\text{R'}}{\overset{\text{OH}}{|}}}
$$

This forms a small concentration of a positively charged intermediate (C), which is far more susceptible to attack by a nucleophile than the ester. Attack by water (step 2) and loss of a proton (step 3) generate intermediate B. The —OR′ group of B can be protonated by the acid (step 4), permitting the ready loss of R′OH (step 5). Loss of a proton (step 6) furnishes the carboxylic acid and regenerates the acid catalyst. All steps in this scheme are reversible; it is also the mechanism of acid-catalyzed esterification of carboxylic acids.

These mechanisms are supported by the demonstration that bond *a* rather than bond *b* is cleaved in most ester hydrolyses.

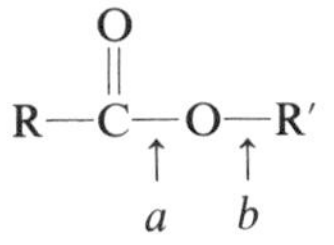

$$
\begin{array}{c}
O \\
\parallel \\
R-C\underset{\uparrow}{-}O\underset{\uparrow}{-}R' \\
ab
\end{array}
$$

One experiment showing the position of cleavage used isotopically labeled ethyl propionate.

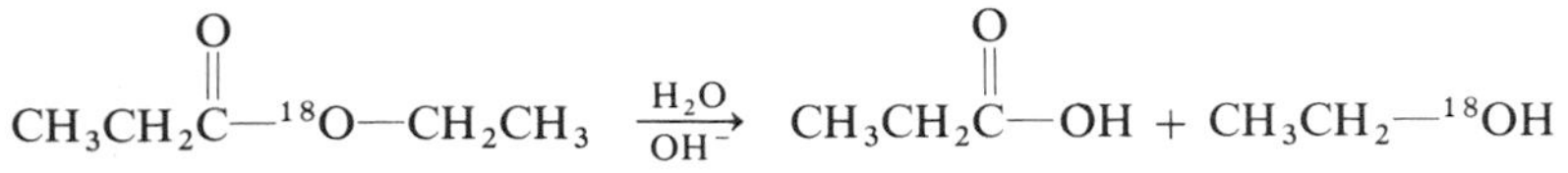

$$
\underset{\substack{\\}}{\overset{\displaystyle O}{\overset{\parallel}{CH_3CH_2C}}}-{}^{18}O-CH_2CH_3 \xrightarrow[OH^-]{H_2O} \underset{\substack{\\}}{\overset{\displaystyle O}{\overset{\parallel}{CH_3CH_2C}}}-OH + CH_3CH_2-{}^{18}OH
$$

Ethyl propionate was synthesized in which the alkoxy oxygen contained a much higher percentage of ^{18}O than the 0.20% in natural sources. After hydrolysis, the excess ^{18}O was found exclusively in the ethanol. This demonstrated that the $^{18}O-CH_2CH_3$ bond had not cleaved.

Other reactions of esters, amides, acyl halides, and acid anhydrides that involve substitution of one nucleophile by another often proceed by mechanisms similar to those for ester hydrolysis. A nucleophilic reagent adds to the carbonyl group to generate a tetrahedral carbon, followed by loss of the nucleophilic group already present.

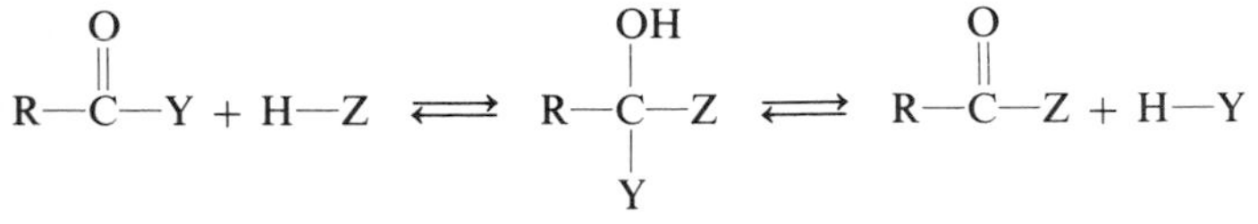

$$
\underset{\substack{\\}}{\overset{\displaystyle O}{\overset{\parallel}{R-C}}}-Y + H-Z \rightleftarrows \underset{\substack{| \\ Y}}{\overset{\displaystyle OH}{\overset{|}{R-C}}}-Z \rightleftarrows \underset{\substack{\\}}{\overset{\displaystyle O}{\overset{\parallel}{R-C}}}-Z + H-Y
$$

$$
Y \text{ and } Z \text{ are } -OH, -OR', -NH_2, -X, \text{ or } \overset{\displaystyle O}{\overset{\parallel}{-OCR'}}
$$

12.8
CLAISEN CONDENSATION

Hydrogens α to the carbonyl groups of esters are considerably more acidic than hydrogens of alkanes. Just as with aldehydes and ketones, the enhanced acidity of α hydrogens is due to formation of a resonance-stabilized anion in which a significant portion of the negative charge resides on oxygen.

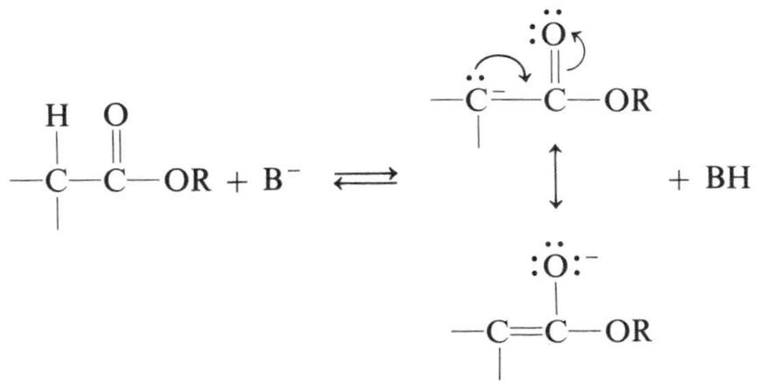

$$
\underset{\substack{| \\ }}{\overset{\displaystyle H \ \ O}{\overset{| \ \ \parallel}{-C-C}}}-OR + B^- \rightleftarrows
$$

The reactions discussed here and in the next section involve such anions.

When ethyl acetate is treated with sodium ethoxide in ethanol, a reaction resembling the aldol condensation (Sec. 9.12) occurs.

$$2 \ CH_3-\overset{\overset{\displaystyle O}{\|}}{C}-OCH_2CH_3 \ \xrightarrow{Na^+ \ ^-OCH_2CH_3} \ CH_3-\overset{\overset{\displaystyle O}{\|}}{C}-CH_2-\overset{\overset{\displaystyle O}{\|}}{C}-OCH_2CH_3 + CH_3CH_2OH$$

Ethyl acetoacetate
"Acetoacetic ester"

This reaction, known as the Claisen condensation, links together two molecules of the ester.

The mechanism involves familiar steps.

Step 1 $\quad CH_3-\overset{\overset{\displaystyle O}{\|}}{C}-OCH_2CH_3 + \ ^-OCH_2CH_3 \ \rightleftharpoons \ ^-CH_2-\overset{\overset{\displaystyle O}{\|}}{C}-OCH_2CH_3 + HOCH_2CH_3$

Step 2 $\quad CH_3-\underset{\underset{\displaystyle OCH_2CH_3}{|}}{\overset{\overset{\displaystyle O}{\|}}{C}} \ + \ ^-CH_2-\overset{\overset{\displaystyle O}{\|}}{C}-OCH_2CH_3 \ \rightleftharpoons \ CH_3-\underset{\underset{\displaystyle OCH_2CH_3}{|}}{\overset{\overset{\displaystyle O^-}{|}}{C}}-CH_2-\overset{\overset{\displaystyle O}{\|}}{C}-OCH_2CH_3$

Step 3 $\quad CH_3-\underset{\underset{\displaystyle OCH_2CH_3}{|}}{\overset{\overset{\displaystyle O^-}{|}}{C}}-CH_2-\overset{\overset{\displaystyle O}{\|}}{C}-OCH_2CH_3 \ \rightleftharpoons \ CH_3-\overset{\overset{\displaystyle O}{\|}}{C}-CH_2-\overset{\overset{\displaystyle O}{\|}}{C}-OCH_2CH_3 + \ ^-OCH_2CH_3$

In the first step ethoxide ion reacts with the ester to generate a low concentration of the ester anion. Just as do other nucleophiles, this anion attacks the carbonyl carbon of another molecule of the ester (step 2). Cleavage of the resulting intermediate (step 3) forms ethyl acetoacetate and regenerates ethoxide ion.

These steps are all reversible and the equilibrium for the overall reaction favors the reactants. Product formation can be favored, however, by taking advantage of the product's acidity. Ethyl acetoacetate is considerably more acidic than ethyl acetate and is converted almost quantitatively to an anion by ethoxide ion.

$$CH_3-\overset{\overset{\displaystyle O^-}{|}}{C}=CH-\overset{\overset{\displaystyle O}{\|}}{C}-OCH_2CH_3$$

$$\updownarrow$$

$$CH_3-\overset{\overset{\displaystyle O}{\|}}{C}-CH_2-\overset{\overset{\displaystyle O}{\|}}{C}-OCH_2CH_3 + \ ^-OCH_2CH_3 \ \longrightarrow \ CH_3-\overset{\overset{\displaystyle O}{\|}}{C}-CH=\overset{\overset{\displaystyle O}{\|}}{C}-OCH_2CH_3 + HOCH_2CH_3$$

Ethyl acetoacetate

$$\updownarrow$$

$$CH_3-\overset{\overset{\displaystyle O}{\|}}{C}-CH=\overset{\overset{\displaystyle O^-}{|}}{C}-OCH_2CH_3$$

The greater acidity is due to formation of a resonance-stabilized anion in which a significant portion of the negative charge is divided between *two* oxygens. Claisen reactions are carried out by using sufficient alkoxide ion to convert all the product to its anion, thereby pushing the equilibrium toward product formation. At the conclusion of the reaction, acid is added to protonate the anion and generate the neutral product.

Diesters react in an intramolecular fashion when doing so results in the formation of five- or six-membered rings.

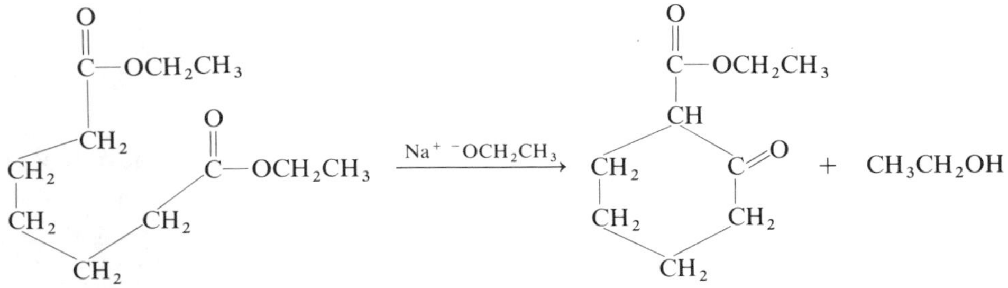

Claisen condensations synthesize esters with ketone functions at the β position. Hydrolysis yields the corresponding β-keto acids.

$$CH_3-\overset{\overset{\displaystyle O}{\|}}{C}-CH_2-\overset{\overset{\displaystyle O}{\|}}{C}-OCH_2CH_3 \xrightarrow[H^+]{H_2O} CH_3-\overset{\overset{\displaystyle O}{\|}}{C}-CH_2-\overset{\overset{\displaystyle O}{\|}}{C}-OH \xrightarrow{\text{heat}} CH_3-\overset{\overset{\displaystyle O}{\|}}{C}-CH_3 + CO_2$$

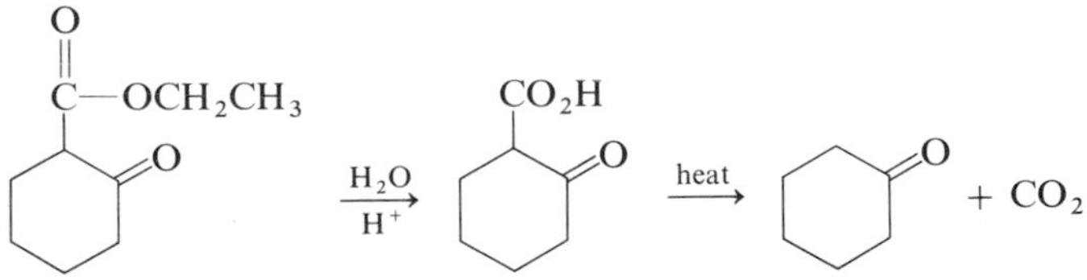

Carboxylic acids that contain β-carbonyl groups lose carbon dioxide on heating. This ready **decarboxylation** involves a cyclic mechanism that produces an enol.

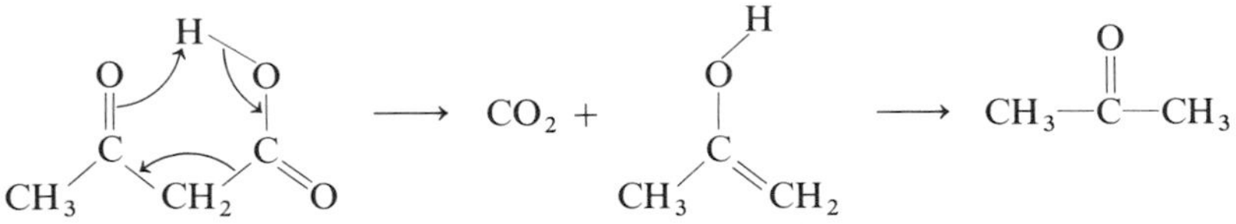

The enol rearranges rapidly to the more stable ketone.

Claisen condensations and related reactions are among the most important ways of forming new carbon-carbon bonds in the laboratory. Such reactions or their reverse are also common in biological syntheses and degradations. One example occurs in the metabolism of long-chain carboxylic acids ("fatty acids"), an important energy source for higher animals. These acids are converted to thiol esters of coenzyme A.

$$RCH_2CH_2\overset{\displaystyle O}{\overset{\|}{C}}-OH \;+\; HS-CoA \;\longrightarrow\; RCH_2CH_2\overset{\displaystyle O}{\overset{\|}{C}}-S-CoA$$

Coenzyme A

Thiol esters are compounds of structure $R-\overset{\displaystyle O}{\overset{\|}{C}}-SR'$. Coenzyme A is a complex coenzyme that contains a thiol function. In a series of enzyme-catalyzed reactions the thiol ester of fatty acid 1 is converted to a β-keto thiol ester.

$$R-CH_2-CH_2-\overset{\displaystyle O}{\overset{\|}{C}}-S-CoA \;\longrightarrow\; \longrightarrow\; \longrightarrow\; R-\overset{\displaystyle O}{\overset{\|}{C}}-CH_2-\overset{\displaystyle O}{\overset{\|}{C}}-S-CoA$$

Thiol ester of acid 1

The β-keto ester is then cleaved in a reverse Claisen condensation to give acetyl coenzyme A and the thiol ester of a new carboxylic acid (2) with two carbons fewer than the original.

$$R-\overset{\displaystyle O}{\overset{\|}{C}}-CH_2-\overset{\displaystyle O}{\overset{\|}{C}}-S-CoA \;+\; HS-CoA \;\longrightarrow\; R-\overset{\displaystyle O}{\overset{\|}{C}}-S-CoA \;+\; CH_3-\overset{\displaystyle O}{\overset{\|}{C}}-S-CoA$$

Coenzyme A   Thiol ester   Acetyl coenzyme A
                     of acid 2

The new thiol ester can undergo the same reaction sequence to lose another two carbons.

Another biological example is the enzyme-catalyzed condensation reaction of acetoacetyl coenzyme A and acetyl coenzyme A.

$$CH_3-\overset{\displaystyle O}{\overset{\|}{C}}-CH_2-\overset{\displaystyle O}{\overset{\|}{C}}-S-CoA \;+\; CH_3-\overset{\displaystyle O}{\overset{\|}{C}}-S-CoA \;\longrightarrow\; CH_3-\overset{\displaystyle OH}{\overset{|}{C}}-CH_2-\overset{\displaystyle O}{\overset{\|}{C}}-S-CoA \;\longrightarrow$$

Acetoacetyl   Acetyl
coenzyme A   coenzyme A

with the branch below the central carbon:
$$\begin{array}{c} CH_2 \\ | \\ C-S-CoA \\ \| \\ O \end{array}$$

$$\longrightarrow \longrightarrow CH_3-\overset{\displaystyle OH}{\overset{|}{\underset{\overset{|}{\underset{CO_2H}{CH_2}}}{C}}}-CH_2-CH_2-OH \longrightarrow \longrightarrow \longrightarrow CH_2{=}\overset{\overset{\textstyle }{}}{\underset{\overset{|}{CH_3}}{C}}-CH_2-CH_2-O-\overset{\displaystyle O}{\overset{\|}{\underset{\overset{|}{OH}}{P}}}-O-\overset{\displaystyle O}{\overset{\|}{\underset{\overset{|}{OH}}{P}}}-OH$$

Mevalonic acid   3-Methyl-3-butenyl diphosphate

Addition of the α carbon of acetyl coenzyme A to the carbonyl group of acetoacetyl coenzyme A resembles a Claisen or aldol condensation. In subsequent steps the condensation product is converted to mevalonic acid, which is ultimately converted to 3-methyl-3-butenyl diphosphate (Sec. 7.9), the precursor of terpenes.

12.9
ALKYLATION OF ESTERS

The anion formed from ethyl acetoacetate gives nucleophilic substitution reactions with alkyl halides.

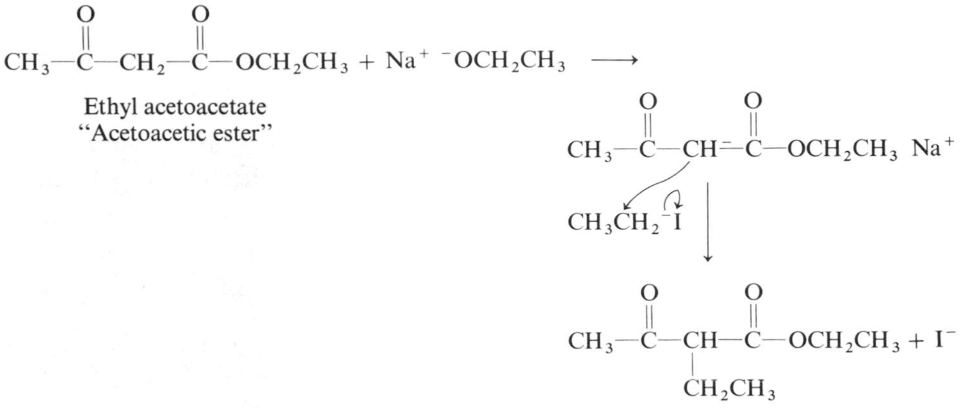

$$CH_3-\underset{\underset{O}{\|}}{C}-CH_2-\underset{\underset{O}{\|}}{C}-OCH_2CH_3 + Na^+ \ ^-OCH_2CH_3 \longrightarrow$$

Ethyl acetoacetate
"Acetoacetic ester"

The result of treating ethyl acetoacetate with base followed by an alkyl halide is to replace a hydrogen by an alkyl group, a reaction called **alkylation**. A general scheme for the alkylation of ethyl acetetoacetate is shown.

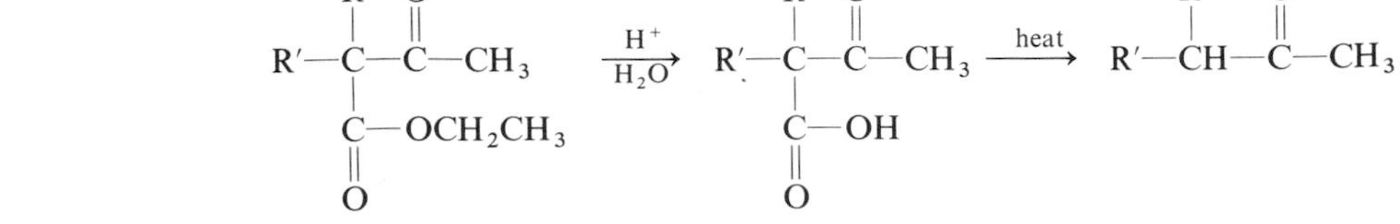

Because the product of alkylation of ethyl acetoacetate still has an acidic α hydrogen, conversion to an anion and reaction with an alkyl halide can be repeated in order to introduce a second alkyl group. The β-keto esters can be hydrolyzed to carboxylic acids that readily lose carbon dioxide. This scheme, sometimes called the acetoacetic ester synthesis, provides a synthesis of ketones having the general structure

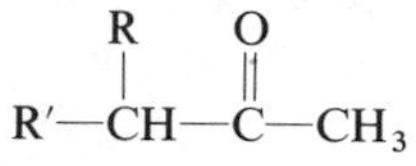

$$
\begin{array}{cc}
R & O \\
| & \parallel \\
R'-CH-C-CH_3 & \text{(R is alkyl, R' is alkyl or H)}
\end{array}
$$

from alkyl halides RX and R'X.

Diethyl malonate also forms a resonance-stabilized anion that reacts with alkyl halides.

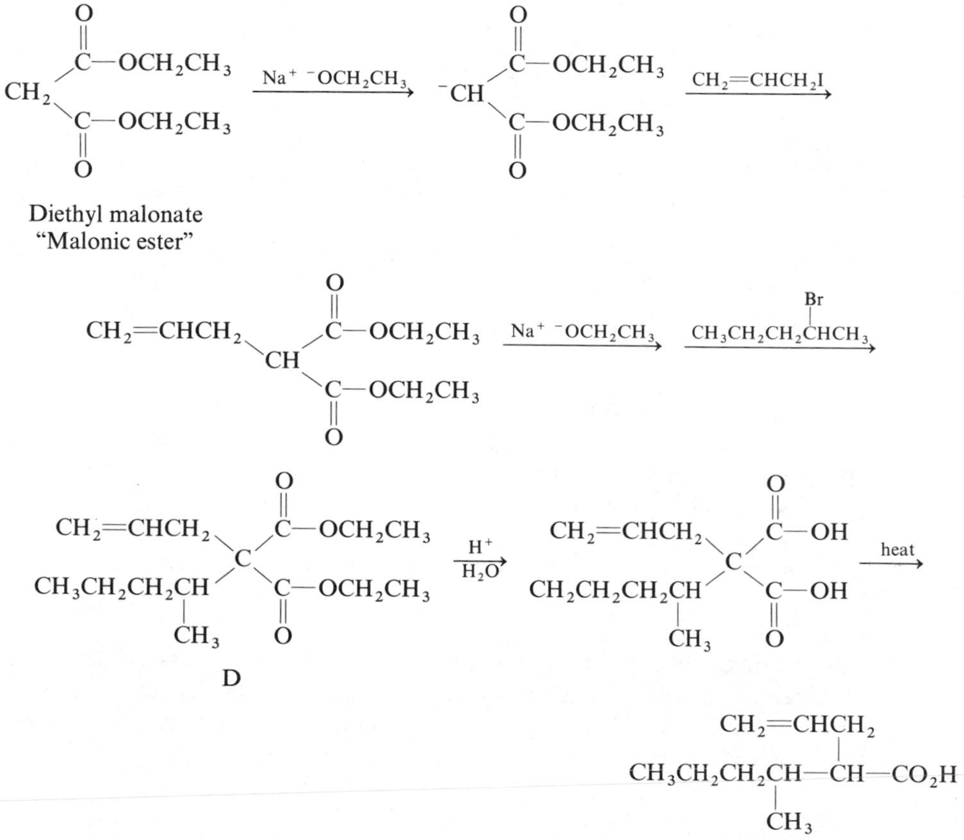

Diethyl malonate
"Malonic ester"

D

$$
\begin{array}{c}
CH_2{=}CHCH_2 \\
| \\
CH_3CH_2CH_2CH-CH-CO_2H \\
| \\
CH_3
\end{array}
$$

Product D obtained from the two alkylation steps is needed for the synthesis of secobarbital (Sec. 12.6). The dicarboxylic acids that form on hydrolysis of malonic esters have a carbonyl group β to a carboxyl group and readily lose a molecule of carbon dioxide. This scheme, sometimes called the malonic ester synthesis, provides a synthesis of carboxylic acids having the structure

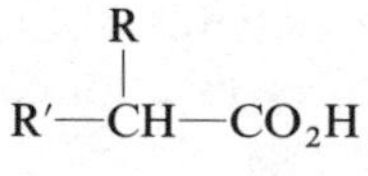

$$
\begin{array}{cc}
R & \\
| & \\
R'-CH-CO_2H & \text{(R is alkyl, R' is alkyl or H)}
\end{array}
$$

from alkyl halides RX and R'X.

Alkylations of ethyl acetoacetate or diethyl malonate are subject to the usual limitations of nucleophilic substitution reactions: tertiary halides ordinarily give elimination rather than substitution products and aryl halides do not react. Nevertheless, such alkylations are among the most significant methods of forming C—C bonds.

12.10
SOME IMPORTANT ESTERS

Di-2-ethylhexyl phthalate and other esters of phthalic acid are used in large amounts as plasticizers, particularly in polyvinyl chloride.

$$\underset{\text{Di-2-ethylhexyl phthalate}}{
\begin{array}{c}
\overset{\displaystyle O}{\overset{\displaystyle \parallel}{}}\qquad \overset{\displaystyle CH_2CH_3}{\overset{\displaystyle |}{}} \\
C-OCH_2CHCH_2CH_2CH_2CH_3 \\
C-OCH_2CHCH_2CH_2CH_2CH_3 \\
\overset{\displaystyle \parallel}{O}\qquad \overset{\displaystyle |}{CH_2CH_3}
\end{array}}$$

Plasticizer molecules fit between polymer chains, reducing attractive forces between them and enabling them to slide more easily over one another. As a result, the polymer is softer and less brittle.

Formaldehyde polymerizes readily to form a linear polymer.

$$n\,CH_2{=}O + H_2O \;\rightleftharpoons\; HO{-}CH_2{-}O{-}CH_2{-}O{-}CH_2{-}O \sim\!\sim\!\sim CH_2{-}O{-}H$$
$$= HO{-}(CH_2{-}O)_{\overline{n}}\,H$$

On heating, however, the polymer is readily degraded by loss of formaldehyde (note the hemiacetal function at each end of the chain). This decomposition is prevented and the stable polymer called Delrin produced by reaction with acetic anhydride to incorporate the terminal hydroxyl groups into ester functions.

$$HO{-}(CH_2{-}O)_{\overline{n}}\,H + 2\,CH_3\overset{\displaystyle O}{\overset{\displaystyle \parallel}{C}}{-}O{-}\overset{\displaystyle O}{\overset{\displaystyle \parallel}{C}}CH_3 \longrightarrow$$

$$\underset{\text{Delrin}}{CH_3\overset{\displaystyle O}{\overset{\displaystyle \parallel}{C}}{-}O{-}(CH_2{-}O)_{\overline{n}}\overset{\displaystyle O}{\overset{\displaystyle \parallel}{C}}CH_3 + 2\,CH_3\overset{\displaystyle O}{\overset{\displaystyle \parallel}{C}}{-}OH}$$

Because of its weather resistance, the polymer of the ester methyl methacrylate is used to manufacture such items as windows, exterior construction panels, outdoor signs, and automobile tail lights.

Methyl methacrylate
Methyl 2-methylpropenoate

$\longrightarrow$

Polymethyl methacrylate
Lucite

The ester linkage occurs in many biologically important molecules. Aspirin, prepared from salicylic acid and acetic anhydride, has analgesic (pain-reducing), antipyretic (fever-reducing), and anti-inflammatory actions.

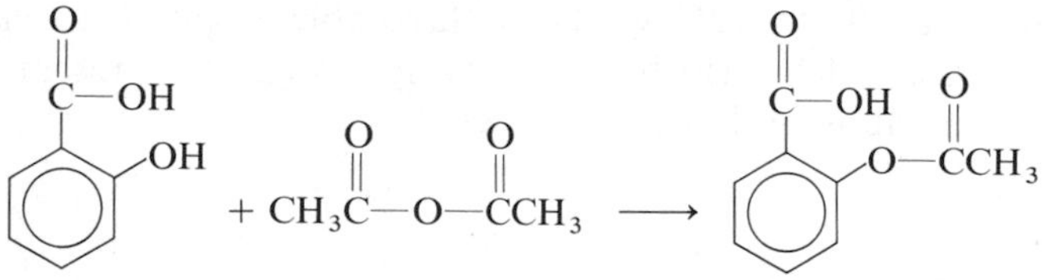

Salicylic acid $+ CH_3\overset{O}{\underset{\|}{C}}-O-\overset{O}{\underset{\|}{C}}CH_3$ $\longrightarrow$ Aspirin

Chrysanthemum flowers contain esters called pyrethrins that have insecticidal properties but are relatively nontoxic to mammals.

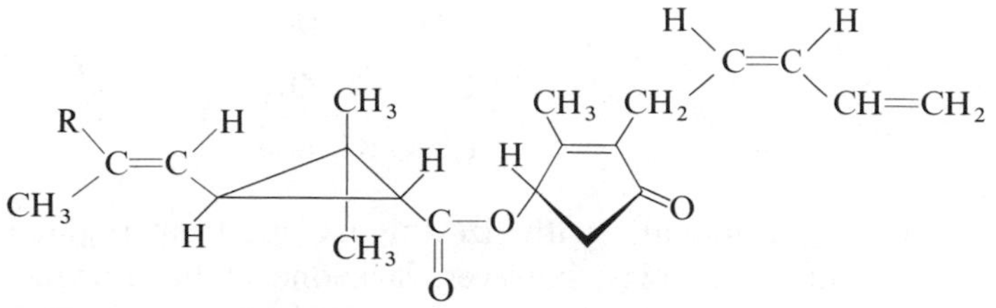

Pyrethrin I $-R = -CH_3$

Pyrethrin II $-R = -\overset{O}{\underset{\|}{C}}OCH_3$

Many pheromones are esters. The female sex attractant of the corn borer is a mixture of *cis*- and *trans*-11-tetradecenyl acetates.

$$CH_3CH_2CH=CH(CH_2)_{10}O-\overset{O}{\underset{\|}{C}}CH_3$$

11-Tetradecenyl acetate

Cyclic esters are called **lactones**. A variety of these heterocyclic compounds are found in living systems. Coumarin, a lactone found in sweet clover and used as a flavoring agent, has the odor of freshly mown hay.

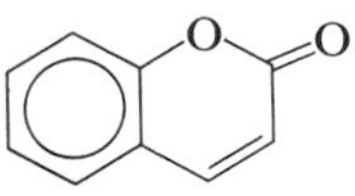

Coumarin

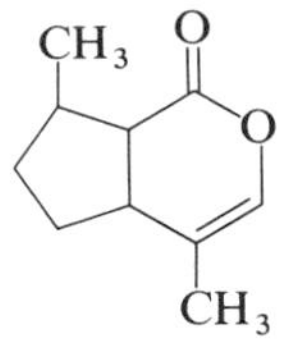

Nepetalactone

Nepetalactone is the constituent in the volatile oil of catnip that at levels as low as one part in 10^9 to 10^{11} evokes a favorable response in cats. L-Ascorbic acid (vitamin C), widely distributed in living systems, is present in particularly large amounts in citrus fruits and tomatoes.

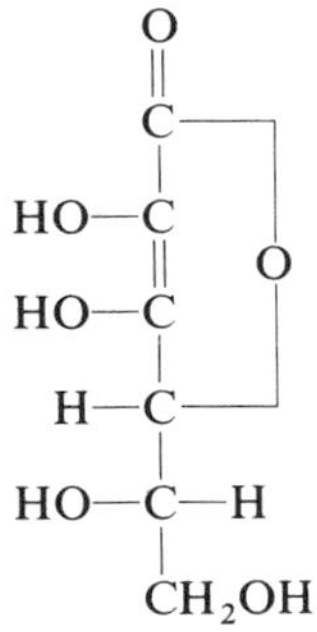

L-Ascorbic acid

Plants and most animals synthesize this lactone from D-glucose. Humans, other primates, and guinea pigs, however, lack one of the enzymes necessary for this synthesis and must obtain vitamin C from their diets.

12.11
AMIDES: NOMENCLATURE
AND PHYSICAL PROPERTIES

Amides are named by replacing the *-ic acid* (trivial) or *-oic acid* (substitutive) ending of the name of the corresponding carboxylic acid by *-amide*.

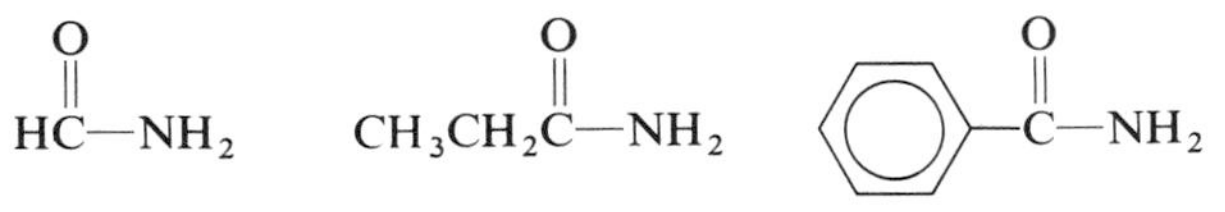

Formamide
Methanamide

Propionamide
Propanamide

Benzamide

Organic groups attached to nitrogen are named as prefixes and labeled with an *N* to identify their position of attachment.

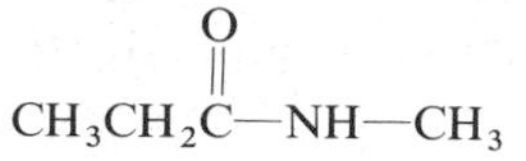

$$CH_3CH_2\overset{\overset{\displaystyle O}{\|}}{C}-NH-CH_3$$

N-Methylpropionamide
N-Methylpropanamide

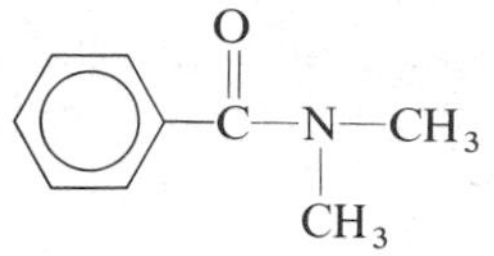

N,N-Dimethylbenzamide

As shown for the following isomeric amides of molecular weight 73, amides tend to have particularly high boiling points for their molecular weights and are often solids.

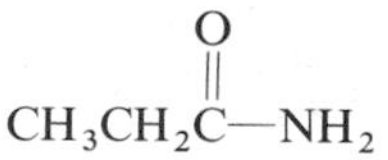

$$CH_3CH_2\overset{\overset{\displaystyle O}{\|}}{C}-NH_2 \qquad CH_3\overset{\overset{\displaystyle O}{\|}}{C}-NH-CH_3 \qquad H\overset{\overset{\displaystyle O}{\|}}{C}-N-CH_3$$

Propionamide	N-Methylacetamide	N,N-Dimethylformamide
(bp 213°	(bp 205°	(bp 152°
mp 81°)	mp 28°)	mp −60°)

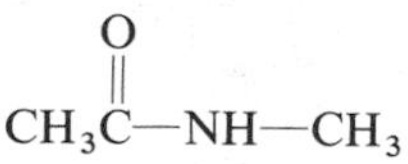

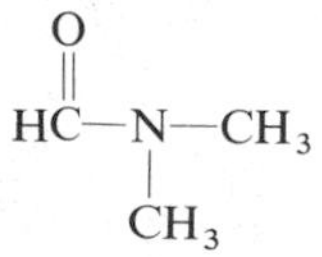

Amides that have N—H bonds hydrogen bond together extensively. Even amides (such as *N,N*-dimethylformamide) that lack N—H bonds, however, have quite high boiling points. Amides are slightly more water soluble than alcohols with the same number of carbons. Most amides with up to four carbons are miscible in water.

Resonance structure B is characteristic of carbonyl compounds.

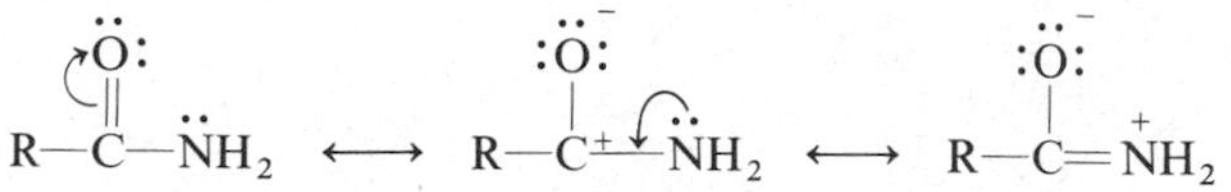

$$R-\overset{\overset{\displaystyle :\ddot{O}:}{\|}}{C}-\ddot{N}H_2 \quad \longleftrightarrow \quad R-\overset{\overset{\displaystyle :\ddot{O}:^-}{|}}{C}{}^+-\ddot{N}H_2 \quad \longleftrightarrow \quad R-\overset{\overset{\displaystyle :\ddot{O}:^-}{|}}{C}=\overset{+}{N}H_2$$

$$A \qquad\qquad\qquad B \qquad\qquad\qquad C$$

The ability of N to donate its unshared electron pair makes resonance structure C also important for amides. As a result, amide functions are unusually polar. This polarity contributes to the high boiling and melting points of amides.

The contribution by resonance structure C is responsible for the geometrical properties of amides. Consistent with the partial double-bond character for the carbon-nitrogen bond, the atoms attached to the C and N lie in a plane and at angles of approximately 120°.

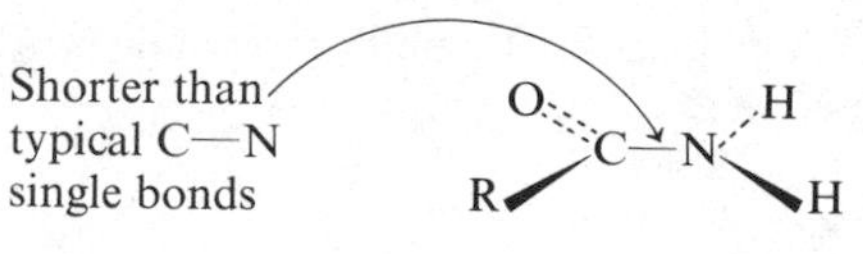

Shorter than
typical C—N
single bonds

Bond angles approximately 120° 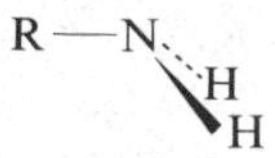   Bond angles approximately 109°

This situation contrasts with the approximately tetrahedral bond angles of amines. Consistent also with partial double-bond character for the carbon-nitrogen bond of an amide, rotation around this bond is much more difficult than around typical single bonds (although not as difficult as around carbon-carbon double bonds). The planar geometry of amide functions is an important component in protein structures. (Chapter 14).

12.12
PREPARATIONS OF AMIDES

Some amides can be prepared directly from carboxylic acids. The salt formed by reaction of a carboxylic acid with ammonia or an amine is heated to a high temperature to form the amide.

$$CH_3CH_2CH_2\overset{\overset{\textstyle O}{\|}}{C}-OH + NH_3 \longrightarrow CH_3CH_2CH_2\overset{\overset{\textstyle O}{\|}}{C}-O^-\,NH_4{}^+ \overset{heat}{\longrightarrow}$$

$$CH_3CH_2CH_2\overset{\overset{\textstyle O}{\|}}{C}-NH_2 + H_2O$$

Butyramide
Butanamide

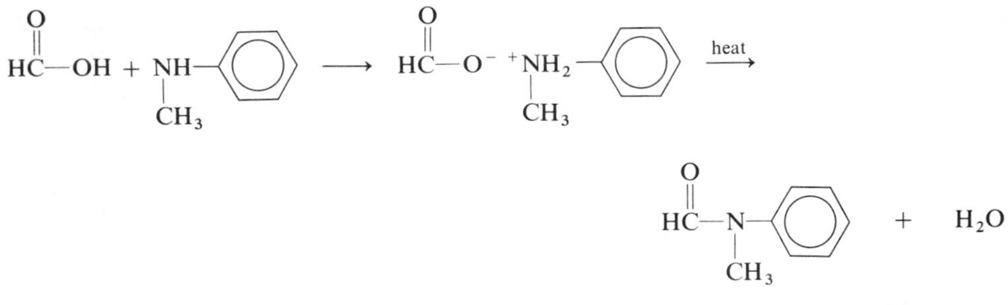

$$\overset{\overset{\textstyle O}{\|}}{HC}-OH + NH-\!\!\!\bigcirc \longrightarrow \overset{\overset{\textstyle O}{\|}}{HC}-O^-\;{}^+NH_2-\!\!\!\bigcirc \overset{heat}{\longrightarrow}$$
$$\qquad\qquad CH_3 \qquad\qquad\qquad\qquad\quad CH_3$$

$$\overset{\overset{\textstyle O}{\|}}{HC}-N-\!\!\!\bigcirc \quad + \quad H_2O$$
$$\qquad CH_3$$

N-Methyl-*N*-phenylformamide

Amides are prepared most frequently by the facile reactions of ammonia and amines with acyl chlorides and acid anhydrides (Sec. 12.3) and occasionally by their more sluggish reactions with esters (Sec. 12.6).

---

problem 12.2   Give the structure and name of the amide formed from
(a) isobutylamine and acetic anhydride.
(b) ammonia and *p*-methoxybenzoyl chloride.
(c) allylamine and acetic anhydride.
(d) diethylamine and hexanoyl chloride.

12.13
REACTIONS OF AMIDES

Even though they have amino groups, amides are far less basic than alkyl or aryl amines. The electron pair of nitrogen is donated partially to the adjoining carbonyl group, and therefore is less available for bonding to a proton.

$$\underset{\substack{R-C-NH_2}}{\overset{\overset{\displaystyle \cdot\cdot O:}{\|}}{}} \quad\longleftrightarrow\quad \underset{\substack{R-C=\overset{+}{N}H_2}}{\overset{\overset{\displaystyle :\overset{\cdot\cdot}{O}:^-}{|}}{}}$$

Because of their considerable resonance stabilization, amides undergo nucleophilic substitution reactions less readily than the other derivatives of carboxylic acids. The increase in such resonance stabilization in the following series is one factor contributing to decreasing reactivity.

increasing resonance stabilization →

$$\underset{RCX}{\overset{\overset{\displaystyle O}{\|}}{}} \qquad \underset{RCOCR}{\overset{\overset{\displaystyle O \cdot O}{\| \,\|}}{}} \qquad \underset{RCOR}{\overset{\overset{\displaystyle O}{\|}}{}} \qquad \underset{RCNH_2}{\overset{\overset{\displaystyle O}{\|}}{}}$$

decreasing reactivity →

Amides can be hydrolyzed to carboxylic acids and amines in the presence of either acid or base. More vigorous conditions—longer time, higher temperature, more catalyst—are required than for hydrolysis of esters, however.

$$\text{cyclopentyl}-\overset{\overset{\displaystyle O}{\|}}{C}-NHCH_3 + H_3O^+ \longrightarrow \text{cyclopentyl}-\overset{\overset{\displaystyle O}{\|}}{C}-OH + CH_3NH_3^+$$

$$CH_3CH_2CH_2CH_2\overset{\overset{\displaystyle O}{\|}}{C}-NH_2 + Na^+\,OH^- \longrightarrow CH_3CH_2CH_2CH_2\overset{\overset{\displaystyle O}{\|}}{C}-O^-\,Na^+ + NH_3$$

Because an ammonium ion (not a nucleophile) is produced in acid and a carboxylate anion in base, hydrolyses in either acid or base are irreversible.

12.14
SOME IMPORTANT AMIDES

The amide linkage occurs in many compounds that have biological activity. Acetaminophen is a common analgesic and antipyretic (Excedrin, Tylenol).

$$HO-\underset{}{\bigcirc}-NH-\overset{\overset{\displaystyle O}{\|}}{C}CH_3 \qquad\qquad \underset{}{\bigcirc}\overset{CH_3}{}-\overset{\overset{\displaystyle O}{\|}}{C}-\underset{\underset{\displaystyle CH_2CH_3}{|}}{N}-CH_2CH_3$$

Acetaminophen
N-(p-Hydroxyphenyl)acetamide

N,N-Diethyl-m-methylbenzamide
N,N-Diethyl-m-toluamide

N,N-Diethyl-m-methylbenzamide is a widely used insect repellent (Off, 6-12).

Lysergic acid diethylamide (LSD) is a potent hallucinogen that produces delusions resembling those of schizophrenia.

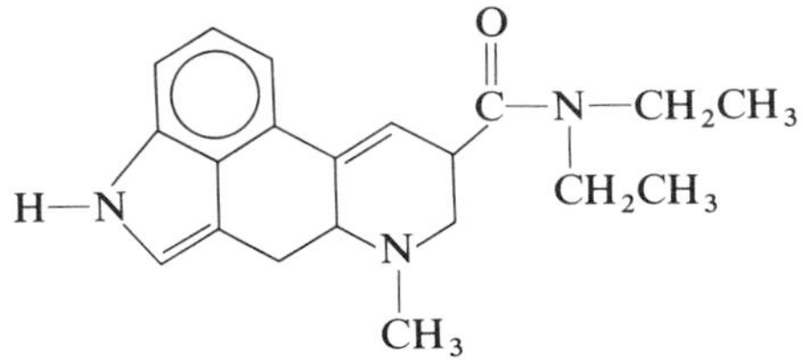

LSD

N,N-Diethyllysergamide
Lysergic acid diethylamide

Less than 50 μg (1 μg = 10^{-6} g) can affect susceptible individuals. Lysergic acid, from which LSD is synthesized, is obtained by the hydrolysis of complex amides known as ergot alkaloids that are isolated from a poisonous fungus infecting rye and other grains.

Antibiotics produced by fermentation using molds of **penicillium notatum** or **chrysogenum** and closely related synthetic compounds are known as penicillins. Penicillins act on microorganisms by interfering with cell-wall development. The sodium and potassium salts of penicillin G are the most commonly used members of this family of compounds.

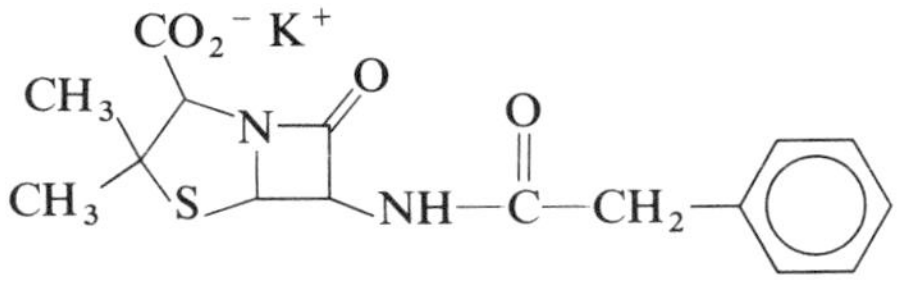

Penicillin G

This molecule contains both acyclic and cyclic amide functions. Cyclic amides are called **lactams**.

Urea is the diamide of carbonic acid.

Urea   Carbonic acid

In most terrestrial vertebrates urea is the normal end product of metabolism of nitrogen-containing compounds. Wöhler's conversion of ammonium cyanate to urea in 1828 was the first indication that synthesis of organic compounds was not unique to living organisms but could also be achieved in the laboratory.

$$NH_4^+ \; {}^-OCN \xrightarrow{\text{heat}} H_2N-\overset{\displaystyle O}{\overset{\|}{C}}-NH_2$$

Ammonium cyanate   Urea

An adult human may excrete as much as 10 kg of urea a year. Even larger amounts—about 27 kg per person every year in the United States—are produced commercially by heating ammonia and carbon dioxide at high temperatures and pressures.

$$CO_2 + 2\,NH_3 \longrightarrow H_2N\!-\!\overset{\displaystyle O}{\overset{\|}{C}}\!-\!NH_2 + H_2O$$

The major use of synthetic urea is as a fertilizer.

12.15
AMIDES OF SULFONIC ACIDS

Just as carboxylic acids have esters and amides, so do sulfonic acids.

$$R\!-\!\overset{O}{\underset{O}{\overset{\|}{\underset{\|}{S}}}}\!-\!OH \qquad R\!-\!\overset{O}{\underset{O}{\overset{\|}{\underset{\|}{S}}}}\!-\!OR \qquad R\!-\!\overset{O}{\underset{O}{\overset{\|}{\underset{\|}{S}}}}\!-\!NH_2 \quad R\!-\!\overset{O}{\underset{O}{\overset{\|}{\underset{\|}{S}}}}\!-\!NHR \quad R\!-\!\overset{O}{\underset{O}{\overset{\|}{\underset{\|}{S}}}}\!-\!NR_2$$

Sulfonic acid   Sulfonate ester   Sulfonamides

$$R\!-\!\overset{\displaystyle O}{\overset{\|}{C}}\!-\!OH \qquad R\!-\!\overset{\displaystyle O}{\overset{\|}{C}}\!-\!OR \qquad R\!-\!\overset{\displaystyle O}{\overset{\|}{C}}\!-\!NH_2 \quad R\!-\!\overset{\displaystyle O}{\overset{\|}{C}}\!-\!NHR \quad R\!-\!\overset{\displaystyle O}{\overset{\|}{C}}\!-\!NR_2$$

Carboxylic acid   Carboxylic acid ester   Carboxylic acid amides

We already encountered sulfonate esters (Sec. 7.9). Sulfonamides can be prepared from reactions of sulfonyl halides with ammonia or primary and secondary amines.

$$\text{C}_6\text{H}_5\!-\!\overset{O}{\underset{O}{\overset{\|}{\underset{\|}{S}}}}\!-\!Cl \;+\; H_2N\!-\!CH_2CH_3 \longrightarrow \text{C}_6\text{H}_5\!-\!\overset{O}{\underset{O}{\overset{\|}{\underset{\|}{S}}}}\!-\!NH\!-\!CH_2CH_3$$

Benzenesulfonyl chloride   *N*-Ethylbenzenesulfonamide

p-Aminobenzenesulfonamide (sulfanilamide) is the parent compound of a family of compounds known as the sulfa drugs.

$$H_2N\!-\!\text{C}_6\text{H}_4\!-\!\overset{O}{\underset{O}{\overset{\|}{\underset{\|}{S}}}}\!-\!NH_2$$

p-Aminobenzenesulfonamide
Sulfanilamide

Sulfadiazine and sulfacetamide are two of the many sulfa drugs in current use.

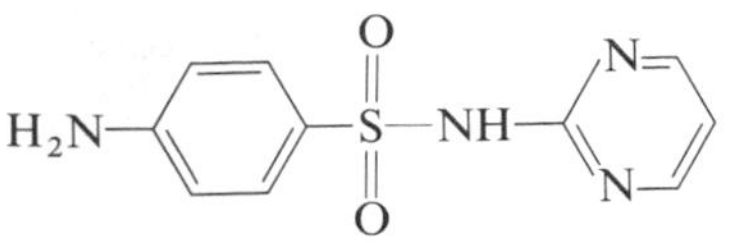

Sulfadiazine 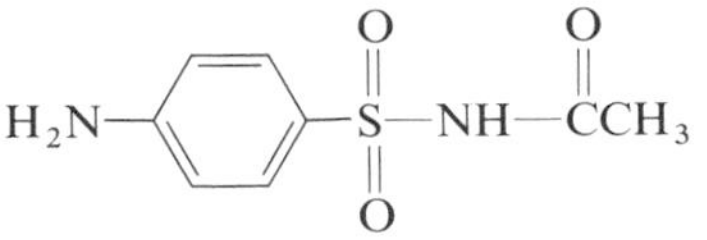   Sulfacetamide

Sulfa drugs inhibit multiplication of many microorganisms. This action is due to their resemblance to *p*-aminobenzoic acid, a compound required in some micro-organisms for synthesis of the coenzyme tetrahydrofolic acid.

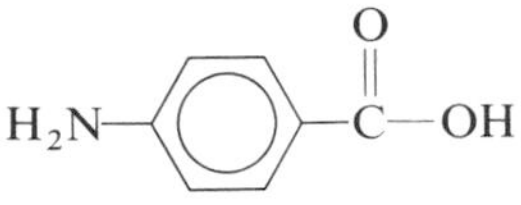

p-Aminobenzoic acid

Sulfa drugs apparently complex with the site on an enzyme that would otherwise accept *p*-aminobenzoic acid and in this way reduce the rate of an enzyme-catalyzed reaction of *p*-aminobenzoic acid required for tetrahydrofolic acid synthesis. Humans obtain tetrahydrofolic acid in another way and are less affected by sulfa drugs.

Saccharin, an amide of both a carboxylic acid and a sulfonic acid, tastes about 500 times as sweet as sugar.

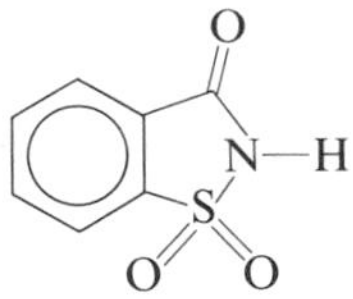

Saccharin

It has been widely used as a sweetening agent by diabetics and in "low-calorie" foods and beverages. Its use, however, has decreased after some indications that it may have weak carcinogenic activity in rats.

12.16
POLYESTERS AND POLYAMIDES

When heated together in the presence of an acid catalyst, dimethyl terephthalate and ethylene glycol react to form the linear polymer known as Dacron or Terylene when used as a fiber or Mylar when used as a film.

$$CH_3O-\overset{O}{\underset{\|}{C}}-\langle\bigcirc\rangle-\overset{O}{\underset{\|}{C}}-OCH_3 \;+\; HO-CH_2CH_2-OH \;\xrightarrow{\;H^+\;}\; CH_3OH\uparrow \;+$$

Dimethyl terephthalate   Ethylene glycol

$$-\overset{O}{\underset{\|}{C}}-\langle\bigcirc\rangle-\overset{O}{\underset{\|}{C}}-O-CH_2CH_2-O-\overset{O}{\underset{\|}{C}}-\langle\bigcirc\rangle-\overset{O}{\underset{\|}{C}}-O-CH_2CH_2-O-\overset{O}{\underset{\|}{C}}-\langle\bigcirc\rangle-\overset{O}{\underset{\|}{C}}-O-CH_2CH_2-O-$$

Dacron

Methanol, the most volatile component, is distilled off as it forms, driving this ester interchange reaction (Sec. 12.6) to completion. Annual production of this most extensively used polyester is about 9 kg per person in the United States. Heating followed by cooling will fix fibers produced from this polymer into particular shapes, introducing permanent creases into garments. Kodel is a polyester made from dimethyl terephthalate and 1,4-di(hydroxymethyl)cyclohexane.

$$CH_3O-\overset{O}{\underset{\|}{C}}-\langle\bigcirc\rangle-\overset{O}{\underset{\|}{C}}-OCH_3 \;+\; HO-CH_2-\langle\bigcirc\rangle-CH_2-OH \;\longrightarrow$$

$$-\overset{O}{\underset{\|}{(C}}-\langle\bigcirc\rangle-\overset{O}{\underset{\|}{C}}-O-CH_2-\langle\bigcirc\rangle-CH_2-O)_n$$

Kodel

Polyesters prepared from nonaromatic dicarboxylic acids have melting points too low to produce satisfactory fibers. Wool or cotton are usually blended with the polyester fibers to give a softer feel and better moisture absorbency to a fabric.

Nylons are polyamides. The first completely synthetic fibers to be marketed were of nylon 6-6.

$$H_2N-(CH_2)_6-NH_2 \;+\; HO-\overset{O}{\underset{\|}{C}}-(CH_2)_4-\overset{O}{\underset{\|}{C}}-OH \;\longrightarrow$$

$$H_3N^{\pm}(CH_2)_6-NH_3{}^+ \;\;{}^-O-\overset{O}{\underset{\|}{C}}-(CH_2)_4-\overset{O}{\underset{\|}{C}}-O^- \;\xrightarrow{\;heat\;}$$

$$-(NH-(CH_2)_6-NH-\overset{O}{\underset{\|}{C}}-(CH_2)_4-\overset{O}{\underset{\|}{C}})_n$$

Nylon 6-6

This polymer is called nylon 6-6 because of its preparation from a six-carbon diamine and a six-carbon dicarboxylic acid. Several other nylons are now produced in significant amounts. Nylon 6 is made from a lactam (cyclic amide).

$$\underset{\substack{\text{CH}_2-\text{CH}_2 \\ | \qquad | \\ \text{CH}_2 \qquad \text{CH}_2 \\ \diagdown \qquad \diagup \\ \text{NH} \qquad \text{CH}_2 \\ \diagdown \!\! \underset{\text{C}}{} \!\! \diagup \\ \overset{\text{O}}{\overset{\|}{}} }}{} \xrightarrow{\text{H}_2\text{O}} \quad \text{H}_2\text{NCH}_2\text{CH}_2\text{CH}_2\text{CH}_2\text{CH}_2\overset{\overset{\text{O}}{\|}}{\text{C}}-\text{OH} \longrightarrow$$

$$-\text{NHCH}_2\text{CH}_2\text{CH}_2\text{CH}_2\text{CH}_2\overset{\overset{\text{O}}{\|}}{\text{C}}-\text{NHCH}_2\text{CH}_2\text{CH}_2\text{CH}_2\text{CH}_2\overset{\overset{\text{O}}{\|}}{\text{C}}-\text{NHCH}_2\text{CH}_2\text{CH}_2\text{CH}_2\text{CH}_2\overset{\overset{\text{O}}{\|}}{\text{C}}-$$

Nylon 6

Nylons have a particularly high tensile strength and resistance to abrasion. Besides their use in fibers for production of fabrics and carpeting, considerable amounts are used in tire cords. Because of their resistance to hydrolysis, nylons are used in fishing nets and marine ropes.

A satisfactory fiber must be flexible, strong, and not easily stretched. An ideal fiber would have polymer chains aligned in a parallel fashion and attracted together sufficiently to retain the fiber shape. Polymer chains that, when stretched out, fit well together and are held in place by strong interchain attractions are required. In polyesters and polyacrylonitrile (Orlon) these attractions are between polar substituent groups. In nylons hydrogen bonds between chains are also involved.

Properties desirable for a fiber contrast with those (Sec. 3.20) needed for an elastomer (rubber). Elastomer chains tend to have a random arrangement; if partially aligned by stretching, they return to their original arrangement when the stretching force is removed. Polymer molecules that make good elastomers have only relatively weak interchain forces and, in some instances, contain bulky groups that prevent the chains from packing closely together.

SUMMARY*

Definitions and Ideas

Carboxylic acids, acyl halides, acid anhydrides, esters, and amides all contain an **acyl group**, $\text{R}\overset{\overset{\text{O}}{\|}}{\text{C}}-$, attached to a nucleophilic group.

Acyl halides have the structure $\text{R}\overset{\overset{\text{O}}{\|}}{\text{C}}\text{X}$, where X is a halogen (usually Cl). They are named by replacing the *-ic acid* ending of a name of the corresponding carboxylic acid by *-yl* and adding the name of the particular halide.

---

* Throughout this summary R can be an alkyl group, aryl group, or hydrogen, and R′ and R″ an alkyl or aryl group.

$$\overset{\displaystyle O}{\underset{\displaystyle \parallel}{}} \overset{\displaystyle O}{\underset{\displaystyle \parallel}{}}$$

Acid anhydrides are compounds of structure $RCOCR$. They are named by adding the word anhydride to the name(s) of the corresponding carboxylic acid(s).

$$\overset{\displaystyle O}{\underset{\displaystyle \parallel}{}}$$

Esters are compounds of structure $RCOR'$. Their names are formed by naming the organic group attached directly to oxygen, followed by the name of the corresponding acid with its *-ic acid* ending changed to *-ate*. Esters have solubilities comparable to those of alcohols with the same number of carbons but boiling points closer to those of hydrocarbons of similar molecular weight.

$$\overset{\displaystyle O}{\underset{\displaystyle \parallel}{}}$$

Amides are compounds of structure $RCNR_2$. They are named by replacing the *-ic acid* or *-oic acid* ending of the name of the corresponding carboxylic acid by *-amide*. Organic groups attached to nitrogen are named as prefixes and labeled with an *N-*. Amides are even somewhat more water soluble than alcohols with the same number of carbons.

Many of the reactions of acyl halides, acid anhydrides, esters, and amides involve substitution of the nucleophilic group attached to the acyl group by another nucleophilic group. In such reactions the order of reactivity is acyl halide > acid anhydride > ester > amide. The reactions ordinarily convert a more reactive compound to a less reactive one. All compounds in this series can be hydrolyzed to a carboxylic acid, although hydrolysis of esters or amides requires the assistance of an acid or base. All but amides react with alcohols to form esters; the reaction of an ester with an alcohol (ester interchange) requires catalysis by an acid or base. All but amides react with ammonia or amines to form amides. These substitution reactions usually proceed by an addition to the carbonyl group, followed by an elimination that regenerates the carbonyl group.

$$\underset{RCY}{\overset{O}{\parallel}} + HZ \; \rightleftharpoons \; \underset{\underset{Z}{|}}{\overset{\overset{OH}{|}}{RCY}} \; \rightleftharpoons \; \underset{RCZ}{\overset{O}{\parallel}} + HY$$

$$\overset{\displaystyle O}{\underset{\displaystyle \parallel}{}}$$

Y and Z are OH, OR′, NR_2, X, or OCR

The α hydrogens of esters (hydrogens bonded to a carbon next to the carbonyl carbon) are considerably more acidic than hydrogens of alkanes. As do other nucleophiles, the anions formed by removing an α hydrogen can (a) add to a carbonyl group or (b) be alkylated by an alkyl halide (a nucleophilic displacement reaction of the halide).

Important Reactions Involving Derivatives of Carboxylic Acids

(a) *Reactions of Carboxylic Acids*

1. *Conversion to acyl halides* $\underset{RCOH}{\overset{O}{\parallel}} \xrightarrow{PX_3 \text{ or } SOCl_2} \underset{RCX}{\overset{O}{\parallel}}$

    X = Cl, Br, or I

2. *Conversion to esters (esterification)* $\underset{\text{RCOH}}{\overset{\text{O}}{\|}} \xrightarrow[\text{H}^+]{\text{R'OH}} \underset{\text{RCOR'}}{\overset{\text{O}}{\|}}$

(b) *Reactions of Acyl Halides*

3. *Conversion to esters (alcoholysis)* $\underset{\text{RCX}}{\overset{\text{O}}{\|}} \xrightarrow{\text{R'OH}} \underset{\text{RCOR'}}{\overset{\text{O}}{\|}}$

4. *Conversion to amides (ammonolysis)* $\underset{\text{RCX}}{\overset{\text{O}}{\|}} \xrightarrow{\text{HNR}_2} \underset{\text{RCNR}_2}{\overset{\text{O}}{\|}}$

(c) *Reactions of Acid Anhydrides*

5. *Conversion to esters (alcoholysis)* $\underset{\text{RCOCR}}{\overset{\text{O O}}{\| \ \|}} \xrightarrow{\text{R'OH}} \underset{\text{RCOR'}}{\overset{\text{O}}{\|}}$

6. *Conversion to amides (ammonolysis)* $\underset{\text{RCOCR}}{\overset{\text{O O}}{\| \ \|}} \xrightarrow{\text{HNR}_2} \underset{\text{RCNR}_2}{\overset{\text{O}}{\|}}$

(d) *Reactions of Esters*

7. *Conversion to carboxylic acids (hydrolysis)*

$$\underset{\text{RCOR'}}{\overset{\text{O}}{\|}} \xrightarrow[\text{H}^+ \text{ or OH}^-]{\text{H}_2\text{O}} \underset{\text{RCOH} + \text{R'OH}}{\overset{\text{O}}{\|}}$$

The acid-catalyzed reaction is the reverse of reaction 2. In the reaction with base, the carboxylic acid is converted to its anion ($\text{RCO}_2{}^-$).

8. *Conversion to other esters (ester interchange)* $\underset{\text{RCOR'}}{\overset{\text{O}}{\|}} \xrightarrow[\text{H}^+ \text{ or R''O}^-]{\text{R''OH}} \underset{\text{RCOR''}}{\overset{\text{O}}{\|}}$

9. *Conversion to amides (ammonolysis)* $\underset{\text{RCOR'}}{\overset{\text{O}}{\|}} \xrightarrow{\text{HNR}_2} \underset{\text{RCNR}_2}{\overset{\text{O}}{\|}}$

10. *Reduction* $\underset{\text{RCOR'}}{\overset{\text{O}}{\|}} \longrightarrow \text{RCH}_2\text{OH} + \text{R'OH}$

The reduction can be carried out by using H_2 and a copper-chromite catalyst (hydrogenolysis) or LiAlH_4, followed by H_2O.

(e) *Reactions of Amides*

11. *Conversion to carboxylic acids (hydrolysis)*

$$\underset{\text{RCNR}_2}{\overset{\text{O}}{\|}} \xrightarrow[\text{H}^+ \text{ or OH}^-]{\text{H}_2\text{O}} \underset{\text{RCOH} + \text{HNR}_2}{\overset{\text{O}}{\|}}$$

In the reaction with acid, the amine is converted to the ammonium ion ($\text{H}_2\text{NR}_2{}^+$). In the reaction with base, the carboxylic acid is converted to its anion ($\text{RCO}_2{}^-$).

PROBLEMS

1. Draw structures for the following compounds.
   (a) methyl acetate
   (b) propionyl chloride
   (c) benzoic anhydride
   (d) ethyl formate
   (e) isopropyl butyrate
   (f) allyl acetate
   (g) butanoyl chloride
   (h) *p*-chlorobenzamide
   (i) phenyl acetate
   (j) methyl hexanoate
   (k) butanamide
   (l) *p*-nitrobenzoyl chloride
   (m) *N,N*-diethylbenzamide
   (n) succinic anhydride
   (o) diethyl malonate
   (p) *tert*-butyl phenylacetate

2. Name the following compounds.

   (a) 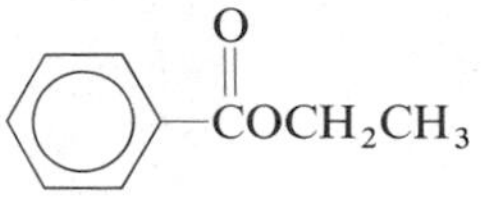
   $$\text{Ph}-\overset{\displaystyle O}{\overset{\displaystyle \|}{C}}OCH_2CH_3$$

   (b)
   $$CH_3\overset{\displaystyle O}{\overset{\displaystyle \|}{C}}\overset{\displaystyle O}{\overset{\displaystyle \|}{C}}CH_3$$

   (c)
   $$CH_3\underset{\underset{\displaystyle CH_3}{|}}{CH}CH_2\overset{\displaystyle O}{\overset{\displaystyle \|}{C}}NH_2$$

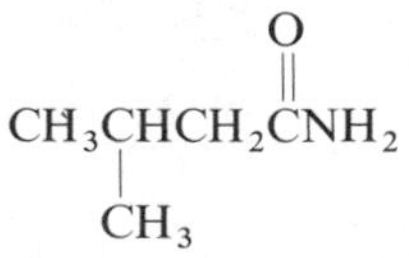

   (d) 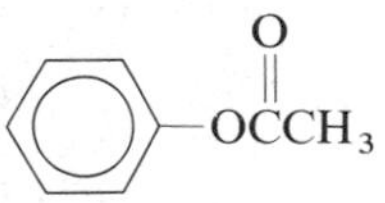
   $$\text{Ph}-O\overset{\displaystyle O}{\overset{\displaystyle \|}{C}}CH_3$$

   (e) 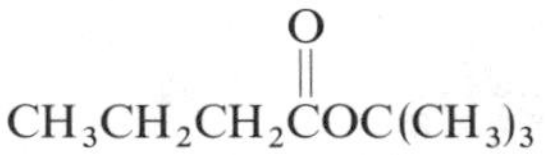
   $$CH_3CH_2CH_2\overset{\displaystyle O}{\overset{\displaystyle \|}{C}}OC(CH_3)_3$$

   (f) 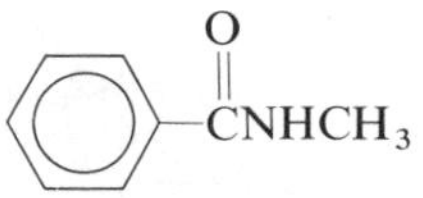
   $$\text{Ph}-\overset{\displaystyle O}{\overset{\displaystyle \|}{C}}NHCH_3$$

   (g) 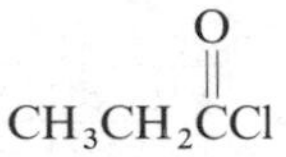
   $$CH_3CH_2\overset{\displaystyle O}{\overset{\displaystyle \|}{C}}Cl$$

   (h) 
   $$H\overset{\displaystyle O}{\overset{\displaystyle \|}{C}}OCH_3$$

3. Draw structures of all compounds that fit the following descriptions.
   (a) amides of molecular formula C_3H_7NO
   (b) esters of molecular formula $C_5H_{10}O_2$

4. Draw structures for any organic products that result from treating benzoic acid with the following reagents.
   (a) KOH
   (b) PCl_5
   (c) CH_3CH_2OH, H^+
   (d) $CH_3CH_2NH_2$
   (e) Br_2, $FeBr_3$
   (f) $NaHCO_3$

   (g) $SOCl_2$
   (h) 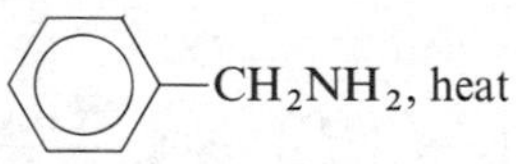 $\text{Ph}-CH_2NH_2$, heat

5. Draw structures for any organic products that result from treating benzoyl chloride with the following reagents.
   (a) CH_3OH
   (b) CH_3NH_2
   (c) H_2O
   (d) NH_3
   (e) benzene, $AlCl_3$

6. Draw structures of the organic products of the following reactions.

   (a) $CH_3CH_2CO_2H \xrightarrow{PCl_3}$

   (b) 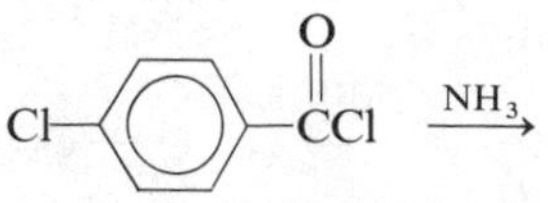 $Cl-\text{Ph}-\overset{\displaystyle O}{\overset{\displaystyle \|}{C}}Cl \xrightarrow{NH_3}$

(c)

$$CH_3CH_2CH_2\overset{O}{\overset{\|}{C}}OCH_3 \xrightarrow{NaOH}$$

(d)

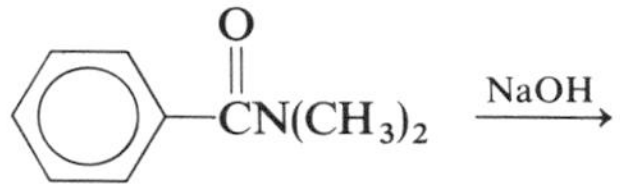

Ph$-\overset{O}{\overset{\|}{C}}N(CH_3)_2 \xrightarrow{NaOH}$

(e)

$$CH_3\overset{O}{\overset{\|}{C}}O\overset{O}{\overset{\|}{C}}CH_3 \xrightarrow{\text{Ph}-NH_2}$$

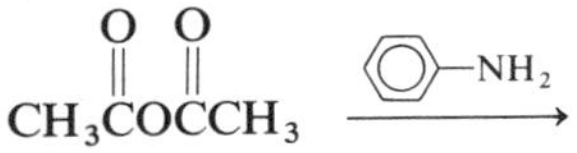

(f) $CH_3CH_2CH_2CO_2H \xrightarrow[H^+]{CH_3OH}$

(g)

$$CH_3CH_2\overset{O}{\overset{\|}{C}}OCH_2CH_3 \xrightarrow{NaOCH_2CH_3}$$

(h)

$(CH_3)_2CH\overset{O}{\overset{\|}{C}}Cl \xrightarrow{CH_3CH_2OH}$

(i)

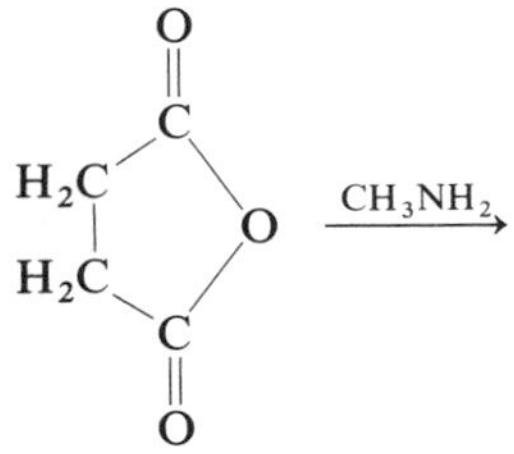

$$\text{(cyclic anhydride)} \xrightarrow{CH_3NH_2}$$

(j)

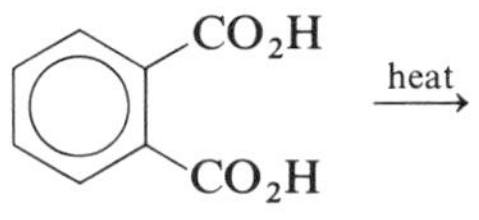

benzene ring with CO_2H, CO_2H $\xrightarrow{heat}$

(k)

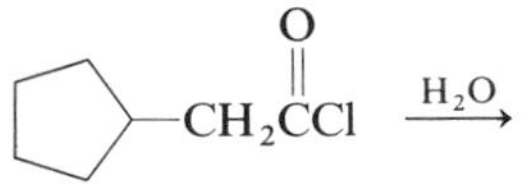

cyclopentyl$-CH_2\overset{O}{\overset{\|}{C}}Cl \xrightarrow{H_2O}$

(l)

Ph$-\overset{O}{\overset{\|}{C}}OCH_2CH_3 \xrightarrow[\text{copper chromite}]{H_2}$

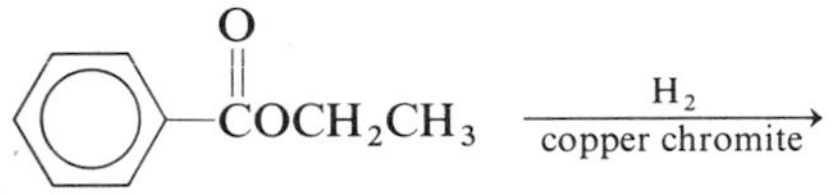

(m)

$$CH_3CH_2O\overset{O}{\overset{\|}{C}}CH_2\overset{O}{\overset{\|}{C}}OCH_2CH_3 \xrightarrow{NaOCH_2CH_3} \xrightarrow{CH_3CH_2Br}$$

(n)

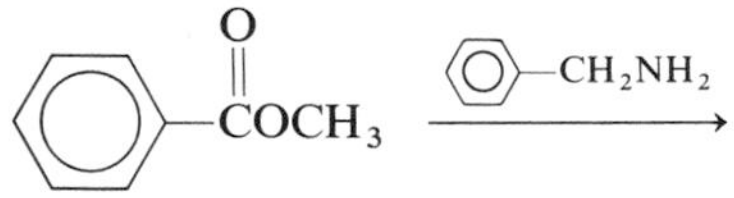

Ph$-\overset{O}{\overset{\|}{C}}OCH_3 \xrightarrow{Ph-CH_2NH_2}$

(o)

$$CH_3CH_2\overset{O}{\overset{\|}{C}}\underset{\underset{CH_3}{|}}{C}HCO_2H \xrightarrow{heat}$$

(p)

$$Cl\overset{O}{\overset{\|}{C}}CH_2CH_2\overset{O}{\overset{\|}{C}}Cl \xrightarrow{H_2NCH_2CH_2NH_2}$$

(q)

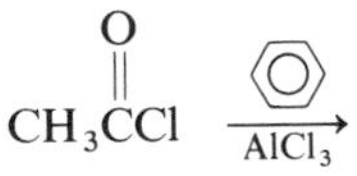

$CH_3\overset{O}{\overset{\|}{C}}Cl \xrightarrow[AlCl_3]{\text{benzene}}$

(r)

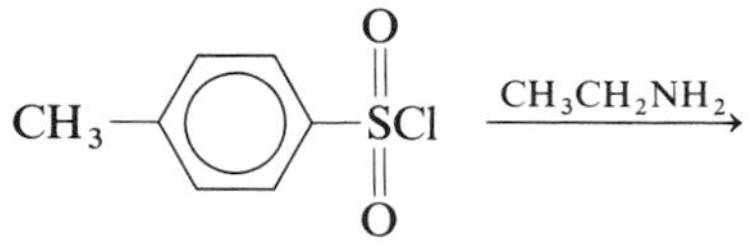

$$CH_3-\text{benzene}-\overset{O}{\underset{O}{\overset{\|}{\underset{\|}{S}}}}Cl \xrightarrow{CH_3CH_2NH_2}$$

(s)

Ph$-\overset{O}{\overset{\|}{C}}OCH_3 \xrightarrow[H^+]{H_2^{18}O}$

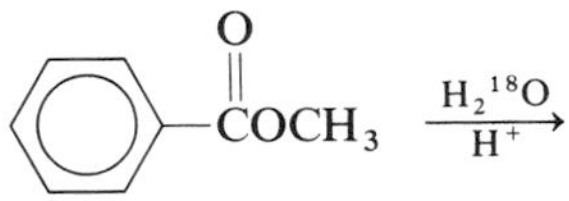

(t)

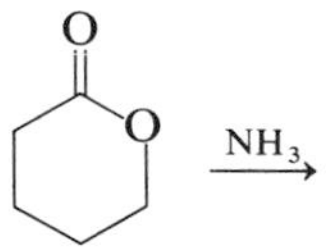

lactone $\xrightarrow{NH_3}$

(u)

Ph$-CO_2H \xrightarrow{SOCl_2} \xrightarrow{CH_3NH_2}$

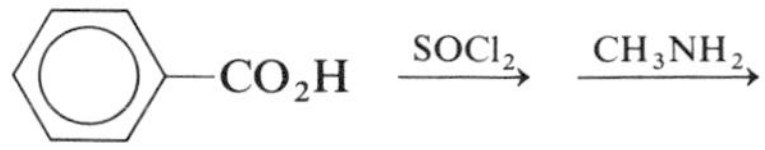

Problems **349**

(v)

$$\underset{\substack{\text{(excess)}}}{CH_3\overset{\displaystyle O}{\overset{\|}{C}}O\overset{\displaystyle O}{\overset{\|}{C}}CH_3} \quad \xrightarrow[\substack{\text{(see structure}\\\text{in Sec. 8.11)}}]{\text{morphine}} \quad \text{(heroin)}$$

(w)

$$CH_3O\overset{\displaystyle O}{\overset{\|}{C}}CH_2\overset{\displaystyle O}{\overset{\|}{C}}OCH_3 \quad \xrightarrow{NaOCH_3} \quad \xrightarrow{CH_3I} \quad \xrightarrow[H^+]{H_2O} \quad \xrightarrow{\text{heat}}$$

(x)

$$CH_3CH_2O\overset{\displaystyle O}{\overset{\|}{C}}CH_2CH_2CH_2CH_2\overset{\displaystyle O}{\overset{\|}{C}}OCH_2CH_3 \quad \xrightarrow{NaOCH_2CH_3} \quad \xrightarrow[H^+]{H_2O} \quad \xrightarrow{\text{heat}}$$

(y)

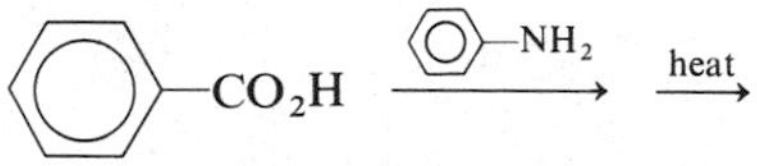

$$\bigcirc\!\!-CO_2H \quad \xrightarrow[\text{heat}]{\bigcirc\!\!-NH_2}$$

(z)

$$\bigcirc\!\!-CH_2CH_2CO_2H \quad \xrightarrow{SOCl_2} \quad \xrightarrow{AlCl_3}$$

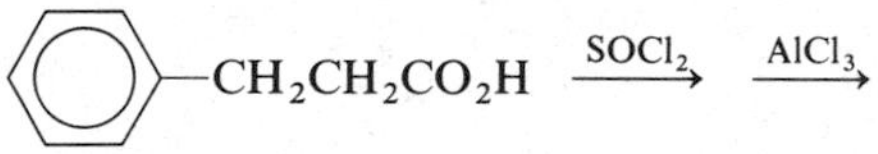

7. Write equations to show how propanoic acid can be converted to each compound.
   (a) methyl propanoate   (b) propanoyl chloride
   (c) propanamide   (d) *N*-methylpropanamide
   (e) propanoic anhydride

8. Write equations to show how the following conversions can be carried out. Some require more than one step.

(a)

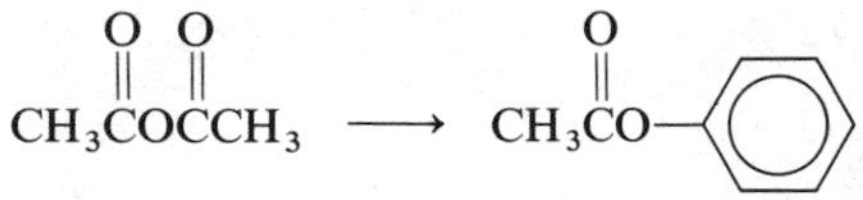

$$CH_3\overset{\displaystyle O}{\overset{\|}{C}}O\overset{\displaystyle O}{\overset{\|}{C}}CH_3 \quad \longrightarrow \quad CH_3\overset{\displaystyle O}{\overset{\|}{C}}O\!\!-\!\!\bigcirc$$

(b)

$$CH_3CO_2H \quad \longrightarrow \quad CH_3\overset{\displaystyle O}{\overset{\|}{C}}N(CH_3)_2$$

(c)

$$CH_3\overset{\displaystyle O}{\overset{\|}{C}}Cl \quad \longrightarrow \quad CH_3\overset{\displaystyle O}{\overset{\|}{C}}NH(CH_2)_5CH_3$$

(d)

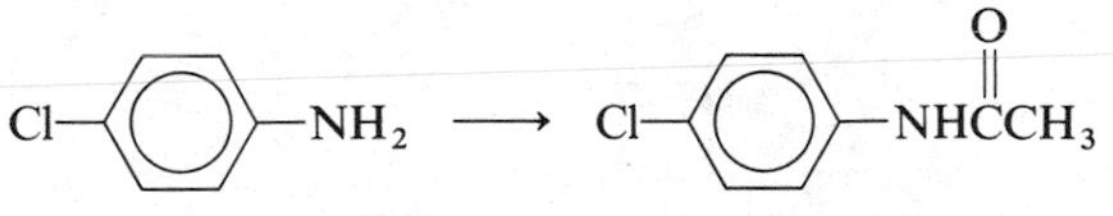

$$Cl\!\!-\!\!\bigcirc\!\!-NH_2 \quad \longrightarrow \quad Cl\!\!-\!\!\bigcirc\!\!-NH\overset{\displaystyle O}{\overset{\|}{C}}CH_3$$

(e)

$$CH_3\!\!-\!\!\bigcirc\!\!-\overset{\displaystyle O}{\overset{\|}{C}}OCH_3 \quad \longrightarrow \quad CH_3\!\!-\!\!\bigcirc\!\!-CH_2OH$$

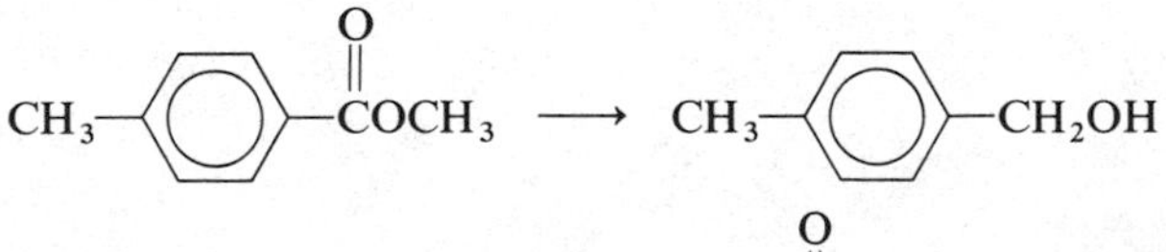

(f)

$$CH_3(CH_2)_{16}CO_2H \quad \longrightarrow \quad CH_3(CH_2)_{16}\overset{\displaystyle O}{\overset{\|}{C}}Cl$$

(g)

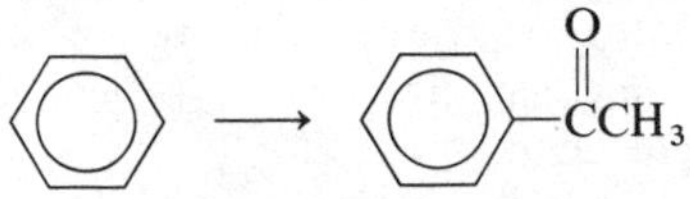

$$\bigcirc \quad \longrightarrow \quad \bigcirc\!\!-\overset{\displaystyle O}{\overset{\|}{C}}CH_3$$

(h)

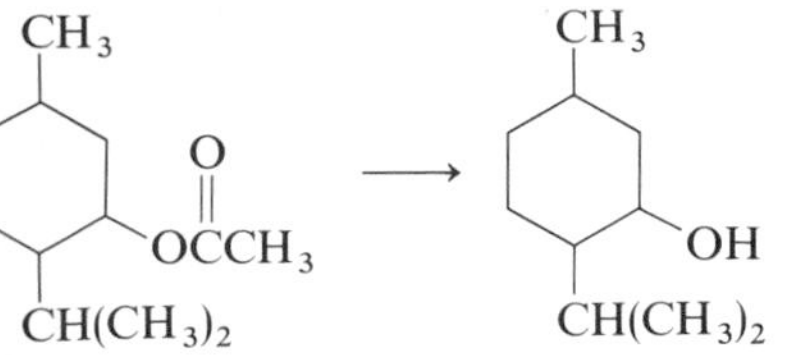

(i)

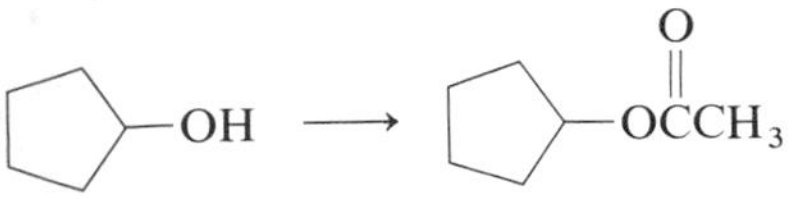

(j)

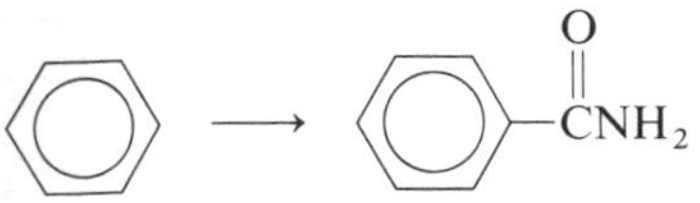

(k)

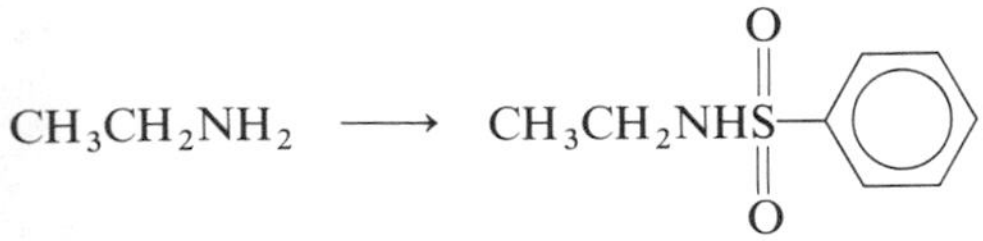

$$CH_3CH_2NH_2 \longrightarrow CH_3CH_2NHS$$

(l)

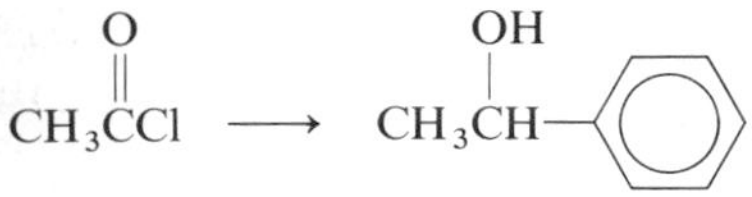

9. Write equations to show how each compound can be synthesized by using a Claisen reaction. Other steps may be needed.

(a)

(b)

10. Write equations to show how each compound can be synthesized from ethyl acetoacetate or diethyl malonate.

(a)   (needed to produce veronal, Sec. 12.6)

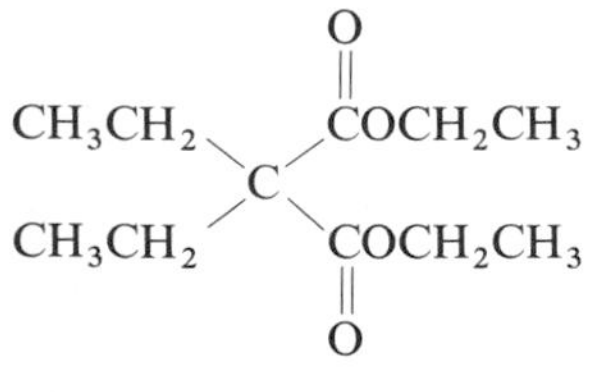

(b)

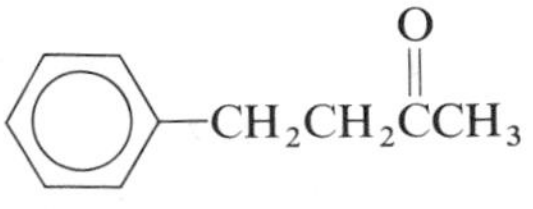

(c) $CH_2{=}CHCH_2CH_2CO_2H$

(d)

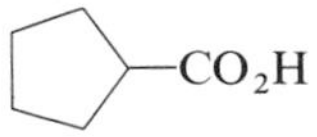

$-CO_2H$ (Hint: from diethyl malonate)

11. What products would result from treating each mixture with sodium ethoxide?
    (a) ethyl acetate and ethyl propionate
    (b) ethyl benzoate and ethyl acetate

12. Draw the structure of a compound of molecular formula $C_6H_{12}O_2$ that on heating with aqueous acid forms a carboxylic acid A and an alcohol B. Oxidation of B with $KMnO_4$ gives A.

13. Draw the structure of a compound of molecular formula $C_5H_{10}O_2$ that on heating with aqueous acid forms carboxylic acid C and alcohol D. D reacts with PBr_3 to form a compound that on treatment with Mg followed by CO_2 gives C.

14. Why isn't sodium hydroxide used instead of an alkoxide salt as the base for the Claisen condensation? Why is sodium ethoxide used instead of sodium methoxide as the base for the Claisen condensation of ethyl acetate?

15. Why are esters and amides less reactive toward nucleophiles than are ketones?

16. Why is *N*-phenylacetamide considerably less reactive toward bromination of the aromatic ring than aniline? Aromatic amines are sometimes converted to their acetyl derivatives because these give more readily controlled electrophilic aromatic substitution reactions.

17. Explain why the reaction of an ester and a Grignard reagent produces a tertiary alcohol.

$$
\underset{\substack{\| \\ RCOR}}{\overset{O}{}} \xrightarrow{R'MgX} \xrightarrow{H_2O} \underset{\substack{| \\ R'}}{\overset{OH}{\underset{}{RCR'}}}
$$

18. Give a mechanism for the ester interchange reaction that occurs in the presence of a catalytic amount of $R''O^- K^+$.

$$
\underset{\substack{\| \\ RCOR'}}{\overset{O}{}} + R''OH \; \rightleftharpoons \; \underset{\substack{\| \\ RCOR''}}{\overset{O}{}} + R'OH
$$

19. Why is the methyl ester of 2,6-dimethylbenzoic acid hydrolyzed only very slowly when heated with aqueous base?

20. The enol forms of most ketones are much less stable than the keto forms and are present only in low concentrations. The enol of ethyl acetoacetate, however, is relatively more stable.

$$
\underset{\substack{\|\quad\| \\ CH_3CCH_2COCH_2CH_3}}{\overset{O\quad O}{}} \; \rightleftharpoons \; \underset{\substack{|\quad\| \\ CH_3C=CHCOCH_2CH_3}}{\overset{OH\quad O}{}}
$$

In hexane solution the equilibrium mixture contains 49% of the enol form and 51% of the keto form. Suggest two factors that stabilize the enol of ethyl acetoacetate.

Lipids

Lipids are important constituents of all plant and animal tissues. The lipid family of compounds includes members belonging to different structural groups. Although they do not share a common structure, however, lipids do share the property of being soluble in nonpolar organic solvents, such as diethyl ether, chloroform, and benzene, but insoluble in water. This behavior contrasts with that of other important groups of biological molecules—carbohydrates, proteins, and nucleic acids—that are insoluble in the same organic solvents. Large alkyl portions found in the structures of all lipids are responsible for their solubility properties.

Compounds commonly known as fats and oils are a major class of lipids. An important food source, they are stored or transported in organisms for ultimate use in metabolic processes. Other major groups of lipids, the glycerophospholipids and sphingolipids, are important components of cell membranes. Waxes are lipids that serve particularly to provide water-impermeable coatings—for instance, to the leaves of plants or the exoskeletons of some insects. All the lipids mentioned so far are esters of carboxylic acids with long alkyl chains. Earlier we discussed other groups of compounds often classified as lipids: terpenes (Sec. 3.21), steroids (Sec. 9.15), and prostaglandins (Sec. 11.7).

13.1
FATS AND OILS: STRUCTURES AND PROPERTIES

Fats and oils, the most abundant lipids, are the major constituents of storage fat cells in animals and plants. Fats and oils, carbohydrates, and proteins are the three major kinds of foodstuffs, but fats and oils provide the most calories per gram. As poor conductors of heat, they also provide thermal insulation in some animals.

Fats and oils are similar in chemical structure. Fat is commonly used to refer to materials that are solid or semisolid at room temperature and oil to materials that are liquid. Fats and oils are complex mixtures of esters of glycerol with a variety of carboxylic acids.

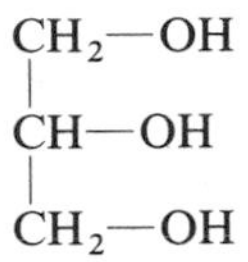

$$CH_2-OH$$
$$CH-OH$$
$$CH_2-OH$$

Glycerol
1,2,3-Propanetriol

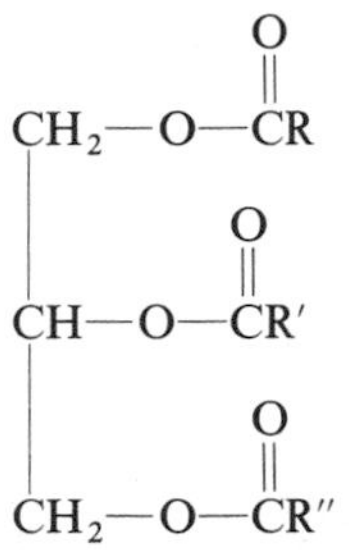

$$CH_2-O-\overset{\overset{\displaystyle O}{\|}}{C}R$$
$$CH-O-\overset{\overset{\displaystyle O}{\|}}{C}R'$$
$$CH_2-O-\overset{\overset{\displaystyle O}{\|}}{C}R''$$

A triacylglycerol

Because of their structures, such molecules are known as **triacylglycerols**.

The carboxylic acids incorporated into fats and oils are often called **fatty acids**. The most commonly encountered fatty acids have long unbranched chains and contain even numbers of carbon atoms. Table 13.1 lists some common fatty

Table 13.1   Some carboxylic acids commonly found in fats and oils

Structure	Trivial name (substitutive name)	Melting point (°C)
$CH_3(CH_2)_{10}CO_2H$	lauric acid (dodecanoic acid)	44
$CH_3(CH_2)_{12}CO_2H$	myristic acid (tetradecanoic acid)	54
$CH_3(CH_2)_{14}CO_2H$	palmitic acid (hexadecanoic acid)	63
$CH_3(CH_2)_{16}CO_2H$	stearic acid (octadecanoic acid)	70
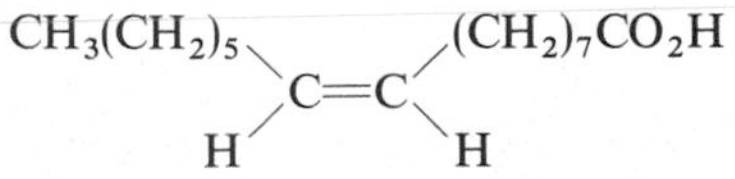 $CH_3(CH_2)_5$ ... C=C ... $(CH_2)_7CO_2H$ (H, H)	palmitoleic acid (*cis*-9-hexadecenoic acid)	0
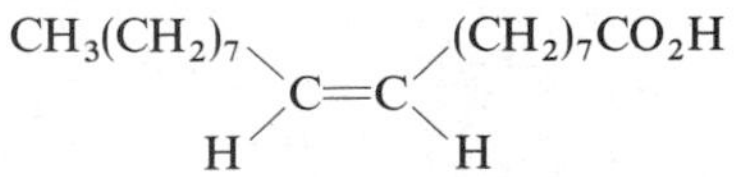 $CH_3(CH_2)_7$ ... C=C ... $(CH_2)_7CO_2H$ (H, H)	oleic acid (*cis*-9-octadecenoic acid)	16
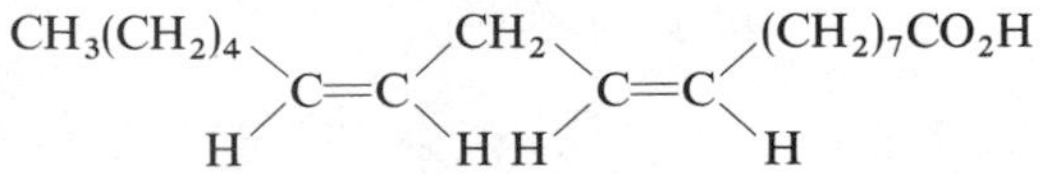 $CH_3(CH_2)_4$ ... C=C ... CH_2 ... C=C ... $(CH_2)_7CO_2H$ (H, H H, H)	linoleic acid (*cis,cis*-9,12-octadecadienoic acid	−5
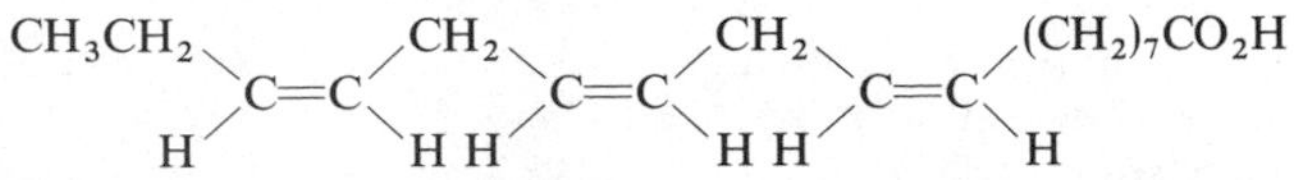 CH_3CH_2 ... C=C ... CH_2 ... C=C ... CH_2 ... C=C ... $(CH_2)_7CO_2H$ (H, H H, H H, H)	linolenic acid (*cis,cis,cis*-9,12,15-octadecatrienoic acid)	−11

acids. Fatty acids with 12, 14, 16, and 18 carbons are the most abundant of the saturated acids. Unsaturated acids with one, two, or three alkene functions are also common; the most abundant have 18 carbons and a cis stereochemistry at each double bond.

Triacylglycerols in which all three hydroxyl groups of glycerol are esterified to the same acid are known as **simple** triacylglycerols.

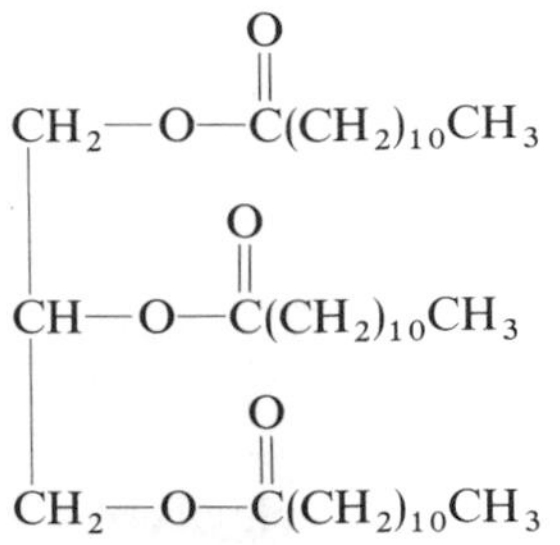

$$
\begin{array}{ll}
\underset{\substack{|\\ \text{CH}_2-\text{O}-\overset{\displaystyle O}{\overset{\|}{\text{C}}}(\text{CH}_2)_{10}\text{CH}_3}}{} & \underset{\substack{|\\ \text{CH}_2-\text{O}-\overset{\displaystyle O}{\overset{\|}{\text{C}}}(\text{CH}_2)_{16}\text{CH}_3}}{}
\end{array}
$$

CH$_2$—O—C(CH$_2$)$_{10}$CH$_3$ CH$_2$—O—C(CH$_2$)$_{16}$CH$_3$

CH—O—C(CH$_2$)$_{10}$CH$_3$ CH—O—C(CH$_2$)$_{16}$CH$_3$

CH$_2$—O—C(CH$_2$)$_{10}$CH$_3$ CH$_2$—O—C(CH$_2$)$_{16}$CH$_3$

Glyceryl trilaurate   Glyceryl tristearate
Trilauroylglycerol   Tristearoylglycerol
(mp 46°)   (mp 73°)

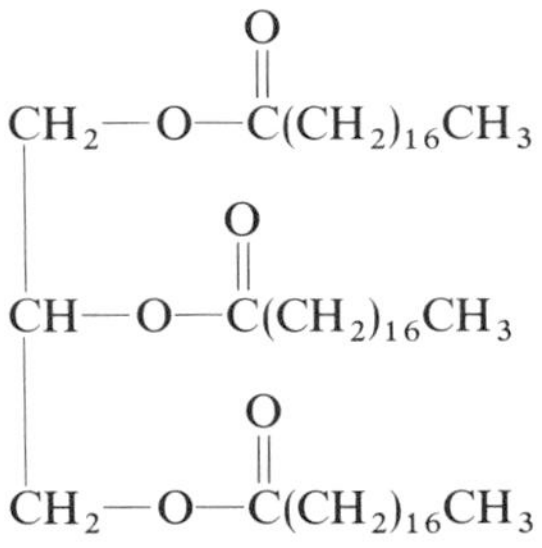

CH$_2$—O—C(CH$_2$)$_7$CH=CHCH$_2$CH=CH(CH$_2$)$_4$CH$_3$

CH—O—C(CH$_2$)$_7$CH=CHCH$_2$CH=CH(CH$_2$)$_4$CH$_3$

CH$_2$—O—C(CH$_2$)$_7$CH=CHCH$_2$CH=CH(CH$_2$)$_4$CH$_3$

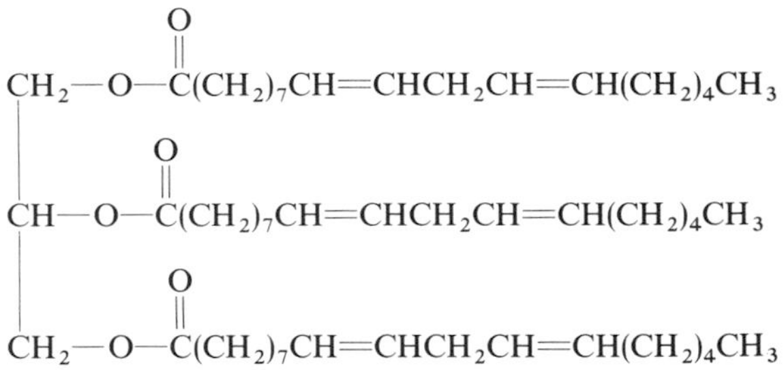

Glyceryl trilinoleate
Trilinoleoylglycerol
(mp −13°)

Mixed triacylglycerols contain two or three different carboxylic acids.

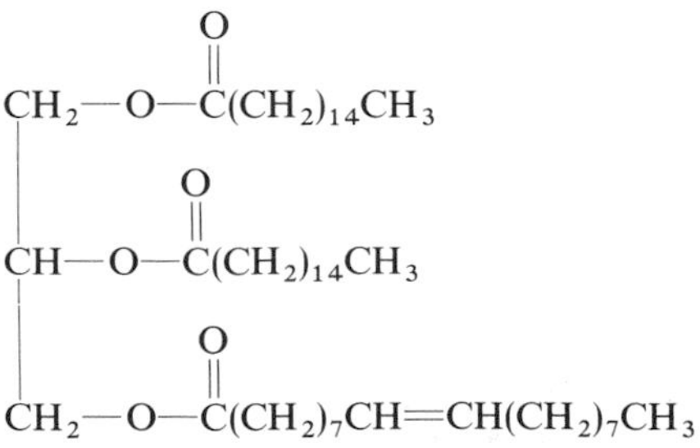

CH$_2$—O—C(CH$_2$)$_{14}$CH$_3$

CH—O—C(CH$_2$)$_{14}$CH$_3$

CH$_2$—O—C(CH$_2$)$_7$CH=CH(CH$_2$)$_7$CH$_3$

Glyceryl 1-oleate 2,3-dipalmitate
1-Oleoyl-2,3-dipalmitoylglycerol
(a mixed triacylglycerol)

Table 13.2   Fatty acid compositions of typical samples of widely used fats and oils (only components present to the extent of 1 to 2% or more are indicated)

	Saturated acids				Unsaturated acids			
					1 C=C		2 C=C	3 C=C
	C_{12}	C_{14}	C_{16}	C_{18}	C_{16}	C_{18}	C_{18}	C_{18}
	Lauric	Myristic	Palmitic	Stearic	Palmitoleic	Oleic	Linoleic	Linolenic
Fat or oil	acid	acid	acid	acid	acid	acid	acid	acid
Animal fats								
Butterfat[a]	4	12	29	11	4	25	2	
Lard		3	24	18	3	42	9	
Tallow (beef fat)		3	26	17	6	43	4	
Vegetable oils								
Coconut[b]	44	18	11	6		7	2	
Corn			13	4		29	54	
Cottonseed		1	29	4	2	24	40	
Linseed			6	4		22	16	52
Olive			14	2	2	64	16	
Palm		1	48	4		38	9	
Peanut[c]			6	5		61	22	
Safflower			8	3		13	75	1
Soybean			11	4		25	51	9

[a] Butterfat also contains butanoic acid (3%), hexanoic acid (1%), octanoic acid (1%), decanoic acid (3%), and 9-tetradecenoic acid (2%).

[b] Coconut oil also contains hexanoic acid (6%) and octanoic acid (6%).

[c] Peanut oil also contains linear, saturated carboxylic acids with 20 carbons (2%), 22 carbons (3%), and 24 carbons (1%).

Natural fats and oils are complex mixtures of simple and mixed triacylglycerols. As in Table 13.2, it is convenient to describe a fat or oil in terms of the composition of the mixture of carboxylic acids obtained on hydrolysis. Many acids are obtained from some fats and oils—more than 60 different carboxylic acids have been isolated from butterfat. The carboxylic acid mixtures obtained from many fats and oils, however, consist mainly of a few compounds. Palmitic, stearic, oleic, and linoleic acids, for instance, constitute more than 95% of the acids obtained from corn, cottonseed, olive, palm, and safflower oils.

The tendency of a mixture of triacylglycerols to be liquid increases with increasing unsaturation (increasing number of carbon-carbon double bonds) and with increasing concentrations of acids with short chains. Note that the simple triacylglycerol of stearic acid melts somewhat higher than that of lauric acid and considerably higher than that of linoleic acid. The same trend is noted in the melting points of the constituent acids themselves (Table 13.1) and has the same origin. Attractive forces between chains in different molecules tend to increase with increasing chain length. Cis double bonds, however, make it more difficult for chains to pack closely together in a regular crystal structure.

The triacylglycerol mixtures obtained from animal sources are often solids. Note that butterfat, lard, and tallow, all solids at room temperature, contain relatively large amounts of saturated acids and only minor amounts of acids with more than one carbon-carbon double bond. In contrast, vegetable oils characteristically have many short chains (coconut oil) or many double bonds. Triacylglycerols from animals that must live at low temperatures tend to have larger amounts of unsaturation. The blubber oil of the Antarctic elephant seal contains more than 80% unsaturated acids, including significant amounts of C_{22} acids containing up to six double bonds.

13.2
FATS AND OILS: ISOLATION AND USES

Chemical processes involving fats and oils include their conversion to components of more expensive foodstuffs, to soaps, and to long-chain acids or alcohols that are raw materials for other chemical products. Fats are obtained from animal tissues by rendering—slow heating, often in the presence of steam. Vegetable oils can be obtained by pressing oil-containing seeds, sometimes after subjecting the seeds to a preliminary heat treatment. Alternatively, oils are extracted (dissolved) out of the seeds with low-boiling organic solvents, which are then evaporated and recovered. Because they are renewable resources, vegetable oils are attracting increased attention as sources of organic raw materials.

13.3
HYDROGENATION OF CARBON-CARBON
DOUBLE BONDS

Hydrogenation of the carbon-carbon double bonds of vegetable oils is carried out on an enormous scale. The use of this reaction stems from the preference of many persons for eating fats, such as butter or lard, instead of oils. Animal fats, however, are not available in sufficient quantities and are more expensive than vegetable oils. Because the higher melting point of an animal fat is related to its smaller number of double bonds, the idea of partially hydrogenating (hardening) vegetable oils to produce substitutes for animal fats has obvious commercial appeal. Such hydrogenation, exemplified by the conversion of glyceryl trioleate to glyceryl tristearate, is usually carried out with a nickel catalyst.

$$
\begin{array}{ccc}
& \overset{\displaystyle O}{\overset{\displaystyle \|}{}} & \\
CH_2-O-C(CH_2)_7CH{=}CH(CH_2)_7CH_3 & & CH_2-O-\overset{O}{\overset{\|}{C}}(CH_2)_{16}CH_3 \\
| & & | \\
CH-O-\overset{O}{\overset{\|}{C}}(CH_2)_7CH{=}CH(CH_2)_7CH_3 & \xrightarrow[Ni]{H_2} & CH-O-\overset{O}{\overset{\|}{C}}(CH_2)_{16}CH_3 \\
| & & | \\
CH_2-O-\overset{O}{\overset{\|}{C}}(CH_2)_7CH{=}CH(CH_2)_7CH_3 & & CH_2-O-\overset{O}{\overset{\|}{C}}(CH_2)_{16}CH_3 \\
\text{Glyceryl trioleate} & & \text{Glyceryl tristearate} \\
(\text{mp } 6°) & & (\text{mp } 73°)
\end{array}
$$

Margarine is produced mainly from cottonseed, soybean, and corn oils.   In the manufacture of margarine the partially hydrogenated vegetable oil is mixed with skim milk that has been slightly fermented.   Salt, artificial color, vitamin A, and other additives are blended in.   Shortenings (such as Crisco and Spry) also are made from partially hydrogenated vegetable oils, sometimes blended with lard or beef fat.

A dietary preference for solids rather than liquids may be unfortunate.   Diets containing high levels of saturated fatty acids seem to have some connection with atheroschlerosis, a thickening and hardening of the inner walls of arteries associated with deposition of cholesterol and other lipids.

13.4
HYDROGENOLYSIS

Like other esters (Sec. 12.6), triacylglycerols are cleaved by catalytic hydrogenation under severe conditions.   As illustrated for a triacylglycerol that is one of many in coconut oil, this reaction produces glycerol and a mixture of primary alcohols.

$$CH_2-O-\overset{\overset{\displaystyle O}{\|}}{C}(CH_2)_{10}CH_3$$
$$CH-O-\overset{\overset{\displaystyle O}{\|}}{C}(CH_2)_{10}CH_3 + 6 H_2 \xrightarrow[\substack{\text{(high temperature} \\ \text{and pressure)}}]{\text{copper chromite}} \text{glycerol} + 2 CH_3(CH_2)_{10}CH_2OH$$
$$CH_2-O-\overset{\overset{\displaystyle O}{\|}}{C}(CH_2)_{12}CH_3 \qquad\qquad + CH_3(CH_2)_{12}CH_2OH$$

Such long-chain alcohols are used for making an important class of detergents (Sec. 13.7) and for other purposes.   The severe hydrogenolysis conditions generally lead to hydrogenation of any carbon-carbon double bonds.

13.5
HYDROLYSIS

Hydrolysis is illustrated for a triacylglycerol that is one of many in beef tallow.

$$CH_2-O-\overset{\overset{\displaystyle O}{\|}}{C}(CH_2)_{14}CH_3$$
$$CH-O-\overset{\overset{\displaystyle O}{\|}}{C}(CH_2)_{16}CH_3 + 3 Na^+ OH^- \longrightarrow \quad CH-OH + CH_3(CH_2)_{14}CO_2^- Na^+$$
$$CH_2-O-\overset{\overset{\displaystyle O}{\|}}{C}(CH_2)_7CH{=}CH(CH_2)_7CH_3$$

$$CH_2-OH$$
$$CH_2-OH$$

$$+ CH_3(CH_2)_{16}CO_2^- Na^+ + CH_3(CH_2)_7CH{=}CH(CH_2)_7CO_2^- Na^+$$

Hydrolysis of fats and oils with aqueous base (Sec. 12.6 and 12.7) is carried out on an enormous scale because the resulting mixtures of carboxylic acid salts are soaps. Alkaline hydrolysis of esters is often called **saponification** (from the Latin *sapo*, soap).

13.6
SOAPS

Sodium salts of unbranched carboxylic acids with 12 to 18 carbons are soaps. Beef tallow and coconut oil are common sources of the salts. Soaps with the C_{16} and C_{18} chains that predominate in beef tallow have excellent cleansing properties. Those with the C_{12} and C_{14} chains that predominate in coconut oil dissolve rapidly and are particularly effective in stabilizing foams. To combine various favorable properties, soaps are sometimes made from mixtures of two or more fats and oils. After the fat or oil has been heated with aqueous base, glycerol is removed and the mixture of sodium salts purified. Alternatively, acid-catalyzed hydrolysis is used, following which the acids are isolated, purified, and converted to salts by treatment with base. Dyes, perfumes, abrasive agents (for scouring soaps), and other additives are mixed in. Then the soap is shaped into bars, cut into flakes, or converted to a powder. Floating soaps are produced by introducing finely dispersed air bubbles to reduce density. K^+ and $NH_4{}^+$ salts are lower melting and more easily dispersed in water than Na^+ salts and so are often used to formulate shaving soaps, shampoos, and liquid soaps.

Soap's cleansing action is related to the solubility properties of the salts of long-chain carboxylic acids. Sodium palmitate is a typical soap molecule.

$$CH_3-CH_2-CH_2-CH_2-CH_2-CH_2-CH_2-CH_2-CH_2-CH_2-CH_2-CH_2-CH_2-CH_2-CH_2-\overset{\overset{\textstyle O}{\|}}{C}-O^-\ Na^+$$

Hydrophobic	Hydrophilic
(tends to be insoluble in water but soluble in organic materials)	(tends to dissolve in water)

The ionic head of this salt has a great tendency to dissolve in water. It is **hydrophilic** (water loving). The long hydrocarbon tail tends to be very insoluble in water. It is **hydrophobic** (water hating). Because the hydrocarbon portion is so large, solubility of sodium palmitate in water to form a true solution is very low. When shaken with water, however, it disperses to form small aggregates called **micelles** that contain hundreds or thousands of palmitate ions. As shown in Figure 13.1, the ionic heads are on the exterior of the micelle, where they can interact with water molecules. The hydrocarbon tails are clustered together away from water. Having many exterior negative charges, micelles repel one another. As a result, they remain suspended in solution rather than coalescing into larger particles that separate from the solution.

Soiled clothes (and hands) are ordinarily coated with water-insoluble oil and grease, which tend to hold dirt particles. These organic materials may be dispersed into the water as small droplets by mechanical agitation during the washing process. Yet they tend to coalesce and separate out of the water to again coat the clothes. In

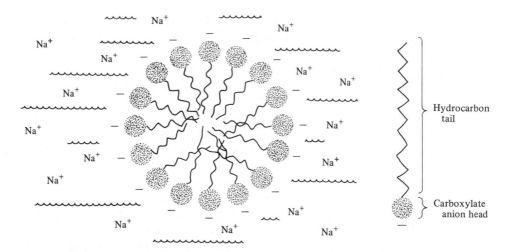

Figure 13.1 A sodium palmitate micelle. Palmitate anions are represented in the diagrammatic fashion shown at the right.

the presence of a soap these organic materials are incorporated into micelles (Fig. 13.2). Because of surface layers of negative charges provided by the carboxylic acid anions, these micelles do not readily coalesce. Therefore the organic materials remain suspended and are removed with the wash water.

In order for a carboxylic acid salt to be an effective soap, its chain cannot be too long or too short. If the chain is too short (less than C_{12}), the salt simply dissolves in

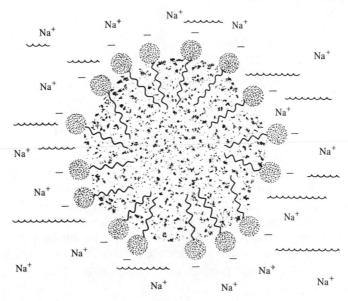

Figure 13.2 A micelle containing long-chain carboxylic acid anions from a soap and incorporating other organic materials, such as fats and oils.

water instead of forming micelles. If the chain is too long (more than C_{18}), the hydrocarbon tail so dominates properties that not much of the salt goes into the solution to form micelles.

A relatively stable suspension of small droplets of one phase in another in which it is inherently insoluble is called an **emulsion**. A soap-stabilized suspension of oil in water is one example. Milk is also an emulsion of an oil in water. A biological example related to the action of soap is the functioning of the bile. Most fats and oils that we ingest pass through the stomach and into the small intestine. There they are mixed with bile, a substance produced in the liver, stored in the gallbladder, and released at intervals through the bile duct into the small intestine. Bile contains compounds that aid in emulsifying fats, oils, and fatty acids and thereby promote both their contact with enzymes catalyzing hydrolysis and their absorption through the intestinal wall. The principal emulsifying compounds in bile are steroids—of which cholic acid is most common—and derivatives in which they are attached by amide linkages to glycine (H_2N—CH_2—CO_2H) and taurine (H_2N—CH_2CH_2—SO_3H).

Cholic acid $-Z = -\overset{\overset{\displaystyle O}{\|}}{C}-OH$

Glycocholic acid $-Z = -\overset{\overset{\displaystyle O}{\|}}{C}-NH-CH_2-CO_2H$

Taurocholic acid $-Z = -\overset{\overset{\displaystyle O}{\|}}{C}-NH-CH_2CH_2-SO_3H$

All these compounds, known collectively as bile acids, contain a carboxyl or sulfonic acid group. In the intestine (pH about 6.5), these acidic functions are essentially completely ionized. The action of a bile acid is related to its soaplike structure—an ionic function ($-CO_2^-$ or $-SO_3^-$) attached to a large hydrocarbon unit.

Conventional soaps function poorly in hard water. Hard waters contain Ca^{2+}, Mg^{2+}, or Fe^{2+} ions, and salts of these ions with anions of carboxylic acids are very insoluble. In hard water, insoluble salts precipitate, removing the cleansing agent from the solution and producing an undesirable scum.

$$2\ CH_3(CH_2)_{14}\overset{\overset{\displaystyle O}{\|}}{C}-O^-\ Na^+ + Ca^{2+} \longrightarrow (CH_3(CH_2)_{14}\overset{\overset{\displaystyle O}{\|}}{C}-O^-)_2\ Ca^{2+} \downarrow + 2\ Na^+$$

(the ring around the tub)

Soaps are also ineffective in acidic solutions. As the salts of weak acids, they are converted by stronger acids to the free carboxylic acids, which are not effective emulsifying agents.

$$R-\overset{\overset{\displaystyle O}{\|}}{C}-O^- + HX \longrightarrow R-\overset{\overset{\displaystyle O}{\|}}{C}-OH + X^-$$

These disadvantages of soaps are one reason for the widespread use of synthetic detergents.

13.7
DETERGENTS

A **detergent** (from the Latin *detergere*, to wipe off) is a molecule that, like a soap, forms micelles in water and emulsifies fats and oils. The most widely used synthetic detergents are sodium alkylbenzenesulfonates.

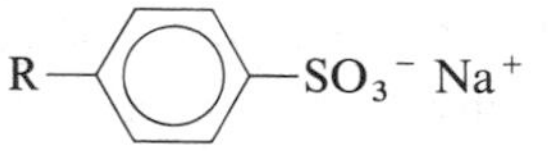

$$R-\underset{}{\bigcirc}-SO_3^- \; Na^+ \qquad\qquad R \text{ has approximately 12 carbons.}$$

A sodium alkylbenzenesulfonate detergent

These compounds are made by sulfonating alkylbenzenes, followed by treatment with base.

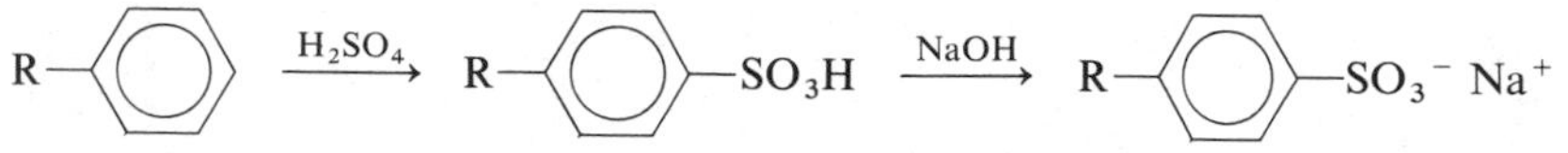

$$R-\bigcirc \xrightarrow{H_2SO_4} R-\bigcirc-SO_3H \xrightarrow{NaOH} R-\bigcirc-SO_3^- \; Na^+$$

Detergents resemble soaps in having an ionic group and a long hydrocarbon tail but offer certain advantages. Their Ca^{2+}, Mg^{2+}, and Fe^{2+} salts do not separate from aqueous solutions. Moreover, as salts of extremely strong acids, they are not converted to the parent acids even in quite acidic solutions.

Several environmental problems have been associated with detergent use. The alkyl chains of alkylbenzenesulfonate detergents were at one time often derived from inexpensive alkenes made by polymerizing propylene.

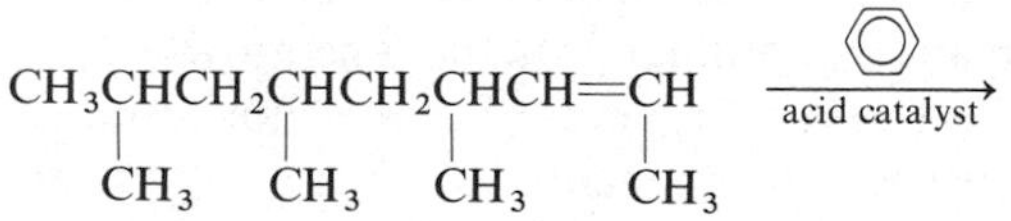

$$\underset{\substack{|\\CH_3}}{CH_3CH}CH_2\underset{\substack{|\\CH_3}}{CH}CH_2\underset{\substack{|\\CH_3}}{CH}CH=CH \xrightarrow[\text{acid catalyst}]{\bigcirc}$$

A tetramer of propylene

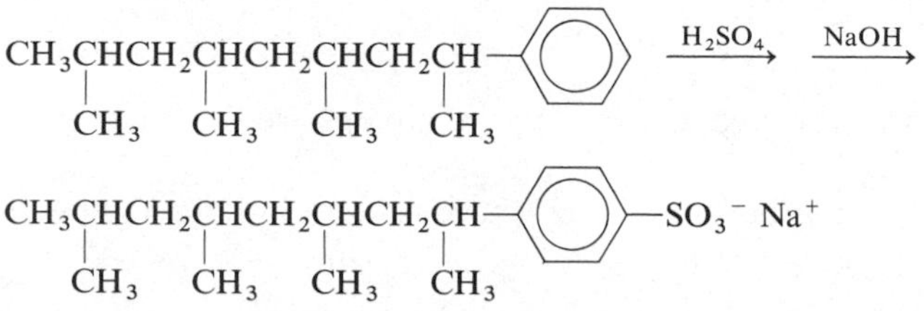

$$\underset{\substack{|\\CH_3}}{CH_3CH}CH_2\underset{\substack{|\\CH_3}}{CH}CH_2\underset{\substack{|\\CH_3}}{CH}CH_2\underset{\substack{|\\CH_3}}{CH}-\bigcirc \xrightarrow{H_2SO_4} \xrightarrow{NaOH}$$

$$\underset{\substack{|\\CH_3}}{CH_3CH}CH_2\underset{\substack{|\\CH_3}}{CH}CH_2\underset{\substack{|\\CH_3}}{CH}CH_2\underset{\substack{|\\CH_3}}{CH}-\bigcirc-SO_3^- \; Na^+$$

The branched chains in these molecules are not rapidly metabolized by most organisms. As a result, these detergents tended to pass unchanged through sewage plants to accumulate in rivers, some of which literally developed a head of foam. Since the mid-1960s, there has been a switch to precursors with straight chains, such as 1-dodecene, that furnish detergents that are more biodegradable.

$$CH_3(CH_2)_9CH{=}CH_2 \xrightarrow[\text{acid catalyst}]{\bigcirc} \xrightarrow{H_2SO_4} \xrightarrow{NaOH} CH_3(CH_2)_9\underset{\underset{\displaystyle CH_3}{|}}{CH}{-}\bigcirc{-}SO_3^-\ Na^+$$

1-Dodecene

The chains of an important group of detergents made by converting alcohols to salts of sulfuric acid esters are even more biodegradable.

$$CH_3(CH_2)_nCH_2OH \xrightarrow[\text{cold}]{H_2SO_4} CH_3(CH_2)_nCH_2OSO_3H \xrightarrow{NaOH} CH_3(CH_2)_nCH_2OSO_3^-\ Na^+$$

$$n = 10, 12, 14, 16$$

Hydrogenolysis of fats and oils (Sec. 13.4) is one way of obtaining the alcohols.

Other environmental problems have stemmed from additives mixed with detergents. Organic molecules rarely constitute as much as 20% of a detergent formulation. Components called "builders" usually constitute more than 50%. Besides adding some desirable properties, builders are also considerably less expensive than detergent molecules. The most widely used builder is pentasodium triphosphate.

$$^-O{-}\overset{\overset{\displaystyle O}{\|}}{\underset{\underset{\displaystyle O_-}{|}}{P}}{-}O{-}\overset{\overset{\displaystyle O}{\|}}{\underset{\underset{\displaystyle O_-}{|}}{P}}{-}O{-}\overset{\overset{\displaystyle O}{\|}}{\underset{\underset{\displaystyle O_-}{|}}{P}}{-}O^-\quad 5\,Na^+ \qquad\qquad ^-O{-}\overset{\overset{\displaystyle O}{\|}}{\underset{\underset{\displaystyle O_-}{|}}{P}}{-}O^-$$

Pentasodium triphosphate   Phosphate ion

One important function of a builder is to form water-soluble complexes with such ions as Ca^{2+} and Mg^{2+}. Although these ions do not form precipitates with detergents, they do adversely affect their performance. The builder also helps to keep the solution alkaline and, by complexing with inorganic ions, aids in the breakdown and suspension of soil. Enormous amounts of phosphate builders end up as phosphate ion, much of which finds its way into rivers and lakes. Phosphate is not toxic; in fact, it is essential for plant and animal growth. The growth of algae in some rivers and lakes is limited by a low natural level of phosphate. Addition of large amounts of phosphate to such waters results in greatly increased algal growth, drastically altering the ecological balance.

13.8 PAINTS

The three principal components of a paint are the following:

1. a *binder* that adheres to a surface, holding the pigment in place and acting as a moisture barrier.

2.   a *pigment* that provides color and often provides much of the covering and protective power of a paint.

3.   a *thinner* (solvent) that aids in applying the paint and then evaporates.

Linseed oil was once the most common binder. This oil contains a large portion of unsaturated acids, particularly linolenic acid, which has three carbon-carbon double bonds. On contact with oxygen in the air, unsaturated acids, especially those with more than one double bond, tend to be converted to a hard, resinous material. This process is complex, involving polymerization of the carbon-carbon double bonds as well as other reactions. Antioxidants (Sec. 5.20) are often added to foodstuffs containing fats and oils to retard such processes.

Alkyd resins are now the usual binders for "oil-based" paints. A typical alkyd resin is prepared by heating glycerol with linseed oil to give principally mono-esters of glycerol.

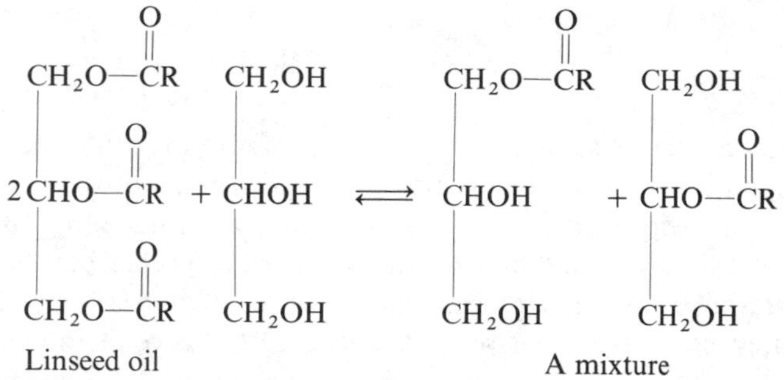

Linseed oil   A mixture

Reaction of this mixture with phthalic anhydride forms ester linkages to the hydroxyl groups, affording a low molecular weight alkyd resin.

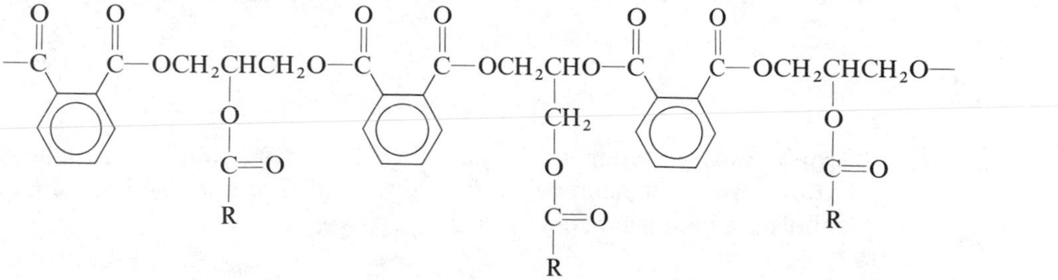

An alkyd resin

R represents long chains, usually unsaturated

When the paint is spread in a thin film in the presence of air, the organic thinner evaporates and the unsaturated groups (R) polymerize to give a highly *cross-linked* polymer.

In water-based paints, often called latex paints, the binder is a stable emulsion of a polymer in water. Generally this is a copolymer prepared from vinyl acetate

$$\begin{matrix} & O \\ & \| \\ (CH_2{=}CHOCCH_3) \end{matrix} \text{ and an ester of acrylic acid } \begin{matrix} & O \\ & \| \\ (CH_2{=}CHCOR). \end{matrix}$$

$$-CH_2-CH-CH_2-CH-CH_2-CH-CH_2-CH-CH_2-CH-CH_2-CH-CH_2-CH-$$

A copolymer of vinyl acetate and an acrylate ester

$$R- \text{ is often } CH_3CH_2CH_2CH_2- \text{ or } CH_3CH_2CH_2CH_2CHCH_2- \\ | \\ CH_2CH_3$$

13.9 WAXES

Waxes are complex mixtures of organic compounds that include alkanes, carboxylic acids, and alcohols. Their major components, however, ordinarily are esters of carboxylic acids and primary alcohols that both have long unbranched chains.

Waxes are found on external portions of higher plants, particularly on leaves, stems, and fruits. Their principal function is often to retard evaporation of water, but they may also have protective functions in reducing mechanical damage and inhibiting fungal and insect attack. Waxes are found on feathers, particularly of waterfowl. Their water repellency prevents water from wetting the feathers, which would greatly increase their weight and reduce their insulating ability.

Spermaceti, a wax that crystallizes from the oil of the sperm whale, has as its major component an ester of palmitic acid with a C_{16} alcohol.

$$\begin{matrix} & & O \\ & & \| \\ CH_3(CH_2)_{14}C & -O(CH_2)_{15}CH_3 \end{matrix}$$

Carnuba wax, a coating on the leaf of the Brazilian palm, is a particularly complex mixture. Esters of unbranched C_{24}, C_{26}, and C_{28} acids with C_{30}, C_{32}, and C_{34} alcohols are significant components.

13.10 GLYCEROPHOSPHOLIPIDS AND SPHINGOLIPIDS

Glycerophospholipids are major components of membranes in all plant and animal cells. Like the triacylglycerols, they are esters of glycerol. Only two of the hydroxyl groups of glycerol are esterified with fatty acids, however—most commonly palmitic, stearic, or oleic. One primary hydroxyl group is esterified to phosphoric acid (see

Sec. 5.7 for phosphate esters). The phosphoric acid is also esterified to an alcohol (ROH) that contains a polar functional group.

$$
\begin{array}{c}
\quad\quad\quad \overset{\displaystyle O}{\overset{\displaystyle \|}{}} \\
CH_2-O-\underset{\underset{\displaystyle OH}{|}}{P}-OR \\
\overset{\displaystyle O}{\overset{\displaystyle \|}{}} \quad\quad | \\
R'C-O-CH \\
\overset{\displaystyle O}{\overset{\displaystyle \|}{}} \quad\quad | \\
R''C-O-CH_2
\end{array}
$$

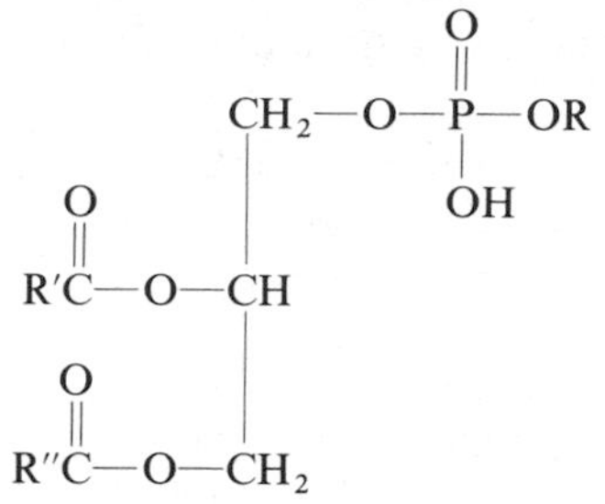

R contains a polar functional group; R' and R'' are long saturated or unsaturated chains of fatty acids.

A glycerophospholipid

In animals and higher plants the most abundant glycerophospholipids have either ethanolamine or choline as this alcohol component.

$$HO-CH_2CH_2-NH_2 \qquad HO-CH_2CH_2-\overset{\overset{\displaystyle CH_3}{|}}{\underset{\underset{\displaystyle CH_3}{|}}{N^+}}-CH_3$$

Ethanolamine   Choline

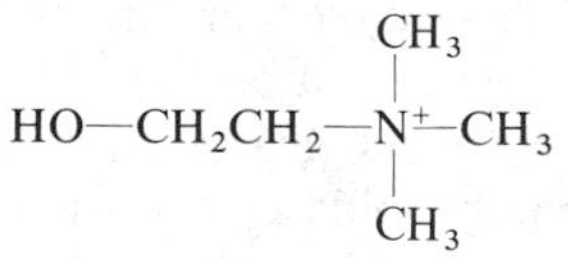

Glycerophospholipids containing ethanolamine are commonly called phosphatidyl-ethanolamines or cephalins and those containing choline are called phosphatidyl-cholines or lecithins.

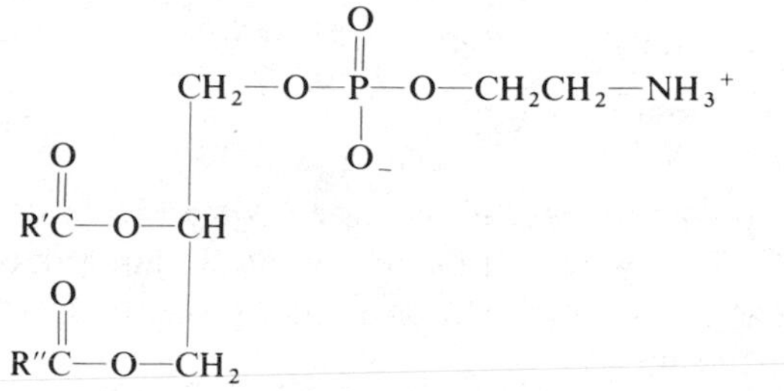

$$
\begin{array}{c}
\quad\quad\quad \overset{\displaystyle O}{\overset{\displaystyle \|}{}} \\
CH_2-O-\underset{\underset{\displaystyle O_-}{|}}{P}-O-CH_2CH_2-NH_3{}^+ \\
\overset{\displaystyle O}{\overset{\displaystyle \|}{}} \quad\quad | \\
R'C-O-CH \\
\overset{\displaystyle O}{\overset{\displaystyle \|}{}} \quad\quad | \\
R''C-O-CH_2
\end{array}
\qquad\qquad
\begin{array}{c}
\quad\quad\quad \overset{\displaystyle O}{\overset{\displaystyle \|}{}} \\
CH_2-O-\underset{\underset{\displaystyle O_-}{|}}{P}-O-CH_2CH_2-N^+(CH_3)_3 \\
\overset{\displaystyle O}{\overset{\displaystyle \|}{}} \quad\quad | \\
R'C-O-CH \\
\overset{\displaystyle O}{\overset{\displaystyle \|}{}} \quad\quad | \\
R''C-O-CH_2
\end{array}
$$

A phosphatidylethanolamine   A phosphatidylcholine or lecithin
or cephalin

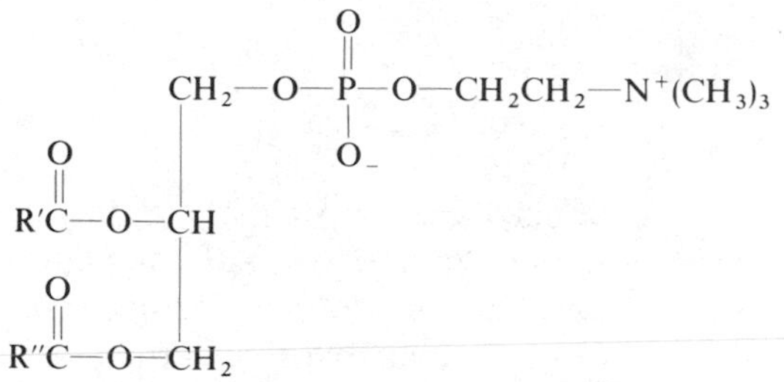

Note the charges on these structures. An —OH group attached to phosphorus in such phosphate esters is relatively acidic (comparable in acidity to phosphoric acid). In biological solutions, most of which are approximately neutral (pH about 7), this group is usually ionized as shown in the preceding structures.* Similarly, the

* Ordinarily we draw acids in their un-ionized forms and amines in their unprotonated forms, although under certain conditions these are not the dominant forms. Because the charges are crucial to their function in a cell, however, glycerophospholipids are drawn here in the forms that predominate in the neutral solutions that are their usual environment.

amino group of an ethanolamine glycerophospholipid is protonated. The quaternary ammonium group of choline, of course, is always charged.

If we draw out more fully the structure of a typical choline glycerophospholipid, the structural resemblance to a soap is evident—both have a polar head attached to a large hydrocarbon portion.

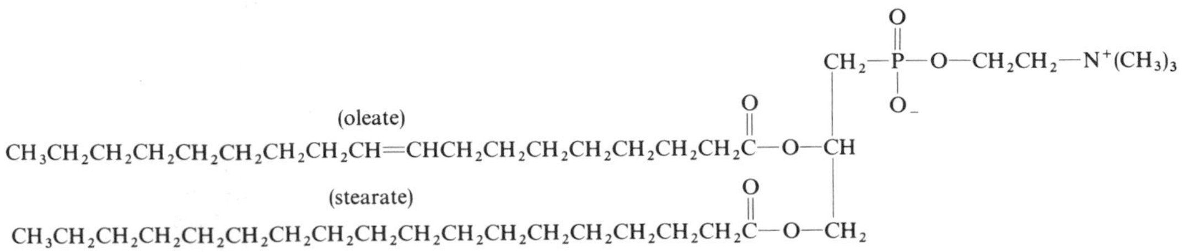

$$CH_2-\overset{\overset{\displaystyle O}{\|}}{P}-O-CH_2CH_2-N^+(CH_3)_3$$

(oleate)

$$CH_3CH_2CH_2CH_2CH_2CH_2CH_2CH_2CH=CHCH_2CH_2CH_2CH_2CH_2CH_2CH_2\overset{\overset{\displaystyle O}{\|}}{C}-O-CH$$

(stearate)

$$CH_3CH_2CH_2CH_2CH_2CH_2CH_2CH_2CH_2CH_2CH_2CH_2CH_2CH_2CH_2CH_2\overset{\overset{\displaystyle O}{\|}}{C}-O-CH_2$$

A typical choline glycerophospholipid

In fact, these molecules are good emulsifying agents. Lecithin, usually obtained from soybeans, is used in the food industry as an emulsifying agent—for example, in margarine and chocolate.

Sphingolipids are also important components of membranes in both animal and plant cells. Particularly large amounts are present in brain and nervous tissue. They derive their name from sphingosine.

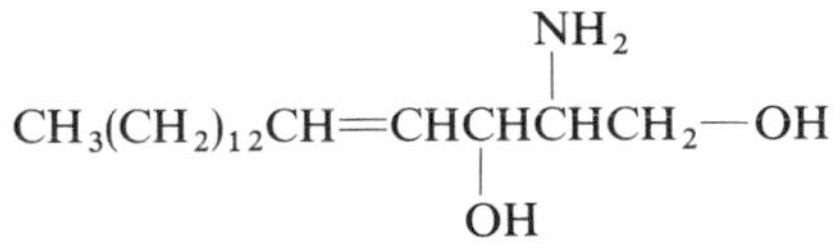

$$CH_3(CH_2)_{12}CH=CHCHCHCH_2-OH$$

Sphingosine

Sphingosine or a closely related compound is one of the three components of a sphingolipid. A second component is a fatty acid, attached by an amide linkage to the amino group of sphingosine. The third component is a polar grouping linked to the primary hydroxyl group of sphingosine.

In sphingomyelins, the most abundant sphingolipids in higher animals, the hydroxyl group is esterified with phosphoric acid, which is also esterified to choline.

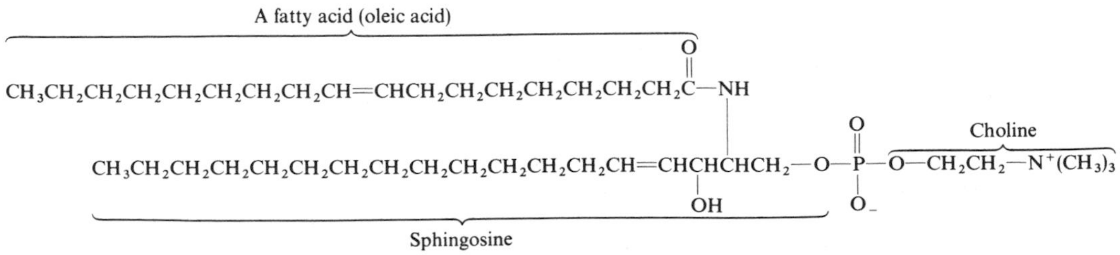

A fatty acid (oleic acid)

$$CH_3CH_2CH_2CH_2CH_2CH_2CH_2CH_2CH=CHCH_2CH_2CH_2CH_2CH_2CH_2CH_2\overset{\overset{\displaystyle O}{\|}}{C}-NH$$

Choline

$$CH_3CH_2CH_2CH_2CH_2CH_2CH_2CH_2CH_2CH_2CH_2CH_2CH_2CH=CHCHCHCH_2-O-\overset{\overset{\displaystyle O}{\|}}{P}-O-CH_2CH_2-N^+(CH_3)_3$$

Sphingosine

A representative sphingomyelin molecule

These lipids are important components of the sheathlike structures that surround nerve fibers.

In glycosphingolipids the polar group is a saccharide. It may be a monosaccharide, often a β-D-galactopyranosyl unit, as shown here in the structure of a simple glycosphingolipid.

$$CH_3CH_2CH_2CH_2CH_2CH_2CH_2CH_2CH_2CH_2CH_2CH_2CH_2CH_2CH_2CH_2CH_2CH_2CH_2CH_2C\overset{\overset{\textstyle O}{\|}}{—}NH$$

$$CH_3CH_2CH_2CH_2CH_2CH_2CH_2CH_2CH_2CH_2CH_2CH_2CH_2CH=CHCHCHCH_2$$
$$\underset{\textstyle OH}{|}$$

A simple glycosphingolipid

The polar group may instead be a polysaccharide, some units of which can be linked to other polar molecules. Glycosphingolipids are often important cell-surface components. Some found on the surface of red blood cells are related to blood-group specificities.

Note that, like the glycerophospholipids, the sphingolipids have structures with two long hydrocarbon chains and a polar head.

13.11
MEMBRANE STRUCTURE

Many cellular components must be confined within the cell and many substances in the external environment must be excluded. Yet nutrients and other essential compounds must readily enter and waste products and compounds to be transferred to other cells must readily leave. Even within a cell, functions are often compartmentalized in ways that require different concentrations of a compound in different parts of the cell.

Membranes, major parts of all cells, act as barriers to the passage of some compounds. In this way, they maintain differences in concentrations of chemical species between a cell and its external environment or between different parts of a cell. They aid and regulate the transport of other substances. Membranes also act as structural elements to which certain metabolic systems are attached.

Membranes are flexible, sheetlike structures. Although composition varies considerably, a typical membrane has 40% lipids and 60% proteins. Membranes contain large amounts of lipids with polar functional groups. In fact, such lipids are found principally in membranes. Glycerophospholipids usually predominate, but sphingolipids are also present.

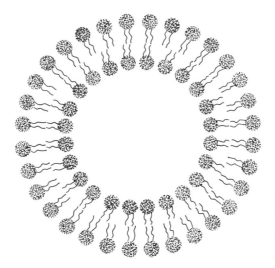

Figure 13.3 A cross-sectional view of a liposome obtained by shaking a polar lipid with water. Water molecules are on both the inside and outside of the bilayer of lipid molecules.

These polar lipids have a polar head and two long hydrocarbon tails.

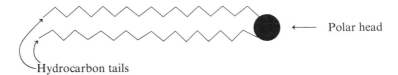

Because of the large hydrocarbon portions, they have little solubility in water. If shaken with water, however, they tend to form structures called **liposomes** that are dispersed through the water. In a liposome (Fig. 13.3) a bilayer (two layers) of lipid surrounds a tiny water droplet. The lipid layers are arranged so that the polar heads are in contact with water, either on the exterior or the interior of the liposome, permitting maximum interaction of the polar groups with water. The hydrocarbon chains are in the interior of the bilayer, hidden from the aqueous environment.

As shown in Figure 13.4,* such a bilayer structure is now thought to provide the framework of a membrane. A variety of proteins are embedded in the membrane. Some are on one side or the other, but others extend completely through the membrane.

Water can pass through a bilayer of polar lipids. Ions and many polar, water-soluble molecules cannot, however. Carbohydrates (such as glucose) and amino acids (Chapter 14), for instance, do not pass through such a bilayer. Some of the proteins embedded in a membrance are involved in selectively transporting some ions and polar molecules across the membrane. The transport of polar materials probably

---

* This picture omits additional components that contribute to membrane structure. Some membranes, for instance, contain a considerable amount of cholesterol, a steroidal lipid (see Sec. 9.15 for its structure); some membrane surfaces have a substantial amount of carbohydrate, some as part of lipids (glycosphingolipids) and some attached to proteins.

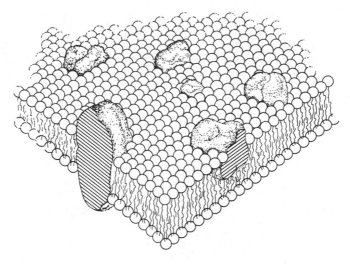

Figure 13.4 A structure proposed for membranes. The framework of the membrane is a bilayer of polar lipid molecules. Proteins are embedded in the membrane, some extending from one side to the other. [From S. J. Singer and G. L. Nicolson, *Science*, **175**, 720 (1972). Copyright 1972 by the American Association for the Advancement of Science.]

requires their conversion to forms more compatible with nonpolar environments. A clue to a way in which ion transport may be facilitated is provided by the observation (Sec. 5.15) that ions can be made more soluble in organic solvents by complexing with cyclic polyethers. In fact, some antibiotics (such as nonactin, Sec. 5.15) with many complexing groups facilitate transport of ions across membranes.

SUMMARY

Definitions and Ideas

Lipids, important compounds in all plants and animals, belong to several structural groups. All, however, share the property of being soluble in nonpolar organic solvents but insoluble in water. All lipids have large alkyl portions responsible for this solubility behavior.

Fats (solids at room temperature) and **oils** (liquids at room temperature) are mixtures of **triacylglycerols**, compounds in which all three hydroxyl groups of glycerol are esterified with carboxylic acids. The carboxylic acids incorporated into fats and oils are often called **fatty acids**. The most abundant fatty acids have long unbranched chains containing even numbers of carbon atoms and often containing one or more alkene functions.

Hydrogenating (H_2 and Ni) the carbon-carbon double bonds of fats and oils raises the melting point. Hydrogenolysis (H_2 and copper chromite, high temperature and pressure) produces long-chain alcohols from the carboxylic acids. Hydrolysis of fats and oils with base (**saponification**) produces salts of long-chain acids.

Salts of long-chain acids function as soaps through their having a long hydrocarbon chain that is **hydrophobic** (water hating) and an ionic portion that is **hydrophilic** (water loving). Synthetic detergents contain similar structural features. Soaps and synthetic detergents are not too soluble in water but do disperse to form **micelles**.

The major constituents of **waxes** are usually esters of carboxylic acids and primary alcohols that both have long unbranched chains.

Glycerophospholipids are esters of glycerol in which two of the hydroxyl groups are esterified with fatty acids. One primary hydroxyl group is esterified to phosphoric acid, which also is esterified to some alcohol containing a polar functional group. **Sphingolipids** are derivatives of sphingosine. The amino group of sphingosine is linked by an amide linkage to a fatty acid and a hydroxyl group is linked to a polar group.

Both glycerophospholipids and sphingolipids have two long hydrocarbon chains and a polar group. These compounds, which are major components of cellular membranes, tend to form bilayer structures in an aqeous solution.

PROBLEMS

1. Define each of the following terms. Where helpful, use structural drawings as illustrations.

   (a) lipid
   (b) fat
   (c) oil
   (d) triacylglycerol
   (e) simple triacylglycerol
   (f) mixed triacylglycerol
   (g) fatty acid
   (h) unsaturated fatty acid
   (i) saponification
   (j) hardening of oils
   (k) soap
   (l) detergent
   (m) micelle
   (n) hydrophilic group
   (o) hydrophobic group
   (p) bile acid
   (q) wax
   (r) paint binder
   (s) paint pigment
   (t) paint thinner
   (u) oil-based paint
   (v) water-based paint
   (w) alkyd resin
   (x) glycerophospholipid
   (y) sphingolipid
   (z) liposome
   (aa) lipid bilayer
   (bb) membrane

2. Draw structures of compounds that fit the following descriptions.
   (a) a triacylglycerol that is a solid at ordinary temperatures
   (b) a triacylglycerol that is a liquid at ordinary temperatures
   (c) three commonly occurring saturated fatty acids
   (d) a commonly occurring unsaturated fatty acid
   (e) a soap
   (f) a detergent with an aromatic group
   (g) a detergent that has no aromatic group

3. Draw the organic products that result from treating glyceryl trioleate (the simple triacylglycerol of oleic acid) with each reagent.
   (a) NaOH
   (b) H_2, Ni
   (c) H_2, copper chromite (high temperature and pressure)

(d) LiAlH$_4$, then H$_2$O

(e) Br$_2$

4. Write equations for what happens to sodium stearate, a typical soap molecule, in the following solutions.

(a) hard water containing Ca^{2+} (b) an acidic aqueous solution

5. When hydrolyzed, a mixture of triacylglycerols gives a mixture of glycerol, palmitic acid, and stearic acid. Draw the structures of all different triacylglycerol molecules that could be present.

6. Explain the ability of soaps and detergents to act as cleansing agents.

7. Why does the following molecule have detergentlike properties?

$$
\begin{array}{c}
\text{O} \\
\| \\
\text{CH}_3(\text{CH}_2)_{14}\text{COCH}_2\text{C(CH}_2\text{OH})_3
\end{array}
$$

8. How could you construct an "invert" soap, a soap in which the organic portion has a positive rather than a negative charge?

9. The "saponification number" is often used as one means of characterizing a fat or oil. It is defined as the number of milligrams of KOH needed to saponify 1 g of the fat or oil.

(a) What is the saponification number of glyceryl tristearate (the simple triacylglycerol of stearic acid), which has a molecular weight of 891?

(b) Which has a higher saponification number, butter or lard?

10. The "iodine number" is another means of characterizing fats and oils. It is defined as the number of grams of I$_2$ absorbed by the carbon-carbon double bonds of 100 g of a fat or oil. Because I$_2$ does not add readily to the alkene functions, a more reactive reagent, usually ICl, is actually used. The result, however is expressed as though I$_2$ had added.

(a) What is the iodine number of glyceryl trioleate (the triacylglycerol of oleic acid), which has a molecular weight of 885?

(b) Which has the highest iodine number: corn, linseed, or soybean oil?

11. Why is it reasonable that metabolism of a fat furnishes more calories than the metabolism of an equal weight of a carbohydrate?

14

Amino Acids and Proteins

A multitude of chemical components are absolutely essential to the existence of even the simplest organisms. None, however, is as diverse in function or interwoven with so many vital processes as the **proteins**, major constituents of all organisms. In fact, most of the genetic information transmitted from an organism to its successors is expressed by the synthesis of appropriate proteins.

The proteins known as **enzymes** catalyze and regulate virtually all biological reactions. Structural proteins are the principal constituents of connective tissue, tendons, cartilage, and ligaments, as well as of skin, scales, feathers, hair, nails, and hoofs. Contractile proteins, the major components of muscles, are responsible for body movements. Proteins acting as protective agents include the antibodies, which combine with and neutralize the effects of foreign substances, and proteins involved in blood clotting. Some proteins—insulin is an example—are hormones and have a role in regulating physiological processes. Others have a storage function: The protein of egg white stores amino acids for future use and the protein known as ferritin stores iron. Proteins involved in transport include hemoglobin, which carries oxygen from the lungs to other tissues throughout the body. Other proteins are toxins: Most snake venoms and the bacterial toxin that causes botulism are examples. Other groups of proteins include the threadlike materials secreted by spiders and silkworms to form their webs or cocoons.

It is estimated that humans contain more than a million different kinds of protein molecules. This large number is needed because of the numerous functions that proteins perform in our bodies. Proteins are large molecules; their molecular weights range from several thousand to several million. Like carbohydrates and nucleic acids, the other high molecular weight biological molecules, proteins are polymers. Proteins are constructed from low molecular weight compounds that contain at least one amino and one carboxyl group and hence are known as **amino acids**. The extraordinary diversity of functions and properties of proteins arises from different numbers and structural combinations of only a small number of amino acids.

14.1
AMINO ACIDS: STRUCTURES

The 20 amino acids that are the principal building blocks of proteins are listed in Table 14.1. The trivial names in the table are accepted by the IUPAC and invariably used instead of more systematic names. These names are often related to a source from which the amino acid was originally obtained. Asparagine, for example, was isolated from asparagus juice. Each is an α-amino acid—the amino function is attached to the carbon next to the carboxyl group.

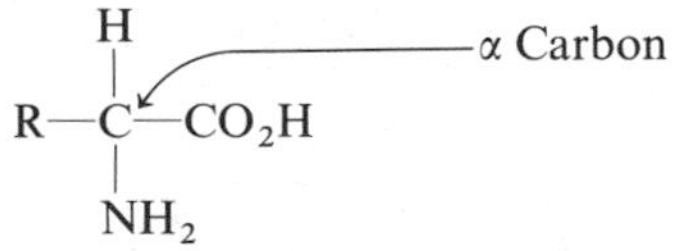

$$R—\underset{\underset{NH_2}{|}}{\overset{\overset{H}{|}}{C}}—CO_2H \qquad \text{— α Carbon}$$

With the exception of glycine (R = H), all the amino acids are chiral and have the same configuration.

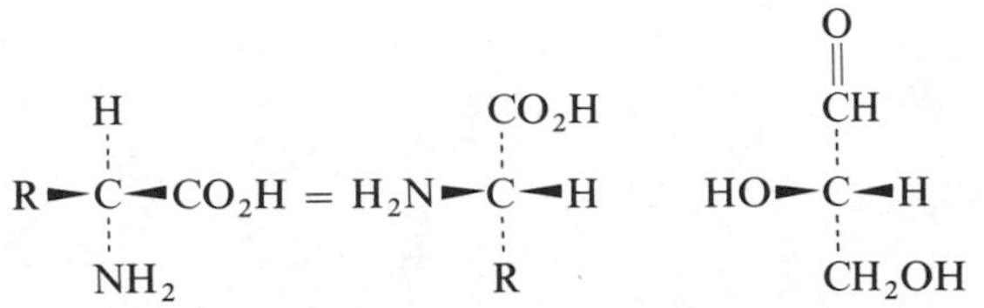

$$R—\underset{\underset{NH_2}{|}}{\overset{\overset{H}{|}}{C}}◄CO_2H = H_2N►\underset{\underset{R}{|}}{\overset{\overset{CO_2H}{|}}{C}}◄H \qquad HO►\underset{\underset{CH_2OH}{|}}{\overset{\overset{CH}{|}}{C}}◄H$$

An L-amino acid   L-Glyceraldehyde

Relative to glyceraldehyde, this configuration is defined as L.

Proteins are linear polymers in which amide linkages (Sec. 12.11) link the α-amino group of one amino acid and the carboxyl group of another.

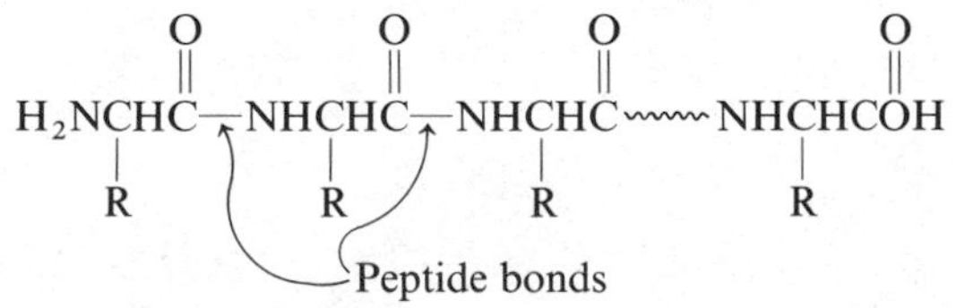

$$H_2N\underset{\underset{R}{|}}{CH}\overset{\overset{O}{||}}{C}—NH\underset{\underset{R}{|}}{CH}\overset{\overset{O}{||}}{C}—NH\underset{\underset{R}{|}}{CH}\overset{\overset{O}{||}}{C}\sim\sim\sim NH\underset{\underset{R}{|}}{CH}\overset{\overset{O}{||}}{C}OH$$

Peptide bonds

Amide linkages between amino acids are often called **peptide bonds**. The variation in **side chains** (R) is responsible for the vast range of properties of proteins.

Table 14.1 The 20 amino acids commonly found in proteins. The eight whose names are italicized are essential for adult humans. Arginine and histidine are needed during periods of growth.

Name	Abbreviation	Formula	Nature of side chain
Glycine	Gly	$\underset{\underset{NH_2}{\vert}}{\overset{\overset{H}{\vert}}{H-C}}-CO_2H$	
Alanine	Ala	$\underset{\underset{NH_2}{\vert}}{\overset{\overset{H}{\vert}}{CH_3-C}}-CO_2H$	
Valine	Val	$\underset{\underset{CH_3\ NH_2}{\vert\ \ \ \vert}}{\overset{\overset{H}{\vert}}{CH_3CH-C}}-CO_2H$	
Leucine	Leu	$\underset{\underset{CH_3\qquad NH_2}{\vert\qquad\vert}}{\overset{\overset{H}{\vert}}{CH_3CHCH_2-C}}-CO_2H$	
Isoleucine	Ile	$\underset{\underset{CH_3\ NH_2}{\vert\ \ \ \vert}}{\overset{\overset{H\ \ \ H}{\vert\ \ \vert}}{CH_3CH_2C\ -\ C}}-CO_2H$	Hydrophobic
Proline	Pro	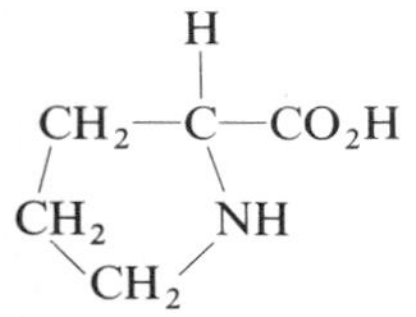 $\underset{\underset{CH_2}{\diagdown}}{\overset{CH_2}{\diagup}} \underset{NH}{\overset{\overset{H}{\vert}}{CH_2-C}}-CO_2H$	
Phenylalanine	Phe	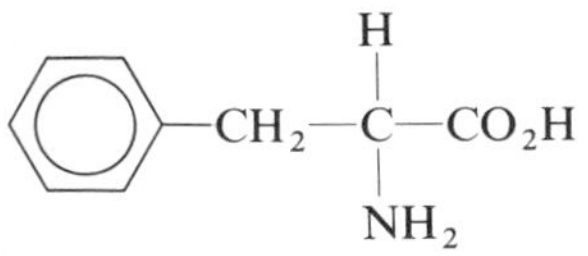 $\bigcirc -CH_2-\underset{\underset{NH_2}{\vert}}{\overset{\overset{H}{\vert}}{C}}-CO_2H$	
Tryptophan	Trp	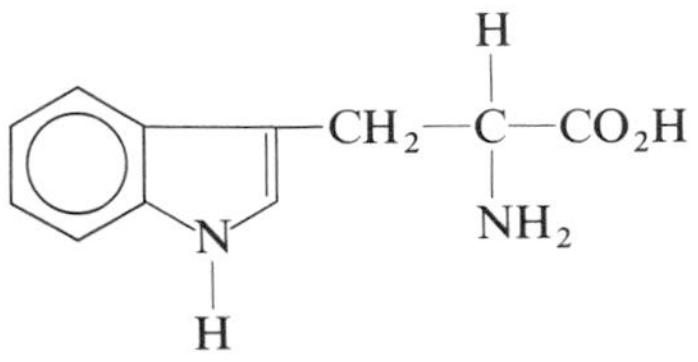 $-CH_2-\underset{\underset{NH_2}{\vert}}{\overset{\overset{H}{\vert}}{C}}-CO_2H$	
Methionine	Met	$CH_3-S-CH_2CH_2-\underset{\underset{NH_2}{\vert}}{\overset{\overset{H}{\vert}}{C}}-CO_2H$	

374

Table 14.1 (Continued)

Name	Abbreviation	Formula	Nature of side chain
Aspartic acid	Asp	$$HO-\overset{\overset{\displaystyle O}{\|\|}}{C}-CH_2-\overset{\overset{\displaystyle H}{\|}}{\underset{\underset{\displaystyle NH_2}{\|}}{C}}-CO_2H$$	Acidic
Glutamic acid	Glu	$$HO-\overset{\overset{\displaystyle O}{\|\|}}{C}-CH_2CH_2-\overset{\overset{\displaystyle H}{\|}}{\underset{\underset{\displaystyle NH_2}{\|}}{C}}-CO_2H$$	
Lysine	Lys	$$H_2N-CH_2CH_2CH_2CH_2-\overset{\overset{\displaystyle H}{\|}}{\underset{\underset{\displaystyle NH_2}{\|}}{C}}-CO_2H$$	Basic
Arginine	Arg	$$H_2N-\overset{\overset{\displaystyle NH}{\|\|}}{C}-NH-CH_2CH_2CH_2-\overset{\overset{\displaystyle H}{\|}}{\underset{\underset{\displaystyle NH_2}{\|}}{C}}-CO_2H$$	
Histidine	His	$$\underset{N\diagdown\quad\diagup NH}{\underset{CH}{CH=C}}-CH_2-\overset{\overset{\displaystyle H}{\|}}{\underset{\underset{\displaystyle NH_2}{\|}}{C}}-CO_2H$$	
Serine	Ser	$$HO-CH_2-\overset{\overset{\displaystyle H}{\|}}{\underset{\underset{\displaystyle NH_2}{\|}}{C}}-CO_2H$$	Hydrophilic (but not very acidic or basic)
Threonine	Thr	$$CH_3-\overset{\overset{\displaystyle OH}{\|}}{\underset{\underset{\displaystyle H}{\|}}{C}}-\overset{\overset{\displaystyle H}{\|}}{\underset{\underset{\displaystyle NH_2}{\|}}{C}}-CO_2H$$	
Tyrosine	Tyr	$$HO-\langle\bigcirc\rangle-CH_2-\overset{\overset{\displaystyle H}{\|}}{\underset{\underset{\displaystyle NH_2}{\|}}{C}}-CO_2H$$	
Cysteine	Cys	$$HS-CH_2-\overset{\overset{\displaystyle H}{\|}}{\underset{\underset{\displaystyle NH_2}{\|}}{C}}-CO_2H$$	
Asparagine	Asn	$$H_2N-\overset{\overset{\displaystyle O}{\|\|}}{C}-CH_2-\overset{\overset{\displaystyle H}{\|}}{\underset{\underset{\displaystyle NH_2}{\|}}{C}}-CO_2H$$	
Glutamine	Gln	$$H_2N-\overset{\overset{\displaystyle O}{\|\|}}{C}-CH_2CH_2-\overset{\overset{\displaystyle H}{\|}}{\underset{\underset{\displaystyle NH_2}{\|}}{C}}-CO_2H$$	

In Table 14.1 amino acids are classified according to side-chain characteristics that are important in determining the structures of proteins. The first category includes amino acids with hydrophobic side chains. When feasible, these side chains tend to be in a nonpolar, organic environment rather than in water. Note that isoleucine has a second chiral carbon. The side chain of proline is also linked to the amine nitrogen. Proline is the only one of the principal amino acids to have a secondary amino group, a feature that has important consequences for protein structure. Although the heterocyclic ring of tryptophan has an N—H bond, its degree of hydrogen bonding to water is insufficient to make this large side chain very hydrophilic. Methionine has a sulfide function, but water hydrogen bonds much less effectively to sulfide than to ether functions.

Aspartic acid and glutamic acid have a carboxyl group in their side chains. Most physiological environments are approximately neutral (pH around 7). We know (Secs. 8.4 and 11.3) that under these conditions carboxyl groups are essentially completely ionized to carboxylate anions. The sodium salt of glutamic acid (monosodium glutamate, MSG) is used in the food industry as a flavor enhancer.

Three amino acids have basic side chains. Lysine contains an amino group in the side chain and we know (Sec. 8.4) that in neutral solutions an amino group is almost completely protonated. Arginine contains a function, known as a guanidino group, that is also sufficiently basic to be mostly protonated in neutral solutions.

$$\underset{\text{A guanidino group}}{H_2N-\overset{\overset{\displaystyle NH}{\|}}{C}-NH- \;+\; H^+ \;\;\rightleftharpoons\;\; H_2N-\overset{\overset{\displaystyle NH_2^+}{\|}}{C}-NH-}$$

The imidazole ring of histidine has one relatively basic nitrogen.* In neutral solutions many histidine side chains are protonated.

$$\begin{array}{cc}
CH\!=\!C & CH\!=\!C \\
N\quad NH & {}^+HN\quad NH \\
CH & CH
\end{array} + H^+ \;\rightleftharpoons$$

The final group contains amino acids with side chains that are not acidic or basic but that participate extensively in hydrogen bonds. When feasible, these hydrophilic side chains tend to be in contact with water molecules rather than with nonpolar organic material. Note that threonine has a second chiral carbon atom. Asparagine and glutamine are the amides of aspartic and glutamic acids.

Humans are unable to synthesize all the amino acids. Those that must be supplied in the diet are known as **essential amino acids**. The essential amino acids

---

* Like the nitrogen in the heterocyclic ring of tryptophan, the other nitrogen of imidazole is not basic because its electron pair is part of the aromatic π cloud (Sec. 8.9).

for adult humans are italicized in Table 14.1. We must obtain these amino acids by ingestion and hydrolysis of proteins of an organism that can synthesize them or that has eaten something else that can. Some sources of protein, such as milk and eggs, are good sources of all the essential amino acids. Others, particularly grains, are not. Corn proteins are deficient in lysine and tryptophan, for instance, and rice proteins in lysine and threonine. D-Amino acids are not incorporated as such into proteins. They are excreted unchanged, isomerized to the corresponding L isomers, or metabolized in other ways.

14.2
AMINO ACIDS: PROPERTIES AND REACTIONS

The properties of amino acids are more typical of simple salts than of ordinary amines or carboxylic acids. Amino acids are crystalline substances that melt or decompose only at high temperatures and are generally very soluble in water but relatively insoluble in organic solvents. These properties occur because solid amino acids are internal salts, often called **zwitterions** (from the German *zwitter*, hybrid) or dipolar ions.

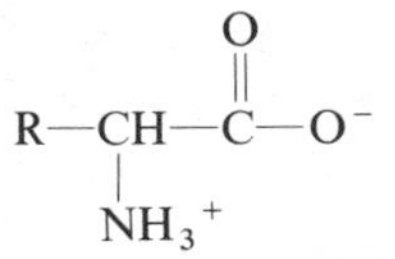

$$\underset{\underset{NH_3^+}{\displaystyle |}}{R-CH}-\overset{\overset{\displaystyle O}{\|}}{C}-O^-$$

Zwitterion form of an amino acid

Even though it has no overall charge, a zwitterion has saltlike properties. A zwitterion structure for an amino acid is reasonable, for we know that an amine and a carboxylic acid react to form a salt.

$$RNH_2 + RCO_2H \; \rightleftharpoons \; RNH_3^+ \; RCO_2^-$$

The dominant form of an amino acid in solution depends on the acidity of the solution.

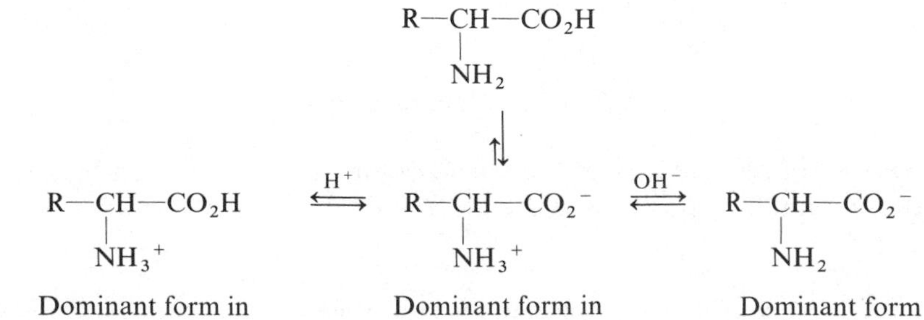

$$\underset{\underset{NH_2}{\displaystyle |}}{R-CH}-CO_2H$$

$$\updownarrow$$

$$\underset{\underset{NH_3^+}{\displaystyle |}}{R-CH}-CO_2H \; \overset{H^+}{\rightleftharpoons} \; \underset{\underset{NH_3^+}{\displaystyle |}}{R-CH}-CO_2^- \; \overset{OH^-}{\rightleftharpoons} \; \underset{\underset{NH_2}{\displaystyle |}}{R-CH}-CO_2^-$$

| Dominant form in strongly acidic solutions | Dominant form in neutral solutions | Dominant form in strongly basic solutions |

In aqueous solutions near neutrality an amino acid exists largely as the zwitterion in equilibrium with a small amount of the uncharged form. In strongly acidic solutions the carboxyl group is protonated to give a positively charged amino acid. In strongly basic solutions the proton is removed from the ammonium group to give a negatively charged amino acid. Acidic or basic groups in the side chains, of course, may also be charged.

Amino acids undergo characteristic amine and carboxylic acid reactions. The carboxyl groups can be converted to esters and the amino groups to amides, for example. Even in neutral solutions, the zwitterion form of the amino acid is in rapid equilibrium with a small amount of a form with uncharged $-NH_2$ and $-CO_2H$ groups, which can undergo these reactions.

The primary amino groups react with nitrous acid in a fashion typical for primary alkyl amines (Sec. 8.6) to produce N_2.

$$\underset{\underset{NH_2}{|}}{R-CH-CO_2H} + HNO_2 \longrightarrow \underset{\underset{OH}{|}}{R-CH-CO_2H} + H_2O + N_2\uparrow$$

Because the volume of N_2 evolved can be easily measured, this reaction provides a method (known as the van Slyke procedure) for determining the number of primary amino groups in a sample containing amino acids or proteins.

Amino acids react with ninhydrin in a complex reaction that generates a deeply purple product.*

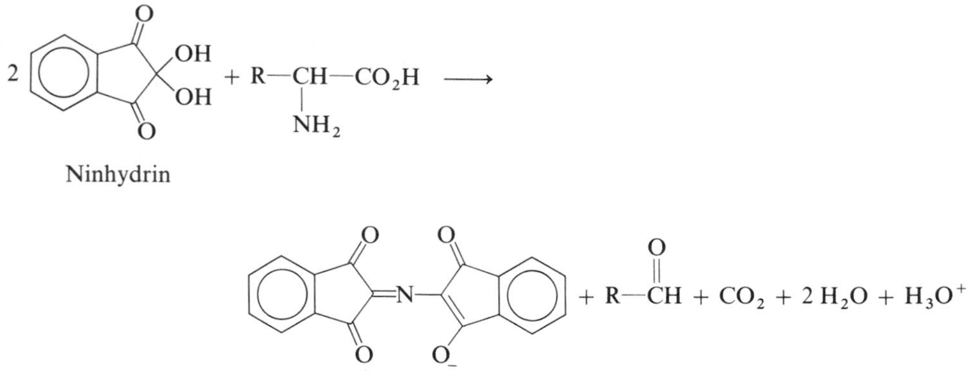

Ninhydrin

A deeply purple-colored anion

Because of the intensity of the purple color, this reaction can be used to determine the amount of amino acid in even a very small sample.

---

* Proline undergoes a different reaction with ninhydrin to generate a yellow product.

The functional groups in side chains of amino acids also undergo their typical reactions. The thiol group of cysteine, for example, is readily oxidized (Sec. 5.16).

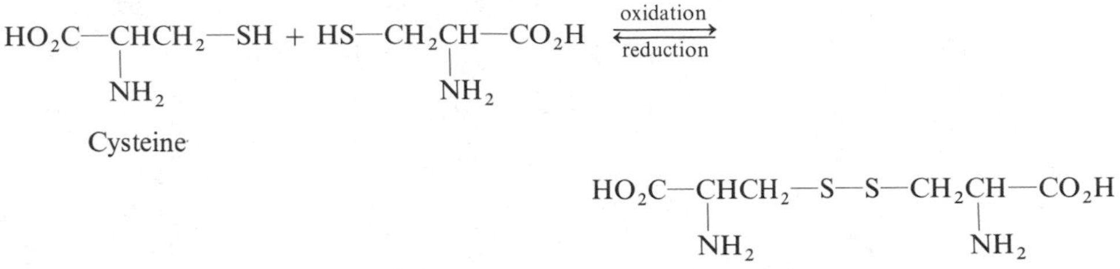

$$HO_2C—CHCH_2—SH + HS—CH_2CH—CO_2H \underset{\text{reduction}}{\overset{\text{oxidation}}{\rightleftarrows}}$$

$$\qquad\quad \underset{NH_2}{|} \qquad\qquad\qquad \underset{NH_2}{|}$$

Cysteine

$$HO_2C—CHCH_2—S—S—CH_2CH—CO_2H$$

$$\qquad\qquad \underset{NH_2}{|} \qquad\qquad\qquad \underset{NH_2}{|}$$

Cystine

This reaction leads to a disulfide known as cystine, which is readily reduced back to cysteine. Disulfide linkages between cysteines are an important feature of protein structures.

14.3
PEPTIDES

The word protein is usually applied only to molecules with molecular weights above 5000 to 10,000, corresponding to 40 to 80 or more amino acid units. Smaller molecules are called **peptides**.

The simplest peptides contain just two amino acids and are called **dipeptides**. Glycylalanine and alanylglycine are examples.

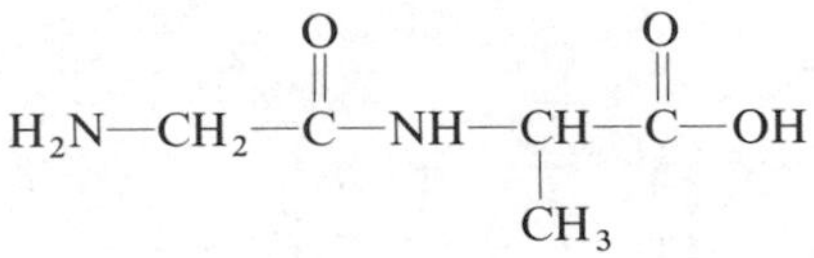

$$\underset{\text{Glycylalanine}\\ \text{Gly-Ala}}{H_2N—CH_2—\overset{O}{\overset{||}{C}}—NH—\underset{\underset{CH_3}{|}}{CH}—\overset{O}{\overset{||}{C}}—OH} \qquad \underset{\text{Alanylglycine}\\ \text{Ala-Gly}}{H_2N—\underset{\underset{CH_3}{|}}{CH}—\overset{O}{\overset{||}{C}}—NH—CH_2—\overset{O}{\overset{||}{C}}—OH}$$

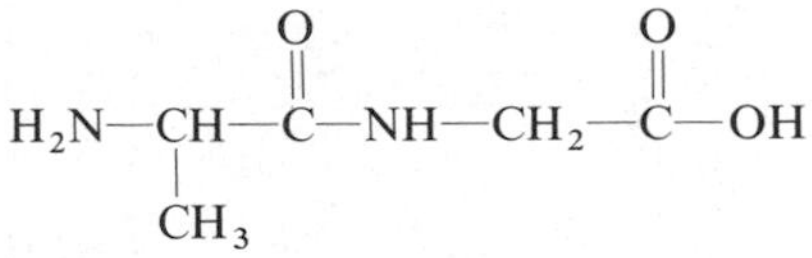

Note that two isomeric dipeptides contain one alanine and one glycine. The peptide bond of the first dipeptide incorporates the carboxyl group of glycine and the amino group of alanine. In the second, the peptide bond incorporates the carboxyl group of alanine and the amino group of glycine.

It is conventional to draw and name peptides with the free α-amino group at the left and the free carboxyl group at the right. Therefore the name valyllysylserine represents the tripeptide in which valine is the **N-terminal amino acid** and serine the **C-terminal amino acid**.

N-terminal end

C-terminal end

$$
\underset{\substack{\displaystyle | \\ CHCH_3 \\ | \\ CH_3}}{H_2N-CH}-\overset{\overset{\displaystyle O}{\|}}{C}-NH-\underset{\substack{\displaystyle | \\ CH_2 \\ | \\ CH_2 \\ | \\ CH_2 \\ | \\ CH_2 \\ | \\ NH_2}}{CH}-\overset{\overset{\displaystyle O}{\|}}{C}-NH-\underset{\substack{\displaystyle | \\ CH_2 \\ | \\ OH}}{CH}-\overset{\overset{\displaystyle O}{\|}}{C}-OH
$$

Valyllysylserine
Val-Lys-Ser (or Val→Lys→Ser)

Peptides are named by listing the amino acids in sequence, starting with the N-terminal amino acid. Except for the C-terminal amino acid, the *-ine* ending of the name of each amino acid is changed to *-yl*.* It is convenient to represent structures of peptides and proteins not by drawings that indicate each atom but simply by listing the three-letter abbreviations (see Table 14.1) for each amino acid. Thus Val-Lys-Ser indicates fully the structure of the preceding tripeptide. In representing a peptide or protein in this abbreviated fashion, it is important that the N-terminal group be at the left. To avoid ambiguity, particularly when a structure cannot be written from left to right, arrows are sometimes used, the head of each arrow pointing toward the C-terminal amino acid (Val→Lys→Ser).

Peptides with four, five, six . . . amino acids are called **tetrapeptides, penta-peptides, hexapeptides** . . . . Peptides with many amino acids are often called **polypeptides**. Just as with amino acids, the functional groups of peptides can be charged.

$$
\underset{\substack{\displaystyle | \\ CHCH_3 \\ | \\ CH_3}}{H_3N^+-CH}-\overset{\overset{\displaystyle O}{\|}}{C}-NH-\underset{\substack{\displaystyle | \\ CH_2 \\ | \\ CH_2 \\ | \\ CH_2 \\ | \\ CH_2 \\ | \\ NH_3^+}}{CH}-\overset{\overset{\displaystyle O}{\|}}{C}-NH-\underset{\substack{\displaystyle | \\ CH_2 \\ | \\ OH}}{CH}-\overset{\overset{\displaystyle O}{\|}}{C}-O^-
$$

Dominant form of Val-Lys-Ser in a neutral solution

Some peptides have significant biological activities. The pentapeptide leucine enkephalin is one of a group of peptides known as endorphins.

* Tryptophan becomes tryptophyl, cysteine becomes cysteinyl, and cystine becomes cystyl.

Tyr-Gly-Gly-Phe-Leu

Leucine enkephalin

The endorphins, isolated from mammalian brain tissue, have opiatelike effects. It is likely that morphine and related drugs exert their effects at receptor sites intended for endorphins.

The nonapeptide vasopressin is a hormone released by the hypothalmus.

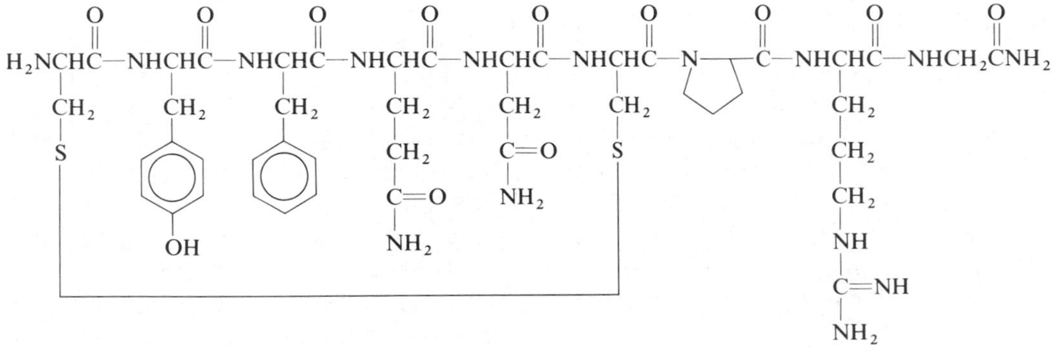

or   Cys-Tyr-Phe-Gln-Asn-Cys-Pro-Arg-GlyNH$_2$

Vasopressin

By affecting the rate of excretion of water from the kidneys, it is important in regulating water balance. Note that the sulfur atoms of two cysteines are incorporated into a disulfide-containing ring. The C-terminal glycine is present as an amide. Oxytocin, a closely related nonapeptide also secreted by the hypothalmus, has quite different physiological activities.

Cys-Tyr-Ile-Gln-Asn-Cys-Pro-Leu-GlyNH$_2$

Oxytocin

It stimulates lactation and, by stimulating uterine contractions, induces labor.

14.4
DETERMINING STRUCTURES OF PEPTIDES
AND PROTEINS

A particular protein or peptide can differ from others not only in amino acid composition but also in the sequence in which the amino acids are arranged. For all but the simplest peptides, the number of possible sequences is enormous. Because 20 different amino acids can potentially occupy each position of a peptide, the number of structurally different tripeptides is $20 \times 20 \times 20 = 20^3 = 8000$. For a decapeptide (10 amino acids), there are $20^{10} = 10,240,000,000,000$ possibilities. There are not sufficient atoms on earth to construct even one molecule of each of the 20^{20} possible peptides having 20 amino acids. Determining the amino acid composition and sequence of a protein or peptide is an important prelude to understanding how it is able to carry out a highly specific function.

14.5
DETERMINING AMINO ACID COMPOSITION

To determine its amino acid composition, a protein or peptide is first hydrolyzed completely to its component amino acids. Then the mixture of amino acids is separated and the quantity of each determined.

Hydrolysis can be effected by extended heating with aqueous hydrochloric acid. Some form of **chromatography** is generally used to separate the amino acids. In some types of chromatography, compounds are separated by taking advantage of the different degrees to which they are adsorbed by some material.

In **column chromatography*** a solution containing compounds to be separated is added to the top of a column packed with granules of a solid adsorbent, often aluminum oxide, silica, or starch. A solvent, poured onto the top of the column, passes through the column and emerges at the bottom. Because the compounds are adsorbed to different degrees by the solid, they move down the column at different rates.

For the separation of amino acids, the solid is often particles of polystyrene that have been sulfonated and treated with base to introduce sulfonate ion functions ($-SO_3^-$).

Polystyrene

The mixture of amino acids is added to the top of the column and an aqueous solution of fixed acidity (pH) allowed to flow down the column. The amino acids move down the column at rates that depend on their structures and basicities and thus emerge at different times. Amino acids that are most positively charged in the solution tend to move most slowly because of strong interactions with the negative sulfonate groups. Ninhydrin is added to the solution emerging from the column.

---

* **Gas chromatography** is another important kind of chromatography. A mixture is vaporized and swept by an inert gas (usually helium) through a narrow tube containing an adsorbent to separate it into its component compounds.

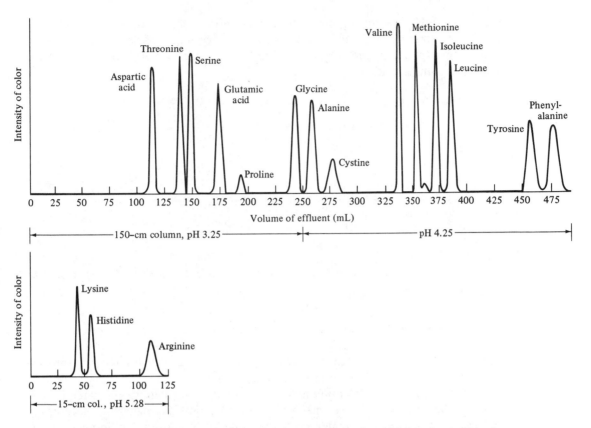

Figure 14.1 Chromatographic separation of a mixture of amino acids on columns of a salt of a sulfonated polystryene. The intensity of the color formed by reaction of ninhydrin and the effluent emerging from the column is plotted against the volume of the effluent. The amino acids with basic side chains take a long time to emerge and were analyzed by using a shorter column. [Reprinted with permission from S. Moore, D. H. Spackman, and W. H. Stein, *Anal. Chem.* **30**, 1185 (1958). Copyright 1958 American Chemical Society.]

The intensity of the color developed in the portion of the solution containing each amino acid indicates how much of that amino acid is present. The entire separation procedure is done automatically in an amino acid analyzer, an instrument that gives a plot like the one in Figure 14.1.

14.6
IDENTIFYING N-TERMINAL AND C-TERMINAL
AMINO ACIDS

Once the amino acid composition of a peptide or protein is known, how can the sequence be determined? It is relatively easy to determine the N-terminal and C-terminal amino acids. One method to determine the N-terminal amino acid involves a reaction with 2,4-dinitrofluorobenzene. Because of its two strongly electron-withdrawing nitro groups, this compound undergoes nucleophilic substitution by an amino group more easily than most aryl halides (Sec. 7.8). The only α-amino group of a peptide or protein available for such a reaction belongs to the N-terminal amino acid.

$$
\underset{R}{H_2NCHC}-\underset{R'}{NHCHC}-\underset{R''}{NHCHC}\cdots\cdots COH \quad \xrightarrow{\;\;O_2N-\langle\bigcirc\rangle-F^{NO_2}\;\;}
$$

$$
O_2N-\langle\bigcirc\rangle^{NO_2}-\underset{R}{NHCHC}-\underset{R'}{NHCHC}-\underset{R''}{NHCHC}\cdots\cdots COH \quad \xrightarrow[H^+]{H_2O}
$$

$$
O_2N-\langle\bigcirc\rangle^{NO_2}-\underset{R}{NHCHCOH} + \underset{R'}{H_2NCHCOH} + \underset{R''}{H_2NCHCOH} + \text{other amino acids}
$$

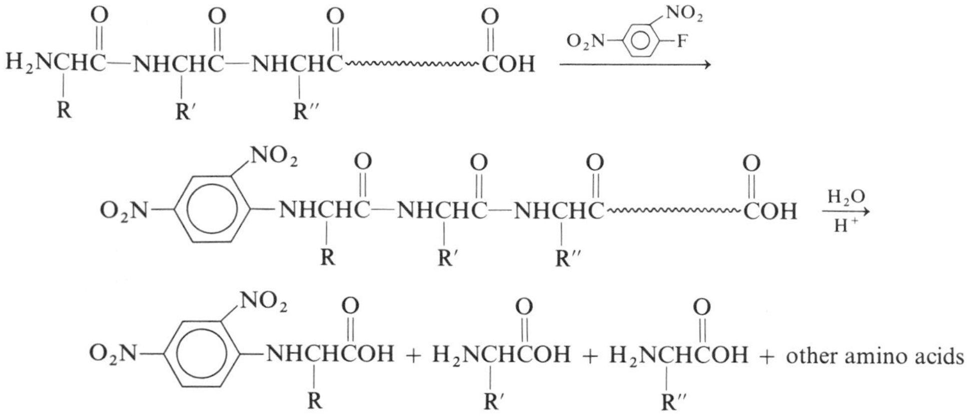

After hydrolysis and separation, the N-terminal amino acid will be present as its *N*-2,4-dinitrophenyl derivative; the other amino acids will be unsubstituted.* The yellow 2,4-dinitrophenyl derivative is easily separated and characterized, thus establishing the identity of the N-terminal amino acid.

One procedure for determining the C-terminal amino acid uses enzymes known as carboxypeptidases. These enzymes, isolated from the pancreas, specifically catalyse hydrolysis of the peptide bond that links the C-terminal amino acid to the chain.

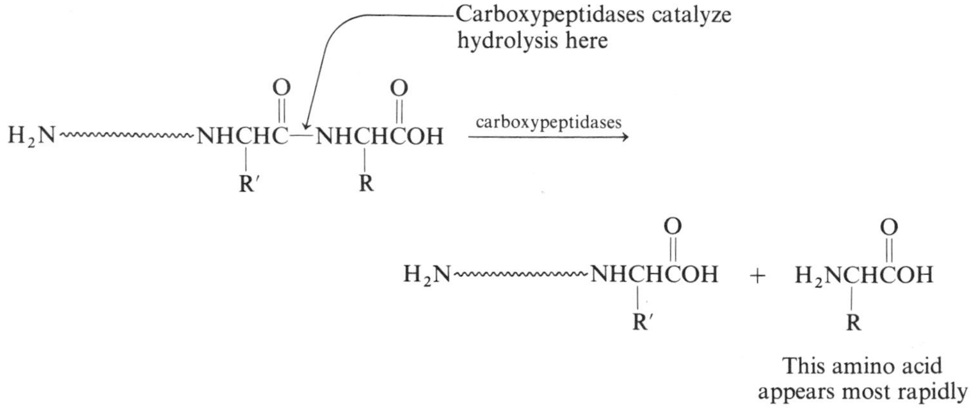

Carboxypeptidases catalyze
hydrolysis here

$$
\underset{R'}{H_2N\cdots\cdots NHCHC}-\underset{R}{NHCHCOH} \quad \xrightarrow{carboxypeptidases}
$$

$$
\underset{R'}{H_2N\cdots\cdots NHCHCOH} + \underset{R}{H_2NCHCOH}
$$

This amino acid
appears most rapidly

---

* The amino group of the side chain of lysine will be substituted, but only if lysine is the N-terminal amino acid does its α-amino group also react.

When hydrolysis of a peptide or protein is catalyzed by carboxypeptidases, the C-terminal amino acid is the first to appear in the solution.

Let's use these procedures to determine the amino acid sequence of a tripeptide that we will call A. By analyzing the amino acid mixture obtained on its hydrolysis, tripeptide A is found to contain one glycine, one alanine, and one serine. Six tripeptides, however, have this amino acid composition (Gly-Ala-Ser, Gly-Ser-Ala, Ala-Gly-Ser, Ala-Ser-Gly, Ser-Gly-Ala, and Ser-Ala-Gly). Reaction of tripeptide A with 2,4-dinitrofluorobenzene, followed by hydrolysis, produces the *N*-2,4-dinitrophenyl derivative of alanine plus free glycine and serine. Therefore the N-terminal amino acid is alanine. On treatment with a carboxypeptidase, glycine appears most rapidly and hence is the C-terminal amino acid. Tripeptide A must be Ala-Ser-Gly.

14.7
DETERMINING AMINO ACID SEQUENCE

In the preceding section we learned that procedures to identify N-terminal and C-terminal amino acids allow us to determine the sequence of a tripeptide. How can we determine the amino acid sequence of a much longer peptide? The trick used can be illustrated with a simple example.

Consider a hexapeptide that we will call Z. Complete hydrolysis of Z, followed by separation and quantitative determination of the amino acids, shows it to have the composition [Ala, Asp, Gly, Phe, Ser, Val].

$$\text{Peptide Z} \xrightarrow[\text{H}^+]{\text{H}_2\text{O}} \text{Ala} + \text{Asp} + \text{Gly} + \text{Phe} + \text{Ser} + \text{Val}$$

Products of complete hydrolysis

The amino acids are written in alphabetical order and separated by commas as their sequence is not known.

Peptide Z is then subjected to **partial hydrolysis**.

$$\text{Peptide Z} \xrightarrow[\text{H}^+]{\text{H}_2\text{O}} \text{[Ala, Gly, Ser]} + \text{[Asp, Gly, Phe]} + \text{[Ala, Val]} + \text{individual amino acids}$$

Peptide A   Peptide B   Peptide C

Products of partial hydrolysis

The hydrolysis is carried out incompletely so that some peptide bonds remain. Partial hydrolysis of peptide Z leads to formation of a mixture that, in addition to individual amino acids, contains two tripeptides (A and B) and a dipeptide (C).

The composition and sequence of each peptide are determined. Complete hydrolysis, followed by separation and quantitative determination of amino acids, for instance, shows that peptide A contains one alanine, one glycine, and one serine. By identifying the N-terminal and C-terminal amino acids, its sequence is found to be Ala-Ser-Gly. (Peptide A was used as an example in the preceding section.) Sequences of the other peptides are similarly determined.

Ala-Ser-Gly   Gly-Phe-Asp   Val-Ala

Peptide A   Peptide B   Peptide C

Each peptide represents a segment of the original chain of peptide Z. Matching the segments

Val-Ala
Ala-Ser-Gly
Gly-Phe-Asp

indicates that the sequence of peptide Z must be

Val-Ala-Ser-Gly-Phe-Asp

The strategy for determining the amino acid sequence of a peptide or protein is to obtain a group of overlapping fragments whose sequences can be determined. The total sequence can be deduced by ordering the fragments. This technique was first used to determine the amino acid sequence of bovine (cow) insulin, a small protein with 51 amino acids (Fig. 14.2). Insulin is a hormone that plays an important role in regulating the glucose level. Note that insulin has two chains of amino acids, held together by disulfide linkages between cysteines. It also has a disulfide linkage between two cysteines in the A chain. To determine which cysteines are joined by disulfide linkages, partial hydrolysis was used to obtain small fragments that had intact disulfide linkages.

Using procedures of the type outlined, sequences have been determined for many larger proteins, including the enzyme bovine glutamate dehydrogenase, which contains 506 amino acids.

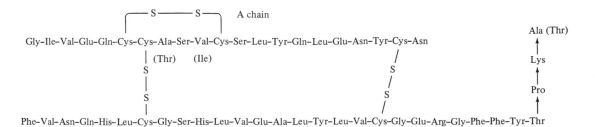

Figure 14.2 The amino acid sequence of bovine insulin. Human insulin differs in having the three amino acid units indicated in parentheses.

14.8
PRIMARY PROTEIN STRUCTURE

The abilities of proteins to function in particular ways are due to specific three-dimensional shapes. These shapes, however, depend ultimately on the sequence of amino acids and the positioning of disulfide linkages between cysteines. The numbers of amino acids of each kind and the order of their attachment by covalent (peptide and disulfide) bonds are referred to as the **primary structure** of a protein. The structure of insulin in Figure 14.2 is its primary structure.

Samples of a given protein isolated from different individuals of the same species usually have identical primary structures, although small differences are sometimes observed. Some small changes in primary structure have relatively little effect on biological activity. Even though it differs from human insulin in the amino acids at three positions, bovine insulin functions satisfactorily in humans. As a fortunate consequence, insulin obtained from a slaughterhouse can be used by diabetics.

Other small changes in primary structure drastically affect normal functioning. Because samples of hemoglobin are easily obtained, such changes in humans have been most extensively studied with this protein. Hemoglobin is the protein in red blood cells responsible for transporting oxygen from the lungs to the rest of the body. It is a **conjugated protein**, one that contains some component in addition to amino acids. The nonamino acid component of a conjugated protein is called a **prosthetic group**. Hemoglobin's prosthetic group is the iron-containing porphyrin heme (Sec. 8.10). Human hemoglobin, a protein of molecular weight 64,500, contains two identical α chains, each with 141 amino acids, and two identical β chains, each with 146 amino acids. One heme is linked to each chain.

The best-known variation in hemoglobin causes sickle-cell anemia. When hemoglobin is not bonded to O_2, the variant hemoglobin molecules tend to aggregate together anomalously and distort red blood cells into a sicklelike shape. Such cells can block capillaries and are also easily ruptured. Ordinary hemoglobin and that responsible for sickle-cell anemia differ in a single amino acid in their β chains. In ordinary hemoglobin glutamic acid is at position 6. In the hemoglobin responsible for sickle-cell anemia, valine is at this position. More than 300 other variant hemoglobins have been detected, some also associated with clinically observable symptoms. Most differ from ordinary hemoglobin in only one amino acid.

14.9
SECONDARY PROTEIN STRUCTURE

We learned that the bond between the carbonyl carbon and the nitrogen of an amide has double-bond character (Sec. 12.11).

A partial double bond

Atoms attached directly to these atoms lie in a plane and rotation around the bond joining them is difficult. Rotation around other bonds of the chain is possible, however.

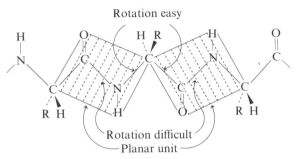

Because of the possibility of rotation around so many bonds, a protein might be expected to assume a large number of shapes (conformations). Ordinarily, however, a protein has one specific shape and this shape is essential for it to fulfill its biological function. Here and in the next section we consider what the three-dimensional shapes of proteins actually are and why one shape may be favored.

On the basis of their shapes and solubilities, most proteins can be placed in one of two broad classes: fibrous proteins and globular proteins. **Fibrous proteins** have relatively extended chains associated together in fiberlike or sheetlike structures. Proteins that have structural functions generally are fibrous. Examples are the **keratins**, the principal constituents of skin, hair, wool, silk, feathers, nails, horns, and hoofs. Consonant with their structural functions, fibrous proteins are often insoluble in water.

Globular proteins have chains that are bent and tightly folded about themselves so that the molecules are compact and have spherical or globular shapes. Proteins whose functions are regulatory, requiring them to interact specifically with particular chemical species, ordinarily are globular. Enzymes, the proteins that catalyze reactions of specific compounds, are globular. So are transport proteins, such as hemoglobin, that bind and transport particular substances. So also are antibodies, proteins that form insoluble complexes with harmful foreign substances in the blood. Globular proteins often are water soluble.

Secondary protein structure refers to regular and repeating conformational arrangements of successive amino acids along the chains of some proteins. Such relatively simple repeating structures are dominant in the fibrous proteins.

The keratins, fibrous structural proteins of animals, fall into two classes: α and β keratins. The β keratins, the less abundant group, include silk fibers and the fibers of a spider's web. As shown in Figure 14.3(a), each chain in a β keratin is extended in a linear fashion and hydrogen bonded to adjoining chains to form a sheetlike structure. Each carbonyl oxygen and amide hydrogen participates in hydrogen bonding. Because it is slightly nonplanar, this structure is often called a **pleated sheet**. Many of these sheets are stacked together to form a β-keratin fiber [Fig. 14.3(b)]. Such stacking brings side chains close together and so those that are bulky

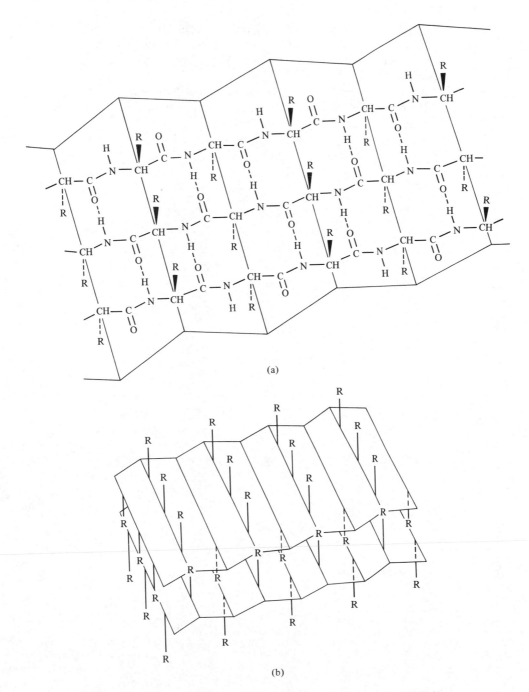

Figure 14.3 (a) A small portion of a pleated-sheet protein structure. The side chains of the amino acids lie alternately above and below the sheet. (b) The pleated sheets are stacked on top of one another to form fibers.

or that have like charges disrupt this arrangement. Consequently, proteins that adopt the pleated-sheet structure are composed principally of amino acids with small, uncharged side chains. Nearly 90% of the amino acids of silk are glycine, alanine, and serine.

In α keratins each chain is coiled in a helical shape that is stabilized by internal hydrogen bonds. Ordinarily the chain is an **α helix**, a right-handed helix (spiraling in the same way as the threads on an ordinary screw or bolt) that has approximately 3.6 amino acid units per turn. As shown in Figure 14.4a, this arrangement is stabilized by hydrogen bonds between successive coils of the helix. The hydrogen bonds are

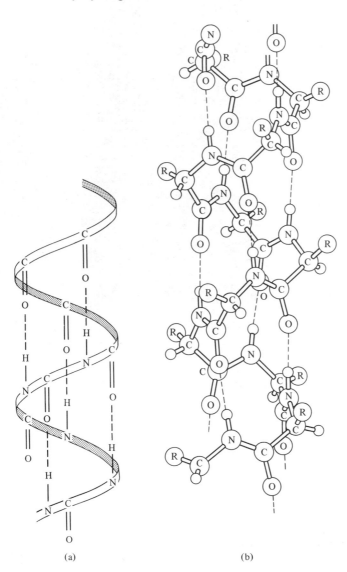

(a) (b)

Figure 14.4 (a) A schematic drawing of a segment of an α helix showing the hydrogen bonds that stabilize the helical structure. (b) A ball-and-stick model of a segment of an α helix. [Figure (b) redrawn from B. W. Low and J. T. Edsall in *Currents in Biochemical Research*, D. E. Green (Ed.), Interscience Publishers, New York, 1956.]

between the carbonyl oxygen of each amino acid and the N—H of the fourth amino acid down the chain.

$$
\begin{array}{c}
\text{H} \quad\quad \text{O} \ \text{H} \quad\quad \text{O} \ \text{H} \quad\quad \text{O} \ \text{H} \quad\quad \text{O} \ \text{H} \quad\quad \text{O} \ \text{H} \quad\quad \text{O} \ \text{H} \quad\quad \text{O} \\
| \quad\quad\ || \ | \quad\quad\ || \ | \quad\quad\ || \ | \quad\quad\ || \ | \quad\quad\ || \ | \quad\quad\ || \ | \quad\quad\ || \\
-\text{N}-\text{CH}-\text{C}-\text{N}-\text{CH}-\text{C}-\text{N}-\text{CH}-\text{C}-\text{N}-\text{CH}-\text{C}-\text{N}-\text{CH}-\text{C}-\text{N}-\text{CH}-\text{C}-\text{N}-\text{CH}-\text{C}- \\
| \quad\quad\quad\ | \quad\quad\quad\ | \quad\quad\quad\ | \quad\quad\quad\ | \quad\quad\quad\ | \quad\quad\quad\ | \\
\text{R} \quad\quad\quad \text{R} \quad\quad\quad \text{R} \quad\quad\quad \text{R} \quad\quad\quad \text{R} \quad\quad\quad \text{R} \quad\quad\quad \text{R}
\end{array}
$$

As shown in the ball-and-stick model in Figure 14.4b, the side chains extend outward from the backbone of the helix. Proteins that contain many amino acids with large side chains tend to have α-helical structures, for the α helix is more able than the pleated sheet to accommodate larger side chains.

One feature that destabilizes an α helix is the presence of many side chains with like charges. Charge repulsion tends to push the side chains farther apart, disrupting the helical coil. As an example, polylysine (a synthetic protein containing only lysine) assumes an α-helical structure in a basic solution in which the amino groups in the side chains are not significantly protonated. In a neutral solution (pH about 7), however, in which the side chains are largely in the form of charged ammonium groups, polylysine has a random structure. Another destabilizing feature is the presence of many proline residues. When incorporated into a protein, proline is unique in having no N—H bond. Therefore a proline introduces a break into the regular internal hydrogen bonding that stabilizes an α helix. Moreover, the —CH_2— group attached to the nitrogen of proline bumps into the next coil of a helix. As a result, a proline causes a kink or bend in a helical structure.

α-Helical chains are often assembled together in a regular fashion. In hair, three α helices are twisted together to form larger units, 11 of which are arranged in a bundle. Large numbers of these bundles are held together in a matrix of additional protein to form a hair fiber. In such assemblies, the α helices are linked together by disulfide bonds between cysteines. In the permanent waving of hair a reducing agent is used to cleave the disulfide bonds. The hair is placed in the desired shape and a mild oxidizing agent is added to form new disulfide bonds to maintain this shape.

Hydrogen bonds maintain the pleated-sheet structure, the α-helix structure, and other structures sometimes encountered. Rotation around the chain bonds is not actually free because a regular hydrogen-bonding pattern imposes strong constraints. Individual hydrogen bonds are weak, but the collective effect of many hydrogen bonds is large. A given protein adopts the conformation in which it is most stable. This conformation maximizes hydrogen bonding consistent with spatial and other requirements of the side chains. Therefore the secondary structure of a protein is determined ultimately by its amino acid sequence. Fibrous proteins are not ordinarily water soluble because most hydrogen bonds involving the N—H and carbonyl groups of the peptide bonds are internal rather than with water molecules. These insoluble proteins also have a high proportion of amino acid side chains that are neutral and not effective participants in hydrogen bonding.

14.10
TERTIARY PROTEIN STRUCTURE

In contrast to the regular, repetitive structures of fibrous proteins, each globular protein has a unique shape. **Tertiary protein structure** refers to the complex conformations that result from the extensive bending and folding of protein chains that are characteristic of globular proteins.

Consider the protein chymotrypsin, an enzyme in the small intestine that catalyzes the hydrolysis of specific peptide bonds. The 241 amino acids of chymotrypsin are in three chains that are linked together by disulfide bonds. The actual geometry of chymotrypsin is shown in Figure 14.5. This complicated shape, essential for the functioning of chymotrypsin, is its most stable shape under normal physiological conditions.

In addition to dispersion forces and constraints due to group size, several factors are important in determining the most stable shape of a given protein.

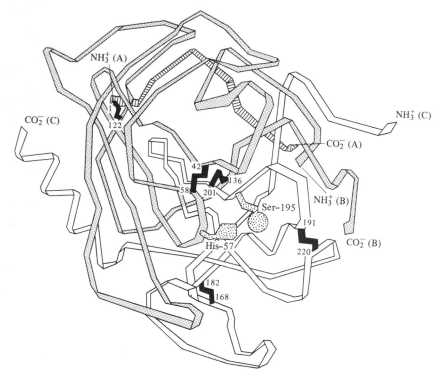

Figure 14.5   The geometrical arrangement of bovine chymotrypsin. The disulfide linkages and the ends of the A, B, and C chains are marked. Amino acids 57 (a histidine in chain B) and 195 (a serine in chain C) are specifically indicated because they play an important part in the catalytic action of this enzyme. [Reprinted by permission from *Nature* **214**, 652 (1967). Copyright 1967 Macmillan Journals Limited.]

1. *Constraints due to disulfide linkages.* Amino acids that occupy widely separated positions on a chain can be forced together by disulfide bonds between cysteines. For example, cysteines at positions 191 and 220 in the C chain of chymotrypsin are linked by a disulfide bond, bending the chain in a way that influences the positions of many other amino acids. As in chymotrypsin, disulfide functions can also link two independent chains.

2. *A tendency for polar groups to be on the exterior.* Hydrogen-bonding or charged side chains and charged end groups ($-CO_2^-$ and $-NH_3^+$) can interact strongly with water. These hydrophilic groups tend to be on the exterior of a globular protein, where the environment is aqueous.

3. *A tendency for nonpolar side chains to be on the interior.* Amino acid side chains that are neutral and do not hydrogen bond efficiently prefer a hydrocarbonlike environment to an aqueous one. Such hydrophobic side chains tend to cluster together in the interior of a globular protein.

4. *Effects of internal hydrogen bonding.* Amide linkages tend to hydrogen bond together. Regions of globular proteins often have regular structures of the types characteristic of fibrous proteins. In chymotrypsin some amino acids at the C-terminal end of chain C are in an α helix. Chymotrypsin also has areas in which chain segments are arranged in a parallel fashion that resembles the pleated-sheet structure. Any hydrogen-bonding groups not hydrogen bonded to water are generally hydrogen bonded internally to other functions in the protein.

5. *Attractions and repulsions between charged groups.* Charged groups with like charges tend to be remote from each other and those with unlike charges to be close together.

The most stable conformation of a protein is the compromise that best satisfies the factors just enumerated.

Globular proteins usually have a larger proportion of polar side chains than do fibrous proteins. Because most of their polar side chains are arrayed on the exterior, globular proteins tend to be water soluble.

Small changes in conditions can reduce or destroy the biological activities of many proteins. Mild heating of proteins (egg white is a familiar example), for instance, may cause a change in properties. So can simply a change in acidity (pH) of the solution or the addition to the solution of urea. Loss of activity that results from conditions too mild to cleave peptide bonds is called **denaturation**. Globular proteins are particularly susceptible to denaturation. Denaturation of a protein results from a breakdown of its characteristic secondary and tertiary structure. Changes in acidity affect structure by changing the number of charged groups. Addition of urea (H_2NCNH_2), a molecule effective at forming hydrogen bonds, disrupts the internal hydrogen bonds required to maintain the structure. Protein structure can also be changed by chemical alteration of side chains. Thiol groups are particularly sensitive. The toxic effects of some mercury compounds are thought to be due to reactions with the thiol groups of cysteines, leading to disruptions in protein structure.

$$R\text{---}SH + CH_3\text{---}Hg\text{---}Cl \longrightarrow R\text{---}S\text{---}Hg\text{---}CH_3 + HCl$$

Some globular proteins contain several chains that are linked not covalently but instead by weaker forces, such as those that contribute to tertiary protein structure. Hemoglobin, which has four independent chains, is among the simplest of such proteins. **Quaternary protein structure** refers to the geometry with which individual chains (subunits) of such proteins are clustered together. Interactions between the subunits of hemoglobin are important in regulating the ability of this protein to bind O_2. Similarly, interactions between subunits in other proteins often play a regulatory role.

14.11
X-RAY CRYSTALLOGRAPHY

Throughout this book we have been concerned with the geometries of molecules. In this chapter we learned that even the shapes of very large proteins are crucial to their functioning. How are molecular geometries determined? One important technique is X-ray crystallography.

Figure 14.6 Electron density of naphthalene in the plane of the molecule. Each line represents a different electron density, the lines nearest the atomic centers (the smallest circles) representing the highest densities. The electron density around hydrogens is much less than around carbons. [Redrawn from S. C. Abrahams, J. M. Robertson, and J. G. White, *Acta Cryst.* **2**, 238 (1949).]

When a crystal of a compound is bombarded with a beam of X rays, the X rays interact with the electrons of the atoms and are scattered in various directions. The scattered X rays, recorded on photographic film or electronically, form a pattern that is characteristic of the arrangement of atoms in the crystal. The pattern can be analyzed mathematically to give the electron density at each location in the crystal. Figure 14.6 shows the electron density through the plane of a molecule in a crystal of naphthalene. From such information the relative locations of the atoms, including accurate bond lengths and angles, can be deduced.

The shape of chymotrypsin (Fig. 14.5) was determined by X-ray crystallography.

14.12
CATALYSIS BY ENZYMES

Virtually all biological reactions are catalyzed by proteins called **enzymes**. Most are globular proteins with molecular weights ranging from 12,000 to several million. Enzymes catalyze few reactions that cannot, in some way, be achieved by chemists in the laboratory. The catalytic activity of enzymes is enormous, however, permitting biochemical reactions to be carried out without such aids as strong acids or bases, powerful oxidizing agents, and high temperatures that are routine in laboratory operations. Even more remarkable is the specificity of enzymes. An enzyme generally catalyzes only a particular type of chemical reaction for a single compound or closely related group of compounds. The compound that undergoes a reaction catalyzed by an enzyme is referred to as a **substrate** for that enzyme.

To get some insight into how enzymes function, we will examine one intensively studied example of enzymatic catalysis—hydrolysis of peptide bonds by the enzyme chymotrypsin. Chymotrypsin catalyzes most readily the hydrolysis of peptide bonds in which the carbonyl group is donated by phenylalanine, tyrosine, or tryptophan.

$$
\begin{array}{cccc}
& O & O & O & O \\
& \parallel & \parallel & \parallel & \parallel \\
-NHCHC & -NHCHC & -NHCHC & -NHCHC- \\
| & | & | & | \\
R & Ar & R & R
\end{array}
\xrightarrow[\text{H}_2\text{O}]{\text{chymotrypsin}}
$$

$$
\begin{array}{cccc}
O & O & & O & O \\
\parallel & \parallel & & \parallel & \parallel \\
-NHCHC & -NHCHCOH + H_2NCHC & -NHCHC- \\
| & | & | & | \\
R & Ar & R & R
\end{array}
$$

Ar is the side chain of phenylalanine, tyrosine, or tryptophan

It is well established that a histidine at position 57 and a serine at position 195 in chymotrypsin play an active role in the hydrolysis. As evident in Figure 14.5, histidine-57 and serine-195 are brought in close proximity by the tertiary structure of the enzyme. The region of the enzyme containing these amino acids contains a cleft into which a segment of a peptide chain can fit. Near histidine-57 and serine-195 is a pocket, surrounded by hydrophobic groups, that has the size to accommodate the aromatic side chain of phenylalanine, tryptophan, or tyrosine. Proteins complex to chymotrypsin in such a manner that one of these aromatic groups is in that pocket. As a result, the carbonyl group of one of these aromatic amino acids is held near histidine-57 and serine-195.

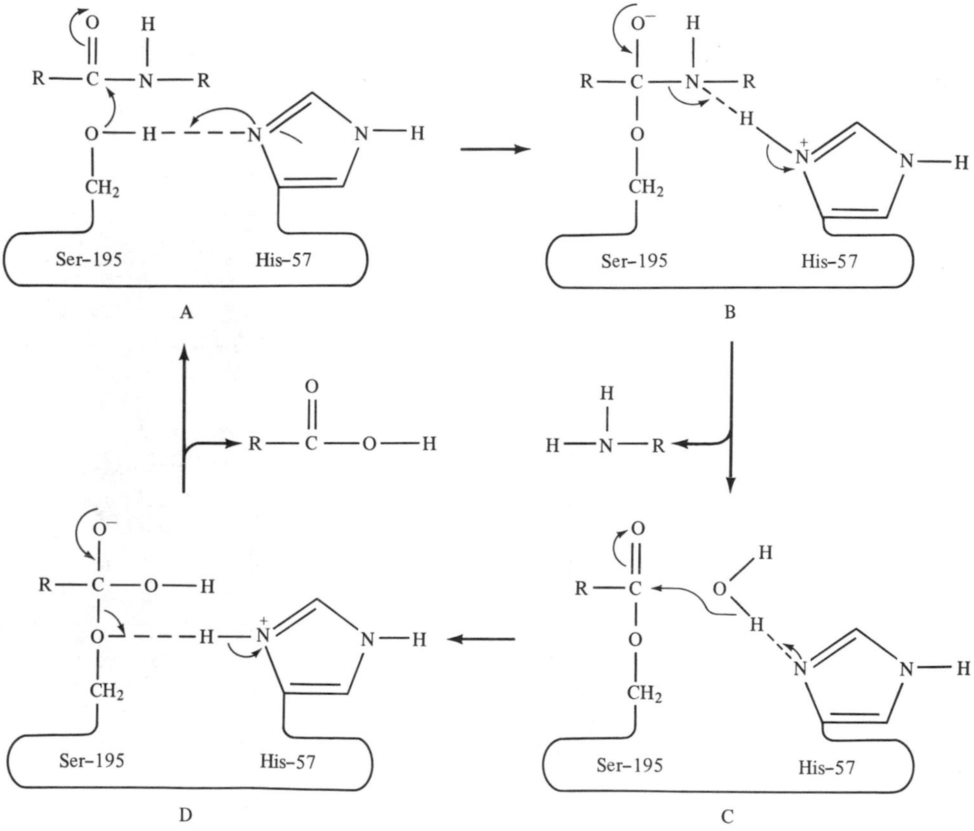

Figure 14.7 A mechanism proposed for the chymotrypsin-catalyzed hydrolysis of proteins. A peptide bond in which the carbonyl group is from a phenylalanine, tyrosine, or tryptophan is held in a region of the enzyme (A) near histidine-57 and serine-195, which are hydrogen bonded together. Nucleophilic attack by the serine hydroxyl group on the peptide carbonyl group to give B is facilitated by increased nucleophilicity of the hydroxyl group due to simultaneous proton removal by histidine. The new N—H of histidine hydrogen bonds to the peptide nitrogen. Cleavage of intermediate B, with simultaneous transfer of a proton from histidine, releases the amino group. This cleavage forms C, in which the acyl group from the peptide is attached by an ester linkage to the serine hydroxyl group. In the cleavage of this ester linkage by water, histidine-57 removes and donates a proton in the same way as in the cleavage of the peptide bond. A water molecule, hydrogen bonded to histidine, attacks the carbonyl group as histidine removes a proton. Cleavage of the resulting intermediate (D), with simultaneous proton transfer from histidine, releases the carbonyl function. This last step regenerates the enzyme (A), which is now available for another catalytic cycle.

A mechanism proposed for the ensuing hydrolysis appears in Figure 14.7. In addition to the details shown, other amino acid residues in the vicinity of histidine-57 and serine-195 are also thought to participate in the catalysis—for example, by providing hydrogen-bonding interactions.

Hydrolysis of peptide linkages in the laboratory is typically achieved by heating for many hours at 100 to 120° with 6 M hydrochloric acid. In contrast, chymotrypsin-catalyzed hydrolysis is rapid at physiological temperatures in neutral solutions. The catalytic activity of enzymes is certainly due to several factors. A large part of the catalysis probably results from binding and positioning the substrate in a favorable orientation with respect to the catalytic groups in the enzyme. In contrast, substrate and catalyst molecules (e.g., hydrochloric acid) in nonenzymatic reactions only rarely approach one another in a favorable orientation. Also important in some cases are the correlated effects of functional groups in the enzyme. In the action of chymotrypsin, for instance, it is proposed that nucleophilic addition reactions of the serine hydroxyl group and of water are facilitated by concurrent transfer of a proton to the histidine side chain.

It is clear why an enzyme must have a complex and very specific structure and why a relatively small change in structure could destroy its functioning. The reactivity of chymotrypsin obviously will be lost if histidine-57 or serine-195 is replaced by another amino acid. Substitution of even a single amino acid in some other region of this enzyme, however, might change the tertiary structure. A resulting change in the spatial relationships of amino acids in the catalytically active region could reduce or destroy the catalytic functioning. Even an amino acid that has a D rather than an L configuration could be damaging, for its side chain would point in a different direction.

It is also clear why the functioning of proteins can be severely affected by reagents that react with one of the amino acids, particularly with one that has a catalytic function. A small amount of diisopropyl fluorophosphate, for example, inactivates chymotrypsin. The hydroxyl group of serine-195 becomes incorporated into a phosphate ester, destroying its ability to participate in catalysis.

$$
\begin{array}{c}
\overset{\displaystyle O}{\underset{\displaystyle \parallel}{}} \\
-NH-CH-C- \\
\mid \\
CH_2 \\
\mid \\
OH
\end{array}
\; + \; (CH_3)_2CHO-\overset{F}{\underset{\parallel O}{P}}-OCH(CH_3)_2 \; \longrightarrow \;
\begin{array}{c}
O \\
\parallel \\
-NH-CH-C- \\
\mid \\
CH_2 \\
\mid \\
O \\
\mid \\
(CH_3)_2CHO-\overset{}{\underset{\parallel O}{P}}-OCH(CH_3)_2
\end{array}
\; + \; HF
$$

Serine-195   Diisopropyl fluorophosphate

Diisopropyl fluorophosphate also combines readily with some serine residues in other enzymes. Its action as a nerve poison (it was developed for this purpose during World War II) is due to its reaction with a critical serine in the enzyme acetylcholine esterase. Acetylcholine is a substance (Sec. 8.12) released by some nerve cells to transmit a nerve impulse to adjoining nerve cells. After triggering a response, acetylcholine is deactivated by a hydrolysis catalyzed by acetylcholine esterase.

$$
\underset{\text{Acetylcholine}}{CH_3\overset{\displaystyle O}{\overset{\|}{C}}-OCH_2CH_2N^+(CH_3)_3} \xrightarrow[\text{H}_2\text{O}]{\text{acetylcholine esterase}} CH_3\overset{\displaystyle O}{\overset{\|}{C}}-OH + \underset{\text{Choline}}{HOCH_2CH_2N^+(CH_3)_3}
$$

Destruction of the catalytic activity of acetylcholine esterase permits the concentration of acetylcholine to increase, thus leading to overstimulation and loss of function of nerve cells.

SUMMARY

Definitions and Ideas

Proteins, major constituents of all organisms, are polymers constructed from a small number of low molecular weight α-amino acids. The **amino acids** differ only in the structures of their **side chains**, which may be only a hydrocarbon group or may include another function, such as amino, carboxyl, hydroxyl, sulfide, or thiol. Proteins that incorporate **prosthetic groups**, components other than amino acids, are known as **conjugated proteins**.

In neutral aqueous solutions, amino acids are present as internal salts called **zwitterions** (or dipolar ions). Amino acids react with ninhydrin to form a deeply purple product. The thiol groups of cysteines are readily oxidized to form a disulfide known as cystine, which is readily reduced back to cysteines ($2\ RSH \rightleftarrows RS\!-\!SR$).

The amino acids in proteins are joined by amide linkages, known as **peptide bonds**, between the α-amino group of one amino acid and the carboxyl group of another. Small molecules formed in this way are called **peptides** (dipeptides, tripeptides . . . polypeptides).

Structures of peptides and proteins are drawn with the **N-terminal amino acid**, the amino acid with the free α-amino group, at the left and the **C-terminal amino acid**, the amino acid with the free carboxyl group, at the right. Structures are often drawn by using three-letter abbreviations for the amino acids. Peptides are named by listing the amino acids in sequence, starting at the N-terminal end. Except for the C-terminal amino acid, the -*ine* ending of the name of each amino acid is changed to -*yl*.

The amino acid composition of a peptide or protein is determined by hydrolyzing it completely to its component amino acids. These are separated by some chromatographic procedure and the quantity of each is determined. The N-terminal and C-terminal amino acids of a peptide or protein can be identified by using reactions that involve only amino acids at the ends of a chain. The amino acid sequence of a peptide or protein can be discovered by partial hydrolysis to form overlapping peptide fragments small enough that their sequences can be determined.

The number of amino acids of each kind and the order of their attachment by covalent (peptide or disulfide) bonds are called the **primary structure** of a protein.

Most proteins can be placed in two broad classes: fibrous and globular. **Fibrous proteins** have relatively extended chains associated together in fiberlike or sheetlike structures. Most proteins that have structural functions are fibrous. **Globular proteins** have chains that are bent and tightly folded about themselves so that the molecules have spherical or globular shapes. Proteins whose functions are regulatory, requiring them to interact specifically with particular chemical species, are usually globular.

Secondary protein structure refers to a regular and repeating conformational arrangement of successive amino acids along the chains of some proteins. One important secondary structure is the **pleated sheet**, in which an amino acid chain is extended in a linear fashion and hydrogen bonded to adjoining chains to form a sheet. Another is the **α helix**, in which a protein chain is coiled in a helical shape. The carbonyl oxygen of each amino acid in an α helix is hydrogen bonded to the N—H of the fourth amino acid down the chain.

Tertiary protein structure refers to complex conformations that result from the extensive bending and folding of protein chains characteristic of globular proteins. In addition to dispersion forces and constraints due to group size, factors important in determining tertiary protein structure are

1. constraints due to disulfide linkages.

2. a tendency for polar groups to be on the exterior.

3. a tendency for nonpolar groups to be on the interior.

4. effects of internal hydrogen bonding.

5. attractions and repulsions of charged groups.

Some globular proteins contain several chains that are not linked covalently. **Quaternary protein structure** refers to the geometry with which the individual chains (subunits) of such proteins are clustered together.

Virtually all biological reactions are catalyzed by proteins called **enzymes**. The compound that undergoes a reaction catalyzed by an enzyme is referred to as a **substrate** for that enzyme. An enzyme can exhibit not only tremendous catalytic activity but also a high specificity for a particular substrate.

PROBLEMS

1. Define each term. Where helpful, illustrate with a structural formula or a drawing.
   (a) α-amino acid
   (b) L-amino acid
   (c) essential amino acid
   (d) zwitterion structure of an amino acid
   (e) peptide bond
   (f) peptide
   (g) tripeptide
   (h) protein
   (i) N-terminal amino acid
   (j) C-terminal amino acid
   (k) partial hydrolysis of a protein
   (l) primary protein structure
   (m) secondary protein structure
   (n) tertiary protein structure
   (o) quaternary protein structure
   (p) conjugated protein
   (q) prosthetic group
   (r) fibrous protein
   (s) globular protein
   (t) α keratin

(u) β keratin   (v) pleated sheet

(w) α helix   (x) enzyme

(y) substrate   (z) denaturation of a protein

2. Draw the structure of at least one amino acid found in proteins that has each feature.
   (a) an acidic side chain
   (b) a basic side chain
   (c) an alkyl group
   (d) an aromatic hydrocarbon ring
   (e) a phenol function
   (f) an alcohol function
   (g) a sulfur-containing function that can form a disulfide linkage
   (h) a sulfur-containing function that cannot form a disulfide linkage
   (i) a heterocyclic ring
   (j) an amide function
   (k) a secondary amino group
   (l) no asymmetric carbon
   (m) two asymmetric carbons

3. Draw complete structures of the following peptides.
   (a) glycylvaline   (b) alanyllysylserine
   (c) Phe-Glu-Gly   (d) an isomer of the peptide in part (a)

4. Name the following peptides.

   (a)

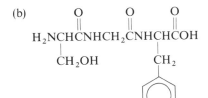

   $$\underset{\substack{|\\CH_3}}{H_2NCHCNHCHCOH}\ \underset{\substack{|\\CH_2CH_2SCH_3}}{}$$

   (b)

   $$H_2NCHCNHCH_2CNHCHCOH$$

5. List all compounds that fit the following descriptions.
   (a) dipeptides that contain one Gly and one Val
   (b) tripeptides that contain one Ala, one Ser, and one Phe
   (c) tripeptides that contain Ala, or Gly, or both
   (d) tetrapeptides that contain three Ala and one Ser

6. How many isomeric peptides fit each description?
   (a) dipeptides constructed from the 20 common amino acids
   (b) tripeptides constructed from the 20 common amino acids
   (c) tripeptides constructed from the 20 common amino acids without using any amino acid more than once

7. Draw the structures of the organic products that result from treating alanine with each reagent.
   (a) a strongly acidic aqueous solution   (b) a strongly basic aqueous solution
   (c) HNO_2 (d) ninhydrin
   (e) CH_3OH, H^+

8. Draw the structures of the organic products that result from treating Gly-Ala-Val with each reagent.

(a) H_2O, H^+

(b) 2,4-dinitrofluorobenzene followed by H_2O, H^+

(c) a carboxypeptidase

9. Draw the structure of the dominant form of valine when it is dissolved in an aqueous solution having a pH value of 2, 7, or 12.

10. Draw the amino acid sequence of a hexapeptide that when hydrolyzed completely gives Ala, Gly, Ile, Met, and Phe. The amount of alanine formed is twice that of the other amino acids. Partial hydrolysis of the hexapeptide gives several smaller peptides that can be shown to have the following structures: Ala-Ala-Ile, Gly-Ala, Ile-Met, and Phe-Gly.

11. Identify each of the following features in the peptide shown below.

Thr-Cys-Ser-Thr-Ser-Thr-Pro-Gly-Val-Tyr-Ala-Arg-Val-Thr-Ala-Leu-Val-Gln-Thr

(a) an N-terminal end

(b) a C-terminal end

(c) a site where a covalent link to another chain might readily form

(d) a site where a bend in an α-helical structure might be found

12. List the smaller peptides that would form most rapidly in hydrolyses of the following peptide catalyzed by each enzyme.

Ala-Glu-Ser-Asp-Phe-Ala-Thr-Lys-Ile-Val-Cys-Tyr-Gly-Asn-Ala-Gly

(a) carboxypeptidase   (b) chymotrypsin

13. How could you use each of the following as an aid in determining end groups of peptides?

(a) aminopeptidases, enzymes that specifically catalyze the hydrolysis of peptide functions having a free amino group alpha to the carbonyl group

(b) lithium borohydride ($LiBH_4$), a reagent that in aqueous solution reduces a carboxyl group to a primary alcohol but does not reduce amide linkages

14. Polyglutamic acid is a protein containing only glutamic acid units. Depending on the pH of the solution in which it is dissolved, it has a highly regular or a random structure. Predict its structure in solutions having a pH of 2 or 7.

15. What would be the result of having one D-phenylalanine in a protein that otherwise is constructed from L-amino acids and has an α-helical structure?

16. α Keratins are easily stretched, particularly when heated in a moist environment. When stretched, they tend to adopt the sheetlike structures characteristic of β keratins, but on cooling they resume their original structures. In contrast, β keratins are not easily stretched. Rationalize the different behavior of the two groups.

17. Briefly describe each technique.

(a) column chromatography   (b) X-ray crystallography

18. Chymotrypsin is formed from chymotrypsinogen, itself not an active catalyst. Chymotrypsinogen, a component of the pancreatic juice secreted into the small intestine, has a single chain of 245 amino acids. A selective hydrolysis of chymotrypsinogen, catalyzed by the enzyme trypsin, excises two dipeptides, forming chymotrypsin. Why may it be advantageous for a creature to store chymotrypsin in the form of chymotrypsinogen and convert it to chymotrypsin only as needed?

Nucleic Acids

15

Polymers called deoxyribonucleic acids (DNA) and ribonucleic acids (RNA) are major components of living cells. They are responsible for the remarkable ability of cells to reproduce themselves. DNA molecules store almost all genetic information. In cells having nuclei, most DNA is in the structures called chromosomes. Each segment of a DNA molecule stores the information necessary to direct the synthesis of the amino acid sequence of a particular protein. By the biosynthesis of specific proteins, the information stored in DNA is ultimately translated into the features that uniquely characterize particular cells and organisms. RNA molecules participate in protein biosynthesis. Few developments in our era have greater potential to affect civilization than our growing understanding of the genetic process, culminating in the recently acquired ability ("genetic engineering") to transfer genetic material from one organism to another.

In this chapter we learn how genetic information is stored in the structure of DNA. Then we briefly examine how this information is transmitted from one cell to another and translated into the synthesis of specific proteins. DNA and RNA molecules must be able to contain and transmit large amounts of information. Yet, just like proteins, they are linear polymers constructed from only a few kinds of small monomeric units. We begin by considering these small units, known as **nucleotides**.

15.1
NUCLEOTIDES

Hydrolysis of nucleic acids leads to nucleotides. Just as amino acids are the monomeric units of proteins, so nucleotides are the monomeric units of nucleic acids. Adenosine 5′-phosphate is one of four nucleotides commonly found in RNA.

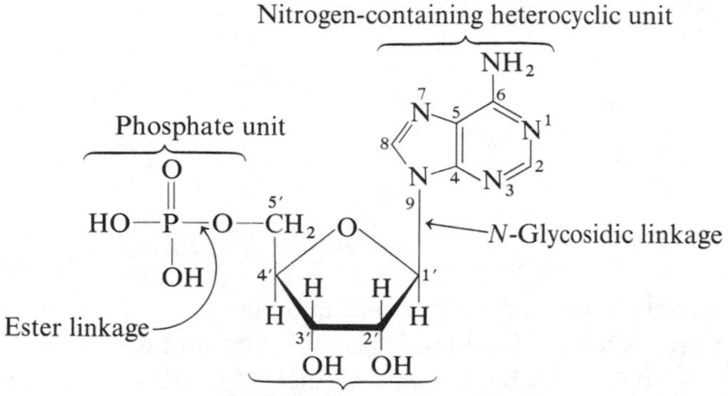

Adenosine 5′-phosphate

As are all nucleotides, it is constructed of three components:

1. a nitrogen-containing heterocyclic compound.

2. a pentose.

3. a molecule of phosphoric acid.

In the nucleotides of RNA the pentose is D-ribose (Sec. 10.1), present as a β-furanoside (Sec. 10.2). The ribose and the nitrogen-containing heterocyclic compound are bonded together by an N-glycosidic linkage (Sec. 10.7). The phosphoric acid and the ribose C-5′ oxygen (called C-5′ to distinguish it from C-5 of the heterocyclic compound) are bonded by an ester linkage. The other nucleotides of RNA differ only in the particular heterocyclic compound that they contain.

Deoxythymidine 5′-phosphate is one of four nucleotides commonly found in DNA.

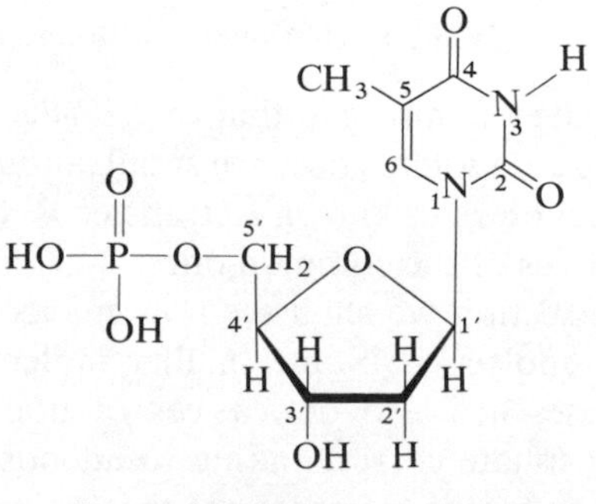

Deoxythymidine 5′-phosphate

In the nucleotides of DNA the pentose is 2-deoxy-D-ribose (ribose lacking the oxygen at C-2) instead of D-ribose. The pentose is linked to a phosphate and a heterocyclic compound in the same way as in RNA nucleotides. The other DNA nucleotides differ only in the heterocyclic compound.

The nitrogen-containing heterocyclic compounds found in nucleotides are derivatives of either purine or pyrimidine.

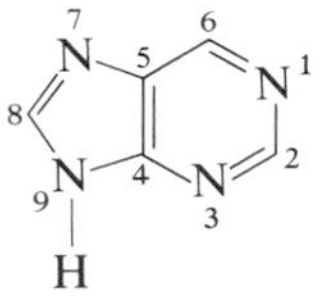

Purine 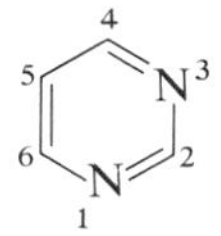   Pyrimidine

The purines adenine and guanine and the pyrimidine cytosine are found both in DNA and RNA nucleotides. The fourth common heterocyclic component is uracil in DNA nucleotides but thymine (5-methyluracil) in RNA nucleotides.

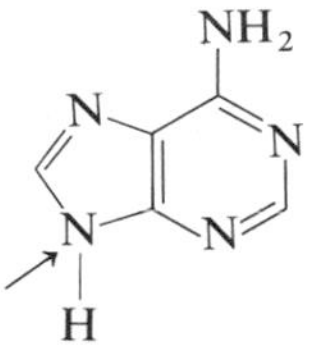

Adenine 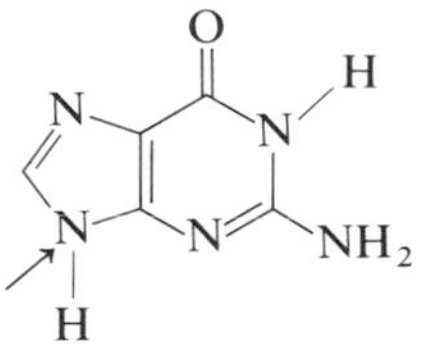   Guanine 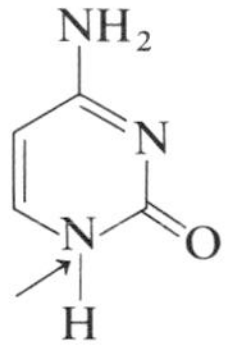   Cytosine

(found in both RNA and DNA)

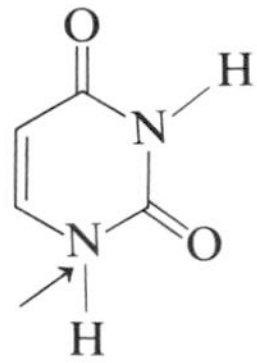

Uracil   Thymine
(found in RNA only)   (found in DNA only)

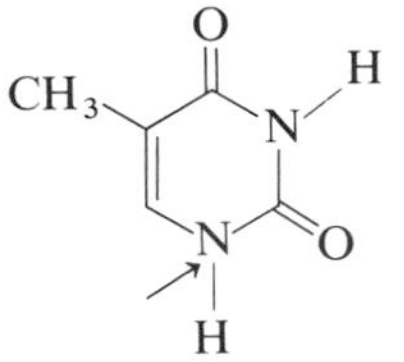

The arrows indicate the nitrogen that in a nucleotide is linked to a pentose. Some other heterocyclic compounds occur in small amounts in some nucleic acids, particularly in the RNA molecules known as transfer RNA's. These rarer nitrogen heterocycles are derivatives of the common ones.

In the next section we shall see how nucleotides are linked together to form RNA and DNA molecules. Some smaller molecules that play important roles in metabolic processes are also derivatives of nucleotides. Adenosine triphosphate (ATP), the triphosphate corresponding to adenosine 5'-phosphate, is the principal carrier of energy to biological processes that require an input of energy.

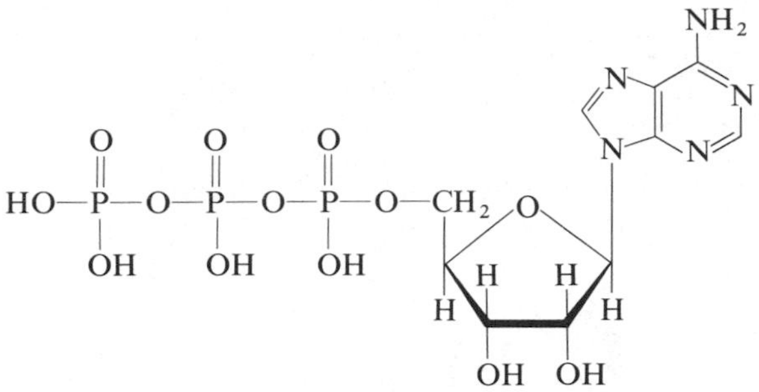

ATP

Energy produced by metabolic processes is used to synthesize ATP from the corresponding diphosphate (ADP). In transferring energy to reactions that require it, ATP is converted back to ADP.

ATP also has other important biochemical roles. One is as a precursor for adenosine 3′, 5′-cyclic monophosphate (cyclic AMP).

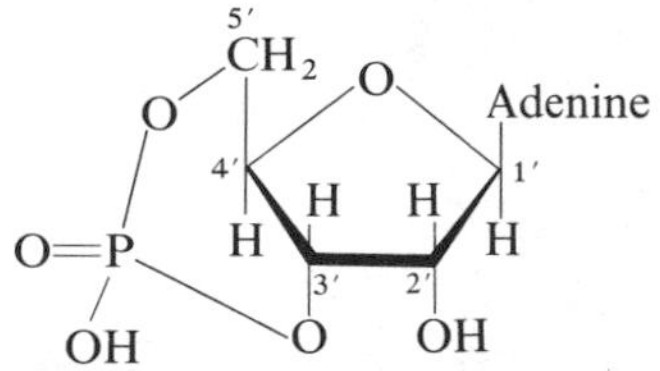

Adenosine 3′,5′-cyclic monophosphate
Cyclic AMP

Delivery of certain hormones through the bloodstream to a cell stimulates conversion of ATP to cyclic AMP, a reaction catalyzed by the enzyme adenylate cyclase in the cell membrane. Cyclic AMP transmits throughout the cell the signal delivered by the hormone.

An adenosine phosphate unit is a portion of the structure of the important biological molecule NADH (Sec. 10.7).

15.2
PRIMARY AND SECONDARY STRUCTURES OF DNA

DNA and RNA are linear polymers in which nucleotide phosphate units are linked together. As shown in Figure 15.1 for a segment of a DNA chain, the polymer backbones contain alternating pentose and phosphate units. Each phosphate is bonded by ester linkages to the C-5′ oxygen of one pentose and to the C-3′ oxygen of another. RNA differs only in the presence of ribose instead of deoxyribose and uracil instead of thymine.

Attached to the pentose-phosphate backbone are nitrogen-containing heterocycles. The sequence of nucleotides in a DNA (or RNA) molecule is its primary

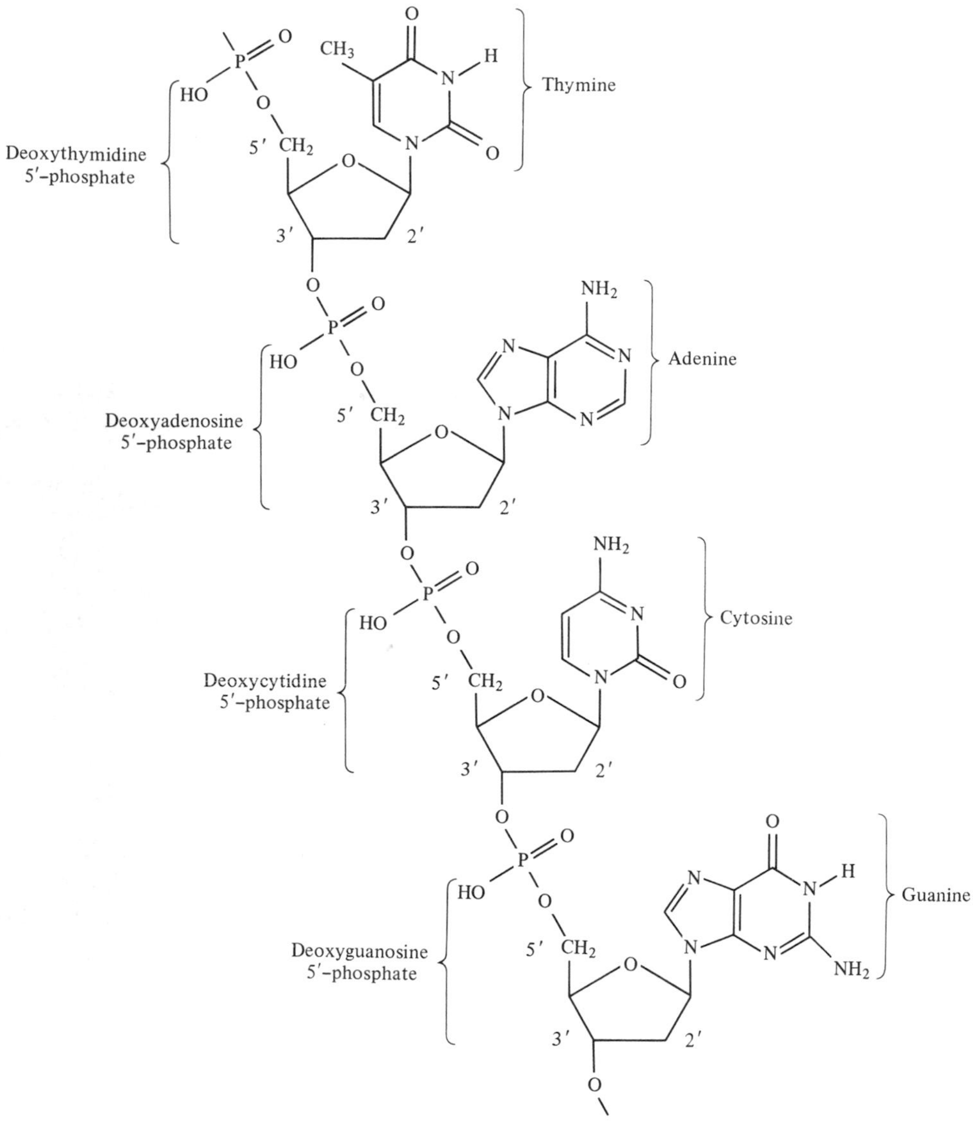

Figure 15.1 A segment of a DNA chain. An RNA chain differs only in having ribose (which has a hydroxyl group at C-2′) as the pentose and uracil instead of thymine (5-methyluracil) as one of the heterocyclic components. The nucleotide containing uracil is named uridine 5′-phosphate.

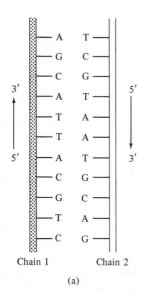

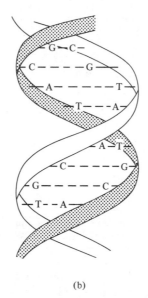

Figure 15.2 (a) An unraveled segment of two complementary polynucleotide chains of DNA. A, C, G, and T represent the four pyrimidines and purines commonly found in DNA. (b) The same segments coiled in a double helix. Along the helix, the pentose carbons are arranged in a $3' \rightarrow 5'$ sequence in one chain but in a $5' \rightarrow 3'$ sequence in the complementary chain.

structure, just as the sequence of amino acids is the primary structure of a protein. And just as differences between proteins are due to different numbers and sequences of 20 amino acids, so differences between DNA molecules (or RNA molecules) are due to different numbers and sequences of nucleotides containing the four heterocycles. One aspect of the composition of DNA is fixed. The amounts of adenine (A) and thymine (T) are equal, as are the amounts of guanine (G) and cytosine (C).* The ratio of A + T to G + C varies considerably, however.

As shown in Figure 15.2, DNA molecules normally consist of two chains. The chains are coiled in a right-handed helical fashion about the same axis, forming what is called the **double helix**. The chains are antiparallel: in a particular direction along the helix, the pentose carbons in one chain are arranged in a $3' \rightarrow 5'$ sequence but in the other chain in a $5' \rightarrow 3'$ sequence. Each chain has 10 nucleotides in a complete turn of the helix. The two chains of the helix are held together by hydrogen bonds between heterocycles attached to one chain and those attached to the other. The polar pentose and phosphate groups (the —OH of the phosphate is ionized to —O⁻ in physiological solutions) are on the exterior of the double helix, exposed to water. The pyrimidines and purines are in the interior. The plane of each heterocyclic ring is perpendicular to the axis (direction) of the helix but parallel to the other rings on its own and the other chain.

In the DNA double helix, A on one chain is always hydrogen bonded to T on the other. Similarly, G and C are always hydrogen bonded together. This specific

* Depending on the context, the one-letter abbreviations will stand either for the heterocycles or for the nucleotide units into which they are incorporated.

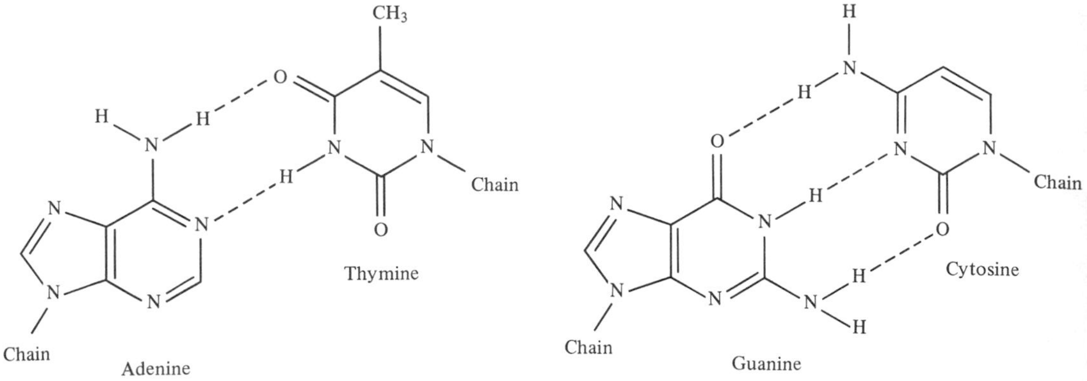

Figure 15.3 Hydrogen bonding between adenine and thymine and between guanine and cytosine located on complementary chains of a DNA double helix.

hydrogen bonding (see Fig. 15.3) is responsible for the 1:1 ratios of A to T and G to C. The hydrogen-bonded A-T and G-C pairs have about the same size and fit comfortably into the helical structure. Hydrogen-bonded pairs that would result from other combinations of the four heterocycles are too large or too small to fit satisfactorily into the helical structure. Moreover, other pairs of the heterocycles do not hydrogen bond together as efficiently. In their antiparallel arrangement in the helix the two chains have complementary sequences. Where A appears in one chain, T appears in the other. G and C are similarly matched.

Just as a pleated sheet or an α-helix structure is considered the secondary structure of a protein, so the double helix is the secondary structure of a DNA molecule.

15.3
DNA REPLICATION

No property of a living cell is more remarkable than its capacity for self-replication. Because the genetic information is contained in DNA, reproduction of a cell requires accurate duplication of its DNA. The process by which DNA is duplicated is called **replication**. During replication a parent DNA molecule is converted to two progeny, both identical to the parent. Much remains to be discovered about replication and some details are not the same for all living organisms. Nevertheless, the essential features are now well established and remarkably simple.

Faithful reproduction of an enormous DNA molecule results from using each of its chains as a template to specify the synthesis of a new, exactly complementary chain. The essential features of replication are illustrated in Figure 15.4. A portion of the DNA helix is unwound. Deoxyribonucleotides (actually triphosphates rather than monophosphates) hydrogen bond to each of the separated chain segments. Hydrogen bonding is specifically between A and T and between G and C. The

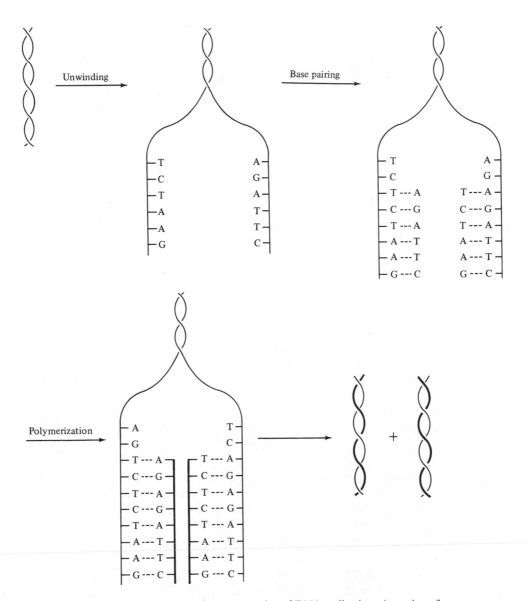

Figure 15.4 A diagrammatic representation of DNA replication. A portion of the helix is unwound. While the unwinding continues, nucleotide derivatives hydrogen bond to each of the separated chains. A bonds specifically to T and G to C. The nucleotide derivatives are linked together in an enzyme-catalyzed reaction. This process forms two double helices, each containing one chain from the parent DNA molecule and one newly synthesized chain.

nucleotide components are enzymatically polymerized (with expulsion of diphosphate) to form a new chain. Extension of this process along the full length of each of the original chains results in the formation of two double helical molecules that are identical to the parent DNA molecule. Each new DNA molecule contains one chain from the parent DNA and one newly synthesized chain.

15.4
RIBONUCLEIC ACIDS

RNA molecules are involved in **translation**, the process in which genetic information stored in DNA leads to the synthesis of specific proteins. There are three types of RNA molecules—messenger, ribosomal, and transfer—each having a different function. All RNA molecules have single chains instead of the double helical structures characteristic of DNA.

Messenger RNA (*m*RNA) molecules carry the genetic message from DNA to the ribosomes. Ribosomes, found in all cells, are particles at which protein synthesis occurs. In a cell with a nucleus, the DNA is confined within it. DNA molecules store the information for many peptides and proteins and often have molecular weights of many billion. Messenger RNA molecules carry the information for the synthesis of only one or at most a few peptides and proteins and have much lower molecular weights.

The process by which *m*RNA molecules are synthesized is called **transcription**. A segment of one chain of a DNA molecule is transcribed into the single chain of a *m*RNA molecule. The RNA chain is synthesized in the reverse direction ($5' \rightarrow 3'$) from that of the DNA chain ($3' \rightarrow 5'$) and is complementary to the DNA chain. RNA has uracil (U) instead of thymine (T) and this serves as the complement of adenine (A) of DNA. After transcription, the *m*RNA moves to a ribosome, where it serves as a template that directs the specific amino acid sequence formed in the biosynthesis of a protein.

Ribosomes are complexes of **ribosomal RNA** (*r*RNA) molecules and proteins. A ribosome has a specific size and three-dimensional shape. Ribosomes contain several characteristic types of *r*RNA molecules, with molecular weights ranging from about 50,000 to a million. Protein synthesis takes place on the surface of ribosomes.

Transfer RNA (*t*RNA) molecules bring the amino acids to the ribosomal site of protein synthesis. They are relatively small, containing only 75 to 90 nucleotide residues (corresponding to molecular weights of about 30,000). Each amino acid has at least one corresponding *t*RNA molecule and some amino acids have several. During protein biosynthesis a free hydroxyl group at the 3' end of the chain of a *t*RNA molecule is attached by an ester linkage to the acyl group of its specific amino acid. As described further in Section 15.6, the *t*RNA molecule for a particular amino acid hydrogen bonds to a segment of *m*RNA that codes for that amino acid. Amino acids bonded to the *t*RNA molecules then are linked together to form a protein chain having the amino acid sequence specified by the *m*RNA molecule.

15.5
THE GENETIC CODE

How does the linear sequence of nucleotides in a *m*RNA molecule specify the synthesis of the unique amino acid sequence of a particular protein? Because there are 20 amino acids but only four nucleotides, one nucleotide cannot code for one amino acid. Messenger RNA has an alphabet of only four letters and must be able to form at least 20 words. A combination of two nucleotides is also insufficient, for there are only $4 \times 4 = 16$ possible sequences of two nucleotides. A combination of three nucleotides provides $4 \times 4 \times 4 = 64$ sequences, more than needed. Various experiments have shown that a sequence of three adjacent nucleotides, called a **codon**, is the coding unit that leads to incorporation of a specific amino acid into a protein chain.

Use of synthetic polyribonucleotides of known composition in place of natural *m*RNA as templates for protein biosynthesis was one important technique for deciphering the "genetic code." For example, poly(U) (-UUUUUUUUUUUUUUU-) led to the synthesis of polyphenylalanine (-Phe-Phe-Phe-Phe-Phe-), indicating that UUU is the codon for phenylalanine. Similarly, poly(C) led to synthesis of polyproline, showing that CCC is the codon for proline.

Using such experiments, the relationships shown in Table 15.1 between amino acids and the 64 codons were established. This code seems to be universal for all organisms, a striking example of the much greater similarity between living organisms at the chemical than the morphological level. All but two of the amino acids are

Table 15.1 The genetic code. The codons of messenger RNA and the amino acid that each specifies during protein biosynthesis.

UUU	Phe	UCU	Ser	UAU	Tyr	UGU	Cys
UUC	Phe	UCC	Ser	UAC	Tyr	UGC	Cys
UUA	Leu	UCA	Ser	UAA	[end]	UGA	[end]
UUG	Leu	UCG	Ser	UAG	[end]	UGG	Trp
CUU	Leu	CCU	Pro	CAU	His	CGU	Arg
CUC	Leu	CCC	Pro	CAC	His	CGC	Arg
CUA	Leu	CCA	Pro	CAA	Gln	CGA	Arg
CUG	Leu	CCG	Pro	CAG	Gln	CGG	Arg
AUU	Ile	ACU	Thr	AAU	Asn	AGU	Ser
AUC	Ile	ACC	Thr	AAC	Asn	AGC	Ser
AUA	Ile	ACA	Thr	AAA	Lys	AGA	Arg
AUG	Met	ACG	Thr	AAG	Lys	AGG	Arg
GUU	Val	GCU	Ala	GAU	Asp	GGU	Gly
GUC	Val	GCC	Ala	GAC	Asp	GGC	Gly
GUA	Val	GCA	Ala	GAA	Glu	GGA	Gly
GUG	Val	GCG	Ala	GAG	Glu	GGG	Gly

specified by more than one codon. Three codons (UAA, UAG, and UGA) do not code for amino acids. Instead they function as periods, terminating the synthesis of a protein chain. In polynucleotide chains, as in peptide chains, there is a problem of direction. We know that Ala-Gly and Gly-Ala are different; Ala is the N-terminal residue in Ala-Gly but the C-terminal residue in Gly-Ala. Similarly, UAG and GAU are different. The codons are written from left to right in the $5' \rightarrow 3'$ direction. In UAG, the $5'$ carbon of the ribose of A is toward U and its $3'$ carbon toward G.

Proteins occasionally contain amino acid residues other than the 20 common amino acids for which there are codons. These unusual amino acids result from enzyme-catalyzed modification of the 20 common amino acids after a protein chain has been synthesized.

15.6
TRANSLATION OF THE GENETIC CODE;
BIOSYNTHESIS OF PROTEINS

Translation of the genetic information contained in the sequences of codons along a *m*RNA molecule into the synthesis of a protein with the proper amino acid sequence involves an intricate series of enzyme-catalyzed steps. Specific hydrogen bonding between complementary nucleotides is the essential feature leading to an appropriate amino acid sequence.

Transfer RNA molecules play a major role in protein biosynthesis. These relatively small polynucleotides have specific three-dimensional shapes. These shapes, determined mainly by internal hydrogen bonds, somewhat resemble a cloverleaf that has been further folded together. One loop of a *t*RNA molecule has an exposed sequence of three nucleotides called an **anticodon**. This sequence forms specific hydrogen bonds with the *m*RNA codon that specifies a particular amino acid. This amino acid is attached by an ester linkage to a hydroxyl group of the ribose of the nucleotide at the $3'$ end of the *t*RNA molecule.

Protein synthesis occurs on the surface of a ribosome. Figure 15.5 shows diagrammatically how the ribosome interacts with *m*RNA and *t*RNA to translate the message of the *m*RNA into the appropriate amino acid sequence.

Figure 15.5   Translation of the information in messenger RNA into synthesis of a specific amino acid sequence. (a) The *m*RNA is held on the ribosome in a highly specific way. At any moment, one codon of the *m*RNA is at the P site. A *t*RNA hydrogen bonded to this site has linked to it the previously synthesized portion of the protein chain. An adjacent codon of *m*RNA is at the A site of the ribosome. A new *t*RNA enters the A site. This *t*RNA (specific for Arg in this example) has an anticodon complementary to the *m*RNA codon. (b) In an enzyme-catalyzed reaction, the amino group of the amino acid linked to the *t*RNA (that is at the A site) becomes attached to the C-terminal carbonyl group of the growing protein chain. This reaction adds one amino acid to the protein chain and cleaves the ester linkage that held the growing protein chain to the end of the old *t*RNA. The old *t*RNA is released and the ribosome moves the length of one codon along the *m*RNA chain. The appropriate *t*RNA (specific for Asp in this example) can then occupy the A site, readying the system for attachment of another amino acid to the growing protein chain. The process continues until a codon is reached that signals for synthesis to end.

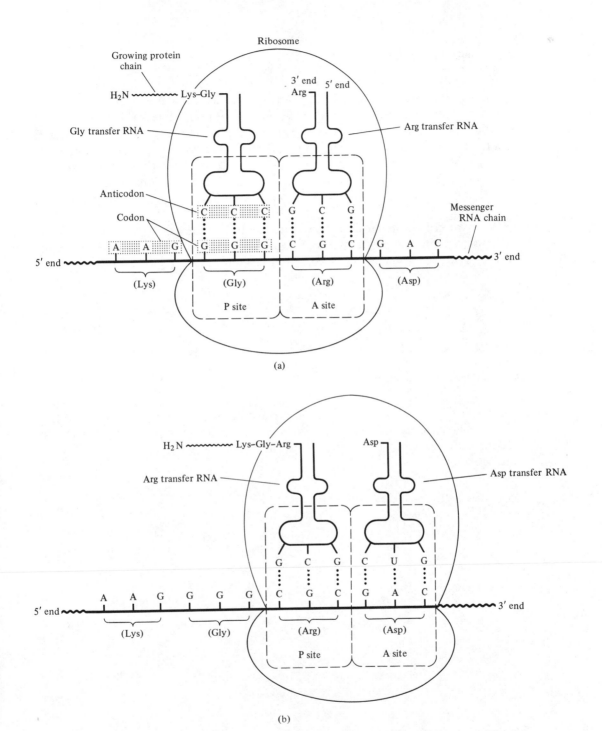

(a)

(b)

A structural change in a DNA molecule that on replication and transcription leads to synthesis of altered DNA and RNA molecules is called a **mutation**. Even a change in one nucleotide can result in a change in the amino acid sequence of a protein and may have serious consequences. The complete amino acid sequences of the α and β chains of normal human hemoglobin and many variant human hemoglobins are known. Most observed variations could be due to an alteration of only a single nucleotide. Sickle-cell anemia, for instance, is due to a substitution of a glutamic acid (known to be coded for by GAG) by a valine. The codons for valine are GUA or GUG, but presumably GUG is involved here.

A variety of chemical and physical agents are known to alter DNA. Some changes to a nitrogen heterocycle lead to its being read as another. As one example, nitrous acid (Sec. 8.6) converts adenine to hypoxanthine.

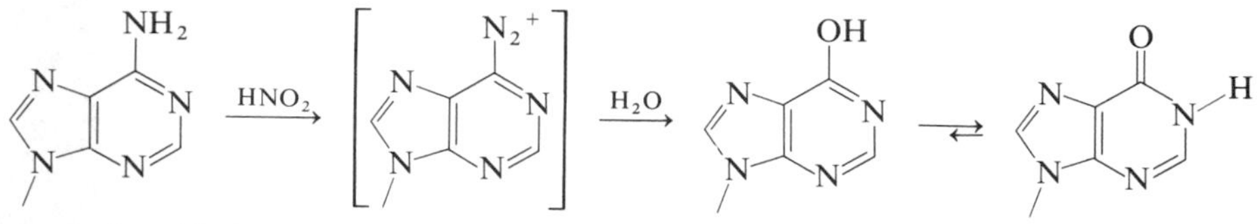

Adenine (pairs with thymine)   Hypoxanthine (pairs with cytosine)

Adenine (A) pairs with thymine (T), but hypoxanthine pairs with cytosine (C).

Other changes to DNA are serious enough to block replication. Ultraviolet radiation, for example, links adjacent thymine rings, stopping replication of DNA at that site.

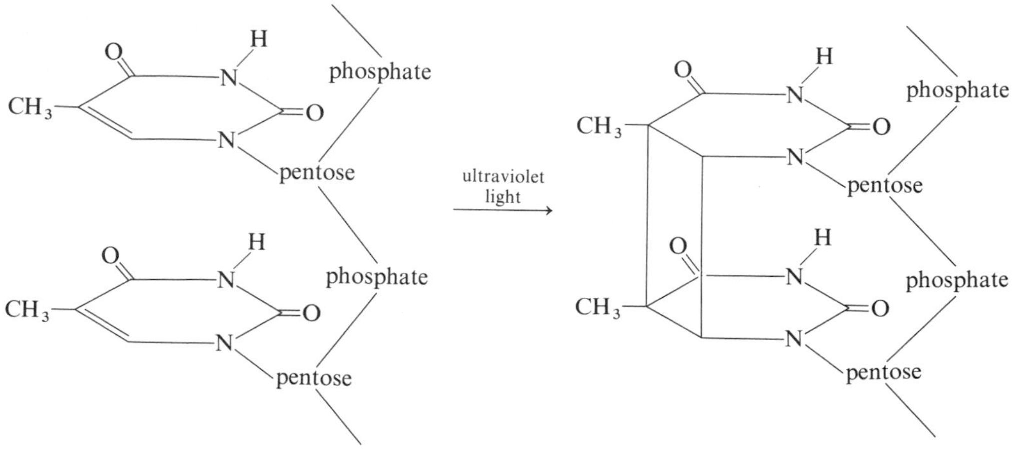

Gamma rays and X rays can break DNA or cause deletion of nucleotides.

An enormous DNA molecule is continuously subjected to chemicals and to radiation that can alter its structure. To help preserve the structure, cells have elaborate mechanisms for the repair of damage to DNA. Some carcinogens certainly initiate cancer by their action on DNA.

Replication, transcription, and translation must occur with great speed and incredible accuracy. Various chemicals, including some antibiotics, interfere with these processes. As an example, some flat molecules, such as the dye proflavine, can fit between successive hydrogen-bonded heterocyclic pairs and interfere with replication and transcription.

Proflavine

SUMMARY

Definitions and Ideas

Deoxyribonucleic acid (DNA) and **ribonucleic acid** (RNA), major components of all living cells, are responsible for the ability of cells to reproduce themselves. They are polymers of **nucleotides**, molecules composed of a purine or pyrimidine derivative, a pentose, and a molecule of phosphoric acid. In RNA, the pentose is D-ribose, present as a β furanoside. It is linked by an N-glycosidic linkage to the heterocyclic compound and by an ester linkage at its C-5' hydroxyl group to the phosphoric acid. An RNA molecule is constructed similarly except that the pentose is 2-deoxy-D-ribose. The heterocyclic compounds in DNA are adenine (A), guanine (G), cytosine (C), and thymine (T); those in RNA are A, G, C, and uracil (U).

DNA and RNA are polymers in which the phosphate unit of one nucleotide is bonded by an ester linkage to the C-3' hydroxyl group of the next. The result is a linear polymer with a backbone of alternating pentose and phosphate units, branching off from which are the heterocyclic components.

DNA molecules usually consist of two chains coiled in a **double helix**. The two chains are arranged in an antiparallel fashion (in a particular direction along the helix, the pentose carbons in one chain are arranged in a 3' → 5' sequence but in the complementary chain in a 5' → 3' sequence). The two chains are held together by hydrogen bonds between the heterocycles attached to one chain and those attached to the other. A is always hydrogen bonded to T and G to C, accounting for the equal amounts of A and T and of G and C found in DNA.

DNA molecules, which store almost all genetic information, are duplicated by a process called **replication**. During replication a portion of the DNA helix is unwound and nucleotide derivatives hydrogen bond to each chain heterocycle, the hydrogen bonding being specifically between A and T and between G and C. The nucleotide derivatives are then enzymatically polymerized to form new chains. This process results in the formation of two new DNA molecules that are identical to the parent DNA molecule. Each contains one chain from the parent and one newly synthesized chain.

The genetic information stored in DNA is transformed into the features that uniquely characterize particular cells and organisms by **translation**, a process resulting in the synthesis of specific proteins. Three kinds of RNA molecules, all having single chains, are involved in this process.

Ribosomes are complexes of proteins and **ribosomal RNA** (*r*RNA) molecules. Protein synthesis takes place on the surface of ribosomes.

Messenger RNA (*m*RNA) molecules carry the genetic message from DNA to the ribosomes. In a process called **transcription**, a segment of one DNA chain is transcribed into a complementary RNA molecule (in which U of RNA serves as the complement of A of DNA). A **codon** is a sequence of three adjacent nucleotides in *m*RNA that specifies a particular amino acid. Of the 64 possible sequences of the nucleotides present in *m*RNA, 61 lead to incorporation of a specific amino acid into a protein chain and 3 terminate protein synthesis.

Each **transfer RNA** (*t*RNA) molecule has an **anticodon**, a sequence of three nucleotides that is complementary to and specifically hydrogen bonds to a codon of messenger RNA. The *t*RNA molecule also has a site to which the amino acid specified by that codon can be attached. During translation a *t*RNA molecule hydrogen bonds to the appropriate codon of *m*RNA, and the amino acid it bears becomes incorporated into the growing amino acid chain.

PROBLEMS

1. Define each term. Where helpful, illustrate with a structural formula or a drawing.
   (a) DNA
   (b) RNA
   (c) nucleotide
   (d) complementary nucleotides
   (e) double helix
   (f) replication
   (g) transcription
   (h) translation
   (i) messenger RNA
   (j) transfer RNA
   (k) ribosome
   (l) ribosomal RNA
   (m) genetic code
   (n) codon
   (o) anticodon
   (p) mutation

2. Draw suitable representations of each of the following.
   (a) purine
   (b) pyrimidine
   (c) β-D-ribofuranose
   (d) β-2-deoxy-D-ribofuranose
   (e) guanosine 5′-phosphate
   (f) deoxycytidine 5′-phosphate

3. What are the two major types of nucleic acids? How do they differ?

4. How are nucleotides linked together in nucleic acids?

5. What are the products of the complete hydrolysis of all ester and *N*-glycosidic links of a DNA molecule? What are the products of similar treatment of an RNA molecule?

6. Explain why in DNA the G:C ratio is 1.00 but the C:A ratio is variable.

7. A segment of one chain of a DNA molecule contains the following sequence (reading from the 5′ to the 3′ end).

-TTCGCATCGTCTGGA-

5′ end   3′ end

(a) What is the sequence of nucleotides in the corresponding segment of the other chain of the DNA molecule? Indicate which end of the sequence is 5′ and which is 3′.

(b) What is the sequence of bases in the mRNA transcribed from the first DNA chain? Indicate which end is 5′ and which 3′.

(c) What is the amino acid sequence coded for by the mRNA sequence that is the answer to part (b)? Assume that a codon begins at the nucleotide at the appropriate end of the segment.

8. What amino acid sequence will be synthesized in response to the following segment of mRNA? Assume that a codon begins with the nucleotide on the left.

-CUGAGCGCAAGAGGGCCG-

9. (a) What amino acid sequence will be synthesized in response to the following segment of mRNA? Assume that a codon begins with the nucleotide on the left.

-GUGGCAUACUGCAGA-

(b) What amino acid sequence will result from the following mRNA segment that differs in one nucleotide from that in part (a)?

-GCGGCAUACUGCAGA-

(c) What amino acid sequence will result from the following mRNA sequence that is identical to that in part (a) except for deletion of one nucleotide?

-UGGCAUACUGCAGA-

10. Which of the following amino acid substitutions resulting in an abnormal protein could be due to the alteration of a single DNA nucleotide?
(a) isoleucine in place of leucine   (b) serine in place of proline
(c) glycine in place of alanine   (d) lysine in place of threonine

11. Why is a mutation involving deletion of one nucleotide pair from DNA likely to have more serious consequences than a mutation involving replacement of one nucleotide pair by another?

16

Spectroscopy of Organic Compounds

How are organic compounds identified? How are their structures determined? The introduction in recent decades of various instrumental methods has greatly simplified such tasks. Particularly important are different types of spectroscopy, methods that measure the interaction of radiation (such as light or radio waves) with molecules. Many instrumental methods are rapid and require only small amounts of material. Unlike chemical tests, they are often nondestructive—the sample of compound is not altered and can still be used for other purposes. Information provided by these procedures is invaluable in identifying and determining structures of organic compounds. The procedures are also useful in detecting and determining the quantities of small amounts of compounds, even in mixtures. Such capabilities are needed to deal with many problems involving biological reactions or environmental pollution. In this chapter we consider four particularly important and widely used instrumental methods.

16.1
ELECTROMAGNETIC RADIATION AND MOLECULES

Light, radio waves, and X rays are all examples of **electromagnetic radiation**. The designation electromagnetic is used because this radiation is associated with electric and magnetic fields that increase and decrease in a periodic fashion. All electromagnetic radiation travels at the same velocity ("the speed of light"), 3.0×10^8 m/sec in a vacuum, but can differ in the frequency with which the electric and magnetic fields increase and decrease. A portion of the range of electromagnetic radiation is

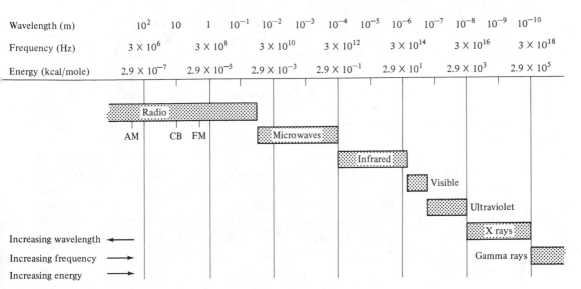

Figure 16.1 Wavelength, frequency, and energy of electromagnetic radiation.

shown in Figure 16.1. Such classifications as X rays or infrared radiation are arbitrary, related to the instruments that produce and detect radiation of particular frequencies. Visible light, for example, is the small portion of electromagnetic radiation that the human eye can detect.

The changes in magnetic and electrical fields as a beam of electromagnetic radiation travels through space are reminiscent of waves on the surface of water and can be described in similar terms. The frequency of the radiation, v, is the number of waves passing a given point in a certain time. Frequency is expressed in cycles per second (generally called hertz, abbreviated Hz). Radiation can also be characterized by its wavelength, the distance between corresponding parts of successive waves. Wavelength and frequency are related by the equation

$$\lambda = \frac{c}{v}$$

where λ = wavelength (in meters)

c = velocity of electromagnetic radiation (3.0×10^8 meters/sec in a vacuum)

v = frequency (in Hz)

The center of the visible light range, for instance, has a frequency of about 5×10^{14} Hz (cycles/sec), which corresponds to a wavelength of

$$\lambda = \frac{c}{v} = \frac{3 \times 10^8 \text{ meters/sec}}{5 \times 10^{14} \text{ cycles/sec}} = 6 \times 10^{-7} \text{ meters}$$

Frequency and wavelength are inversely proportional—radiation having a high frequency has a short wavelength.

When interacting with molecules, electromagnetic radiation behaves as though it consists of discrete, indivisible packets of energy. These smallest units of electromagnetic radiation are called **photons**. Photons of each frequency (wavelength) have a characteristic energy that is proportional to frequency.

$$E = \text{constant} \times v$$

The values of energy in Figure 16.1 are for one mole (Avogadro's number) of photons.

Quantum mechanics indicates that molecules of a given compound can only have certain energies. A particular molecule absorbs only radiation that has certain characteristic frequencies (wavelengths). These frequencies correspond to energy differences between the molecule's present energy state and other permitted states of higher energy. Consider a molecule presently having energy state E_1, for which E_2 and E_3 are also possible energies. This molecule can absorb radiation whose photons have energies equal to $E_2 - E_1$ or $E_3 - E_1$ but not other radiation.

On absorbing radiation, a molecule is raised in energy to a new energy state. Each molecule has a different set of permissible energy states and hence a unique pattern of absorption of radiation.

In **absorption spectroscopy** a sample is placed in a beam of electromagnetic radiation. The frequency (wavelength) of the radiation is varied and the absorption intensity (the extent to which the sample absorbs radiation of each frequency) is determined. A plot of absorption intensity at different frequencies is called an **absorption spectrum**.

16.2
INFRARED SPECTROSCOPY

Figure 16.2 shows a typical infrared spectrum. An experienced chemist would recognize at a glance that it is the spectrum of a simple alcohol. In this section we consider what is shown in an infrared spectrum and what we can infer about the nature of the compound causing it.

The region of infrared radiation ordinarily used extends from about 2.5 to 16 microns (1 micron = 1 micrometer = 10^{-6} meter). It is more customary, however, to use wavenumbers to specify the positions of absorptions. The wavenumber of a particular radiation is the number of waves per centimeter, which is simply one divided by the wavelength (λ) in centimeters.

$$\text{wavenumber} = \frac{1}{\text{wavelength in cm}}$$

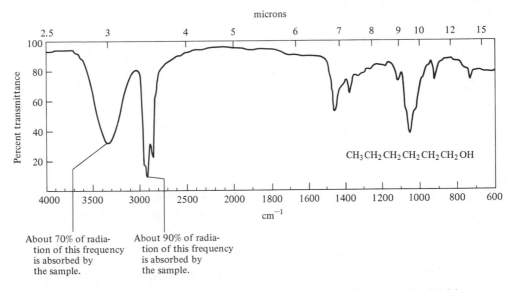

About 70% of radia-
tion of this frequency
is absorbed by
the sample.

About 90% of radia-
tion of this frequency
is absorbed by
the sample.

Figure 16.2   Infrared spectrum of 1-hexanol. [© Sadtler Research Laboratories, Division of Bio-Rad Laboratories, Inc. (1983).]

An infrared spectrum is usually presented as a plot of the percentage of radiation of each wavelength transmitted through the sample (0% transmission corresponds to total absorption by the sample).

An infrared spectrometer, an instrument whose essential components are shown in Figure 16.3, is used to obtain the spectrum. The infrared spectra shown in this chapter are of thin layers of liquid organic compounds held between two

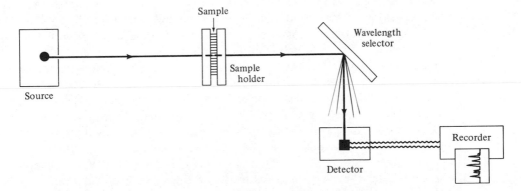

Figure 16.3   A schematic diagram of the essential elements of an infrared spectrometer. The source emits infrared radiation spanning a wide frequency range. A wavelength selector permits only a particular wavelength of the radiation transmitted by the sample to fall on a detector at a given time. The detector measures the intensity of radiation, and a plot of this intensity for each wavelength is traced on paper by a recorder.

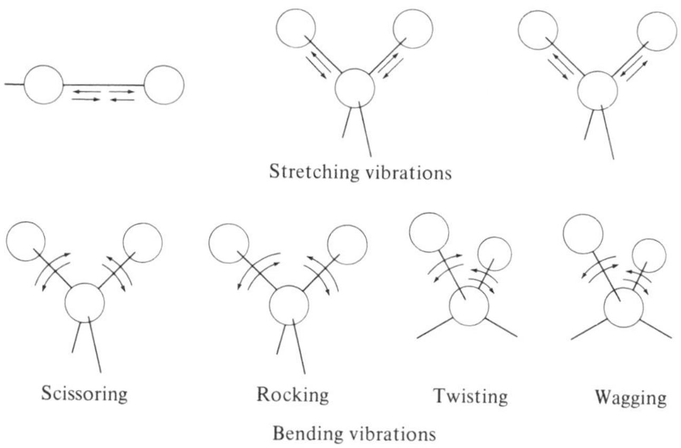

Stretching vibrations

Scissoring   Rocking   Twisting   Wagging

Bending vibrations

Figure 16.4 Some molecular vibrations involving only a few atoms.

plates of sodium chloride (sodium chloride does not absorb in this spectral region). Spectra can also be taken of gases or of compounds dissolved in a solvent. When a solvent is used, absorptions due to the solvent must be subtracted. Spectrometers ordinarily do this by measuring the difference between the absorption of radiation by two sample cells—one containing the solution of the compound, the other containing only solvent. A routine infrared spectrum takes only a few minutes and a few milligrams of sample. When necessary, much smaller amounts of sample can be used.

What happens to a molecule when it absorbs infrared radiation? The bond lengths and angles that we presented for molecules are average values. The atoms of a molecule are actually in constant motion relative to one another. Those motions that stretch bonds and deform angles between bonds are known as molecular vibrations. Figure 16.4 illustrates some of these vibrations. **Stretching** vibrations change distances between adjacent atoms. **Bending** vibrations change bond angles.

On absorption of a photon of infrared radiation, a molecule is excited to a higher energy level in which vibrational motions are more energetic. To some extent, vibrational motions involve all parts of a molecule. Some vibrations, however, are relatively localized, involving mainly two or, at most, a few atoms. As a result, some structural features lead to infrared absorptions that appear in a narrow frequency range, regardless of the structure of the remainder of the molecule. Such characteristic absorptions make infrared spectra useful in providing structural information about compounds of unknown structure.

Some typical frequency regions are shown in Figure 16.5. Particularly useful are the stretching vibrations of bonds involving the very light hydrogen atom (C—H, O—H, N—H) and of multiple bonds (C=C, C=O, C≡C, C≡N). Stretching vibrations involving single bonds between two heavy atoms (e.g., C—C, C—N, C—O) tend to be more associated with movements involving many atoms and to appear in broader frequency ranges. Moreover, these absorptions fall in regions also occupied by bending vibrations. The region from about 1400 to 600 cm^{-1} is

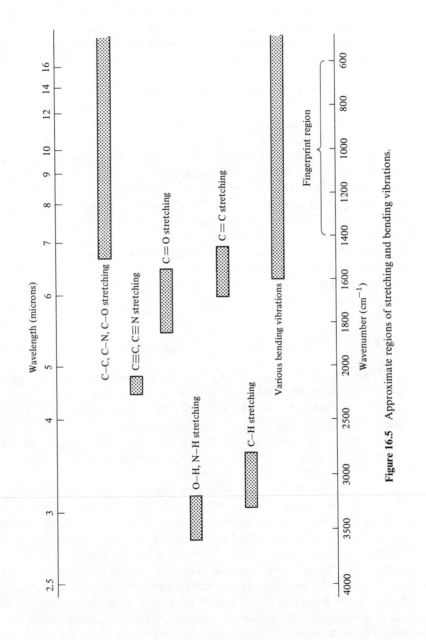

Figure 16.5 Approximate regions of stretching and bending vibrations.

called the "fingerprint" region. It is often difficult to assign all absorptions in this region to particular structural features. Yet, like the fingerprint of an individual, the absorption of an organic molecule in this region is complex and unique.

Table 16.1 lists in more detail some important stretching vibrations. With this information, we can begin to relate some of the features of spectra to the structural features responsible for them.

The first two spectra in Figure 16.6 lack significant absorptions between 2800 and 1500 cm^{-1}. Therefore multiple bonds are absent. The fairly sharp absorptions in the 2970- to 2850-cm^{-1} region are characteristic of C—H stretching vibrations of a saturated carbon. Although these spectra of isomeric alkanes have many similarities, they are unmistakably different, particularly in the fingerprint region.

Return now to the spectrum in Figure 16.2. It resembles the spectra of the alkanes but also has a strong, broad band centered at 3340 cm^{-1}. This absorption is typical of the O—H stretching vibration of a hydroxyl group. The particularly great width of this absorption is due to the presence of a number of different hydrogen-bonded species, each having a somewhat different absorption frequency.

The spectra in Figure 16.6(c) and (d) are of compounds that form 2,4-dinitrophenylhydrazone derivatives (Sec. 9.7) and therefore are probably aldehydes or ketones. Consistent with this assumption, both spectra have very strong absorptions characteristic of a carbonyl group at about 1700 cm^{-1}. The first compound can be identified readily as an aldehyde because of the absorption at 2720 cm^{-1}

$$\overset{\displaystyle O}{\overset{\displaystyle \|}{}}$$

typical of the C—H stretching of a C—H grouping. The other compound must be a ketone.

A sharp absorption at 2220 cm^{-1} in the spectrum in Figure 16.6(e) is characteristic of a triple bond. The presence of an aromatic ring is indicated by the group of absorptions between 1600 and 1450 cm^{-1} and is consistent with the absorption at 3060 cm^{-1} typical of =C—H stretching vibrations.

The very broad absorption in the 2500- to 3500-cm^{-1} region of the spectrum in Figure 16.6(f), along with the strong carbonyl absorption at 1710 cm^{-1}, is representative of a carboxylic acid.

In attempting to deduce structural information about an unknown molecule from its infrared spectrum, our approach is to look for the presence or absence of absorptions having the frequencies characteristic of certain functional groupings. Infrared spectroscopy is also used extensively to identify samples of already known compounds. Many compounds have a boiling point of about 110° or a solubility of about 2 g in 100 g of water (just as many people resemble the author in weighing about 135 pounds and being 5 feet 7 inches tall). In contrast to such single-valued properties as boiling point and solubility, however, an infrared spectrum corresponds to a multitude of properties—the absorption of radiation at each of a large number of frequencies. Identical infrared spectra are identical in a large number of features (just as are identical fingerprints or faces) and almost certainly are of the same compound.

Table 16.1 Some characteristic absorption frequencies due to stretching vibrations

Bond	Functional class		Absorption range (cm^{-1}) (intensity of absorption)[a]		Notes
C—H	alkane (—C—H)		2850–2970	(m–s)	
	alkene (C=C⟨H)		3010–3100	(m)	
	alkyne (C≡C—H)		3290–3320	(s)	
	aromatic (Ar—H)		3000–3100	(m)	
	aldehyde (C—H) with O double bond		2720–2850	(m)	sometimes two absorptions in this region
O—H	alcohol	free	3600–3640	(s)	sharp
		hydrogen bonded	3200–3600	(s)	generally broad
	carboxylic acid	hydrogen bonded	2500–3500	(m)	several broad, overlapping absorptions
N—H	amines, amides		3300–3500	(m)	
C=C	alkenes		1600–1680	(v)	
	aromatic rings		1450–1600		3 to 4 absorptions in this region
			near 1600	(v)	these absorptions vary considerably in intensity
			near 1580	(v)	
			near 1500	(v)	
			near 1450	(s)	overlaps with C—C bending vibrations
C≡C			2100–2260	(v)	
C≡N			2210–2260	(m)	
C—O	alcohols, ethers, carboxylic acids, esters		1050–1350	(s)	
C=O	aldehydes, ketones carboxylic acids, esters, amides, acyl halides		1550–1850	(s)	
C—N	amines		1020–1360	(s)	
C—Cl			800–600	(s)	
C—Br			600–500	(s)	

[a] The following abbreviations are used to indicate the intensity of absorption: s, strong; m, medium; w, weak; v, variable.

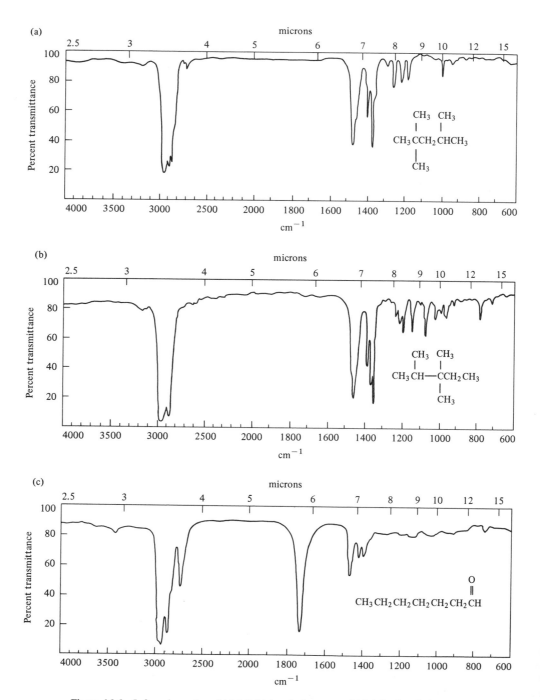

Figure 16.6 Infrared spectra of (a) 2,2,4-trimethylpentane; (b) 2,3,3-trimethylpentane; (c) heptanal; (d) 2-methylcyclohexanone; (e) benzonitrile; and (f) 2-methylbutanoic acid. [© Sadtler Research Laboratories, Division of Bio-Rad Laboratories, Inc. (1983).]

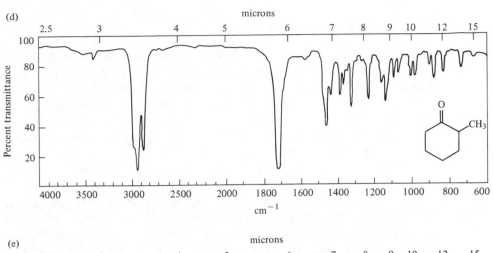

(d)

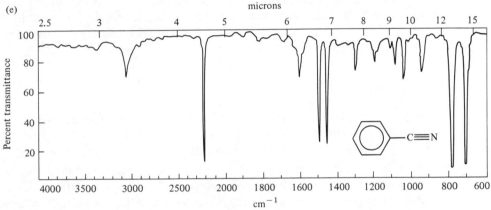

(e)

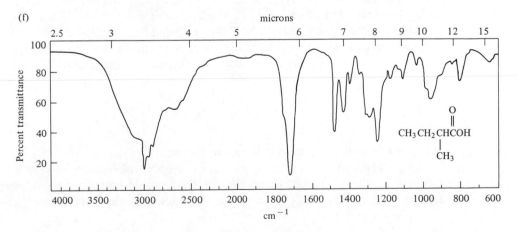

(f)

Figure 16.6 (*Continued*).

427

16.3
ULTRAVIOLET-VISIBLE SPECTROSCOPY

Figure 16.7 shows a typical ultraviolet-visible spectrum. Wavelengths in ultraviolet-visible spectra are usually given in nanometers, abbreviated nm (1 nm = 10^{-9} m). Wavelengths ordinarily used extend from 200 to 800 nm, a range that includes all visible radiation and a portion of ultraviolet radiation. Most compounds, if they absorb significantly at all in this range, absorb only at the shorter wavelength (ultraviolet) end. For such compounds, the spectrum usually shows only the region in which there is significant absorption.

The line tracing an ultraviolet-visible spectrum is generally shown moving toward the top of the paper as absorption of radiation increases. Therefore absorptions are upside down from the way they appear in infrared spectra. The extent to which radiation is absorbed is ordinarily plotted as a quantity called absorbance (A).* Absorbance is proportional to the number of molecules of the compound in the path of the beam of radiation and hence to the length of the cell holding the sample. If the compound is dissolved in a solvent, as is usually the case, then absorbance is also proportional to the concentration of the solution.

$$A = \varepsilon cl$$

where ε = extinction coefficient or molar absorbance of the compound
$\quad\quad c$ = concentration of the solution (in moles/liter)
$\quad\quad l$ = length of the sample cell (in cm)

Absorbance is also directly proportional to the quantity ε, called the extinction coefficient or molar absorbance. A compound has a characteristic value of ε at each wavelength. Depending on the compound and the wavelength, ε ranges from zero to more than 10^5. The larger the value of ε, the larger the portion of the radiation that is absorbed. The spectrum in Figure 16.7 is of a 5.3×10^{-5} M solution in a cell 1 cm in length. From these values of c and l and the value of A that can be read from the spectrum, we can calculate ε for 4-methyl-3-buten-2-one at each wavelength. At the point of greatest absorption at 237 nm, A is 0.61 and

$$\varepsilon = \frac{A}{cl} = \frac{0.61}{(5.3 \times 10^{-5}) \times 1} = 11,500$$

A scale of ε for this compound is shown in Figure 16.7.

In comparison with infrared spectra, most ultraviolet-visible spectra are relatively simple in appearance. Generally they have no more than a few absorption peaks and often only one. As a result, ultraviolet-visible spectra are frequently

---

* Absorbance = log (I_o/I), where I_o is the intensity of radiation that enters the sample and I the intensity of radiation that emerges.

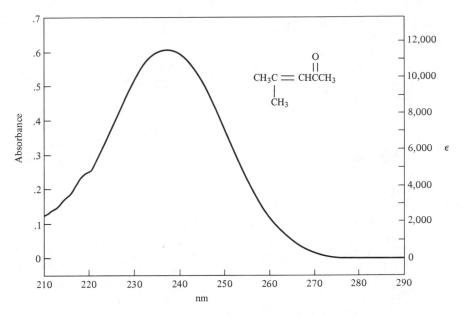

Figure 16.7 The ultraviolet spectrum of a methanol solution of 4-methyl-3-buten-2-one. [©Sadtler Research Laboratories, Division of Bio-Rad Laboratories, Inc. (1983).]

described by giving the wavelength (λ_{max}) and ε of the point of highest absorption (the "maximum") of each absorption peak. The spectrum of 4-methyl-3-buten-2-one has λ_{max} 237 nm, ε 11,500.

The essential features of an ultraviolet-visible spectrometer are similar to those of an infrared spectrometer. The solvent used to dissolve the compound must not absorb strongly in the wavelength region to be studied. Ethanol, water, and alkanes are commonly used as solvents. The sample cells are usually made of quartz; unlike glass, quartz does not absorb significantly at wavelengths longer than 200 nm.

Energies associated with ultraviolet-visible radiation are large, ranging from 35.8 kcal/mole for photons at 800 nm to 143 kcal/mole at 200 nm. These energies are comparable to those of typical covalent bonds (about 50 to 120 kcal/mole). Energies associated with infrared radiation are smaller: 2 to 11 kcal/mole of photons for the usual infrared range (600 to 4000 cm^{-1}). While absorption of infrared radiation increases the energies of internal vibrations, absorption of ultraviolet-visible radiation increases the energy of electrons. Electrons in a molecule normally occupy the orbitals of lowest energy. Absorption of ultraviolet-visible radiation excites an electron from an occupied orbital to a normally unoccupied orbital of higher energy.

Excitation of σ electrons (Sec. 4.17) to higher energy levels requires absorption of radiation having very high energies. The major absorptions of alkanes, alcohols, and ethers are all at wavelengths below (less than) 200 nm. As exemplified by 1-butene and acetone,* the major absorptions of compounds with isolated multiple bonds are also usually below 200 nm.

$$CH_3CH_2CH{=}CH_2 \qquad CH_3\overset{\overset{\displaystyle O}{\|}}{C}CH_3$$

λ_{max} 175 nm, ε 11,200 $\qquad\quad$ λ_{max} 188 nm, ε 1900

Molecules with two or more conjugated double bonds have strong absorptions at longer wavelengths. As in the following examples of polyenes, molecules absorb at progressively longer wavelengths as the number of conjugated double bonds increases.

Two conjugated C=C $\qquad$ $CH_2{=}CH{-}CH{=}CH_2$ $\qquad$ $CH_3{-}CH{=}CH{-}CH{=}CH_2$

$\qquad\qquad\qquad\qquad\qquad$ λ_{max} 217 nm, ε 21,000 $\qquad$ λ_{max} 224 nm, ε 26,000

Three conjugated C=C $\quad$ $CH_2{=}CH{-}CH{=}CH{-}CH{=}CH_2$ $\qquad$ vitamin D_3 (structure in Sec. 9.15)

$\qquad\qquad\qquad\qquad\qquad$ λ_{max} 258 nm, ε 35,000 $\qquad\qquad\qquad$ λ_{max} 265 nm, ε 20,000

Five conjugated C=C $\qquad$ vitamin A (structure in Sec. 3.21)

$\qquad\qquad\qquad\qquad\qquad$ λ_{max} 326 nm, ε 50,000

Eleven conjugated C=C $\quad$ β-carotene (structure in Sec. 3.21)

$\qquad\qquad\qquad\qquad\qquad$ λ_{max} 478 nm, ε 120,000

Although 1,4-pentadiene has two double bonds, they are not conjugated and its spectrum is similar to that of 1-butene.

$$CH_2{=}CH{-}CH_2{-}CH{=}CH_2$$

λ_{max} 178 nm, ε 17,000

Aldehydes and ketones also absorb at wavelengths that increase with the number of conjugated double bonds.

$$CH_2{=}CH{-}\overset{\overset{\displaystyle O}{\|}}{C}{-}CH_3 \qquad CH_3{-}\overset{\overset{\displaystyle CH_3}{|}}{C}{=}CH{-}\overset{\overset{\displaystyle O}{\|}}{C}{-}CH_3 \qquad CH_3{-}CH{=}CH{-}\overset{\overset{\displaystyle O}{\|}}{C}H$$

λ_{max} 212 nm, ε 7100 $\qquad$ λ_{max} 237 nm, ε 11,500 $\qquad$ λ_{max} 220 nm, ε 15,000

$$CH_3{-}CH{=}CH{-}CH{=}CH{-}\overset{\overset{\displaystyle O}{\|}}{C}H \qquad CH_3{-}CH{=}CH{-}CH{=}CH{-}CH{=}CH{-}\overset{\overset{\displaystyle O}{\|}}{C}H$$

λ_{max} 271 nm, ε 25,000 $\qquad\qquad$ λ_{max} 315 nm, ε 37,000

---

* Acetone has an absorption at 279 nm, but it is weak (ε 15). Other aldehydes and ketones have comparable absorptions.

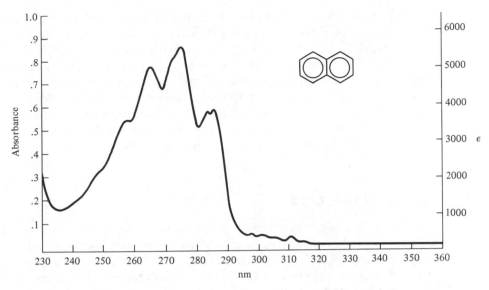

Figure 16.8   The ultraviolet spectrum of a methanol solution of naphthalene.
[© Sadtler Research Laboratories, Division of Bio-Rad Laboratories, Inc. (1983).]

Figure 16.8 shows the ultraviolet spectrum of naphthalene. Absorptions of aromatic compounds move to longer wavelengths with increasing size of the conjugated system.

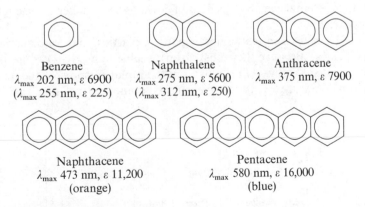

Benzene
λ_{max} 202 nm, ε 6900
(λ_{max} 255 nm, ε 225)

Naphthalene
λ_{max} 275 nm, ε 5600
(λ_{max} 312 nm, ε 250)

Anthracene
λ_{max} 375 nm, ε 7900

Naphthacene
λ_{max} 473 nm, ε 11,200
(orange)

Pentacene
λ_{max} 580 nm, ε 16,000
(blue)

Ultraviolet-visible spectroscopy is of some use in structure determination by revealing information about the nature of conjugated portions of molecules. Although differing greatly in complexity, 1,3,5-hexatriene and vitamin D_3 share the same system of three conjugated carbon-carbon double bonds and hence have similar ultraviolet absorptions. If something is known about the molecule, then details of the ultraviolet-visible spectrum may reveal finer points of structure. Alkyl substituents on a diene or an α,β-unsaturated ketone, for instance, increase the wavelength of absorption. Therefore the exact frequency of an absorption known to have one of these conjugated systems reveals the extent of substitution by alkyl groups.

Ultraviolet-visible spectroscopy is also useful as an analytical tool. If the ε of a compound at some wavelength is known, then the measured absorbance of a solution provides the concentration of that compound (assuming, of course, the absence of other compounds absorbing at that wavelength). When ε is large, concentrations of very dilute solutions can be measured. Ultraviolet-visible spectroscopy is particulary useful for analyses of clinical and biological samples, in which the quantities of important compounds are frequently small.

16.4
COLORED MOLECULES; DYES

The human eye responds to radiation from about 375 to 750 nm, which is therefore called visible. As indicated in Figure 16.9, different wavelengths of visible radiation give sensations of different colors. Light—such as that from the sun or a light bulb—that spans the range of visible frequencies appears white to us. Compounds that absorb certain frequencies of white light appear colored. For example, β-carotene (which helps impart the color to carrots) and naphthacene are orange and pentacene is blue. We see these colors because the light reflected from these compounds is deficient in the absorbed frequencies. Figure 16.9 shows the apparent or complementary colors of objects that absorb light in particular wavelength regions. β-Carotene and naphthacene absorb light at the blue end of the visible range. Light reflected from these compounds is deficient in blue light and appears orange. Pentacene absorbs light of considerably longer wavelengths and, as a result, appears blue.

Dyes used in textile fibers, papers, plastics, and foodstuffs are usually organic molecules. Because they must absorb visible light, dyes ordinarily possess extended conjugated systems. Earlier we saw examples of some of the major classes of dyes. Alizarin (Sec. 5.20) is a member of the "quinone dyes" and Tyrian purple (Sec. 7.10)

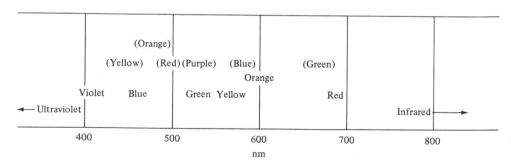

Figure 16.9   Colors associated with different wavelengths in the visible region of electromagnetic radiation. The complementary color (in parentheses) is the apparent color of an object that absorbs light in a particular wavelength region.

of the "indigoid dyes." Many dyes belong to the family of "azo dyes" (Sec. 8.7). Malachite green is an example of the "triphenylmethane dyes," which are derivatives of the triphenylmethyl cation.

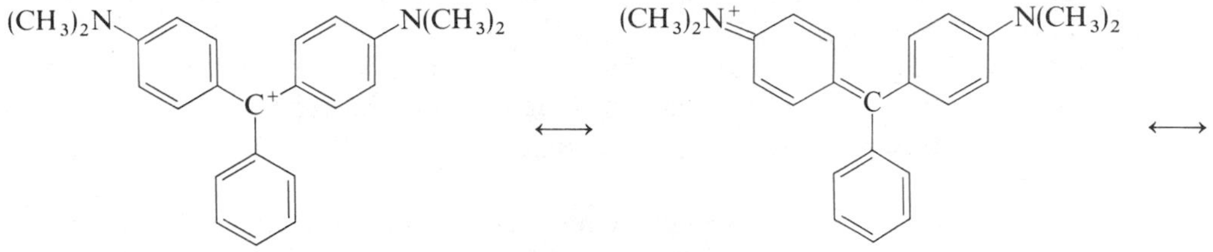

Malachite green

Many resonance structures can be drawn for this particularly stable carbocation.

16.5
CLINICAL CHEMISTRY

Biological molecules for which testing is important are often present in exceedingly small amounts. As a result, many analytical procedures in clinical and biochemical laboratories must be sensitive. Such procedures often involve determining the ultraviolet-visible absorption of a suitable species. Because biological samples, such as a specimen of blood serum, generally contain numerous components, an analytical procedure must also be highly specific for a particular component.

A good example of a modern clinical chemistry procedure is the "GOT" test. This test determines the concentration in blood serum of the enzyme aspartate aminotransferase, also known as glutamic oxaloacetic transaminase (GOT). This enzyme catalyzes the following reversible transformation.*

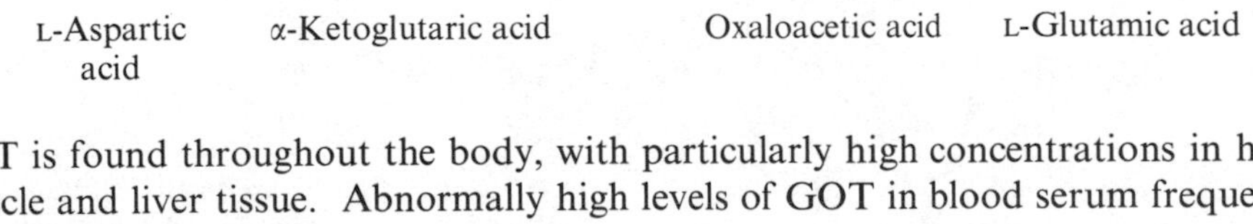

L-Aspartic acid   α-Ketoglutaric acid   Oxaloacetic acid   L-Glutamic acid

GOT is found throughout the body, with particularly high concentrations in heart muscle and liver tissue. Abnormally high levels of GOT in blood serum frequently

---

* This transamination reaction occurs by the process involving pyridoxamine phosphate and pyridoxal phosphate shown in an example in Section 9.7.

indicate damage to heart or liver cells that has resulted in leakage of their enzymes into the blood. The GOT level is high for example, in the blood serum of a person having infectious hepatitis or cirrhosis of the liver. Following blockage of a coronary artery (myocardial infarction), the GOT level in serum rises due to the destruction of some heart cells.

In analytical procedures for GOT a sample of blood serum is added to a solution containing aspartic and α-ketoglutaric acids. The rate of their conversion to oxaloacetic and glutamic acids is proportional to the GOT concentration in the serum.

In one widely used test procedure, formation of oxaloacetic acid is monitored by a second reaction that is conveniently followed by ultraviolet spectroscopy. In this procedure the solution to which the blood serum sample is added also contains NADH and the enzyme malate dehydrogenase. Malate dehydrogenase catalyzes the reversible reduction of oxaloacetic acid by NADH to yield L-malic acid and NAD$^+$ (see Sec. 6.9).

$$
\begin{array}{c}
\mathrm{CO_2H} \\
|\\
\mathrm{C{=}O} \\
|\\
\mathrm{CH_2} \\
|\\
\mathrm{CO_2H}
\end{array}
\;+\; \mathrm{NADH + H^+}
\;\underset{\xrightarrow{\text{malate dehydrogenase}}}{\rightleftharpoons}\;
\begin{array}{c}
\mathrm{CO_2H} \\
|\\
\mathrm{HO{-}C{-}H} \\
|\\
\mathrm{CH_2} \\
|\\
\mathrm{CO_2H}
\end{array}
\;+\; \mathrm{NAD^+}
$$

Oxaloacetic acid   L-Malic acid

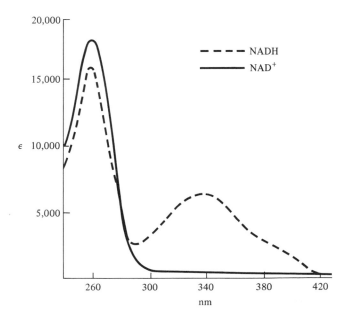

Figure 16.10 Ultraviolet spectra of NADH and NAD$^+$. Partial structures of these compounds appear in Section 9.9 and a full structure of NADH is shown in Section 10.7.

As shown in Figure 16.10, NADH has a strong absorption at 340 nm, but NAD^+ absorbs only weakly at that wavelength. Disappearance of absorption at 340 nm occurs at the same rate as reduction of oxaloacetic acid. Sufficient malic dehydrogenase and NADH are used to ensure that oxaloacetic acid is reduced essentially as rapidly as it forms. Therefore the easily measured rate of NADH disappearance is related to the concentration of GOT. Because of the high catalytic activity of GOT, the amounts of oxaloacetic acid produced and NADH consumed are much larger than the amount of GOT present in the serum. A concentration of NADH large enough to be easily measured can be used. Moreover, the absorption at 340 nm in the test solution is due mainly to the relatively large amount of NADH rather than to various components in the serum sample that also absorb at that wavelength.

Sophisticated tests, such as the GOT test, are carried out in very large numbers in clinical laboratories. Instruments that automatically perform all operations needed for such tests and provide a printed report of the results are now commonplace.

16.6
PHOTOCHEMISTRY

Photochemistry refers to chemistry in which absorption of ultraviolet-visible radiation is involved. The absorption of this radiation by a molecule excites an electron and generates a new, higher-energy chemical species. This electronically **excited** molecule often has very different chemical properties and greater reactivity than the original ground-state molecule.

Some excited alkenes react with another molecule of alkene to form products containing cyclobutane rings.

$$\| {}^{*} + \| \longrightarrow \square \qquad (\text{*indicates an excited molecule})$$

Dimerization of thymine rings (Sec. 15.7), the most common damage caused to DNA by exposure to ultraviolet radiation, is an important example. In an electronically excited alkene the double bond is weaker and cis-trans isomerization is more facile. The isomeric 1,2-diphenylethylenes, for example, are interconverted by irradiation with ultraviolet light.

The isomerization of 7-dehydrocholesterol to vitamin D_3 (Sec. 9.15) caused by ultraviolet light is a reaction of an electronically excited cyclic diene.

Electronically excited molecules are involved in important biological processes, including photosynthesis and vision, and in commercial photographic and photoduplication processes.

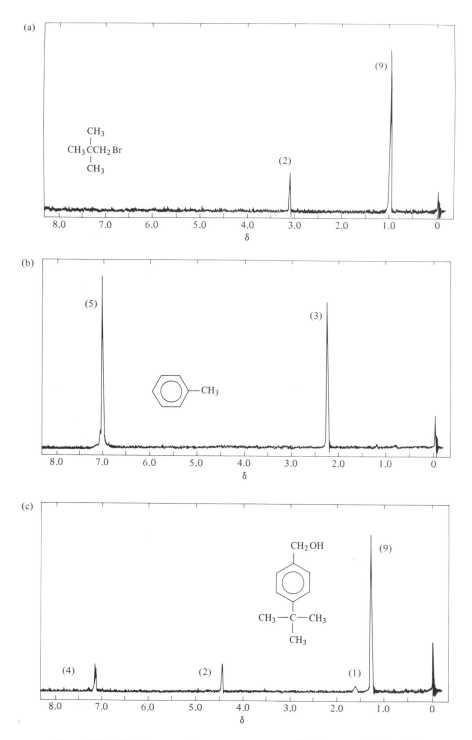

Figure 16.11 Nuclear magnetic resonance spectra of (a) 1-bromo-2,2-dimethyl-propane; (b) toluene; and (c) *p-tert*-butylbenzyl alcohol. Relative areas of absorptions are indicated by the numbers in parentheses. [© Sadtler Research Laboratories, Division of Bio-Rad Laboratories, Inc. (1983).]

Some simple nuclear magnetic resonance (nmr) spectra are shown in Figure 16.11. Absorptions are recorded as in ultraviolet spectroscopy, the line moving to the top of the paper with increasing absorption of electromagnetic radiation. The horizontal frequency scale is an arbitrary one, to be explained later. The absorption that appears at 0 on this scale in each spectrum is due to tetramethylsilane, $(CH_3)_4Si$, added as a standard.

By examining a large number of such spectra, you would soon decipher the following relationships between molecules and their spectra:

1. Absorptions are due to the hydrogens.

2. Hydrogens in different molecular environments absorb at different positions that are characteristic of each environment.

3. The intensity of an absorption (the area under the absorption peak) is proportional to the number of hydrogens responsible for the absorption.

These generalizations can be illustrated with the three spectra in Figure 16.11. The spectrum of 1-bromo-2,2-dimethylpropane has two absorptions with relative areas of 2:9, corresponding to the two methylene and nine *tert*-butyl hydrogens. Toluene has two absorptions with relative areas of 5:3, corresponding to the aryl and methyl hydrogens. Finally, *p-tert*-butylbenzyl alcohol has four absorptions in the ratio 4:2:1:9, corresponding to the aryl, methylene, hydroxyl, and *tert*-butyl hydrogens. Note that the aryl hydrogens of the two aromatic compounds absorb at about 7 on the scale and the hydrogens of the two *tert*-butyl groups at approximately 1.

In a nuclear magnetic resonance spectrometer (Fig. 16.12), the compound is placed between the poles of a strong magnet. Some atomic nuclei have a property called a nuclear spin that makes them behave as though they are tiny bar magnets. In a strong magnetic field these nuclei are aligned in certain ways relative to the field. A proton, the nucleus of the abundant ^1H isotope of hydrogen, has a nuclear spin. In a strong magnetic field a proton has only two possible orientations: in the direction of the field or opposed to the field. The difference in energy between protons having these orientations is proportional to the magnetic field strength.

$$E = \text{constant} \times H$$

where E = the energy difference between the spin orientations of the nucleus
the constant is characteristic of the particular atom (^1H in our examples)
H = the strength of the magnetic field

As shown in Figure 16.13, a proton can be shifted to the higher energy orientation by absorption of radiation. All protons have equal probabilities of absorbing, so the size of an absorption band is related to the number of protons giving rise to it. Nmr spectrometers generally have magnetic field strengths of 14,092 or 23,487 gauss. In comparison, the earth's magnetic field is only about 0.3 to 0.7 gauss. These strong magnetic fields lead to absorptions for protons at 60 or 100 megahertz (MHz, million cycles per second), frequencies in the radio range of electromagnetic radiation.

Spectra are ordinarily taken of compounds in solution, although they can be of pure liquids. Commonly used solvents include those like CCl_4 that have no hydrogens to give interfering absorptions. Because deuterium (^{2}H or D) absorbs in a

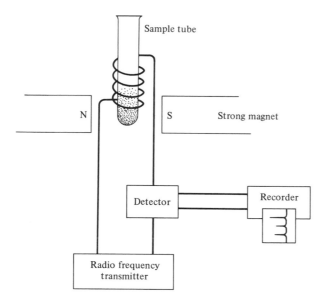

Figure 16.12 The essential features of a simple nmr spectrometer. The sample is placed between the poles of a powerful magnet and is surrounded by a coil connected to a radio frequency transmitter. In effect, the spectrometer measures and records the amount of radio frequency radiation absorbed by the sample as the magnetic field strength is altered.

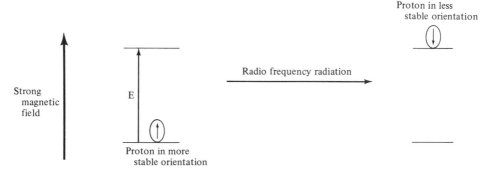

Figure 16.13 Protons in a magnetic field have two possible orientations relative to the field. A proton having the more stable orientation in the direction of the field is changed to the orientation opposed to the field by absorption of radiation having energy E.

very different region than hydrogen, such solvents as $CDCl_3$, in which all H's have been replaced by D's, are also widely used. A few tenths of a milliliter of a 5 to 10% solution suffices for routine spectra. With special techniques and spectrometers, much smaller amounts can be used.

16.8
NUCLEAR MAGNETIC RESONANCE SPECTROSCOPY: CHEMICAL SHIFTS

From the discussion so far it would seem that, at a given radio frequency, all hydrogens would absorb at the same magnetic field strength. We saw from the sample spectra, however, that hydrogens in different environments have different absorptions. Differences arise because electron motion in a molecule produces small, localized magnetic fields (H_{local}). The strength of H_{local} differs from one region of a molecule to another, depending on the electron environment. Therefore the actual or effective strength (H_{eff}) of the magnetic field at a hydrogen is the sum of the large magnetic field supplied by the spectrometer (H_0) and H_{local}.

$$H_{eff} = H_0 + H_{local}$$

At a given radio frequency, all hydrogens absorb at the same H_{eff}. Because H_{local} varies, however, hydrogens in different environments in a compound absorb at slightly different values of H_0. In an nmr spectrometer the magnetic field (H_0) is changed slightly to give the absorption of each group of hydrogens.

The alterations in H_0 required to give absorptions of different protons are called **chemical shifts**. Chemical shifts are small. For a spectrometer operating at a frequency of 60 MHz, most shifts fall within a range of only 0.15 gauss at about 14,092 gauss. The actual values of H_0 are proportional to the frequency at which the spectrometer operates and so are different for a spectrometer operating at 100 MHz. Therefore it is customary to express chemical shifts as δ values, which are the same for all spectrometer frequencies. These values describe the positions of absorptions in parts per million (ppm) of the magnetic field away from a particular absorption chosen as a standard.

$$\delta = \frac{H_{reference} - H_{compound}}{H_{reference}} \times 10^6$$

where $H_{compound} = H_0$ at which an absorption of the compound appears
$\ H_{reference} = H_0$ at which the reference substance absorbs

The factor of 10^6 places most absorptions in the convenient range of 0 to 10 δ. Tetramethylsilane, the usual reference compound ($\delta = 0$), has only a single absorption that comes at one boundary of this range. Moreover, it is soluble in most organic solvents and chemically inert to most compounds.

How different must the environments of hydrogens be for them to give different absorptions? Hydrogens that are not fully equivalent usually absorb in at least slightly different positions. Deciding if hydrogens are equivalent is actually a structural and stereochemical problem that we encountered much earlier. One way of determining if several hydrogens are equivalent is to consider replacing each by a chlorine. Hydrogens whose replacement would give different products are not equivalent and generally have different nmr absorptions. The number of absorptions expected (and observed) for some simple compounds are listed below.

One absorption

$$CH_3CH_3 \quad CH_3Br \quad CH_3\overset{\overset{\displaystyle CH_3}{|}}{\underset{\underset{\displaystyle CH_3}{|}}{C}}CH_3 \quad CH_3OCH_3 \quad CH_3\overset{\overset{\displaystyle O}{||}}{C}CH_3 \quad CH_2{=}CH_2$$

Two absorptions (hydrogens indicated by the same letter contribute to the same absorption)

$$\overset{a \quad b}{CH_3CH_2Br} \qquad \overset{a \quad b \quad b \quad a}{CH_3CH_2CH_2CH_3} \qquad \overset{a \quad b}{CH_3\underset{\underset{\displaystyle a}{|}}{\underset{\displaystyle CH_3}{CH}}}{-}O{-}\overset{b \quad a}{\underset{\underset{\displaystyle a}{|}}{\underset{\displaystyle CH_3}{CHCH_3}}} \qquad \overset{a \quad b}{CH_3\underset{\underset{\displaystyle a}{|}}{\underset{\displaystyle CH_3}{C}}{=}CH_2}$$

Three absorptions

$$\overset{a \quad b \quad c}{CH_3CH_2CH_2Br} \qquad \overset{a \quad b \quad c \quad b \quad a}{CH_3CH_2CH_2CH_2CH_3} \qquad \overset{a \quad b \quad c}{CH_3CH_2OH} \qquad \begin{array}{cc} \overset{a}{H} & \overset{}{Br} \\ & \diagdown C{=}C \diagup \\ \underset{b}{H} & \underset{c}{H} \end{array}$$

Note that even hydrogens that differ stereochemically give different absorptions. Occasionally absorptions are so similar in position that they appear as a single peak. Absorptions of the hydrogens ortho, meta, and para to the methyl group in toluene are an example.

Table 16.2 lists chemical shifts of hydrogens in some commonly encountered structural situations. Chemicals shifts of hydrogens bonded to oxygen and nitrogen are particularly variable in position, for they depend on the extent of hydrogen bonding, which is related to the structure of the compound, its concentration, and the solvent in which it is dissolved.

Three generalizations are useful in estimating absorption positions.

1. Values for absorptions of —CH_3 hydrogens can be used to estimate values for —CH_2R and —CHR_2 hydrogens in similar environments. Absorptions of —CH_2R hydrogens are usually 0.2 to 0.5 δ and of —CHR_2 hydrogens 0.6 to 1.0 δ greater than of —CH_3 hydrogens. Note, for example, the first three entries in Table 16.2 listing absorptions of such hydrogens in

Table 16.2 Characteristic chemical shifts of hydrogens

Location of hydrogen	Chemical shift (δ)	Notes
$R\text{—}CH_3$	0.8–1.0 ⎫	
$R\text{—}CH_2\text{—}R$	1.2–1.4 ⎬ alkanes (R = saturated alkyl)	
$R\text{—}CH\text{—}R$ with R	1.5–1.7 ⎭	
$\begin{array}{c}\diagdown\\ \diagup\end{array}C{=}C\begin{array}{c}\diagup H\\ \diagdown H\end{array}$	4.6–5.0	
$\begin{array}{c}\diagdown\\ \diagup\end{array}C{=}C\begin{array}{c}\diagup H\\ \diagdown R\end{array}$	5.1–5.8	(R = saturated alkyl)
$\text{—}C{\equiv}C\text{—}H$	1.7–3.3	
$Ar\text{—}H$	6.5–8.0	
$\begin{array}{c}\diagdown\\ \diagup\end{array}C{=}C\begin{array}{c}\diagup CH_3\\ \diagdown R\end{array}$	1.6–1.8	(R = H or saturated alkyl)
$Ar\text{—}CH_3$	2.1–2.5	
$\underset{\text{—C—}CH_3}{\overset{O}{\|\|}}$	1.9–2.6	aldehydes, ketones, esters, acetic acid
$\diagdown N\text{—}CH_3$	2.2–2.9	amines, amides
$\text{—}O\text{—}CH_3$	3.3–3.9	methanol, ethers, esters
$Cl\text{—}CH_3$	3.1	
$Br\text{—}CH_3$	2.7	
$\underset{\text{—CH}}{\overset{O}{\|\|}}$	9.2–10.2	aldehydes
ROH	0.5–5.5	alcohols
$ArOH$	4–8	phenols
$\underset{\text{—COH}}{\overset{O}{\|\|}}$	10–13	carboxylic acids
$\diagdown NH$	0.5–5.0	amines

In similar environments, —CH_2R absorbs at values approximately 0.2 to 0.5 greater and —CHR_2 at values 0.6 to 1.0 greater.

alkanes or the following values for alkyl chlorides

$$CH_3Cl \qquad RCH_2Cl \qquad R_2CHCl$$

$$3.1\ \delta \qquad \text{about } 3.5\ \delta \qquad \text{about } 4.1\ \delta$$

2. Even a substituent one or two carbons away exerts some effect on a hydrogen absorption. The absorptions of the methyl groups of ethyl chloride and propyl chloride, for instance, have slightly greater chemical shifts than do absorptions of methyl groups of alkanes (0.8 to 1.0 δ).

$$CH_3Cl \qquad CH_3CH_2Cl \qquad CH_3CH_2CH_2Cl$$

$$3.1\ \delta \qquad 1.3\ \delta \qquad 1.1\ \delta$$

3. The effects of substituents are roughly additive. In the following examples the —CH_2—absorptions are shifted considerably more than they would be by either substituent group alone.

$$ClCH_2Cl \qquad CH_2{=}CHCH_2Cl \qquad \text{⬡—}CH_2Cl$$

$$5.3\ \delta \qquad\qquad 4.0\ \delta \qquad\qquad 4.5\ \delta$$

16.9
NUCLEAR MAGNETIC RESONANCE SPECTROSCOPY:
SPIN-SPIN SPLITTINGS

Each compound whose spectrum is shown in Figure 16.14 has hydrogens in two environments and might be expected to give two absorptions. Each spectrum has absorptions in the expected regions, but the absorptions are sets of peaks or **multiplets**. The spectrum of ethyl bromide, for example, has four peaks centered at 3.3 δ where the —CH_2— is expected to absorb and three peaks at 1.7 δ where the —CH_3 is expected to absorb. Such multiplets are not unusual. In fact, they are far more common than single absorption peaks.

The splitting of the absorption of a group of hydrogens into a multiplet is due to the spins of nearby protons, usually those on adjoining carbons, and is called **spin-spin splitting**. In simple cases, the number of peaks in a multiplet is one more than the number of hydrogens on adjoining carbons. The spectra in Figure 16.14 are consistent with this simple rule.

CH_3—CH_2—Br		Cl—$\overset{\underset{\displaystyle\vert}{Cl}}{CH}$—$CH_3$		Br—$\overset{\underset{\displaystyle\vert}{Br}}{CH}$—$CH_2$—Br	
3 peaks	4 peaks	4 peaks	2 peaks	3 peaks	2 peaks
(triplet)	(quartet)	(quartet)	(doublet)	(triplet)	(doublet)

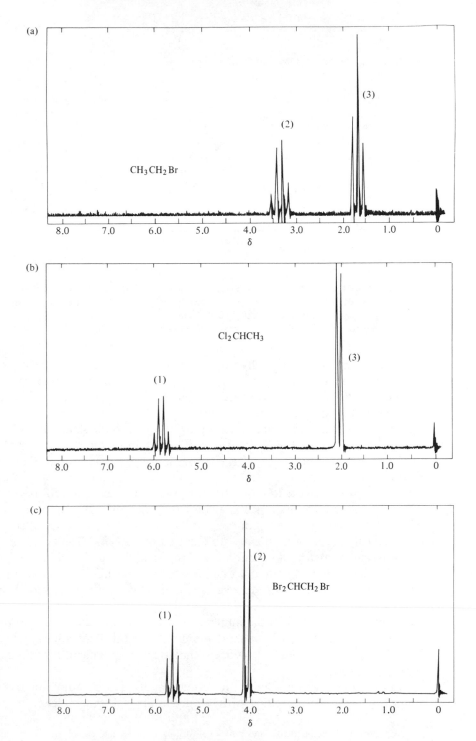

Figure 16.14 Nuclear magnetic resonance spectra of (a) ethyl bromide; (b) 1,1-dichloroethane; and (c) 1,1,2-tribromoethane. Relative areas of absorptions are indicated by the numbers in parentheses. [© Sadtler Research Laboratories, Division of Bio-Rad Laboratories, Inc. (1983).]

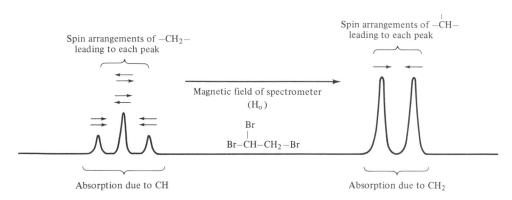

Figure 16.15 The nmr spectrum of 1,1,2-tribromoethane. Each peak in the CH_2 doublet is due to molecules in which the adjoining CH hydrogen has one of the two possible spin orientations. The peaks in the CH triplet arise from molecules having different spin arrangements for the two CH_2 hydrogens.

The multiplet due to the CH_3 hydrogens of ethyl bromide, for instance, has three peaks, one more than the number of hydrogens on the adjoining carbon. The multiplet due to the CH_2 hydrogens has four peaks, one more than the number of hydrogens of the adjoining methyl group.

Splitting of an absorption into a multiplet is due to the small magnetic fields of the nuclei of nearby hydrogens. Consider the CH_2 absorption of 1,1,2-tribromoethane. The magnetic field at the CH_2 hydrogens is slightly increased in molecules in which the nuclear spin of the tertiary hydrogen on the adjoining carbon is aligned with the spectrometer's field (H_0). It is decreased, however, in molecules in which the spin of the tertiary hydrogen is against the field. About 50% of the molecules have each of these spin arrangements. As shown in Figure 16.15, absorptions of equal intensity are seen at slightly different values of H_0. The spins of the CH_2 hydrogens can be arranged in four ways. However, the two combinations in which one hydrogen is aligned with and one against the field have the same effect. As a result, the CH_2 hydrogens give three absorptions, the middle absorption being twice as strong as the others.

Table 16.3 summarizes the numbers and relative areas of peaks in multiplets that arise from splitting by different numbers of equivalent adjacent hydrogens. When absorptions are split by hydrogens that are not equivalent, then the multiplets are more complex.

Splitting is generally significant between hydrogens on adjoining carbons but unimportant between more distant hydrogens.* It is also significant between two

---

* Hydrogens bonded to oxygen or nitrogen ordinarily exchange rapidly from one molecule to another. In a brief period of time, such hydrogens are in molecules having all possible spin combinations of any hydrogens on the adjoining carbons. As a result, OH and NH hydrogens usually give unsplit absorptions (that represent the average of all these environments) and do not split the absorptions of hydrogens on adjoining carbons.

Table 16.3   Number and relative areas of peaks in multiplets due to splitting by different numbers of equivalent hydrogens[a]

Number of hydrogens causing the splitting	Number of peaks	Relative areas of peaks
1	2 (doublet)	1:1
2	3 (triplet)	1:2:1
3	4 (quartet)	1:3:3:1
4	5 (quintet)	1:4:6:4:1
5	6 (sextet)	1:5:10:10:5:1
6	7 (septet)	1:6:15:20:15:6:1

[a] The information here is accurate only when the chemical shift difference between the absorptions of the interacting hydrogens is large. When it is small, the relative areas will be different and more peaks will sometimes be observed.

hydrogens on the same carbon *if* they have different chemical shifts. Splitting is not observed for hydrogens that have the same chemical shift, however. Because all hydrogens of $ClCH_2CH_2Cl$, benzene, or cyclopentane are equivalent, only single absorptions are seen for these molecules. Splitting of the absorptions of ortho, meta, and para hydrogens of toluene is not observed, for these absorptions have almost identical chemical shifts.

16.10
NUCLEAR MAGNETIC RESONANCE SPECTROSCOPY: APPLICATIONS

Nuclear magnetic resonance spectra give a wealth of information useful for identifying molecules. The number of absorptions is related to the number of hydrogen environments in a molecule. Areas of absorptions indicate the relative numbers of hydrogens in each environment. Chemical shifts give information about the nature of each hydrogen environment. Splitting into multiplets is related to the number of nearby hydrogens.

The preceding nmr spectra are due to ^1H nuclei (protons). Many nuclei, including ^{12}C and ^{16}O which are so abundant in organic compounds, have no nuclear spins. Some other nuclei, however, have spins and give nmr absorptions, although at very different radio frequencies (or magnetic field strengths) than protons. Some spectrometers are now sufficiently sensitive to make it routine to obtain spectra due to ^{13}C nuclei, whose normal abundance in C is only 1.1%. The other nuclei most used for spectra of organic compounds are ^{19}F and ^{31}P, the only naturally occurring isotopes of F and P.

In mass spectrometry a compound is bombarded with rapidly moving electrons. Collision of an energetic electron with a molecule often knocks off a valence electron to form a positive ion called the molecular ion.

At least some of the molecular ions fragment to smaller species, many of which are also positively charged. A plot of the relative amounts of positive ions of different masses formed from a compound on electron impact is called a **mass spectrum**. Unlike the other instrumental methods considered previously, mass spectrometry does not involve interaction of a compound with electromagnetic radiation.

In the instrument called a mass spectrometer a small amount of a compound is vaporized and bombarded with an electron beam. By application of electric and magnetic fields, the resulting positive ions are sorted according to mass.* The abundance of ions of each mass is measured by a detector and recorded. As shown in Figure 16.16, mass spectra are often presented as bar graphs.

Every compound has a typical pattern of fragment ions. Note the significant differences between the spectra of isomeric alkanes in Figure 16.16. Matching the mass spectrum of an unknown sample with that of an already known compound is an important way to identify an organic compound. Mass spectrometry is particularly useful for this purpose because it requires only very small samples. As little as 10^{-8} to 10^{-11} g of a compound suffices with sensitive mass spectrometers. Because of its sensitivity, mass spectrometry is used extensively for identifying environmental contaminants or biological compounds whose effects may be important even though they are present in very small amounts.

Mass spectra also give information about the structures of unknown compounds. Fragmentation depends on structure in ways that are partly predictable. Fragmentation, for example, tends to produce the more stable of several possible carbocations. Note that two prominent peaks in the mass spectrum of 2-methylpentane correspond to secondary carbocations that can be formed by cleavage of carbon-carbon bonds.

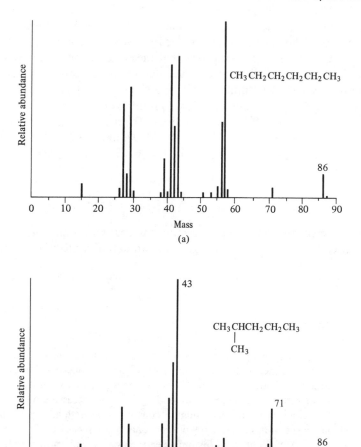

Figure 16.16   Mass spectra of (a) hexane and (b) 2-methylpentane.

Cleavage of carbon-carbon bonds of hexane can produce only primary ions and less selectivity is observed in its fragmentation.

The peak of greatest mass in a mass spectrum generally is due to the molecular ion and thus has the molecular weight of the compound. The observation that the peak of greatest mass in each of the mass spectra in Figure 16.16 is 86 suggests that

---

* A mass spectrometer sorts ions according to the ratio of their mass to the number of charges. Most ions have only a single charge, however, and hence are separated strictly according to mass.

the compounds have that molecular weight.* Because they undergo complete fragmentation, however, the molecular ions of some compounds are not observed.

Using mass spectrometers that measure masses with great precision, not only
molecular weights but also molecular formulas can be determined. Compounds
with a variety of formulas, including C_6H_{14}, $C_5H_{10}O$, and $C_4H_{10}N_2$, have molecular weights of 86. The atomic weight of each isotope (except ^{12}C, which is defined
as exactly 12.00000), however, differs from an integral value by a small amount.
For example, ^1H is 1.00078, ^{14}N 14.00307, and ^{16}O 15.99492. Therefore the molecular weights corresponding to the following formulas are not exactly 86.

$$
\begin{array}{ll}
C_6H_{14} & 86.10948 \\
C_5H_{10}O & 86.07311 \\
C_4H_{10}N_2 & 86.08434
\end{array}
$$

Mass spectrometers that determine masses with sufficient accuracy to distinguish
between such values permit assigning an actual molecular formula.

SUMMARY

Definitions and Ideas

Electromagnetic radiation behaves in some respects as a wave. It can be defined in
terms of either its frequency or its wavelength, properties that are inversely proportional.
Frequency is the number of waves passing a given point in a certain time, and **wavelength**
is the distance between corresponding parts of successive waves. Electromagnetic radiation can also behave as though it consists of discrete indivisible packets of energy called
photons that have an energy proportional to frequency.

A molecule absorbs only electromagnetic radiation having frequencies that correspond in energy to differences between the molecule's present energy state and permitted
higher energy states. A plot of the extent to which radiation is absorbed (absorption
intensity) as the frequency of radiation is changed is called an **absorption spectrum**.

Absorption of infrared radiation excites molecules to higher energy levels in which
vibrational motions are more energetic. Some vibrations are relatively localized, involving
mainly two or, at most, a few atoms. As a result, some structural features lead to infrared
absorptions that appear in a narrow frequency range, regardless of the structure of the
remainder of the molecule. Stretching vibrations of bonds involving a hydrogen (C—H,
O—H, N—H) and of multiple bonds are particularly important. Positions in infrared
spectra are often expressed in **wavenumbers** (1/wavelength in cm).

Absorption of ultraviolet-visible radiation by a molecule excites an electron from
an occupied orbital to a normally unoccupied orbital of higher energy. Ultraviolet-visible
absorptions in the wavelength region usually studied are found chiefly for compounds
with conjugated multiple bonds. Ultraviolet-visible spectra are often described by giving

---

* Note that a mass spectrometer sorts ions according to the mass of each individual ion, not according to the average mass of many. Most hexane molecules have mass 86, but due to the presence
of isotopes other than ^{12}C and ^1H, some have other masses. For example, about 7% contain one
^{13}C ($^{12}C_5{}^{13}CH_{14}$) and have a mass of 87. The mass spectrum of hexane shows an absorption at 87 that
is about 7% as large as that at 86.

the wavelength (λ_{max}) and ε (extinction coefficient or molar absorbance) of the point of highest absorption (the "maximum") of each absorption peak.

Nuclear magnetic resonance (nmr) spectra record the absorption of radio frequency electromagnetic radiation by a molecule placed in a strong magnetic field as the strength of that field is varied. The spectra are due to nuclei, usually of ^{1}H, that have a property called nuclear spin and behave like tiny bar magnets. In a strong magnetic field a proton (the nucleus of ^{1}H) can be aligned either in the direction of or opposed to the field. Absorption of the radio frequency radiation changes a proton from the more stable orientation in the direction of the field to the less stable orientation opposed to it.

Hydrogens that are not fully equivalent generally absorb at slightly different spectrometer field strengths typical of their molecular environments. The alterations in spectrometer magnetic field strength required to give absorptions of different protons are called **chemical shifts**. Chemical shifts occur because the spectrometer magnetic field is supplemented within a molecule by small magnetic fields generated by the motion of electrons. Chemical shifts are small and often expressed as δ **values**, parts per million of the magnetic field strength from the absorption of a standard (usually tetramethylsilane). Intensities of absorptions are proportional to the numbers of hydrogens responsible for them.

Due to **spin-spin splitting**, an nmr absorption often appears as a group of peaks known as a **multiplet**. The splitting of an absorption into a multiplet is due to the spins of nearby protons, usually those on adjoining carbons. In simple cases, the number of peaks in a multiplet is one more than the number of equivalent hydrogens responsible for the splitting. Hydrogens having the same chemical shift do not split each other's absorptions.

Collision of a compound with a rapidly moving electron often knocks off a valence electron to form a positive ion called a **molecular ion**. At least some molecular ions fragment to smaller species, many of which are also positively charged. A **mass spectrum** is a plot of the relative abundance of positive ions of different masses. Every compound has a characteristic pattern of fragment ions. Some mass spectrometers measure mass so accurately that they can determine molecular formulas.

PROBLEMS

1. How can the members of each pair of compounds be distinguished by a glance at their infrared spectra?

   (a) $CH_3CH_2OCH_2CH_3$ and $CH_3CH_2CH_2CH_2OH$

   (b) and $CH_3CH_2CH_2CH=CH_2$

   (c) and

   (d) $(CH_3)_3N$ and $CH_3CH_2CH_2NH_2$

   (e) $CH_3CH_2CH_2\overset{\overset{\displaystyle O}{\|}}{C}OH$ and $CH_3CH_2\overset{\overset{\displaystyle O}{\|}}{C}OCH_3$

   (f) $CH_3CH_2C{\equiv}CH$ and $CH_3CH_2CH=CH_2$

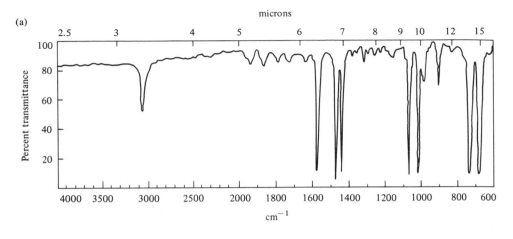

(a)

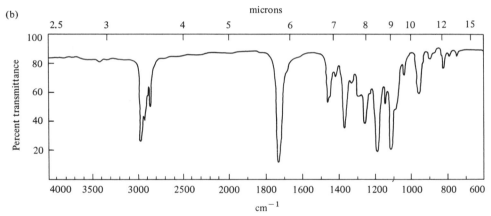

(b)

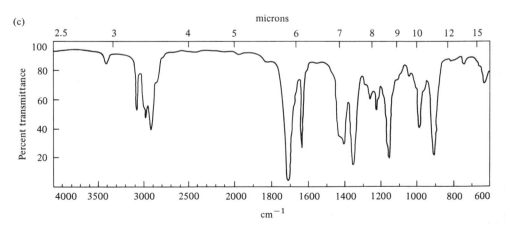

(c)

Figure 16.17 Infrared spectra to accompany problem 2. [© Sadtler Research Laboratories, Division of Bio-Rad Laboratories, Inc. (1983).]

2. From each group of compounds, select the one that has the indicated infrared spectrum. Assign as many of the absorptions as you can to specific structural features of the molecule.

(a) Figure 16.17(a)

$$(CH_3)_2CHNH_2, \quad CH_3CO_2H, \quad CH_3CH_2OCH_2CH_3, \quad CH_3CHCH_2CH_3,$$
$$\overset{|}{CH_3}$$

[benzene ring]—OH,   [benzene ring]—Br,   [benzene ring]—$\overset{\overset{\displaystyle O}{\|}}{C}CH_3$,   [cyclopentane ring]

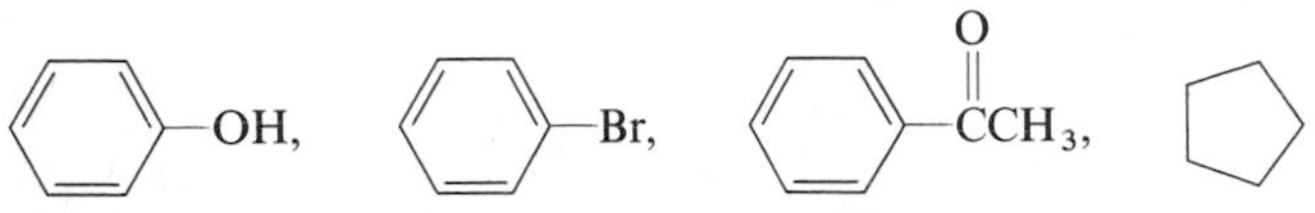

(b) Figure 16.17(b)

$$CH_3CH_2CH_2\overset{\overset{\displaystyle O}{\|}}{C}OCH(CH_3)_2, \quad CH_2{=}CHCH_2\overset{\overset{\displaystyle O}{\|}}{C}CH_3, \quad CH_3\overset{\overset{\displaystyle O}{\|}}{C}CH_2CH_2OH,$$

$$CH_2{=}CHCH_2OCH_2CH_3, \quad CH_3CH_2\overset{\overset{\displaystyle O}{\|}}{C}NH_2, \quad CH_3CH_2CH_2C{\equiv}CH$$

(c) Figure 16.17(c)

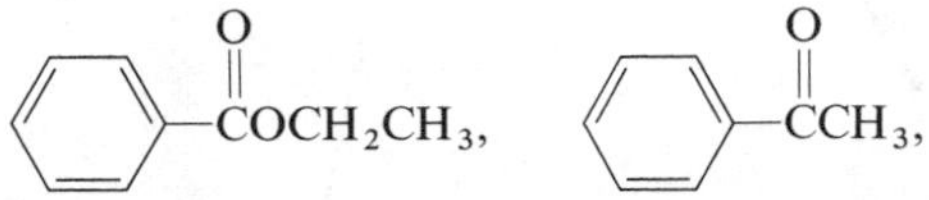

[benzene ring]—$\overset{\overset{\displaystyle O}{\|}}{C}OCH_2CH_3$,   [benzene ring]—$\overset{\overset{\displaystyle O}{\|}}{C}CH_3$, $CH_3CH_2\overset{\overset{\displaystyle O}{\|}}{C}CH_2CH_3$,

$$CH_2{=}CHCH_2CH_2\overset{\overset{\displaystyle O}{\|}}{C}CH_3, \quad CH_2{=}CHCH_2\overset{\overset{\displaystyle O}{\|}}{C}OH, \quad CH_3CH_2\overset{\overset{\displaystyle O}{\|}}{C}OCH(CH_3)_2$$

3. Explain why the infrared spectrum of glucose shows no significant absorption in the C=O stretching vibration region.

4. Select from each group the compound that absorbs ultraviolet light at the longest wavelength.

(a) $CH_2{=}CHCH{=}CHCH_2OH,$ $CH_2{=}CHCH{=}CH\overset{\overset{\displaystyle O}{\|}}{C}H,$

$$CH_2{=}CHCHCH{=}CH_2$$
$$\overset{|}{OH}$$

(b) [benzene ring]—$CH_2CH{=}CH_2$,   [benzene ring]—$CH{=}CHCH_3$

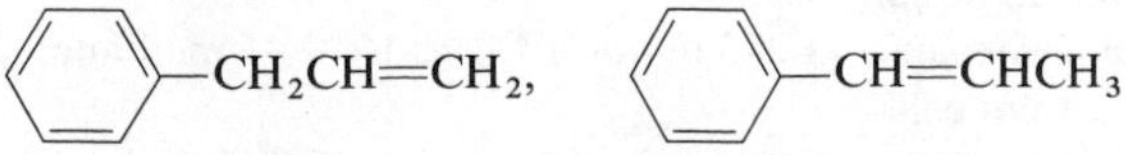

(c) [benzene ring]—CH_2—[benzene ring]   [naphthalene ring]

5. Which one of the following compounds has the ultraviolet spectrum shown in Figure 16.18?

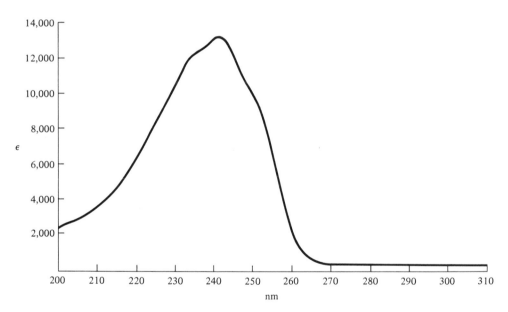

$$CH_3CH_2\overset{\overset{\displaystyle O}{\|}}{C}CH_2CH_3, \quad CH_3CH_2\overset{\overset{\displaystyle OH}{|}}{C}HCH_2CH_3, \quad CH_3\overset{|}{C}=CHCH=\overset{|}{C}CH_3,$$

$$\begin{array}{c} H \\ \diagdown \\ CH_3CH_2 \end{array} C=C \begin{array}{c} H \\ \diagup \\ \diagdown CH_2CH_3 \end{array} \quad CH_3\overset{|}{C}HCH_2\overset{\overset{\displaystyle CH_3}{|}}{C}CH_3,$$

Figure 16.18   Ultraviolet spectrum of a methanol solution of an unknown compound to accompany problem 5. [© Sadtler Research Laboratories, Division of Bio-Rad Laboratories, Inc. (1983).]

6. Use the data in the ultraviolet spectrum of naphthalene in Figure 16.8 to calculate the following quantities.
   (a) the absorbance at 275 nm of a 1.2×10^{-4} M methanol solution of naphthalene in a 1-cm cell
   (b) the concentration of a methanol solution of naphthalene that in a 1-cm cell exhibits an absorbance of 0.15 at 275 nm

7. A mixture of the terpenes α-ionone and β-ionone has an odor resembling cedar wood and is used in perfumery. Which one of these isomeric compounds has an ultraviolet spectrum with λ_{max} 294 nm, ε 8700? Predict a reasonable wavelength for an absorption maximum of the other isomer.

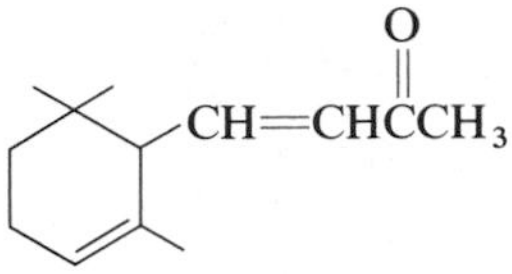

α-Ionone 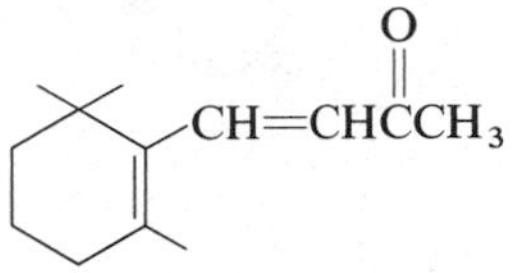   β-Ionone

8. Which of the 20 amino acids commonly found in proteins (Table 14.1) have a significant ultraviolet absorption at a wavelength longer than 210 nm?

9. How would you expect the ultraviolet-visible absorptions of a diamond (Sec. 2.15) and a piece of graphite (Sec. 4.10) to differ?

10. Give the maximum number of nmr absorptions that each compound can exhibit (ignoring any splittings) and the relative areas of the absorptions.

(a) $CH_3CH_2CH_3$

(b) CH_3CHCH_3
$\quad\quad\quad\;\;|$
$\quad\quad\quad\;\;CH_3$

(c) $CH_3CHCH_2CH_3$
$\quad\quad\;\;|$
$\quad\quad\;\;CH_3$

(d) $\quad\quad CH_3$
$\quad\quad\quad\;|$
$\quad\;CH_3CCH_2CH_3$
$\quad\quad\quad\;|$
$\quad\quad\quad CH_3$

(e) $\square$

(f) $CH_3CH{=}CH_2$

(g) 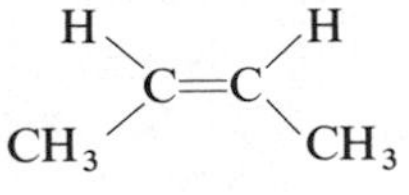 $\underset{CH_3}{H}\!\!\diagdown\!\!\underset{}{\overset{}{C}}{=}\underset{CH_3}{\overset{H}{C}}$

(h) $\square$

(i) $CH_3{-}\!\!\bigcirc\!\!{-}CH_3$

(j) CH_3CHCH_3
$\quad\quad\quad|$
$\quad\quad\quad Cl$

(k) $CH_3CH_2CH_2CH_2Cl$

(l) 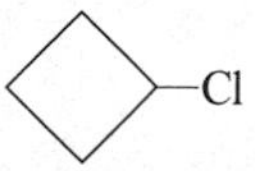 $\diamondsuit{-}Cl$

(m) $CH_2{=}CHCl$

(n) 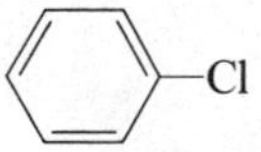 $\bigcirc{-}Cl$

(o) $CH_3CH_2\overset{\overset{\displaystyle O}{\|}}{C}H$

(p) 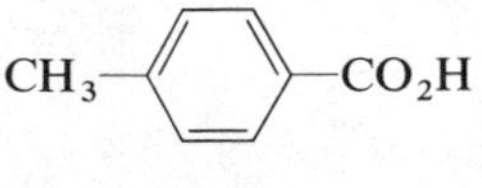 $CH_3{-}\!\!\bigcirc\!\!{-}CO_2H$

11. Indicate how the members of each group of isomeric compounds can be distinguished by a quick glance at their nuclear magnetic resonance spectra.

(a)

$$CH_3\overset{\overset{\displaystyle O}{\|}}{C}CH_3 \text{ and } CH_3CH_2\overset{\overset{\displaystyle O}{\|}}{C}H$$

(b) $CH_3CH_2OCH_2CH_3$ and $CH_3CH_2CH_2CH_2OH$

(c) $BrCH_2CH_2Br$ and CH_3CHBr_2

(d)

⬠ and ◇—CH_3

(e) the three dibromobenzenes
(f) the three tribromobenzenes

12. Predict the approximate chemical shifts of the nmr absorptions due to each group of hydrogens in the following compounds.

(a) CH_3CO_2H

(b)
$$CH_3\overset{\overset{\displaystyle CH_3}{|}}{\underset{\underset{\displaystyle CH_3}{|}}{C}}CO_2H$$

(c)
$$CH_3\overset{\overset{\displaystyle O}{\|}}{C}CH_2CH_3$$

(d) $CH_3\underset{\underset{\displaystyle Cl}{|}}{C}HCH_3$

(e) 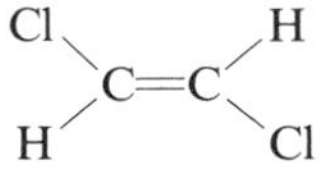
$$\overset{Cl}{\underset{H}{}}\!\!\diagdown\!\!\diagup^{\!\!\!\!\diagup}_{\!\!\!\!\diagdown}\overset{H}{\underset{Cl}{}}\\ C\!=\!C$$

(f) $ClCH_2CH_2CH_2Cl$

13. Sketch out the approximate appearance of the nuclear magnetic resonance spectra of the compounds listed in problem 12. Include any effects due to splitting of absorptions.

14. Select from each group of compounds the one that has the indicated nmr spectrum. Assign each absorption to a specific group of hydrogens.
(a) Figure 16.19(a)

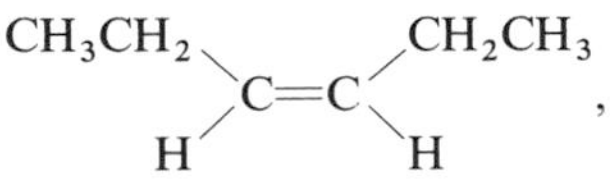

$$\overset{CH_3CH_2}{\underset{H}{}}\!\!\diagdown\!\!\!\diagup\overset{CH_2CH_3}{\underset{H}{}}\\ C\!=\!C$$, $CH_2\!=\!CHCH_2CH_2CH_2CH_3$,

$$CH_2\!=\!\underset{\underset{\displaystyle CH_3}{|}}{C}CH_2CH_2CH_3,$$ $$CH_3CH\!=\!\underset{\underset{\displaystyle CH_3}{|}}{C}CH_2CH_3,$$

$$(CH_3)_2C\!=\!CHCH_2CH_3,$$ $$\overset{CH_3}{\underset{H}{}}\!\!\diagdown\!\!\!\diagup\overset{CH_2CH_3}{\underset{CH_3}{}}\\ C\!=\!C$$

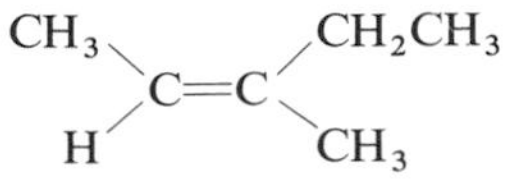

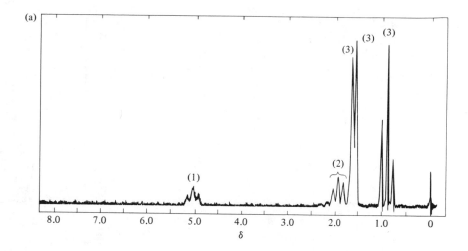

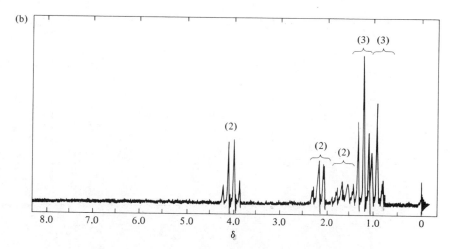

Figure 16.19 Nuclear magnetic resonance spectra to accompany problem 14. Relative areas of absorptions are indicated by the numbers in parentheses. [© Sadtler Research Laboratories, Division of Bio-Rad Laboratories, Inc. (1983).]

(b) Figure 16.19(b)

$$(CH_3)_2CHCOCH_2CH_3, \quad CH_3CH_2CCH_2OCH_2CH_3,$$

$$CH_3CH_2CH_2COCH_2CH_3, \quad CH_3CH_2COCH_2CH_2CH_3,$$

$$CH_3CH_2CH_2CH_2COCH_3, \quad CH_3COCH_2CH_2CH_2CH_3$$

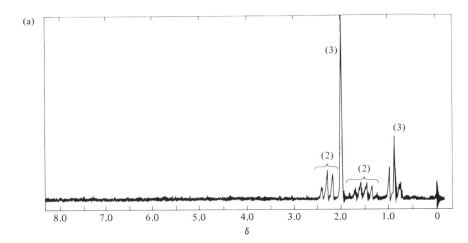

(a)

(3)

(3)

(2)

(2)

8.0 7.0 6.0 5.0 4.0 3.0 2.0 1.0 0

δ

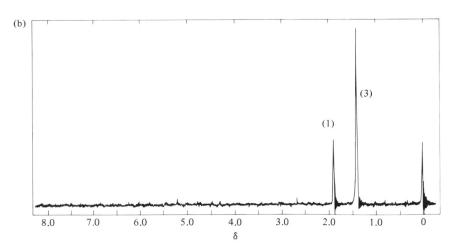

(b)

(3)

(1)

8.0 7.0 6.0 5.0 4.0 3.0 2.0 1.0 0

δ

Figure 16.20 Nuclear magnetic resonance spectra to accompany problem 15. Relative areas of absorptions are indicated by the numbers in parentheses. [© Sadtler Research Laboratories, Division of Bio-Rad Laboratories, Inc. (1983).]

15. Draw the structure of the compound responsible for the nmr spectrum, and assign each absorption to a specific group of hydrogens.
    (a) Figure 16.20(a)   (b) Figure 16.20(b)
        $C_5H_{10}O$ $C_6H_{12}O_2$

16. Draw the structure of each compound whose molecular formula and nmr spectrum are indicated. (The numbers in parentheses are the relative areas of absorptions.)
    (a) $C_2H_3Cl_3$ $2.7\ \delta$ singlet
    (b) $C_4H_8Br_2$ $1.9\ \delta$ singlet   (3)
                         3.9   singlet   (1)

(c) C_4H_8O 1.0 δ triplet   (3)
                          2.0   singlet   (3)
                          2.4   quartet   (2)

(d) C_4H_9Cl 1.6 δ singlet

(e) $C_5H_{10}Cl_2$ 1.1 δ singlet   (9)
                          5.4   singlet   (1)

(f) $C_5H_{10}O$ 1.0 δ triplet   (3)
                          2.4   quartet   (2)

(g) $C_5H_{12}O$ 0.9 δ triplet   (3)
                          1.2   singlet   (6)
                          1.5   quartet   (2)
                          2.9   singlet   (1)

(h) $C_6H_4Br_2$ 6.9 δ triplet   (1)
                          7.3   doublet   (2)
                          7.6   singlet   (1)

(i) $C_6H_4Cl_2$ 7.2 δ singlet

(j) C_6H_{12} 1.4 δ singlet

(k) C_6H_{12} 1.0 δ triplet   (3)
                          2.0   quartet   (2)
                          4.6   singlet   (1)

(l) $C_6H_{12}O$ 1.1 δ singlet   (3)
                          2.1   singlet   (1)

(m) $C_7H_{14}O$ 1.0 δ doublet   (6)
                          2.6   septet   (1)

(n) $C_7H_{14}O$ 0.9 δ triplet   (3)
                          1.6   sextet   (2)
                          2.3   triplet   (2)

(o) C_8H_{10} 1.2 δ triplet   (3)
                          2.5   quartet   (2)
                          7.1   singlet   (5)

(p) $C_9H_{10}O$ 2.0 δ singlet   (3)
                          3.6   singlet   (2)
                          7.2   singlet   (5)

(q) $C_{10}H_{13}Cl$ 1.4 δ singlet   (6)
                          3.5   singlet   (2)
                          7.2   singlet   (5)

17. Extraction of the cuticle of female houseflies with hexane or diethyl ether gives a mixture of compounds. One component, particularly attractive to male houseflies, has molecular formula $C_{23}H_{46}$ and the following nmr spectrum:

                                       0.9 δ triplet   (3)
                                       1.3   complex   (17)
                                       2.0   complex   (2)
                                       5.3   triplet   (1)

Describe as much as you can about the structure of the compound. What remains uncertain?

18. A compound isolated from the sweat of some schizophrenic patients has molecular formula $C_7H_{12}O_2$ and the following nmr spectrum:

0.9 δ	triplet	(3)
1.6	sextet	(2)
2.1	singlet	(3)
2.2	triplet	(2)
5.6	singlet	(1)
12.4	singlet	(1)

Draw a structure of a compound that might have this spectrum. What remains uncertain?

19. If the effects of splitting are ignored, the various isomeric dichlorocyclopropanes have one, two, or three nmr absorptions. Indicate the number of absorptions exhibited by each isomer.

20. Explain why the two methylene hydrogens of 1,2-dibromo-1-phenylethane ($Ph-CHBr-CH_2Br$) exhibit different chemical shifts.

21. The largest peak in the mass spectrum of each of the following compounds is due to a fragment ion rather than the molecular ion. Predict for each compound the mass of that largest peak.

(a) $(CH_3)_4C$ (b)  [cyclohexyl]$-CH_2CH_3$ (c) $CH_2=CHCH_3$

22. Explain the origin in the mass spectrum of methane of a peak at mass 17 that is about 1.1% as intense as the peak at mass 16.

23. Describe how you could use a specific instrumental method to readily distinguish between the members of each pair of compounds.

(a) 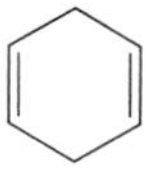 [cyclohexene structure] and [cyclohexadiene structure]

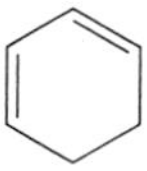

(b) $ClCH_2CH_2Cl$ and CH_3CHCl_2

(c) $CH_3(CH_2)_8CH_3$ and $CH_3(CH_2)_9CH_3$

(d) 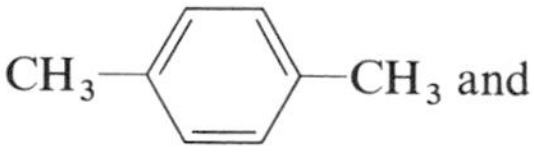 CH_3-[benzene ring]$-CH_3$ and [benzene ring]$-CH_2CH_3$

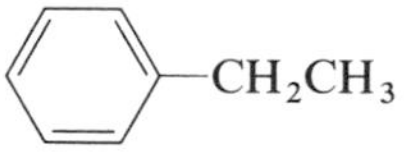

(e)
$$CH_3CH_2OCH_2CH_3 \text{ and } CH_3CH_2\overset{\overset{\displaystyle OH}{|}}{C}HCH_2CH_3$$

(f) 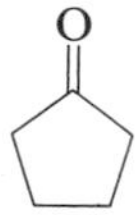 [cyclopentanone structure] and $CH_3CH=CHCCH_3$ (with $\overset{O}{\|}$)

24. What is the wavelength of radiation that has a frequency of 3×10^{14} Hz? What is the frequency of radiation with a wavelength of 3×10^{-8} meter?

Answers
to In-Text Problems

CHAPTER 1

1.1 H:C:::N:

This structure uses the number of valence electrons contributed by each atom (one from H, four from C, and five from N) and places a completed valence shell around each atom (two valence electrons around H and eight around C and N).

1.2 (a)

$$\underset{\substack{\text{H}\overset{\displaystyle \text{O}}{\underset{|}{\overset{||}{-}}}\text{C}\text{—H}}}{} \quad \text{or} \quad \underset{\substack{\text{H}\overset{\displaystyle :\text{O}:}{\underset{|}{\overset{||}{-}}}\text{C}\text{—H}}}{} \quad \text{or} \quad \text{H}:\overset{..}{\underset{..}{\text{C}}}:\text{H}\;\; :\!\overset{\displaystyle :\text{O}:}{}$$

(b) H—C＝C—H   or   H:C::C:H
  | | .. ..
  H H   H H

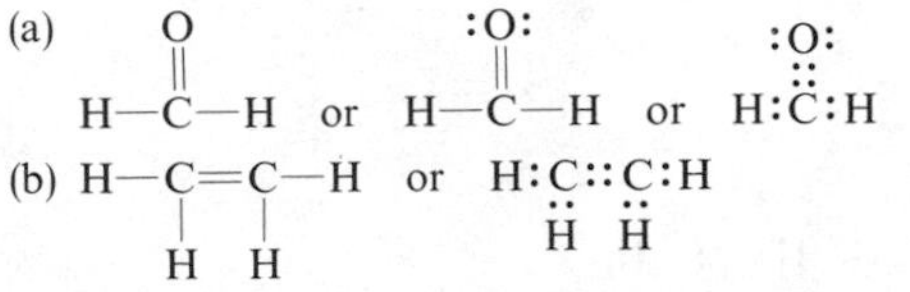

1.3 Formulas (a), (c), (e), and (f) represent butane. Note that each has its four carbons linked in a continuous chain. Formulas (b) and (d) represent isobutane. Each has only three carbons attached in a continuous chain.

CHAPTER 2

2.1

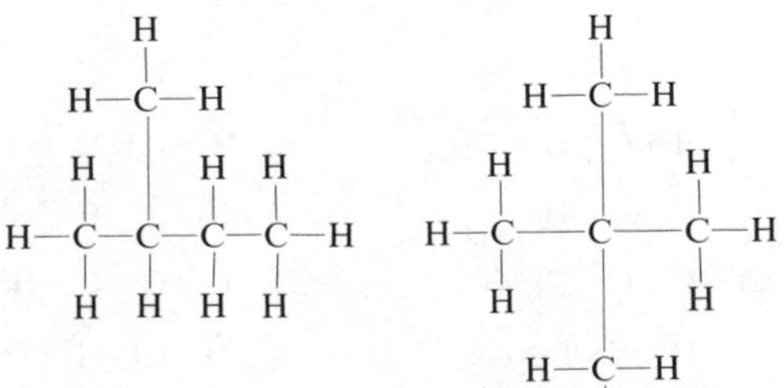

Isopentane   Neopentane

2.2 When examining a structure that you have drawn, look carefully at the attachments between the carbon atoms. If the five carbons are linked in a continuous chain, then the structure represents pentane. If only four carbons are in a continuous chain (the other carbon being linked to one of the interior carbons), then the structure represents isopentane. If the longest chain contains only three carbons, then the structure represents neopentane.

2.3 (a) This structural formula represents the third hexane isomer pictured in Section 2.2. Note that it has a continuous chain of five carbons with the remaining carbon attached to the middle carbon of the chain.

(b) This structural formula represents the second hexane isomer pictured in Section 2.2. It has a continuous chain of five carbons with the remaining carbon attached to a chain carbon that is next to an end carbon.

(c) This structural formula represents the first hexane isomer pictured in Section 2.2. Its carbons are linked in a continuous chain.

2.4 (a)

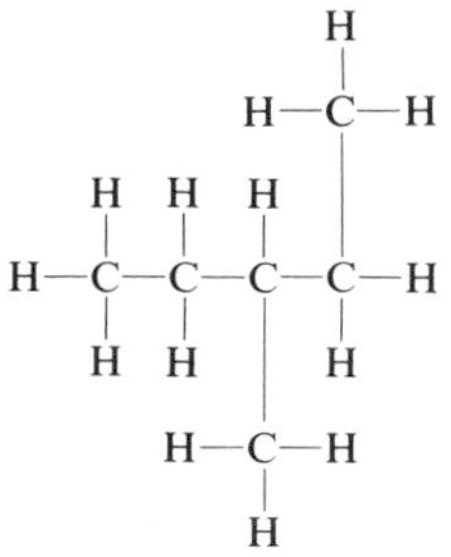

(b)

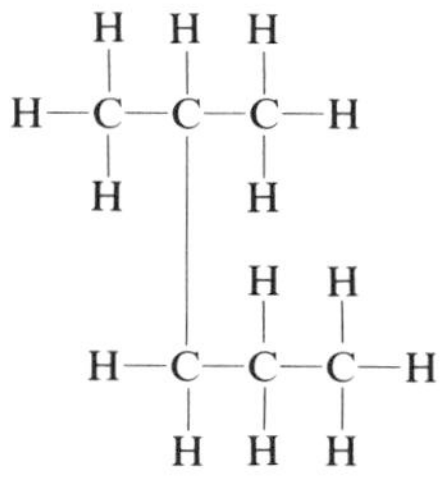

(c)

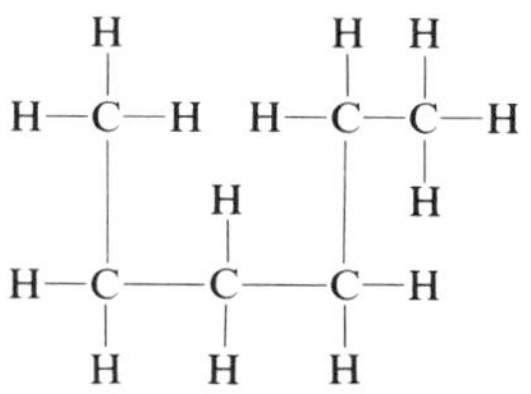

2.5 (a) 3-chlorohexane

(b) 2-chloro-5-methylheptane

Don't be fooled by the way the chain is drawn. The longest continuous chain has seven carbons.

2.6 5-*sec*-butylnonane

2.7

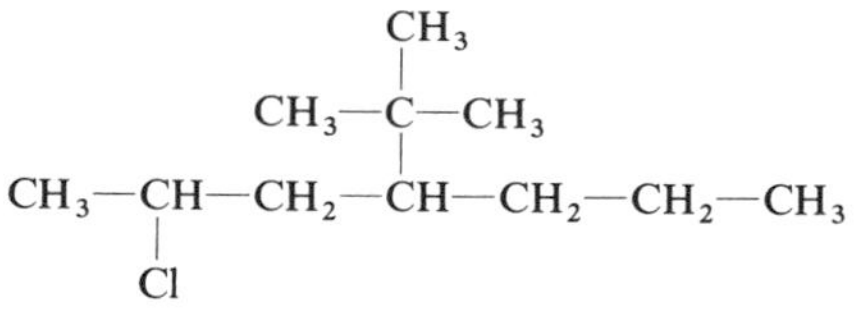

2.8 From butane:

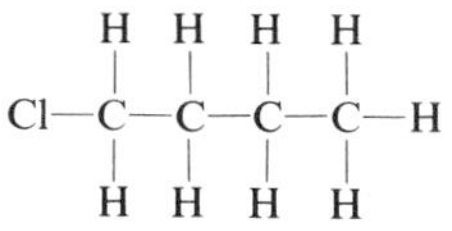

Butyl chloride
1-Chlorobutane

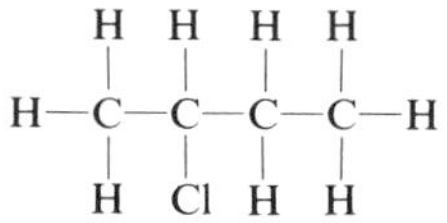

sec-Butyl chloride
2-Chlorobutane

From isobutane:

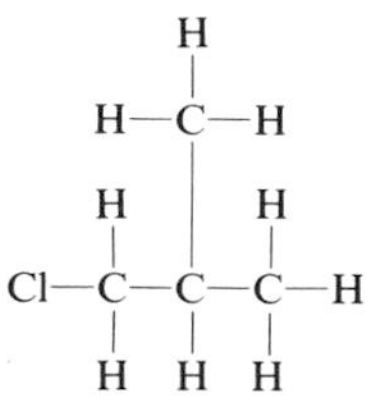

Isobutyl chloride
1-Chloro-2-methylpropane

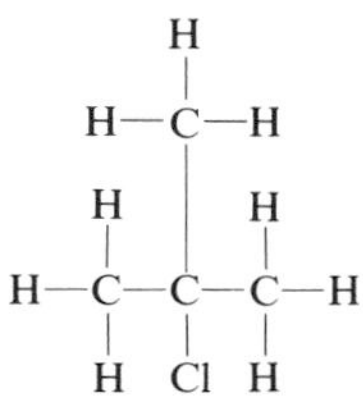

tert-Butyl chloride
2-Chloro-2-methylpropane

2.9 (a)

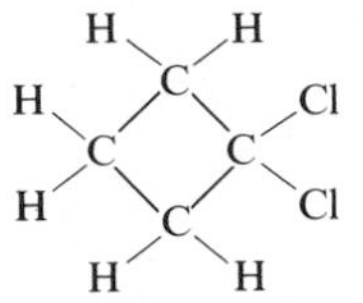

1,1-Dichlorocyclobutane

(b)

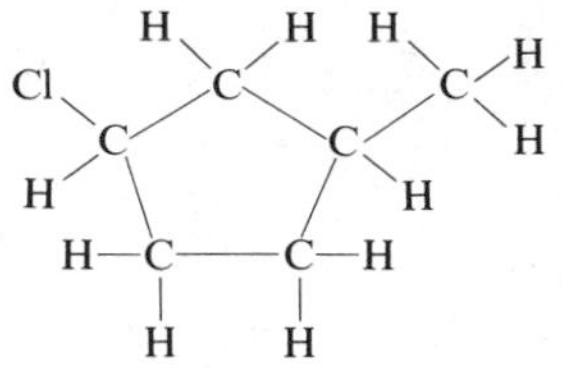

1-Chloro-3-methylcyclopentane

(c)

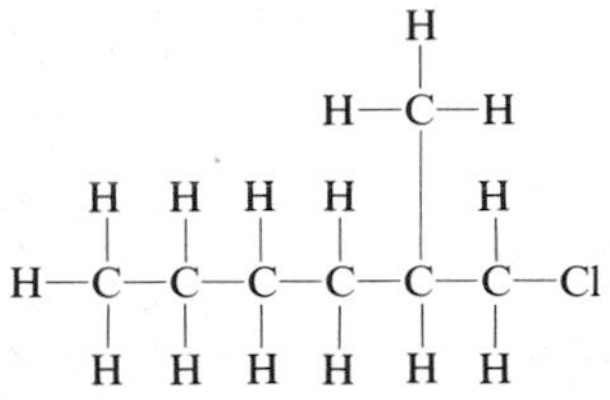

1-Chloro-2-methylhexane

2.10

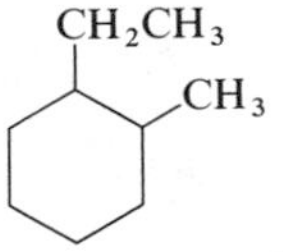

CH_2CH_3

CH_3

2.11

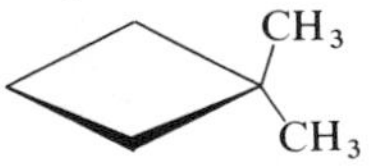

1,1-Dimethylcyclobutane

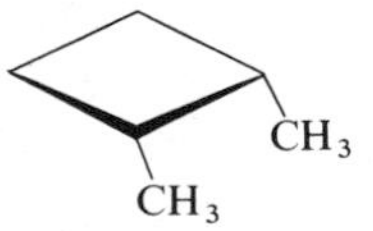

cis-1,2-Dimethylcyclobutane

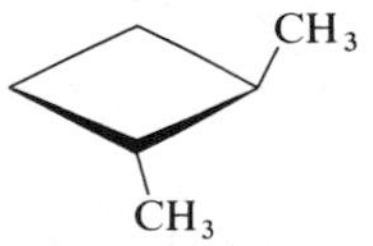

trans-1,2-Dimethylcyclobutane

cis-1,3-Dimethylcyclobutane

trans-1,3-Dimethylcyclobutane

CHAPTER 3

3.1 $CH_2{=}C{-}CH_2{-}CH_2{-}CH_3$ (Note that this is the same as $CH_3{-}C{-}CH_2{-}CH_2{-}CH_3$)
$\quad\quad\quad\quad |$ $\quad\quad\quad\quad\quad\quad\quad\quad\quad\quad\quad\quad\quad\quad\quad\quad\quad\quad\quad\quad\quad ||$
$\quad\quad\quad\quad CH_3$ $\quad\quad\quad\quad\quad\quad\quad\quad\quad\quad\quad\quad\quad\quad\quad\quad\quad\quad\quad\quad\quad CH_2$

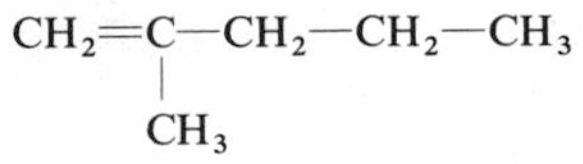

$CH_3{-}C{=}CH{-}CH_2{-}CH_3$
$\quad\quad\quad\quad |$
$\quad\quad\quad\quad CH_3$

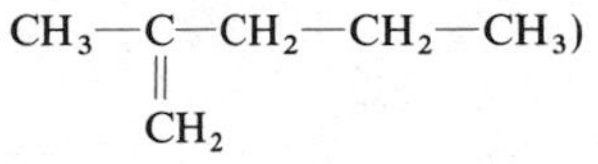

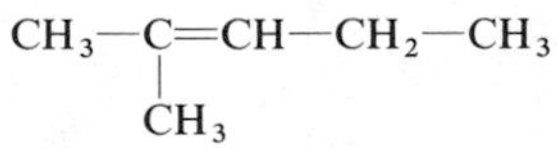

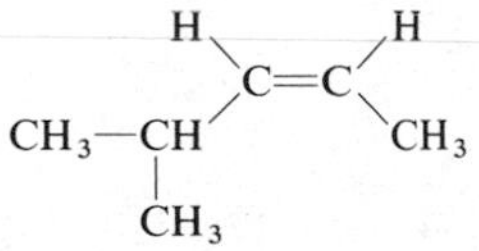

Cis isomer

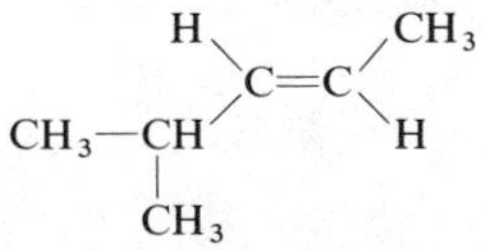

Trans isomer

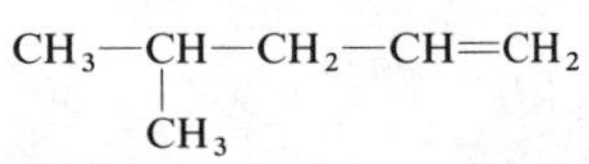

$CH_3{-}CH{-}CH_2{-}CH{=}CH_2$
$\quad\quad\quad |$
$\quad\quad\quad CH_3$

Because the double bond can often be in more than one position and sometimes can be cis or trans, there are generally more alkenes than alkanes with a given skeleton.

3.2   (a) 2-isopropyl-3-methyl-1-pentene

      Remember that the parent chain must include the double bond even though there may be a longer chain in the molecule.

      (b) *trans*-4-chloro-2-pentene

3.3   (a) $CH_2{=}CH{-}CH_2{-}\underset{\underset{CH_3}{|}}{CH}{-}CH_3$ (b) $\underset{ClCH_2}{\overset{H}{\diagdown}}C{=}C\underset{\diagdown CH_2Cl}{\overset{H}{\diagup}}$

3.4   A compound with a carbon-carbon double bond has two hydrogens less than the corresponding saturated compound. For example, propene is C_3H_6 and propane is C_3H_8. Similarly, a compound with a ring has two hydrogens less than the corresponding acyclic compound. Cyclopropane is C_3H_6 and propane is C_3H_8, for instance. In fact, for each ring or double bond that it has, a compound has two hydrogens less than an acyclic alkane (C_nH_{2n+2}) with the same number of carbons. A compound of molecular formula C_6H_{10} has four hydrogens less than hexane (C_6H_{14}). Therefore it has two double bonds, two rings, or one of each. The observation that on hydrogenation it reacts with only one mole of H_2 (to form C_6H_{12}) indicates that it has only one double bond. Therefore it must also have one ring.

3.5   The compound must not contain double bonds. Because it has four hydrogens less than the corresponding acyclic alkane ($C_{10}H_{22}$), the compound must have two rings.

3.6   (a) $CH_3{-}CH_2{-}CH_2{-}CH_2{-}CH_2{-}CH_3$ (b) $CH_3{-}\underset{\underset{CH_3}{|}}{\overset{\overset{Br}{|}}{C}}{-}\overset{\overset{Br}{|}}{CH}{-}CH_3$

    (c) $CH_3{-}\underset{\underset{CH_3}{|}}{\overset{\overset{Cl}{|}}{C}}{-}CH_3$ (d) cyclopentane with OH and CH_3 substituents

3.7   The carbocation intermediate formed by the attack of bromine on ethylene can react with any nucleophile in the solution.

$$CH_2{=}CH_2 \xrightarrow{Br_2} \overset{Br}{\underset{}{CH_2}}{-}\overset{+}{CH_2}$$

with products:

$\underset{\underset{Br}{|}}{\overset{\overset{Br}{|}}{CH_2}}{-}CH_2$ (via Br^-)

$\underset{\underset{Cl}{|}}{\overset{\overset{Br}{|}}{CH_2}}{-}CH_2$ (via Cl^-)

$\underset{\underset{\overset{+}{OH_2}}{|}}{\overset{\overset{Br}{|}}{CH_2}}{-}CH_2 \longrightarrow \underset{\underset{OH}{|}}{\overset{\overset{Br}{|}}{CH_2}}{-}CH_2$ (via H_2O)

Sodium chloride and water are not sufficiently electrophilic to react with ethylene to produce a carbocation intermediate.

3.8   Hydrogen chloride adds more rapidly to 2-methyl-1-propene, for that addition proceeds by formation of a tertiary carbocation.

$$CH_2{=}\underset{\underset{CH_3}{|}}{C}{-}CH_3 \xrightarrow{HCl} CH_3{-}\underset{\underset{CH_3}{|}}{\overset{+}{C}}{-}CH_3 \qquad CH_2{=}CH_2 \xrightarrow{HCl} CH_3{-}\overset{+}{CH_2}$$

                                 Tertiary   Primary

3.9   (a) *cis*-2-butene and *trans*-2-butene

      (b) cyclohexene

      The double bond is cleaved in exactly the same way as the double bonds of acyclic alkenes, but the carbon chain linking the carbons of the double bond is unaffected by the reaction.

3.10 (a) CH₃—CH=CH—CH₃ + CH₃—CH₂—CH=CH₂

(cis + trans)
Major product   Minor product

(b) CH₃—CH=CH—CH₂—CH₂—CH₃ + CH₃—CH₂—CH=CH—CH₂—CH₃

(cis + trans)   (cis + trans)
2-Hexene   3-Hexene

Elimination in either direction forms an alkene with two alkyl substituents. In fact, approximately equal amounts of the 2-hexenes and 3-hexenes are formed.

3.11 It is necessary to use a sequence of two reactions.

$$CH_3—CH_2—CH_2—OH \xrightarrow[\text{heat}]{H^+} CH_3—CH=CH_2 \xrightarrow{HCl} CH_3—\underset{\underset{Cl}{|}}{CH}—CH_3$$

3.12

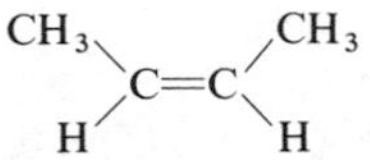

$$\underset{H}{\overset{CH_3}{\diagdown}}C=C\underset{H}{\overset{CH_3}{\diagup}}$$

Remember that in considering the addition of hydrogen to alkenes (Sect. 3.5), we learned that the two hydrogen atoms generally add from the same side of the multiple bond.

CHAPTER 4

4.1 (a) 1-chloro-2-nitrobenzene or *o*-chloronitrobenzene
Remember that substituents are usually arranged alphabetically.
(b) 4-bromotoluene or *p*-bromotoluene
The names 1-bromo-4-methylbenzene and *p*-bromomethylbenzene are not generally used.
(c) 4-ethylphenol or *p*-ethylphenol
The names 1-ethyl-4-hydroxybenzene and *p*-ethylhydroxybenzene are not generally used.
(d) cyclopentylbenzene or phenylcyclopentane

4.2 (a) NO₂ (b) NH₂ (c) NO₂ (d)

4.3 (a) 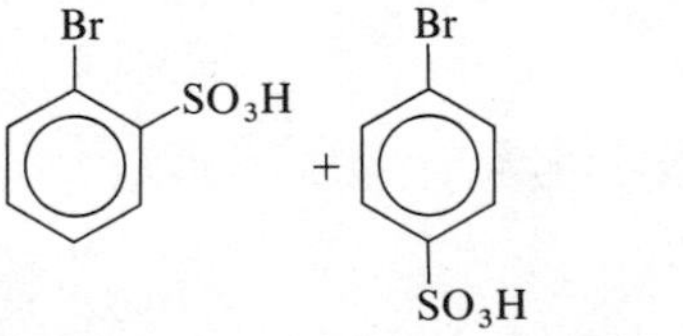 Br   (b) Br

+

The bromine is ortho-para directing.

SO₃H

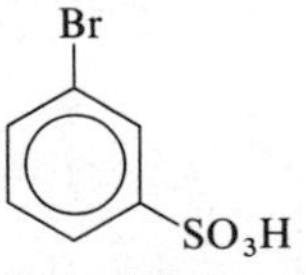

The sulfonic acid group is meta directing.

4.4 (a)

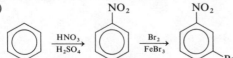

If the steps were reversed, bromobenzene would be formed in the first step. Because a halogen is ortho-para directing, nitration of bromobenzene produces mainly ortho and para products.

(b)

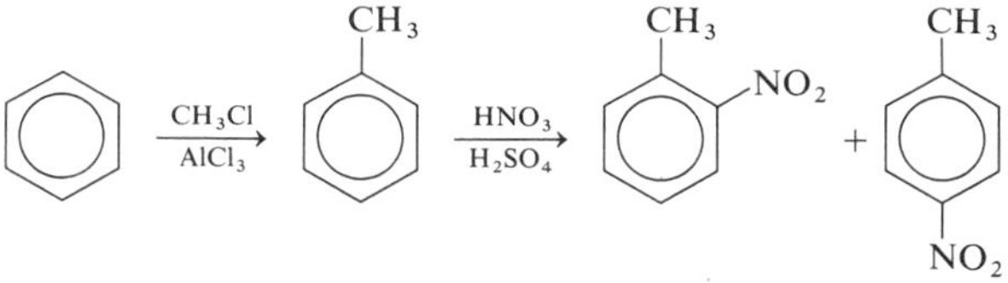

The para isomer must be separated from the accompanying ortho isomer.

4.5

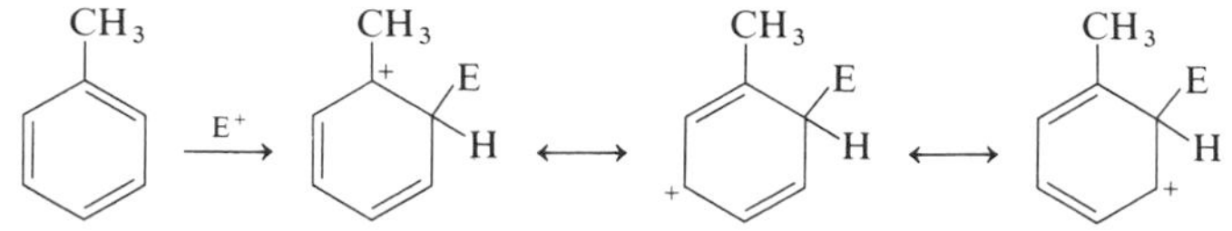

4.6

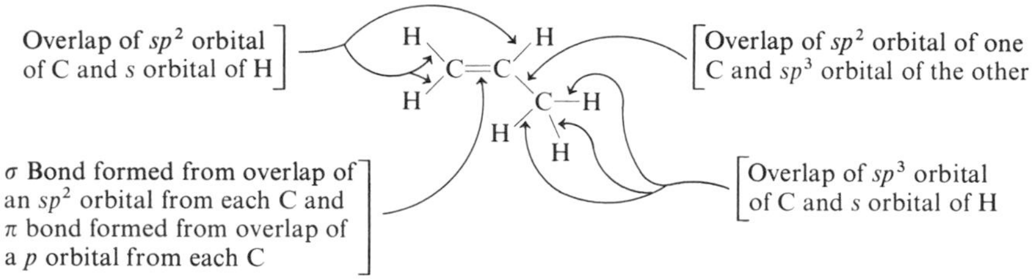

$$\text{Overlap of } sp^2 \text{ orbital of C and } s \text{ orbital of H}$$

$$\left[\text{Overlap of } sp^2 \text{ orbital of one C and } sp^3 \text{ orbital of the other}\right]$$

$$\begin{bmatrix}\sigma \text{ Bond formed from overlap of} \\ \text{an } sp^2 \text{ orbital from each C and} \\ \pi \text{ bond formed from overlap of} \\ \text{a } p \text{ orbital from each C}\end{bmatrix}$$

$$\left[\text{Overlap of } sp^3 \text{ orbital of C and } s \text{ orbital of H}\right]$$

The following atoms lie in a plane with each bond angle approximately $120°$.

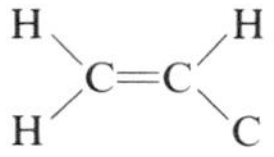

The four atoms attached to the methyl carbon, however, are arranged in a tetrahedral fashion with bond angles of approximately $109°$.

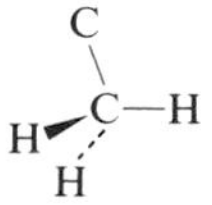

CHAPTER 5

5.1   (a) 2-pentanol   (b) 1-methylcyclohexanol
        (c) 3-phenyl-1-propanol   (d) diisopropyl ether or 2-isopropoxypropane

5.2   (a) CH_3—CH—CH_3 (b) CH_3—CH—CH_2—OH
              | |
              OH   Br
          Secondary   Primary

        (c)   OH   (d) CH_2=CH—CH_2—O—CH_3
                           |
        CH_3—CH_2—C—CH_2—CH_3
                           |
                          CH_3
              Tertiary

5.3   (a) $(CH_3)_2CHO^- \ Na^+$ (b)   O
                                                              ||
                                              $(CH_3)_2CHCCH_3$

(c)

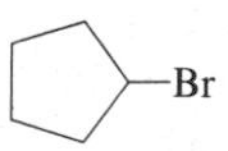

—Br

(d)

$$\text{(CH}_3)_2\text{CHCOH}$$

with O double bonded above C.

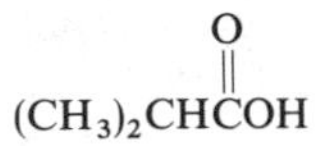

5.4 $(\text{CH}_3)_3\text{COH} \xrightarrow{\text{Na}} (\text{CH}_3)_3\text{CO}^- \text{Na}^+ \xrightarrow{\text{CH}_3\text{CH}_2\text{Br}} (\text{CH}_3)_3\text{COCH}_2\text{CH}_3$

Because the alkyl halide is tertiary, the following combination of reagents leads mainly to elimination rather than ether formation.

$$\text{CH}_3\text{CH}_2\text{OH} \xrightarrow{\text{Na}} \text{CH}_3\text{CH}_2\text{O}^- \text{Na}^+ \xrightarrow{(\text{CH}_3)_3\text{CBr}} \text{CH}_2{=}\text{C}(\text{CH}_3)_2$$

CHAPTER 6

6.1   Structures (a), (c), and (d) represent chiral molecules. Each has one asymmetric carbon (indicated below with an asterisk).

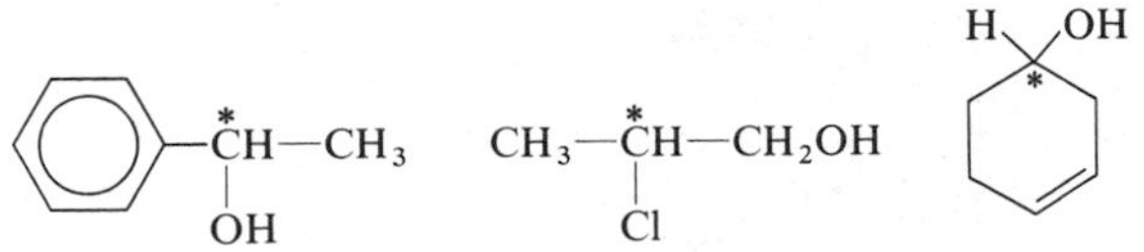

(a)   (c)   (d)

Remember that (d) does have four bonds to each carbon. It is easy to forget when looking at simplified structures that all bonds not otherwise indicated are to hydrogens. Although the asymmetric carbon of (d) is bonded directly to two methylene (CH_2) groups, these substituents become different when we travel on to the next carbons.

6.2   (a)

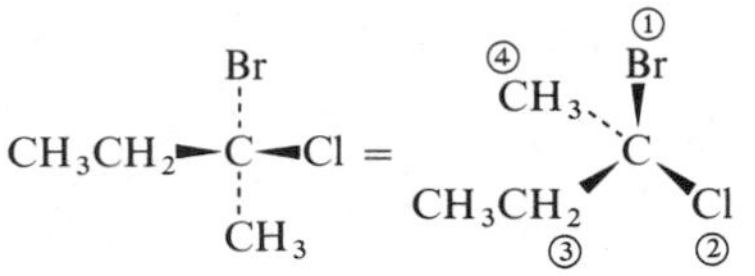

(*R*)-2-Bromo-2-chlorobutane

(b)

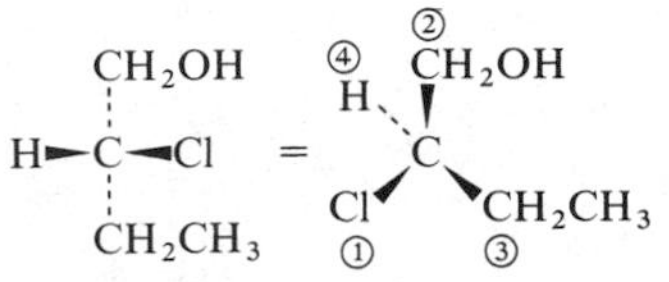

(*R*)-2-Chloro-1-butanol
—C—O outranks —C—C

6.3   (a) This structure has two nonidentical asymmetric carbons and so corresponds to four stereoisomers (two pairs of enantiomers).

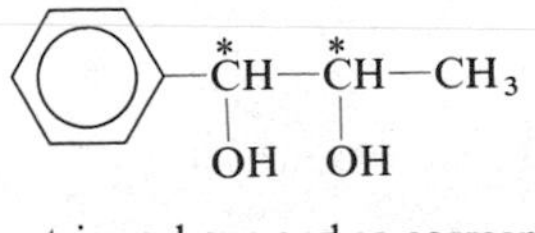

(b) This structure has two identical asymmetric carbons and so corresponds to three stereoisomers (a pair of enantiomers and a meso compound).

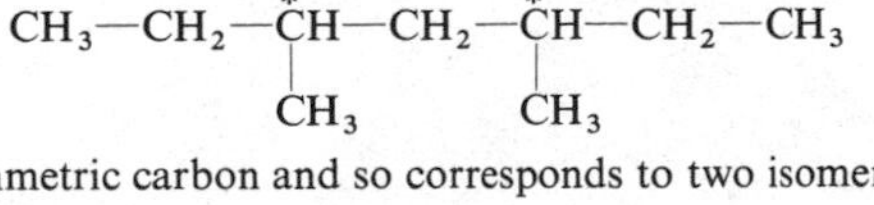

(c) This structure has one asymmetric carbon and so corresponds to two isomers (a pair of enantiomers).

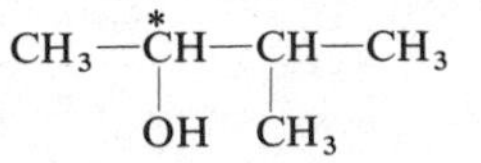

(d) Both cis and trans configurations are possible for this structure.

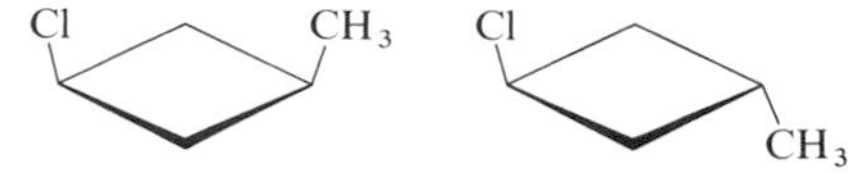

Cis   Trans

Neither the cis nor the trans isomer, however, has chiral carbons and thus only one cis isomer and one trans isomer exist. Note that because of the particular placement of substituents, the two CH_2 groups in each isomer are identical. Another way of noting the lack of chirality is to prepare a perspective drawing (or a model) of each isomer and note that it is identical with its mirror image.

CHAPTER 8

8.1   (a) isopropylamine or 2-propanamine (primary)
    (b) triethylamine or *N,N*-diethylethanamine (tertiary)
    (c) cyclopentylmethylamine or *N*-methylcyclopentanamine (secondary)
    (d) 2-phenyl-1-propanamine (primary)

8.2 $H_2NCH_2CH_2CH_2NH_2$ > $CH_3CH_2CH_2CH_2NH_2$ > $CH_3CH_2N(CH_3)_2$

    bp 136° (mw 74)   bp 78° (mw 73)   bp 37° (mw 73)

These compounds have similar molecular weights and so that is not a factor. The highest boiling compound has two amino groups and hence more opportunities for hydrogen bonding between molecules than exist for the middle compound. The lowest boiling compound is a tertiary amine and so its molecules cannot form hydrogen bonds to one another.

8.3   (a) $(CH_3CH_2)_2NH_2^+$ Cl^- (b) $(CH_3CH_2)_3NH^+$ Br^-
    (c) 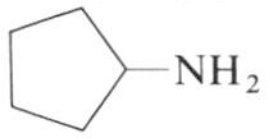   (d) $CH_3CH_2CH_2N(CH_3)_3^+$ I^-

8.4   (a)

The following is another possible synthesis.

(b)

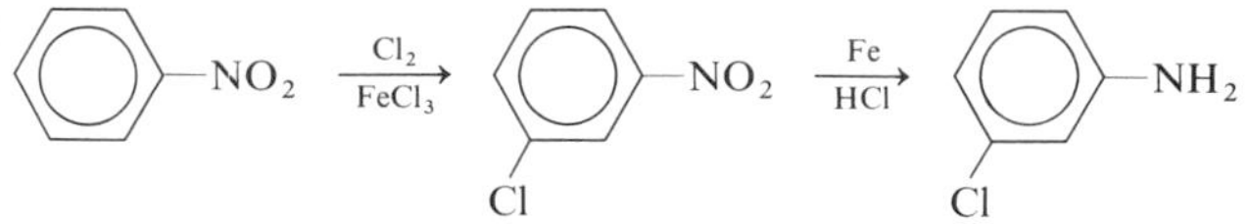

The steps cannot be reversed. Reduction of nitrobenzene gives aniline, and chlorination of aniline produces mainly *ortho*- and *para*-chloroanilines.

CHAPTER 9

9.1   (a) 2-bromobutanal   (b) 3-methyl-1-phenyl-2-butanone or benzyl isopropyl ketone

9.2   (a) 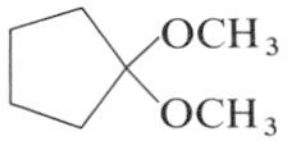   (b)

(c)

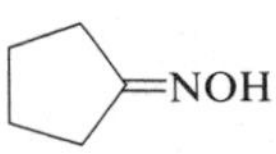

=NOH

(d) **no reaction under normal conditions**

9.3   (a) Either of the following routes would work.

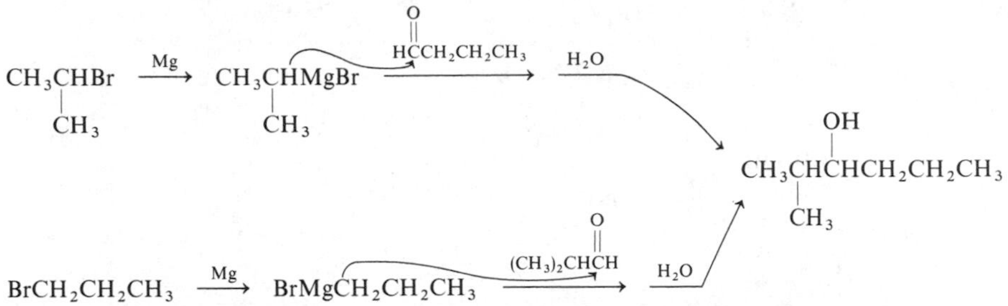

$$CH_3CHBr \xrightarrow{Mg} CH_3CHMgBr \xrightarrow{\underset{\text{O}}{\overset{\text{O}}{HCCH_2CH_2CH_3}}} \xrightarrow{H_2O}$$

with CH₃ groups, leading to

$$\underset{CH_3}{CH_3CHCHCH_2CH_2CH_3}^{OH}$$

$$BrCH_2CH_2CH_3 \xrightarrow{Mg} BrMgCH_2CH_2CH_3 \xrightarrow{\underset{\text{O}}{\overset{\text{O}}{(CH_3)_2CHCH}}} \xrightarrow{H_2O}$$

(b) The carbon skeleton can be assembled from the reaction of an appropriate Grignard reagent with cyclopentanone, but then some additional tinkering is necessary to get rid of the hydroxyl group.

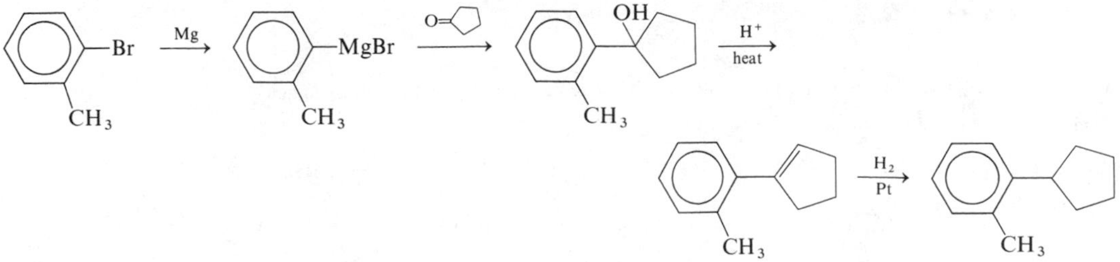

9.4

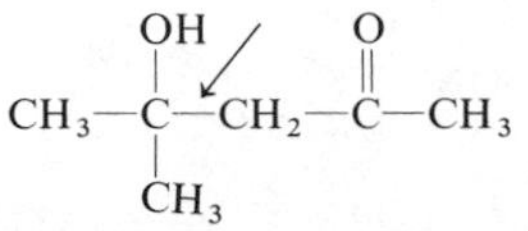

$$\underset{\overset{|}{CH_3}}{\overset{OH}{CH_3-\underset{}{C}-CH_2}}-\overset{O}{\overset{||}{C}}-CH_3$$

The arrow indicates the new carbon-carbon bond.

9.5   The framework can be synthesized by an aldol condensation of butanal.

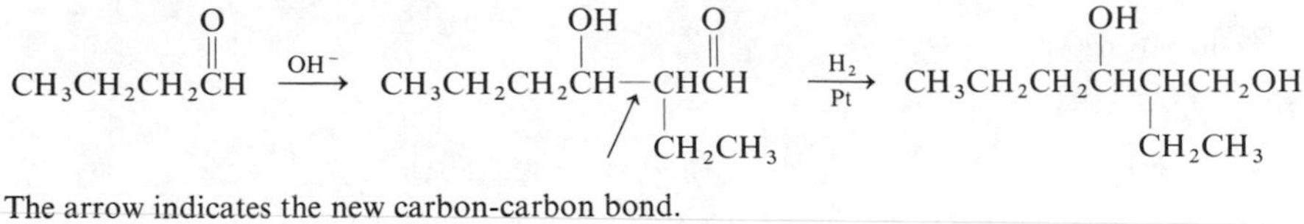

$$CH_3CH_2CH_2\overset{O}{\overset{||}{CH}} \xrightarrow{OH^-} CH_3CH_2CH_2\overset{OH}{\underset{}{CH}}-\overset{O}{\overset{||}{CHCH}} \xrightarrow[Pt]{H_2} CH_3CH_2CH_2\overset{OH}{\underset{CH_2CH_3}{CH}}CHCH_2OH$$

The arrow indicates the new carbon-carbon bond.

CHAPTER 10

10.1

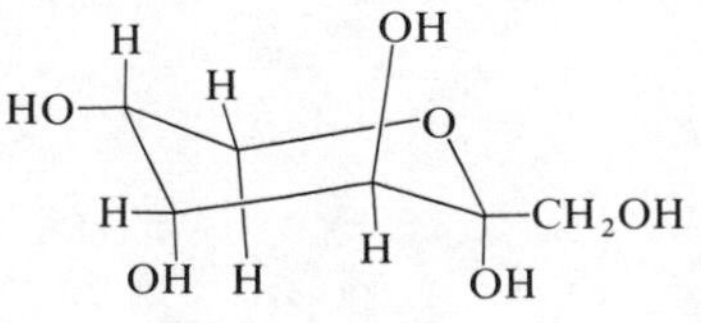

α-D-Fructopyranose

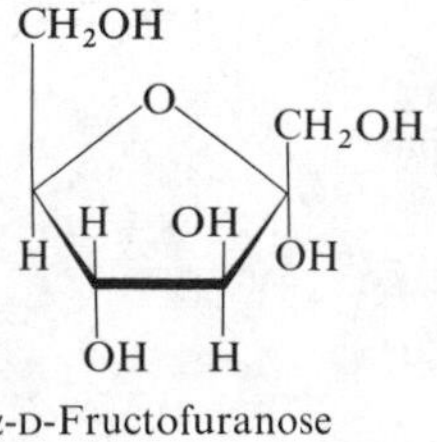

α-D-Fructofuranose

10.2

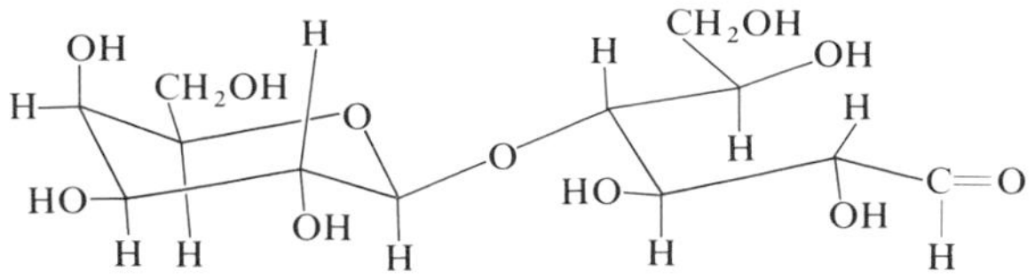

Acyclic form

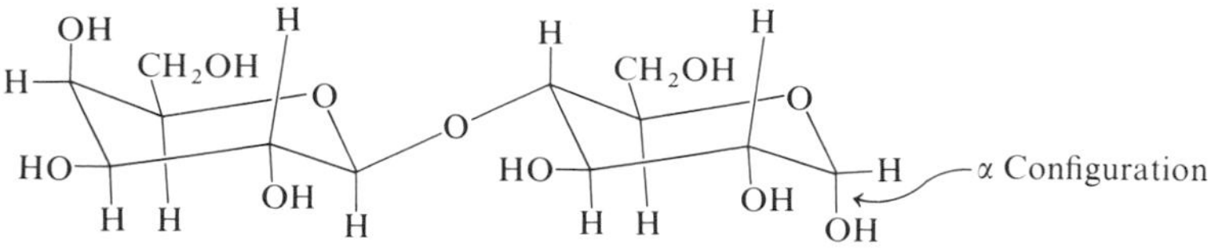

α-D-Glucopyranose form

CHAPTER 11

11.1 (a) 2-chlorohexanoic acid or 2-chlorocaproic acid   (b) *p*-bromobenzoic acid

11.2 (a)

$$Cl_2CHCOH$$ (with C=O)

(b)

$(CH_3)_3C-\bigcirc-COH$ (with C=O)

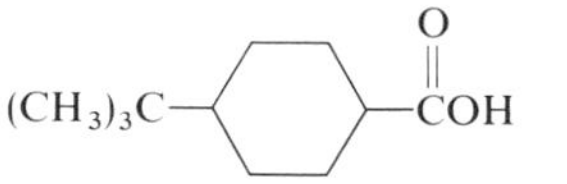

This could be either cis or trans.

(c)

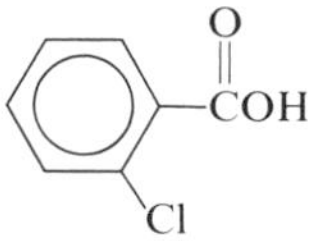

(benzene ring)$-COH$ (with C=O), with Cl substituent

(d)

$$CH_2{=}CHCH_2CHCOH$$ (with C=O), with Br substituent

11.3 (a)

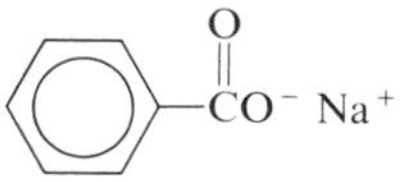

(benzene ring)$-CO^-\,Na^+$ (with C=O)

Sodium benzoate

(b)

$$CH_3COH$$ (with C=O)

Acetic acid or ethanoic acid

(c)

$$CH_3CH_2CO^-\,(CH_3)_3NH^+$$ (with C=O)

Trimethylammonium propanoate or trimethylammonium propionate

11.4 From an alkyl halide:

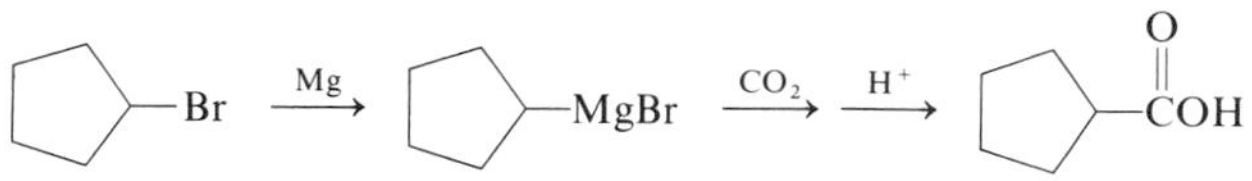

$\bigcirc{-}Br \xrightarrow{Mg} \bigcirc{-}MgBr \xrightarrow[]{CO_2,\ H^+} \bigcirc{-}COH$ (with C=O)

From an alcohol:

$\bigcirc{-}CH_2OH \xrightarrow{K_2Cr_2O_7} \bigcirc{-}COH$ (with C=O)

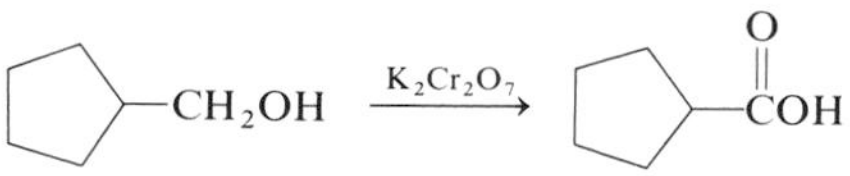

CHAPTER 12

12.1   (a)

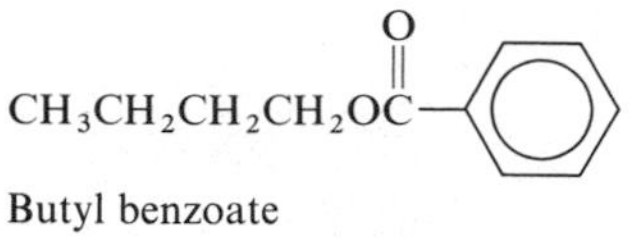

$$CH_3CH_2CH_2CH_2O\overset{\displaystyle O}{\overset{\|}{C}}-\bigcirc$$

Butyl benzoate

(b)

$$\bigcirc-CH_2O\overset{\displaystyle O}{\overset{\|}{C}}CH_2CH_2CH_3$$

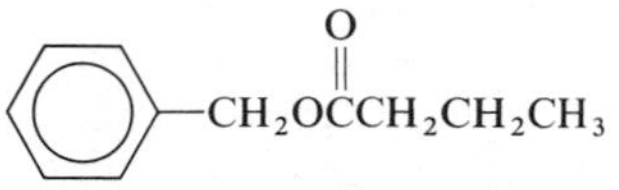

Benzyl butyrate or benzyl butanoate

(c)

$$CH_2{=}CHCH_2O\overset{\displaystyle O}{\overset{\|}{C}}CH_3$$

Allyl acetate or allyl ethanoate

(d)

$$(CH_3)_2CHO\overset{\displaystyle O}{\overset{\|}{C}}CH_2CH_3$$

Isopropyl propionate or isopropyl propanoate

12.2   (a)

$$(CH_3)_2CHCH_2NH\overset{\displaystyle O}{\overset{\|}{C}}CH_3$$

N-Isobutylacetamide or *N*-isobutylethanamide

(b)

$$H_2N\overset{\displaystyle O}{\overset{\|}{C}}-\bigcirc-OCH_3$$

p-Methoxybenzamide

(c)

$$CH_2{=}CHCH_2NH\overset{\displaystyle O}{\overset{\|}{C}}CH_3$$

N-Allylacetamide or *N*-allylethanamide

(d)

$$(CH_3CH_2)_2N\overset{\displaystyle O}{\overset{\|}{C}}CH_2CH_2CH_2CH_2CH_3$$

N,N-Diethylhexanamide or *N,N*-diethylcaproamide

Index

---

*Boldface numbers are used to indicate one or two pages that, by providing a brief definition or explanation, may be useful for quick reference.